Intelligent Adaptive Control with System Constraints

Kapil Sachan • Ahmad Jobran Al-Mahasneh •
Sreenatha Anavatti • Radhakant Padhi

Intelligent Adaptive Control with System Constraints

Evolving Neural Network Approaches for Robust Adaptive Control

Springer

Kapil Sachan
Flight Mechanics and Control Division
CSIR-National Aerospace Laboratories
Bengaluru, India

Ahmad Jobran Al-Mahasneh
Mechatronics Engineering Department
Philadelphia University
Amman, Jordan

Sreenatha Anavatti
School of Engineering and Information Technology
University of New South Wales at Canberra
Canberra, NSW, Australia

Radhakant Padhi
Department of Aerospace Engineering
Indian Institute of Science
Bengaluru, Karnataka, India

ISBN 978-981-95-8219-8 ISBN 978-981-95-8220-4 (eBook)
https://doi.org/10.1007/978-981-95-8220-4

This Springer imprint is published by the registered company Springer Nature Singapore Pte Ltd.
The registered company address is: 152 Beach Road, #21-01/04 Gateway East, Singapore 189721, Singapore

Foreword by Maruthi R. Akella

Learning and adaptation are central traits for any intelligent system. Successful theory-to-practice transition of any nonlinear control system lies in its ability to: (a) learn and improve the system model using measurements; (b) the capacity for making online adjustments and adaptations to predetermined control policies subject to physical constraints and model uncertainty; and (c) meet all the user-imposed stability requirements and performance specifications. Salient to note here is that even when the governing dynamics models admit satisfactory linear approximations, adaptation and learning introduce nonlinearities into the control loops that are unavoidable and must be fully addressed as part of the certification and validation process. Viewed in this context, the book *Intelligent Adaptive Control with System Constraints* provides an excellent basis both as a text for a two-course graduate sequence and as reference material for practicing engineers in the field. The authors are active and recognized researchers in the fields of nonlinear control, optimization, and intelligent systems.

The topics presented in the book are logically arranged, starting with the mathematical and technical foundations for nonlinear stability, controls, constrained systems, and adaptation before transitioning toward model-free and neuro-adaptive systems and concluding with capstone presentations of important practical application examples. The layout is such that Chapters 1–6 can ideally be the basis for a first graduate course on nonlinear adaptive control, followed by Chapters 7–12 that can form the core for a second course specializing in intelligent control systems and their practical applications. The mathematical descriptions are well-organized and accessible for any student with working knowledge of ordinary differential equations, linear algebra, and linear system theory. An exhaustive set of references is provided for each chapter that can be useful for background reading of the foundational results and approaches presented across the book.

In summary, this book serves as an excellent source for both students and professionals in the fields of intelligent systems and adaptive control. By bridging theoretical foundations with practical applications, it equips readers with the necessary tools to design and analyze controllers for systems with or without accurate model descriptions. The overall subject matter expertise of the authors and the logical progression of topics ensure that this text will prove to be a valuable resource, fostering deeper understanding and innovations in the realm of intelligent adaptive control.

Cockrell Family Chair in Engineering,
Department of Aerospace Engineering and
Engineering Mechanics, University of Texas at
Austin (UT Austin), Texas, USA

Prof Maruthi R. Akella

Foreword by B. N. Suresh

Inherently, all systems are nonlinear, and control of such systems requires appropriate control techniques having the capability to adapt to the nonlinearity and provide the required performance characteristics. Current trends demand new and effective control procedures, particularly in complex aerospace applications. Therefore, research publications that provide the latest updates and novel developments which can be applied by the practicing engineers involved in the development of such systems are the need of the hour. In this connection, the book *Intelligent Adaptive Control with System Constraints* by the renowned researchers, encompassing practical aspects of applications in various walks of engineering, is a laudable effort.

I have been fortunate to closely associate with the works of Professor Radhakant Padhi for the last 15 years in varying aspects of aerospace missions, namely, Moon Landing missions, Space Vehicle Reentry dynamics, Orbit and Attitude control of Lagrange Point 'L1' ADITYA Mission, Hypersonic Vehicle Guidance, and many more cross-field applications. I have no doubt that Professor Radhakant Padhi, with his ability to profess optimal control techniques with passion and his understanding of nonlinear control techniques to meet challenges of modern-day constraint requirements, is one of the researchers who is best suited to write a book on this topic. Professor Sreenatha Anavatti is well-versed in modern nonlinear control methods and has contributed significantly to this area for the last two decades. Dr. Kapil Sachan and Dr. Ahmad Jobran Al-Mahasneh, who have been the students of these researchers, have contributed significantly to the enhancement of the knowledge base in this research area. Their contribution in unravelling very fine aspects of the subject like 'Intelligent adaptive control with system constraints' is indeed noteworthy and is no doubt an excellent gift to the world of nonlinear control techniques.

The chronology of the chapters for this book has been carefully chosen and elegantly presented to make a big impact on readers. After an introduction to the basic theory of nonlinear systems and classifications, the theories that ensure the stability of nonlinear systems have been presented lucidly. The subsequent chapter provides all classical nonlinear control theories. Hence, the first three chapters provide a basis for the understanding of upcoming intelligent, adaptive, nonlinear control theories for the student community. After this, the next three chapters deal with advanced control theories to include (*i*) physical, operational, and user-imposed state constraints, (*ii*) adaptive control designs that account for system uncertainties, and (*iii*) a two-step neuro-adaptive control design. The first six chapters together could be considered for a one-semester introductory course on adaptive nonlinear control. Armed with basic adaptive control design knowledge, the subsequent five chapters are dedicated to (*i*) state-constrained neuro-adaptive control designs, (*ii*) system transformation-based state-constrained control methods, (*iii*) neural networks and their functionality and adaptability for complex systems, (*iv*) model-free adaptive control designs focusing on general regressive neural networks, and (*v*) model-free control following evolving regression-based control approaches, all of which are the most advanced control strategies currently being developed and employed in practical applications. The last chapter demonstrates many of the control methods in practical applications so that the interested researchers and practicing engineers can develop and practice the efficacy of the presented methods.

I am quite delighted to note that this book will be a boon to young researchers and practicing engineers to learn the state-of-the-art, adaptive, intelligent nonlinear control techniques effectively. While the first six chapters could form a course on fundamentals of adaptive nonlinear control theory, the next six chapters form an advanced adaptive intelligent nonlinear control course. The gratifying factor is that the authors have enlisted all the relevant references in a consolidated manner with practical applications across the fields of engineering. I am certain that the teachers and students involved in this specialized area would benefit vastly by this elegantly curated book on various aspects of nonlinear controls with exposure to several complex practical systems.

Chancellor, Indian Institute of Space Science and Technology, Thiruvananthapuram, India
Dr. B. N. Suresh

Foreword by Matthew Garratt

Intelligent Adaptive Control with System Constraints is a useful resource textbook for advanced undergraduate and graduate students, particularly those studying nonlinear control systems. The organization and the contents of the book provide not only basic insight into nonlinear control systems but also advanced methods for adaptive intelligent control of these systems.

The word nonlinear holds a passion for control researchers who deal with real-world dynamic problems where adaptiveness is a necessity rather than a choice. There is a vast number of resources available for linear systems due to their simplicity and the ease of application of the available linear mathematics. However, the same is not true of nonlinear systems, wherein each one can exhibit its own characteristic behavior. In addition, the mathematics associated with nonlinear systems makes them more challenging. Thus, there is a necessity for additional textbooks where the basics are introduced along with more advanced techniques for the control of nonlinear systems. This book definitely provides one such useful textbook.

My background is in experimental implementation and validation of guidance and control algorithms for various systems, starting with autonomous ground vehicles to uncrewed aerial vehicles, with a particular focus on vertical take-off and landing capable platforms. I have experienced the challenges of the effects of nonlinear dynamics on the design and validation of controllers for these types of vehicles. This has made researchers like me think of alternate solutions based on the unique systems at hand, leading to the evolution of new design techniques. This book will be a good text for many of these advanced techniques, with a particular focus on the nuances of nonlinear systems. Solutions are discussed with examples.

The organization of the book provides a clear insight into the authors' work. The first six chapters are dedicated to the introduction of nonlinear systems, classical nonlinear and adaptive control techniques, with numerical examples. The concepts from linear techniques for analyzing stability using linearization are also discussed. The foundation presented in the first chapter serves as a basis for the advanced topics in the remaining chapters of the book. Powerful nonlinear design techniques like dynamic inversion, Lyapunov-based control, backstepping control, and sliding mode control are introduced with simple examples and comprehensive theoretical foundations. Constraints play a major role in the design and operation of dynamic systems. These constraints may arise from physical as well as operational limitations. Understanding these constraints and their effects leads to better-controlled systems. Chapter 4 introduces these very well. Basic adaptive control, like direct and indirect adaptive control, along with model-reference adaptive control and neuro-adaptive control design, is discussed in Chapter 6.

The later Chapters 7 to 11, provide comprehensive methodologies for Barrier Lyapunov Function-based design, transformation-based adaptive control, along with model-free adaptive control methods using general neural regression approaches. Neural networks and evolving systems, the foundations of modern intelligent control strategies, are introduced here as well. Here, emphasis is placed on systems that evolve with time, making them intelligent as well.

Case studies provide a good platform to appreciate the theoretical work. The book includes Chapter 12 on case studies as an application of various control methods in practical scenarios. Thus, the book aims to provide not only theoretical knowledge to the students but also their practical utility.

In summary, this book is an informative, self-contained, and reader-friendly one that introduces nonlinear systems and their characteristics along with methods for their control under constraints. The authors deserve high appreciation for their passion and efforts in bringing out a very useful book backed by their vast combined experience.

School of Engineering and Technology,
University of New South Wales,
UNSW@Canberra, Canberra, Australia

Prof. Matthew Garratt

Preface

The nonlinear control system has an aura, both for researchers and engineers in this field. This is due to the nature of the promise it offers in the closed loop, and the fascinating mathematics surrounding it making it possible. We, the authors, have been fascinated by the different types of control system design techniques for nonlinear systems in general, and adaptive control in particular. Although many design techniques have been reported in the literature for adaptive control of dynamical systems, and the field continues to grow, there is a general hesitation to apply these to practical systems to harness their true potential. In our understanding and conviction, perhaps two key reasons for this are: (i) lack of lucidity in the associated math reported in the literature, making it esoteric to have a good understanding by all, and (ii) its inability to explicitly account and honor the system constraints, namely the constraints in both its inputs and its states and outputs. Unless this is addressed, the system may either fail to adapt or, worse, may fail during the process of adaptation. This book is an attempt to address these issues.

We, the last two authors of this book, have been working in this field for more than the last two decades, especially in the context of intelligent adaptive control using artificial neural networks. However, while working toward research publications, we started realizing the necessity for a good and lucid textbook that can help graduate students to quickly understand the topic and move ahead with their research. At the same time, it should also give sufficient confidence to the practicing engineers to confidently apply these wonderful techniques to solve real-world problems.

During this journey, we were very fortunate to have two very brilliant students, namely, Kapil and Ahmad (the first two authors of this book), join us, Radhakant and Sreenatha, respectively, for their PhD degrees about a decade ago. While this development at both places was unknown to each other (we are continents apart), a golden opportunity came at the time of their graduation. Even though examiners from abroad for thesis evaluation are usually selected randomly, by coincidence, we happened to be the examiners of students of each other. It turned out that both had done fantastic work, and a combination of their thesis would serve as a great reference book. The idea was proposed by Radhakant and agreed on by all immediately. That is how we started working on this project after Kapil and Ahmad graduated. Kapil and Ahmad have their PhDs in nonlinear and evolutionary control design techniques. Both have the enthusiasm to pass on their knowledge to the budding students and researchers in this area. Kapil has undertaken substantial research on the design of nonlinear adaptive control techniques, with a special focus on addressing system constraints. In parallel, Ahmad has conducted extensive research on neural data-driven control, with a particular emphasis on modifying neural architectures to enhance the adaptive control of nonlinear systems. This enthusiasm was clearly visible, and we could combine the two theses rather quickly. However, there was a problem with the systematic flow of ideas. Also, a book needs to be clear and flawless in its presentation, and the development of ideas should be systematic as well. Once again, we started to address these concerns. However, a realization came at this point in time that instead of a reference book, perhaps it would serve better as a graduate-level textbook so that one need not worry too much about his/her prior knowledge. We wanted to have a book that looks different from the available ones. The revised goals were:

1. To write a book that is useful for beginners as well as advanced students
2. Can be used as a graduate text as well as a reference book for advanced research
3. Can be used for the teaching process, both by the teacher and the students, by having examples and exercises

With these objectives in mind, we started upgrading the content of the book. However, it took us some time to put everything in an organized manner. We are very happy about the outcome, and we hope the readers will like the content and flow as well. We have tried to organize the book in such a way that the beginners are provided with a framework of mathematics and background material. In fact, the first few chapters (six) provide the basics for advanced material in the remaining chapters, including the important nonlinear controller design methods that include evolutionary techniques like neural network systems. Stability analysis, which forms the foundation for any controller to operate, is brought out clearly using different techniques like Lyapunov methods. The later chapters provide advanced materials that are useful both toward teaching and referencing. Numerical problems are included for every concept introduced in the book for clarity. In addition, a few exercise problems are provided to enhance the learning capability of the reader. This will also facilitate the teachers to use this book as a text/reference book in their courses.

We would also like to take this opportunity to express our sincere gratitude to our family members for their patience and cooperation throughout this journey. It is amazing to see the advancement of technology as well, which facilitated real-time collaboration among us from three different countries, both for meeting at regular intervals through video conferencing and manuscript preparation. Our sincere thanks to the Publisher as well, who kept on giving us numerous extensions as we were not able to meet the deadlines frequently due to various reasons.

Once again, we believe that the book is comprehensive and will be a useful source for teaching as well as for research. Happy reading.

Bengaluru, India — Kapil Sachan
Amman, Jordan — Ahmad Jobran Al-Mahasneh
Canberra, NSW, Australia — Sreenatha Anavatti
Bengaluru, Karnataka, India — Radhakant Padhi

Competing Interests The authors have no competing interests to declare that are relevant to the content of this manuscript.

About the Book

This book, *Intelligent Adaptive Control with System Constraints*, has been developed as a comprehensive and invaluable resource for advanced undergraduate and graduate students studying adaptive and intelligent nonlinear control systems. This book also contains all the necessary fundamentals and, hence, is self-contained.

The book is organized into four parts: a summary of classical nonlinear and adaptive control design techniques; advanced nonlinear adaptive control with a particular focus on system constraints; intelligent model-free adaptive control design techniques using various neural networks; and finally, some applications to real-world problems through a few case studies.

Introduction to nonlinear systems and classical nonlinear and adaptive control design approaches are introduced in Chaps. 1–6. Chapter 1 begins with a thorough introduction to nonlinear systems, offering detailed explanations of their classification, characteristics, and behaviors. It also covers the mathematical preliminaries necessary for understanding these systems, including linearization techniques, which are essential for analyzing the stability and control of nonlinear dynamics. The foundational knowledge presented in this chapter serves as a critical basis for the advanced topics explored in the subsequent chapters.

In Chap. 2, a detailed stability analysis of nonlinear systems using the Lyapunov method is introduced. This chapter explores stability concepts for both autonomous and non-autonomous nonlinear systems, presenting a theoretical framework and practical insights for ensuring system stability in various contexts.

Chapter 3 focuses on the core philosophies of classical nonlinear control design. It introduces powerful techniques such as nonlinear dynamic inversion, Lyapunov-based control, backstepping control, and sliding mode control. These methodologies are explained comprehensively, with theoretical foundations accompanied by numerous examples to enhance understanding and illustrate their practical implementation.

Chapter 4 carries forward the discussion in Chap. 3 to constrained control design, an essential area for systems with physical or operational limitations. This chapter offers a comprehensive overview of system constraints and control strategies specifically tailored for handling state constraints. It focuses on methods like Barrier

Lyapunov Function-based techniques, dynamic inversion-based approaches, and transformation-based strategies to stabilize systems without violating user-imposed state constraints.

Chapter 5 extends the concepts of adaptive control by exploring various design philosophies for handling system uncertainties. It covers a wide range of classical adaptive control strategies, including direct and indirect approaches, basic adaptive control, and model-reference adaptive control. The chapter emphasizes both their theoretical foundations and practical implementation in real-world scenarios.

Building on the backdrop of Chaps. 1–5, Chap. 6 presents a neuro-adaptive control design applicable to various classes of nonlinear systems. This two-step control design process first constructs an approximate system to capture modeling uncertainties of the unknown function (including its gradient) through neural network application, followed by the design of a tracking controller. This adaptive method is known for its smooth learning capabilities and is applied further in state-constrained control designs.

State-constrained neuro-adaptive control designs are presented in Chaps. 7 and 8. Chapter 7 introduces Barrier Lyapunov Function-based state-constrained adaptive control for various types of nonlinear systems. This advanced methodology addresses the challenge of ensuring system stability while adhering to user-imposed state constraints. In Chap. 8, the focus shifts to transformation-based state-constrained adaptive control for different classes of nonlinear systems. This chapter provides an in-depth discussion on how transformations can simplify and resolve constrained control problems, emphasizing both their theoretical principles and practical applications.

Chapter 9 shifts to the domain of neural networks and self-evolving neural networks, foundational to many modern control strategies. It introduces the basic concepts of neural networks, discussing their structure, functionality, and applicability in adaptive control for complex systems. Additionally, the evolving systems framework is explored, emphasizing its capacity to adapt to varying conditions and uncertainties.

Model-free adaptive control methods are presented in Chaps. 10 and 11. Chapter 10 introduces model-free adaptive control methods, with a particular focus on general regression neural networks. This chapter demonstrates how these networks can be utilized to design adaptive control systems that do not depend on explicit system models, enhancing their versatility and applicability across various contexts. In Chap. 11, model-free control is developed following general regression neural networks-based evolving control approaches. This chapter explores how these methods can be employed to develop adaptive systems that evolve over time, thereby improving their performance and reliability in dynamic and uncertain environments.

Finally, Chap. 12 consolidates all the concepts discussed throughout the book by presenting a series of case studies. These case studies illustrate the application of various control methods in practical scenarios, enhancing the confidence of the readers to apply.

By addressing both the theoretical and practical aspects of intelligent and adaptive control, this book equips readers with the knowledge and skills necessary to design efficient control systems. It emphasizes the importance of respecting system constraints, and offers innovative solutions for overcoming their challenges. Whether one is a graduate student new to the field or an experienced practitioner seeking to deepen your expertise, this book serves as an invaluable resource for mastering adaptive and intelligent control systems, including honoring the system constraints.

Contents

About the Authors

Kapil Sachan is currently a Senior Scientist in the Flight Mechanics and Control Division at National Aerospace Laboratories, Bangalore. He obtained both his Master of Engineering in Aerospace Engineering and his PhD from the Indian Institute of Science (IISc), Bangalore, with a specialization in control systems theory.

His research encompasses both theoretical and applied aspects of aerospace control systems. His core areas of expertise include guidance and control of aircraft and space vehicles, nonlinear adaptive control, optimal control, and safety-constrained adaptive control. He has published in reputed international journals and conferences and serves as a reviewer for several leading scientific journals.

As an integral member of the National Control Law Team, Dr. Sachan is actively involved in projects of national significance, including the development of flight control laws for commercial and high-performance aircraft and high-aspect-ratio wing aircraft with significant structural flexibility. He has also conducted research on hypersonic vehicle control and on precision lunar soft-landing technologies.

Ahmad Jobran Al-Mahasneh earned his B.Sc. and M.Sc. degrees in Mechatronics/Mechanical Engineering from the Jordan University of Science and Technology in 2010 and 2014, respectively. He received his Ph.D. in Mechanical Engineering from the University of New South Wales, Australia, in 2020. He earned the UNSW Dean's Award for Outstanding PhD thesis.

He is currently an Assistant Professor in the Department of Mechatronics Engineering at Philadelphia University, Jordan. He has taught a wide range of courses, including System Dynamics, Machine Intelligence, Digital Control, Applied Nonlinear Control, Robotics, and Programmable Logic Controllers (PLC). He has also supervised numerous undergraduate projects and graduate theses in the areas of control systems, robotics, energy management systems, and mobile robotics.

His research interests span adaptive and intelligent control, evolving systems, neural networks and deep learning, nonlinear control, energy management systems, and autonomous systems. He has published many papers in reputed journals and conferences such as *IEEE Transactions on Cybernetics*, *IEEE Transactions on Systems, Man, and Cybernetics*, and *Energy Conversion and Management*. He is a regular reviewer in highly ranked journals.

Sreenatha Anavatti is an academic with the University of New South Wales (UNSW) at Canberra, Australia, in the School of Engineering and Technology. Before joining UNSW, he was an Associate Professor at the Indian Institute of Technology, Mumbai, India. He has his Ph.D. from Indian Institute of Science, Bengaluru.

His research interests include guidance and control of autonomous systems as applied to Uncrewed Aerial Vehicles, Uncrewed Ground Vehicles, and Underwater Vehicles. His works include both theoretical and experimental aspects applied to autonomous systems. He has been working with Fuzzy and Neural systems, Evolutionary control systems, Nonlinear dynamic systems and their applications, and Physics Informed Neural Networks as applied to control. He has co-authored more than

300 publications in these areas. He has supervised more than 25 Ph.D. scholars in these domains. He has given a number of keynote and motivational speeches. His collaborations include Seoul National University (under the Brain-Korea fellowship twice), Institute of Technology, Surabaya, NTU Singapore, and Indian Institute of Science, Bengaluru.

During his tenure at the Indian Institute of Technology, he has undertaken a number of projects of National importance from the Indian Space Research Organisation, the Defence Research and Development Organisation, and the Aeronautical Development Agency (ADA). At UNSW, his research has been well supported by the Defence Science Technology (DST) group including a number of projects related to autonomous systems.

Radhakant Padhi is currently working as a full Professor (HAG scale) and as an associate faculty at the Centre for Cyber-Physical Systems, Indian Institute of Science (IISc), Bengaluru, India. His research interests include optimal and nonlinear control synthesis algorithms, as well as their applications to challenging practical problems in aerospace, biomedical, and mechanical engineering. The elegant nonlinear and optimal guidance and control algorithms developed by him and his team have been globally well-received for their simplicity and effectiveness.

As of 2026, he has co-authored approximately 300 publications in peer-reviewed international journals and conferences, as well as a book on *Satellite Formation Flying*, published in 2021. Besides this book on *Intelligent Adaptive Control with System Constraints*, he is preparing two additional books: one on *Applied Optimal Control and State Estimation* and the other on *Flight Mechanics of Aerospace Vehicles*. He has also co-authored a motivational book for young minds titled "*The Gap Year Advantage: Efforts Are in Your Hands, Results Are Not*," published in 2025. Currently, he serves as an Associate Editor of the *Unmanned Systems* journal and has previously been an Associate Editor of two more international journals in the control and automation field.

He is a Fellow of the Indian National Academy of Engineering, Aeronautical Society of India, Astronautical Society of India, Institute of Electronics and Telecommunication Engineers, and Institute of Engineers India. He is an Associate Fellow of the American Institute of Aeronautics and Astronautics (AIAA) and a Senior Member of the Institute of Electrical and Electronics Engineers (IEEE). During 2014-2020, he was a member of the Council of the International Federation of Automatic Control (IFAC).

His recent passion is his two deep-tech startups incubated at IISc (VAPL.tech and VTPL.tech), one focusing on the Advanced Guidance and Control of aerospace vehicles for challenging missions, and the other focusing on diabetes care. Through these, he intends to focus on utilizing his knowledge and experience for the real benefit of the country and, by extension, the mankind.

Chapter 1
Introduction and Motivation

This chapter lays the foundation for the rest of the book by introducing fundamental concepts of nonlinear systems and essential mathematical tools. It provides a framework for understanding the complexities inherent in nonlinear systems and the challenges involved in designing effective controllers for such systems.

1.1 Introduction of Nonlinear Systems

The interest in analyzing nonlinear systems originates from the fact that almost all real-world systems exhibit nonlinear characteristics, making them more realistic representations of real-life problems. Nonlinear systems are commonly found in fields such as aerospace, automobiles, robotics, and industrial processes. Hence, the study of nonlinear systems is crucial for robustly stabilizing control performance in various engineering applications. Therefore, an understanding of a few properties and the stability behavior of nonlinear systems is crucial before delving into control design for such systems.

As the formal definition, *nonlinear systems* are systems that are "not necessarily linear," whereas linear systems are defined as the systems that satisfy the "*principle of superposition*", which has two main properties: (i) scaling, where multiplying the input by a constant α results in the output being scaled by α, and (ii) additivity, where the response to multiple inputs applied simultaneously is the sum of the individual responses to each input applied separately. For example, consider a static system described by the equation of a straight line, $y = mx + c$, where y is the output, m is the slope, and c is a constant. This system is linear only if $c = 0$ (i.e., the system passes through the origin), as otherwise, the principle of superposition does not hold, as shown in Fig. 1.1.

K. Sachan et al., *Intelligent Adaptive Control with System Constraints*,
https://doi.org/10.1007/978-981-95-8220-4_1

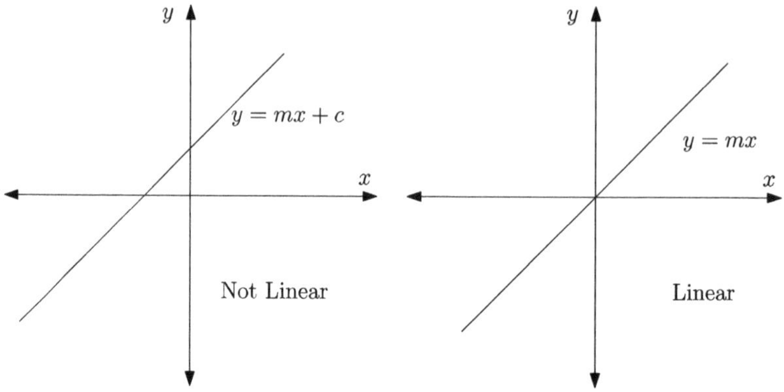

Fig. 1.1 Difference between linear and nonlinear static systems

Example 1.1 Consider the dynamical system $\dot{x} = 2x$. This system is a linear system because it satisfies

(i) scaling, i.e., $\alpha\dot{x} = \alpha(2x) = 2(\alpha x)$
(ii) additivity, i.e., $\frac{d}{dt}(x_1 + x_2) = \dot{x}_1 + \dot{x}_2 = 2x_1 + 2x_2 = 2(x_1 + x_2)$

Example 1.2 Consider the dynamical system $\dot{x} = 2x + 3$. The principle of superposition can be applied as

(i) scaling, i.e., $\alpha\dot{x} = \alpha(2x + 3) \neq 2(\alpha x) + 3$
(ii) additivity, i.e., $\frac{d}{dt}(x_1 + x_2) = \dot{x}_1 + \dot{x}_2 = (2x_1 + 3) + (2x_2 + 3) \neq 2(x_1 + x_2) + 3$

Since this dynamical system does not satisfy the principle of superposition, it is a nonlinear system.

Example 1.3 Consider the dynamical system $\dot{x} = 2\sin x$. The principle of superposition can be applied as

(i) scaling, i.e., $\alpha\dot{x} = \alpha(2\sin x) \neq 2\sin(\alpha x)$
(ii) additivity, i.e., $\frac{d}{dt}(x_1+x_2) = \dot{x}_1+\dot{x}_2 = 2\sin x_1+2\sin x_2 \neq 2\sin(x_1+x_2)$

Since this dynamical system does not satisfy the principle of superposition, it is a nonlinear system.

A typical nonlinear system can be represented by

$$\dot{\boldsymbol{x}} = \boldsymbol{f}(\boldsymbol{x}, \boldsymbol{u}, t) \tag{1.1}$$

subject to initial condition $\boldsymbol{x}_0$, where $\boldsymbol{x} = [x_1, x_2, \ldots, x_n]^T \in \Re^n$ and $\boldsymbol{u} = [u_1, u_2, \ldots, u_m]^T \in \Re^m$ are state and control inputs of the system, respectively, and $\boldsymbol{f} = [f_1, f_2, \ldots, f_n]^T \in \Re^n$ is a nonlinear function of state $\boldsymbol{x}$, control input $\boldsymbol{u}$, and time t.

The generic output response of the system (1.1) can be written as

$$\boldsymbol{y} = \boldsymbol{h}(\boldsymbol{x}, \boldsymbol{u}, t) \tag{1.2}$$

where $\boldsymbol{y} = \left[y_1, y_2, \ldots, y_p\right]^T \in \Re^p$ represents the system output and $\boldsymbol{h} = \left[h_1, \ldots, h_p\right]^T \in \Re^p$ is a nonlinear function of $\boldsymbol{x}$, $\boldsymbol{u}$, and time t.

Note that the nonlinearities in a nonlinear system can be broadly classified into *hard nonlinearities* and *strong nonlinearities*. Hard nonlinearities are characterized by abrupt or discontinuous behavior, often resulting in discontinuous and/or non-smooth system responses. These include phenomena such as saturation, dead zones, hysteresis, and other discontinuities in derivatives. In contrast, strong nonlinearities refer to systems where nonlinear effects are predominant over linear ones, such as higher-order terms in a Taylor series. This book deals only with strongly nonlinear systems. For discussions on hard nonlinearities and associated concepts, the reader is referred to Slotine et al. (1991).

1.2 Classification of Nonlinear Systems

The analysis of a generic nonlinear system presents considerable challenges, prompting researchers to explore classes of nonlinear systems for analysis. This categorization provides a structured approach to examining the complexities of nonlinear systems and uncovering the fundamental principles that govern their behavior. It also facilitates the development of effective control strategies for various classes of nonlinear systems. Nonlinear systems can be classified in various forms as follows.

1.2.1 Time-Dependent System Classification

The dynamics of many practical systems vary over time due to changes in their parameters. Therefore, nonlinear systems can be classified based on their explicit dependency on time, as follows.

1. *Autonomous systems:* An autonomous system is defined as a system whose dynamics do not depend explicitly on time. Note that this explicit time dependency typically comes because of time-varying parameters in the system dynamics or the forcing function itself becoming an explicit function of time (open-loop control). In other words, if the parameter variation in system dynamics is either nonexistent or sufficiently slow, then the system can be classified as an autonomous system.

An autonomous system can be represented as

$$\dot{\boldsymbol{x}} = \boldsymbol{f}(\boldsymbol{x}, \boldsymbol{u})$$

where, along with system dynamics, the forcing function $\boldsymbol{u}$ is also not an explicit function of time. For example, consider a spring-mass-damper system represented by the following equations:

$$\begin{aligned}\dot{x}_1 &= x_2, \\ \dot{x}_2 &= -\frac{k}{m}x_1 - \frac{c}{m}x_2 + f,\end{aligned}$$

where k is the spring constant, c is the damping constant, m is the mass of the block, and f is the externally applied force. This is an autonomous system, as its parameters do not vary with time.

2. *Nonautonomous systems:* A nonautonomous system is a type of nonlinear system whose dynamics vary explicitly with time. In other words, when the parameter variation over time is fast and significant and/or when the forcing function is an explicit function of time, which influences the behavior and properties of the system, the system is classified as nonautonomous.
A nonautonomous system can be represented as

$$\dot{\boldsymbol{x}} = \boldsymbol{f}(\boldsymbol{x}, \boldsymbol{u}, t)$$

For example, consider a more general spring-mass-damper system defined by the equations:

$$\begin{aligned}\dot{x}_1 &= x_2, \\ \dot{x}_2 &= -\frac{k(t)}{m}x_1 - \frac{c}{m}x_2 + f,\end{aligned}$$

where the spring constant $k(t)$ is a time-dependent parameter, changing over time. Thus, this spring-mass-damper system is nonautonomous.

1.2.2 Control-Affine Nonlinear Systems

Control-affine systems are a class of nonlinear systems that are nonlinear with respect to the state but linear (or "affine") in control. A good advantage of such a system is that it often facilitates an analytic (closed-form) solution of control. These systems have broad applications in various areas, including robotics, aerospace engineering, and process control, where *control-oriented modeling* is carried out. These systems can be autonomous or nonautonomous in nature. Mathematically, a control-affine autonomous nonlinear system can be represented as follows:

$$\dot{\boldsymbol{x}} = \boldsymbol{f}(\boldsymbol{x}) + \boldsymbol{G}(\boldsymbol{x})\boldsymbol{u} \tag{1.3}$$

$$\boldsymbol{y} = \boldsymbol{h}(\boldsymbol{x}) \tag{1.4}$$

where $\boldsymbol{x} \in \Re^n$, $\boldsymbol{u} \in \Re^m$, and $\boldsymbol{y} \in \Re^p$ represent the state, input, and output vectors of the system, respectively, $\boldsymbol{f}(\boldsymbol{x})$ and $\boldsymbol{h}(\boldsymbol{x})$ are a nonlinear functions and, $\boldsymbol{G}(\boldsymbol{x}) \in \Re^{n \times m}$ is the control input gain matrix. Note that if the dimensions of the output vector and control input are the same, i.e., $m = p$, the nonlinear system is also *square* in the input-output sense (also known as the control-affine square nonlinear system).

Control-affine square nonlinear autonomous systems exhibit a diverse range of applications, leading to their classification into distinct categories. These categories play a vital role in the design of effective control strategies, including feedback linearization, backstepping, and sliding mode techniques. The specific categories associated with control-affine nonlinear systems are given as follows:

1. *Nonlinear relative degree one system:* If the output of a nonlinear system is required to be differentiated once to obtain an explicit relationship between the input and output, the system is said to have a relative degree of one. A typical mathematical representation of a nonlinear relative degree one system is given by

$$\dot{\boldsymbol{x}} = \boldsymbol{f}(\boldsymbol{x}) + \boldsymbol{G}(\boldsymbol{x})\boldsymbol{u} \tag{1.5}$$

$$\boldsymbol{y} = \boldsymbol{h}(\boldsymbol{x}) \tag{1.6}$$

   where $\boldsymbol{x}$, $\boldsymbol{u}$, and $\boldsymbol{y}$ represent the state, control input, and output vectors of the system, respectively. Note that the dimensions of the state, input, and output of such systems are equal.

   These types of nonlinear systems appear in several control design applications. For instance, while designing a three-loop aircraft autopilot, aircraft equations are represented in the form of a relative degree one system. The following set of nonlinear equations represents aircraft attitude dynamics (Stevens et al. 2015), where P, Q, and R are required aircraft attitude rate inputs to achieve desired aircraft attitudes ϕ, θ, and ψ, thereby having an equal number of inputs, states, and outputs:

$$\begin{aligned}\dot{\phi} &= P + Q \sin\phi \tan\theta + R \cos\phi \tan\theta \\ \dot{\theta} &= Q \cos\phi - R \sin\phi \\ \dot{\psi} &= Q \sin\phi \sec\theta + R \cos\phi \sec\theta\end{aligned}$$

2. *Nonlinear relative degree two system:* If the output of a nonlinear system is required to be differentiated twice to obtain an explicit relationship between the

input and output, the system is said to have a relative degree of two. A typical nonlinear relative degree two system is described by the differential equations of the following form:

$$\begin{bmatrix} \dot{\boldsymbol{x}}_1 \\ \dot{\boldsymbol{x}}_2 \end{bmatrix} = \begin{bmatrix} \boldsymbol{x}_2 \\ \boldsymbol{f}(\boldsymbol{x}, t) + \boldsymbol{G}(\boldsymbol{x}, t)\boldsymbol{u} \end{bmatrix} \tag{1.7}$$

$$\boldsymbol{y} = \boldsymbol{h}(\boldsymbol{x}_1) \tag{1.8}$$

where state variables $\boldsymbol{x}_1$ and $\boldsymbol{x}_2$ are defined as $\boldsymbol{x}_1 \triangleq \left[x_{1_1},\ x_{1_2},\ \ldots,\ x_{1_n}\right]^T \in \Re^n$, and $\boldsymbol{x}_2 \triangleq \left[x_{2_1},\ x_{2_2},\ \ldots,\ x_{2_n}\right]^T \in \Re^n$, respectively, and complete state vector is defined as $\boldsymbol{x} \triangleq \left[\boldsymbol{x}_1^T, \boldsymbol{x}_2^T\right]^T$. The output of the system is selected as a function of $\boldsymbol{x}_1$, i.e., $\boldsymbol{y} = \boldsymbol{h}(\boldsymbol{x}_1)$. The control input is represented by $\boldsymbol{u}$, and $\boldsymbol{f}(\boldsymbol{x}, t)$ is a nonlinear function. The matrix $\boldsymbol{G}(\boldsymbol{x}, t)$ represents the control effectiveness matrix. In this book, it is assumed that $\boldsymbol{G}(\boldsymbol{x}, t) \in \Re^{n \times n}$ is nonsingular $\forall t$, which facilitates easy control synthesis.

Any system that can be modeled using Newton's laws leads to this class of nonlinear systems. For instance, the dynamics of an ith agent of a multi-agent system is represented as

$$\dot{x}_i = v_i \tag{1.9}$$

$$\dot{v}_i = \gamma_i \phi_i(x_i, v_i, t) + g_i(x_i, v_i)u_i, \quad i = 1, \ldots, n \tag{1.10}$$

where n is the number of agents, x_i and v_i are position and velocity of the ith agent, g_i is the control direction (or effectiveness) and u_i represents the control input to the agent. The term $\gamma_i \phi_i$ is a part of system dynamics.

3. *Euler-Lagrange systems:* The Euler-Lagrange (EL) system dynamics can be categorized as a subset of nonlinear systems with a relative degree of two, as defined in Eq. (1.7). However, there are distinct differences in terms of explicit time dependence and the characteristics of the control effectiveness matrix $\boldsymbol{G}(\boldsymbol{x}) \in \Re^{n \times n}$. The EL systems are autonomous in nature, with a symmetric positive definite control effectiveness matrix.

   Note that many robotic systems are represented by EL systems. For example, the dynamics of the rigid robotic manipulator is represented as follows:

$$\boldsymbol{M}(\boldsymbol{q})\ddot{\boldsymbol{q}} + \boldsymbol{C}(\boldsymbol{q}, \dot{\boldsymbol{q}})\dot{\boldsymbol{q}} + \boldsymbol{G}(\boldsymbol{q}) = \boldsymbol{\tau}$$

   where $\boldsymbol{q} \in \Re^n$, $\dot{\boldsymbol{q}} \in \Re^n$, and $\ddot{\boldsymbol{q}} \in \Re^n$ denote the vector of joint position, velocity, and acceleration, $\boldsymbol{M}(\boldsymbol{q}) \in \Re^{n \times n}$ denotes the positive definite inertia matrix, $\boldsymbol{C}(\boldsymbol{q}, \dot{\boldsymbol{q}}) \in \Re^{n \times n}$ denotes the Coriolis torques, $\boldsymbol{G}(\boldsymbol{q})$ denoted gravitational forces, and $\boldsymbol{\tau}$ represents the control torque input.

4. *Strict-feedback systems:* A nonlinear system is said to be a strict-feedback nonlinear system if it can be described by differential equations of the following form:

$$\dot{x}_i = f_i(\overline{\boldsymbol{x}}_i) + g_i(\overline{\boldsymbol{x}}_i)x_{i+1}, \quad i = 1, \ldots, n-1$$
$$\dot{x}_n = f_n(\overline{\boldsymbol{x}}_n) + g_n(\overline{\boldsymbol{x}}_n)u, \quad n \geq 2$$

where $\overline{\boldsymbol{x}}_i = [x_1, \ldots, x_i]^T \in \Re^i$, $i = 1, \ldots, n$. The state vector is denoted as $\boldsymbol{x} = [x_1, x_2, \ldots, x_n]^T \in \Re^n$. The control input and system output are represented as $\boldsymbol{u} \in \Re$ and $y = x_1 \in \Re$, respectively. The control coefficient matrix is represented by $g_i(\overline{\boldsymbol{x}}_i), \forall i$, and $f_i(\overline{\boldsymbol{x}}_i), \forall i$ is a nonlinear function of state. Many practical systems are represented as strict-feedback nonlinear systems. For instance, the dynamic model of an electrohydraulic servo system is represented as follows:

$$\dot{x}_1 = x_2$$
$$\dot{x}_2 = x_3$$
$$\dot{x}_3 = f(x_1, x_2, x_3) + b\,u$$

where x_1, x_2, and x_3 represent the position, velocity, and acceleration of the hydraulic cylinder piston motion, respectively. Control input is represented by u, term $f(x_1, x_2, x_3)$ is a nonlinear function of states, and b is a system dynamics-related parameter.

5. *Feedback linearizable systems:* A nonlinear system is said to be a feedback linearizable MIMO nonlinear system if it can be described by differential equations of the following form:

$$\begin{bmatrix} x_1^{(n_1)} \\ \vdots \\ x_m^{(n_m)} \end{bmatrix} = \begin{bmatrix} f_1(\boldsymbol{x}) + g_{1_1}(\boldsymbol{x})u_1 + \ldots + g_{1_m}(\boldsymbol{x})u_m \\ \vdots \\ f_m(\boldsymbol{x}) + g_{m_1}(\boldsymbol{x})u_1 + \ldots + g_{m_m}(\boldsymbol{x})u_m \end{bmatrix}$$

where $x_i^{(n_i)} \triangleq \frac{d^{n_i} x_i}{dt^{n_i}}$, $\boldsymbol{x} \triangleq [x_1, \ldots, x_1^{(n_1-1)}, \ldots, x_m, \ldots, x_m^{(n_m-1)}]^T \in \Re^n$ with $n = n_1 + \ldots + n_m$ is the overall state vector; x_i and u_i, $\forall i = 1, \ldots, m$ represent output and input to the system, respectively; and f_i is a nonlinear function.
The nonlinear system can also be written in the following compact form:

$$\boldsymbol{x}^{(n)} = \boldsymbol{f}(\boldsymbol{x}) + \boldsymbol{G}(\boldsymbol{x})\boldsymbol{u} \tag{1.11}$$

where $\boldsymbol{x}^{(n)} = \left[x_1^{(n_1)}, \ldots, x_m^{(n_m)}\right]^T \in \Re^m$, $\boldsymbol{f}(\boldsymbol{x}) = [f_1(\boldsymbol{x}), \ldots, f_m(\boldsymbol{x})]^T \in \Re^m$,

and $\boldsymbol{G}(\boldsymbol{x}) = \begin{bmatrix} g_{1_1}(\boldsymbol{x}) & \ldots & g_{1_m}(\boldsymbol{x}) \\ \vdots & \ddots & \vdots \\ g_{m_1}(\boldsymbol{x}) & \ldots & g_{m_m}(\boldsymbol{x}) \end{bmatrix} \in \Re^{m \times m}$.

This is a generic type of nonlinear system, and many complex and coupled practical systems are represented by this type of nonlinear system dynamics. For example, the following nonlinear system dynamics represent a two-link robotic manipulator:

$$\begin{bmatrix} \ddot{\theta}_1 \\ \ddot{\theta}_2 \end{bmatrix} = [H]^{-1} \begin{bmatrix} h\dot{\theta}_2 & h\left(\dot{\theta}_1 + \dot{\theta}_2\right) \\ -h\dot{\theta}_1 & 0 \end{bmatrix} \begin{bmatrix} \dot{\theta}_1 \\ \dot{\theta}_2 \end{bmatrix} + [H]^{-1} \begin{bmatrix} \tau_1 \\ \tau_2 \end{bmatrix}$$

where

$$\begin{aligned} H &= \begin{bmatrix} H_{11} & H_{12} \\ H_{21} & H_{22} \end{bmatrix} \\ H_{11} &= a_1 + 2a_3 \cos\theta_2 + 2a_4 \sin\theta_2 \\ H_{12} &= H_{21} = a_2 + a_3 \cos\theta_2 + a_4 \sin\theta_2 \\ H_{22} &= a_2, \quad h = a_3 \sin\theta_2 - a_4 \cos\theta_2 \end{aligned}$$

with

$$\begin{aligned} a_1 &= I_1 + m_1 {l_{c1}}^2 + I_e + m_e {l_{ce}}^2 + m_e {l_1}^2 \\ a_2 &= I_e + m_e {l_{ce}}^2 \\ a_3 &= m_e l_1 l_{ce} \cos\delta_e \\ a_4 &= m_e l_1 l_{ce} \sin\delta_e \end{aligned}$$

Here, $\theta_1 \in \Re$ and $\theta_2 \in \Re$ represent joint positions, and $\dot{\theta}_1 \in \Re$ and $\dot{\theta}_2 \in \Re$ represent joint velocities of the manipulator. The torque τ_j $(j = 1, 2)$ are control inputs to the system. All parameters used in the dynamics have their standard dimensions and usual meanings and can be found in Slotine et al. (1991).

1.2.3 Input-Output Dimension-Dependent Classification

Based on the number of control inputs and outputs, nonlinear systems can be classified into the following four categories:

1. *Single-input-single-output systems (SISOs):* As the name implies, a nonlinear system is referred to as a *single-input-single-output* system when it has a scalar input and scalar output. For example, by substituting $m = p = 1$ in Eqs. (1.1) and (1.2), the nonlinear system can be transformed into a SISO system. An illustrative

example of a SISO nonlinear system can be as follows:

$$\begin{aligned}\dot{x}_1 &= x_1^2 + x_2 \\ \dot{x}_2 &= x_1 + (1 + x_2)^2 + u_1 \\ y &= x_1\end{aligned}$$

where $\boldsymbol{x} = [x_1, x_2]^T$ is the state of the system. The system has a single input u_1 and a single output x_1.

2. *Single-input-multi-output systems (SIMOs):* A *single-input-multi-output* system is classified as a nonlinear system when it consists of a scalar input and multiple outputs. For instance, when $m = 1$ in Eq. (1.1), the nonlinear system is transformed into a SIMO system. A representative example of a SIMO nonlinear system is

$$\begin{aligned}\dot{x}_1 &= x_1^2 + x_2 \\ \dot{x}_2 &= x_1 + (1 + x_2)^2 + u_1 \\ y &= [x_1, x_2]^T\end{aligned}$$

where the system has a single input u_1 and two outputs x_1 and x_2.

3. *Multi-input-single-output systems (MISOs):* A nonlinear system is categorized as a *multi-input-single-output* system when it consists of multiple inputs and a scalar output. For example, by substituting $p = 1$ in Eq. (1.2), the nonlinear system is transformed into a MISO system. An example of a MISO nonlinear system is

$$\begin{aligned}\dot{x}_1 &= x_1^2 + x_2 + u_1 \\ \dot{x}_2 &= x_1 + (1 + x_2)^2 + u_2 \\ y &= h(x_1, x_2)\end{aligned}$$

where the system has multiple inputs u_1 and u_2 and a single output y.

4. *Multi-input-multi-output systems (MIMOs):* A nonlinear system is categorized as a *multi-input-multi-output* (MIMO) system when it comprises multiple inputs and multiple outputs. As an illustration, the system described by Eqs. (1.1) and (1.2) represents a nonlinear MIMO system. A typical example of a MIMO nonlinear system can be exemplified as follows:

$$\begin{aligned}\dot{x}_1 &= x_1^2 + x_2 + u_1 \\ \dot{x}_2 &= x_1 + (1 + x_2)^2 + u_2 \\ y &= [x_1, x_2]^T\end{aligned}$$

where the system has two inputs u_1 and u_2 and two outputs x_1 and x_2.

Table 1.1 Input-output dimension-dependent classification for nonlinear systems

Nature of the system	Output dimension	Input dimension
SISO	1	1
SIMO	p	1
MISO	1	m
MIMO	p	m

The above definitions of input-output dimension-dependent classification of nonlinear systems are tabulated in Table 1.1.

1.3 Linearization: A Brief Overview

Analyzing the behavior of nonlinear systems presents significant challenges. In addition to advanced methods that deal with nonlinear systems in their original form, a commonly employed approach is linearization. This involves creating linearized models around specific operating points, allowing for simplified analysis.

Linearization is a method to approximate nonlinear systems around an "operating point," which leads to a linear system around that point. Linear systems hold great significance in practical applications due to their simplicity and ease of control design. A typical linear system can be written as

$$\dot{\boldsymbol{x}} = \boldsymbol{A}\boldsymbol{x} + \boldsymbol{B}\boldsymbol{u} \tag{1.12}$$

$$\boldsymbol{y} = \boldsymbol{C}\boldsymbol{x} + \boldsymbol{D}\boldsymbol{u} \tag{1.13}$$

where $\boldsymbol{x} \in \Re^n$, $\boldsymbol{y} \in \Re^p$, and $\boldsymbol{u} \in \Re^m$ are the state, output, and control input of the system, respectively. The matrices $\boldsymbol{A} \in \Re^{n\times n}$, $\boldsymbol{B} \in \Re^{n\times m}$, $\boldsymbol{C} \in \Re^{p\times n}$, and $\boldsymbol{D} \in \Re^{p\times m}$ define the overall behavior of the system, known as system matrix, control matrix, output matrix, and output control matrix, respectively. It can be noted that the matrices $\boldsymbol{A}$, $\boldsymbol{B}$, $\boldsymbol{C}$, and $\boldsymbol{D}$ are time-independent, which can happen if the system is continuously linearized around an operating trajectory. Details of these are discussed next.

1.3.1 *Operating Point and Equilibrium Point*

A linear system is constructed by approximating a nonlinear system around an operating point. An operating point is defined as a point through which the system trajectory passes, including the initial and final points.

Consider the nonlinear system:

$$\dot{\boldsymbol{x}} = \boldsymbol{f}(\boldsymbol{x}, \boldsymbol{u}) \tag{1.14}$$

where state $\boldsymbol{x} \in \Re^n$ and control input $\boldsymbol{u} \in \Re^m$. Assume that $(\boldsymbol{x}_0, \boldsymbol{u}_0)$ is one of the operating points of the system. Then, for the operating point, the system (1.14) is expressed as $\dot{\boldsymbol{x}}_0 = \boldsymbol{f}(\boldsymbol{x}_0, \boldsymbol{u}_0)$.

On the other hand, if a system starts at an equilibrium point, it will remain there indefinitely. This implies that an equilibrium point is a constant solution of the nonlinear system. Therefore, while all equilibrium points are operating points, the converse is not necessarily true. Consider $(\boldsymbol{x}^*, \boldsymbol{u}^*)$ to be one of the equilibrium points of the system. Then, the system (1.14) is denoted as $\boldsymbol{f}(\boldsymbol{x}^*, \boldsymbol{u}^*) = 0$ in the equilibrium point (as $\dot{\boldsymbol{x}}^* = 0$). Note that linear systems can have only a single equilibrium point, whereas nonlinear systems may possess multiple equilibrium points. The local stability behavior of the nonlinear system can be studied by linearizing it around an equilibrium point.

1.3.2 Linearization

The nonlinear system (1.14) can be linearized around the operating point $(\boldsymbol{x}_0, \boldsymbol{u}_0)$. Taking $\boldsymbol{x} = \boldsymbol{x}_0 + \Delta\boldsymbol{x}$ and $\boldsymbol{x} = \boldsymbol{u}_0 + \Delta\boldsymbol{u}$, using the Taylor series expansion, the system in (1.14) can be written as

$$\dot{\boldsymbol{x}}_0 + \Delta\dot{\boldsymbol{x}} = \boldsymbol{f}(\boldsymbol{x}_0, \boldsymbol{u}_0) + \left[\frac{\partial \boldsymbol{f}}{\partial \boldsymbol{x}}\right]_{(\boldsymbol{x}_0,\boldsymbol{u}_0)} \Delta\boldsymbol{x} + \left[\frac{\partial \boldsymbol{f}}{\partial \boldsymbol{u}}\right]_{(\boldsymbol{x}_0,\boldsymbol{u}_0)} \Delta\boldsymbol{u} + HOT \quad (1.15)$$

where *HOT* represents the collection of higher-order terms. The matrices $\left[\frac{\partial \boldsymbol{f}}{\partial \boldsymbol{x}}\right]$ and $\left[\frac{\partial \boldsymbol{f}}{\partial \boldsymbol{u}}\right]$ are evaluated as

$$\left[\frac{\partial \boldsymbol{f}}{\partial \boldsymbol{x}}\right]_{(\boldsymbol{x}_0,\boldsymbol{u}_0)} = \begin{bmatrix} \frac{\partial f_1}{\partial x_1} & \cdots & \frac{\partial f_1}{\partial x_n} \\ \vdots & \ddots & \vdots \\ \frac{\partial f_n}{\partial x_1} & \cdots & \frac{\partial f_n}{\partial x_n} \end{bmatrix}_{(\boldsymbol{x}_0,\boldsymbol{u}_0)} \triangleq \boldsymbol{A} \quad (1.16)$$

$$\left[\frac{\partial \boldsymbol{f}}{\partial \boldsymbol{u}}\right]_{(\boldsymbol{x}_0,\boldsymbol{u}_0)} = \begin{bmatrix} \frac{\partial f_1}{\partial u_1} & \cdots & \frac{\partial f_1}{\partial u_m} \\ \vdots & \ddots & \vdots \\ \frac{\partial f_n}{\partial u_1} & \cdots & \frac{\partial f_n}{\partial u_m} \end{bmatrix}_{(\boldsymbol{x}_0,\boldsymbol{u}_0)} \triangleq \boldsymbol{B} \quad (1.17)$$

Since $\dot{\boldsymbol{x}}_0 = \boldsymbol{f}(\boldsymbol{x}_0, \boldsymbol{u}_0)$, neglecting the *HOT*, the approximated form of (1.15) can be expressed as

$$\Delta\dot{\boldsymbol{x}} \approx \left[\frac{\partial \boldsymbol{f}}{\partial \boldsymbol{x}}\right]_{(\boldsymbol{x}_0,\boldsymbol{u}_0)} \Delta\boldsymbol{x} + \left[\frac{\partial \boldsymbol{f}}{\partial \boldsymbol{u}}\right]_{(\boldsymbol{x}_0,\boldsymbol{u}_0)} \Delta\boldsymbol{u} \quad (1.18)$$

At this stage, for convenience, with a slight abuse of notation, $\Delta \boldsymbol{x}$ and $\Delta \boldsymbol{u}$ are redefined as $\Delta \boldsymbol{x} \triangleq \boldsymbol{x}$ and $\Delta \boldsymbol{u} \triangleq \boldsymbol{u}$. With this, the local deviation dynamics of the nonlinear system (1.14) around $(\boldsymbol{x}_0, \boldsymbol{u}_0)$ can be written as

$$\dot{\boldsymbol{x}} = \boldsymbol{A}\boldsymbol{x} + \boldsymbol{B}\boldsymbol{u} \tag{1.19}$$

This is how a linearized system dynamics is obtained for a nonlinear system.

Example 1.4 Consider the system defined by

$$\ddot{x} + 0.6\dot{x} + 3x = 0$$

To obtain the equilibrium points of the system, it is first represented in the state-space form by selecting $x_1 = x$ and $x_2 = \dot{x}$. This results in

$$\begin{bmatrix} \dot{x}_1 \\ \dot{x}_2 \end{bmatrix} = \begin{bmatrix} x_2 \\ -0.6x_2 - 3x_1 \end{bmatrix} \tag{1.20}$$

The equilibrium points can be found by fixing $\dot{x}_1 = 0$ and $\dot{x}_2 = 0$. By solving, the equilibrium point is obtained as $x_1^* = 0$ and $x_2^* = 0$. It can be noticed that the given system is a linear system. Thus, it has only one equilibrium point.

Example 1.5 Consider the system

$$\ddot{x} + 0.6\dot{x} + 3x + x^2 = 0$$

Again, to determine the equilibrium points, the system is represented in state-space form by selecting $x_1 = x$ and $x_2 = \dot{x}$. This results in

$$\begin{bmatrix} \dot{x}_1 \\ \dot{x}_2 \end{bmatrix} = \begin{bmatrix} x_2 \\ -0.6x_2 - 3x_1 - x_1^2 \end{bmatrix} \tag{1.21}$$

Again, the equilibrium points can be found by fixing $\dot{x}_1 = 0$ and $\dot{x}_2 = 0$. By solving, the equilibrium point is obtained as $x_1^* = 0, -3$ and $x_2^* = 0$. It can be noticed that this nonlinear system has multiple equilibrium points, namely, $(0, 0)$ and $(-3, 0)$.

(continued)

Example 1.5 (continued)
To linearize the system, define $f_1 = x_2$ and $f_2 = -0.6x_2 - 3x_1 - x_1^2$. Hence, the system in (1.21) can be linearized as $\begin{bmatrix} \dot{x}_1 \\ \dot{x}_2 \end{bmatrix} = A \begin{bmatrix} x_1 \\ x_2 \end{bmatrix}$, where A is

$$A_{2\times 2} = \begin{bmatrix} \frac{\partial f_1}{\partial x_1} & \frac{\partial f_1}{\partial x_2} \\ \frac{\partial f_2}{\partial x_1} & \frac{\partial f_2}{\partial x_2} \end{bmatrix}_{(x_{1_0}, x_{2_0})} = \begin{bmatrix} 0 & 1 \\ -3 - 2x_{1_0} & -0.6 \end{bmatrix}$$

Hence, around equilibrium point (0, 0), the linear system is obtained as

$$\begin{bmatrix} \dot{x}_1 \\ \dot{x}_2 \end{bmatrix} = \begin{bmatrix} 0 & 1 \\ -3 & -0.6 \end{bmatrix} \begin{bmatrix} x_1 \\ x_2 \end{bmatrix}$$

and around equilibrium point (−3, 0), the linear system is obtained as

$$\begin{bmatrix} \dot{x}_1 \\ \dot{x}_2 \end{bmatrix} = \begin{bmatrix} 0 & 1 \\ 3 & -0.6 \end{bmatrix} \begin{bmatrix} x_1 \\ x_2 \end{bmatrix}$$

While linear systems provide valuable approximations in many practical scenarios (especially facilitating easier control system design), it is crucial to note that the validity of these linearized models diminishes when the system operates beyond its nominal operating range. Moreover, linear models fail to capture certain frequently observed phenomena in nonlinear systems, such as multiple equilibrium points, limit cycles, chaos, finite-time escape, and other intricate dynamic behaviors, which are explained in the next section.

1.4 Characteristics of Nonlinear Systems

A typical nonlinear system exhibits several unique characteristics in its response, which makes it difficult to analyze. The major frequently observed characteristics are given as follows:

1. *Multiple equilibrium points:* Nonlinear systems can have multiple equilibrium points. Moreover, based on the initial condition, a steady state trajectory for a stable system can converge to one of them.

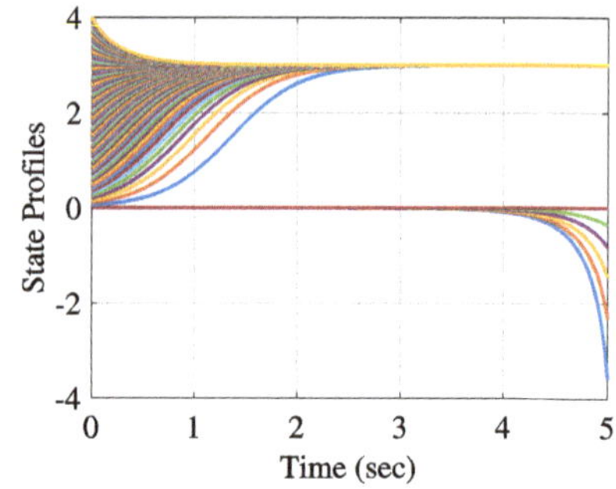

Fig. 1.2 Equilibrium points and their characteristics

Example 1.6 Consider the nonlinear system:

$$\dot{x} = 3x - x^2$$

where system has two equilibrium points $x^* = \{0,\ 3\}$. Based on the initial condition, the trajectory will converge to the equilibrium points. It can be seen in Fig. 1.2 that if the initial condition is the same as the equilibrium point, then the system trajectory remains at that point. However, for all other positive initial conditions, the system trajectories $x(t)$ converge to 3. Thus, the equilibrium point 3 is a stable equilibrium point. On the other hand, since system trajectories are diverging in nature, if the initial conditions are chosen around the equilibrium point 0, it is an unstable equilibrium point.

Example 1.7 The nonlinear system

$$\dot{x} = 3x - x^3$$

where system has three equilibrium points $\boldsymbol{x}^* = \{0,\ \pm\sqrt{3}\}$. The trajectory will converge to the equilibrium points based on the nature of the equilibrium points. Since trajectories are converging to $\boldsymbol{x}^* = \{\sqrt{3}, -\sqrt{3}\}$, both equilibrium points are stable, whereas $x^* = 0$ is an unstable equilibrium point. The system trajectories are shown in Fig. 1.3.

2. *Limit cycle:* Nonlinear systems can display oscillations for a fixed time period without external excitation. These oscillations are called limit cycles, or self-excited oscillations. A limit cycle can be classified as stable or unstable depending on the converging or diverging behavior of the trajectories of the solution in its vicinity, respectively.

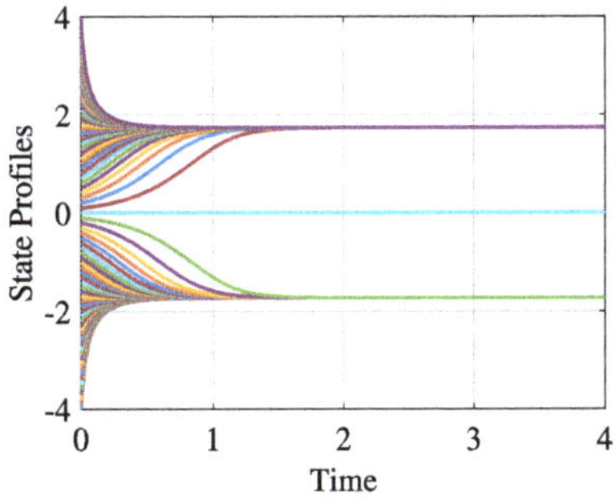

Fig. 1.3 Equilibrium points and their characteristics

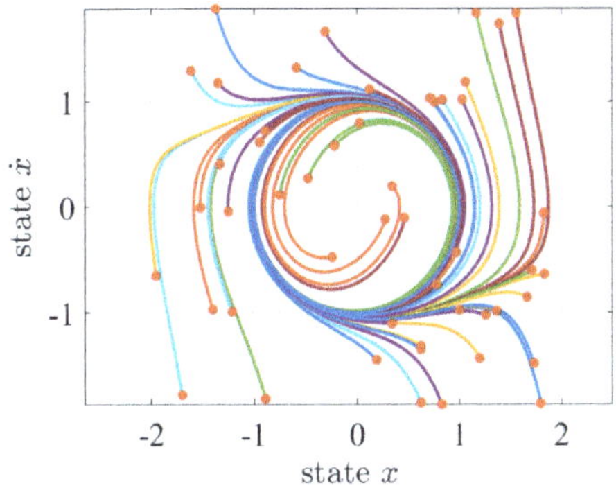

Fig. 1.4 Limit cycle of nonlinear system

Example 1.8 The dynamics of a nonlinear system described by

$$\ddot{x} = -x + \left(1 - x^2 - \dot{x}^2\right)\dot{x}$$

has a stable limit cycle. The time history of the system trajectories (as shown in Fig. 1.4) is converging to the limit cycle.

3. *Chaos:* For stable linear systems, small differences in initial conditions can only cause small differences in output. On the other hand, nonlinear systems can display a phenomenon called chaos, where the system output is extremely sensitive to initial conditions. This chaotic behavior of the nonlinear system can lead to a more complicated behavior. The essential feature of chaos is the unpredictability of the system trajectory. The truncation error of computers also makes system trajectory prediction inaccurate. That is why special higher-order numerical integration schemes with high bit computation are generally used to mitigate this error to a good extent. Even if the model of a nonlinear system is exactly known, the response of the system in the long run still cannot be well predicted.

Fig. 1.5 Chaotic behavior of the nonlinear system

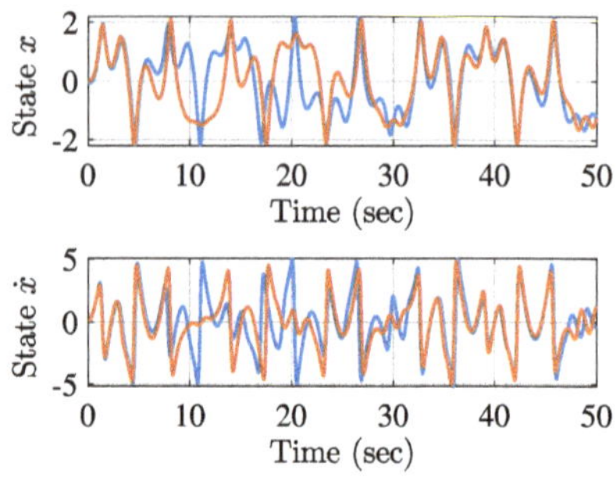

Example 1.9 The nonlinear system

$$\ddot{x} + 0.1\dot{x} + x^5 = 6\sin(t)$$

has a chaotic behavior due to the presence of strong nonlinearity x^5. By changing the initial condition, the behavior becomes chaotic. As can be seen from Fig. 1.5, a slight change in initial condition leads to a large difference in output response.

Note that weather modeling typically results in a chaotic system. Hence, even though short-duration weather prediction is fairly accurate in general, long-duration prediction becomes a challenge even with supercomputers.

4. *Finite-time escape :* In the case of linear systems, an unstable response can reach infinity as time approaches infinity. However, in the case of nonlinear systems, a phenomenon occurs where a system can become unbounded within a finite-time interval. This property of nonlinear systems is known as finite-time escape.

Example 1.10 Consider the nonlinear system:

$$\dot{x} = x^2$$

Subject to $x_0 = 1$, it has the solution

$$x(t) = \frac{1}{t-1}$$

As $t \to 1$, $x(t) \to \infty$. Hence, the solution is valid for $t \in [0, 1)$. Thus, the solution has a finite-time escape at $t = 1$.

Table 1.2 Comparison between nonlinear and linear systems

S.N.	Nonlinear systems	Linear systems
1	More realistic	Approximation to reality
2	Need not obey the principle of superposition (not necessarily linear)	Obeys the principle of superposition
3	Usually difficult to analyze and design	Usually simpler to analyze and design
4	Tools are under development	A lot of tools are well-developed
5	Can have multiple equilibrium points	Only single equilibrium point
6	Non-uniqueness of solution	Unique solution
7	System stability depends on the initial condition	Stability nature is independent of the initial condition (Hence, "zero" initial conditions are assumed in the transfer function approach)
8	Limit cycles (self-sustained oscillations)	No limit cycles
9	Bifurcations (number of equilibrium points and their stability nature can vary with parameter values)	No bifurcation
10	Finite-time escape	No finite-time escape
11	Chaos (very small difference in I.C. can lead to a large difference in output as time increases. That is why predicting the weather for a long time is inaccurate)	No chaos

A brief comparison between the characteristics of nonlinear and linear systems is given in Table 1.2.

Following the introduction of nonlinear systems and their distinctive characteristics, the next section presents the necessary mathematical preliminaries. Throughout the subsequent chapters, these foundational concepts are applied to analyze nonlinear systems and to examine the design and analysis of closed-loop control systems.

1.5 Mathematical Preliminaries

A few selected fundamental mathematical principles used in control system design and analysis are presented in this section for completeness so that the reader need not refer to other literature for the same.

1.5.1 Vector and Vector Norm

A vector $\boldsymbol{x} = [x_1, \ldots, x_n]^T \in \Re^n$ belongs to a set of real numbers. For example, $\boldsymbol{x} = [1, 1]^T$ is a two-dimensional vector, represented as $\boldsymbol{x} \in \Re^2$, having two real numbers as its elements.

The norm of a vector $\boldsymbol{x}$ is a real-valued function often represented as $\|\boldsymbol{x}\|$, which has the following properties:

1. $\|\boldsymbol{x}\| > 0 \; \forall \boldsymbol{x} \in \Re^n$ and $\|\boldsymbol{x}\| = 0$ if and only if $\boldsymbol{x} = 0$
2. $\|\alpha \boldsymbol{x}\| = |\alpha| \|\boldsymbol{x}\|$ for any scalar $\alpha \in \Re$
3. $\|\boldsymbol{x} + \boldsymbol{y}\| \leq \|\boldsymbol{x}\| + \|\boldsymbol{y}\| \quad \forall \boldsymbol{x}, \boldsymbol{y}$ (triangle inequality)

The concept of norm is introduced to invoke the notion of distance between two vectors. The distance can be measured as the absolute sum of vector elements, as the square root of the quadratic sum of vector elements, or in many other similar ways. In concise form, the pth norm of a vector, known as l_p norm, is defined as follows:

$$\|\boldsymbol{x}\|_p = \left(\sum_{i=1}^{n} |x_i|^p \right)^{\frac{1}{p}}, \quad \text{for } 1 \leq p < \infty$$

$$\|\boldsymbol{x}\|_\infty = \max_{1 \leq j \leq n} |x_i|, \quad \text{for } p = \infty$$

The commonly used vector norms are

$$\|\boldsymbol{x}\|_1 = |x_1| + |x_2| + \cdots + |x_n| \qquad (l_1 \text{ norm})$$
$$\|\boldsymbol{x}\|_2 = \left(|x_1|^2 + |x_2|^2 + \cdots + |x_n|^2\right)^{1/2} \qquad (l_2 \text{ norm})$$
$$\|\boldsymbol{x}\|_\infty = \left(|x_1|^\infty + |x_2|^\infty + \cdots + |x_n|^\infty\right)^{1/\infty} = \max_i |x_i| \; (l_\infty \text{ norm})$$

where $\|\boldsymbol{x}\|_1$ is a sum of the absolute value of the elements of the vector elements; $\|\boldsymbol{x}\|_2$, also known as Euclidean norm, is positive square root of square sum of vector elements; and $\|\boldsymbol{x}\|_\infty$ represents the largest absolute element of a vector.

To illustrate the concept of the vector norm more clearly, consider $\boldsymbol{x} = [x_1, x_2]^T$, where x_1 and x_2 are chosen such that norms $\|\boldsymbol{x}\|_1 = \|\boldsymbol{x}\|_2 = \|\boldsymbol{x}\|_\infty = 1$. Then, the relation between norms $\|\boldsymbol{x}\|_1$, $\|\boldsymbol{x}\|_2$, and $\|\boldsymbol{x}\|_\infty$ is graphically presented in Fig. 1.6, where l_1 norm is bounded by l_2 norm and l_∞ norm is the upper bound of all norms.

Norm Inequalities Vector norm inequalities are frequently utilized in mathematical analysis to show the stability and boundedness of a dynamical system response, besides numerous other applications. In a finite-dimensional vector space, all norms are equivalent; i.e., any two vector norms are related to one another by the following inequality:

$$k_1 \|\boldsymbol{x}\|_a \leq \|\boldsymbol{x}\|_b \leq k_2 \|\boldsymbol{x}\|_a, \quad \forall \boldsymbol{x} \in \Re^n \tag{1.22}$$

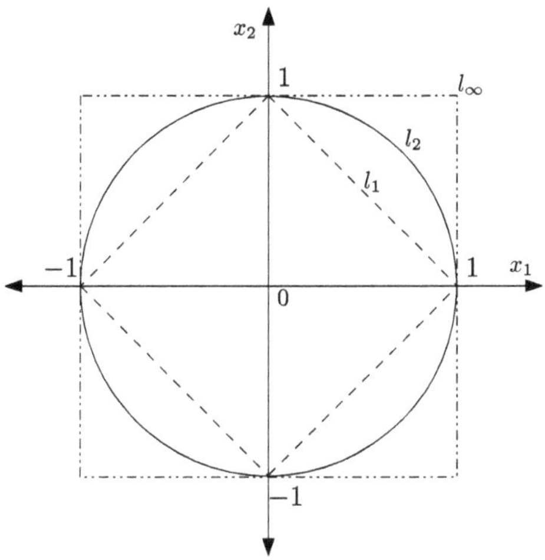

Fig. 1.6 Representation of norms $\|\boldsymbol{x}\|_1 = \|\boldsymbol{x}\|_2 = \|\boldsymbol{x}\|_\infty = 1$

where a and b represent norm and k_1 and k_2 are positive constants related as $k_2 \geq k_1$. Especially, the following inequalities can be proven for an n-dimensional vector, which are useful in the context of this book:

$$\begin{aligned}
\|\boldsymbol{x}\|_2 &\leq \|\boldsymbol{x}\|_1 \leq \sqrt{n}\,\|\boldsymbol{x}\|_2, && \forall \boldsymbol{x} \in \Re^n \\
\|\boldsymbol{x}\|_\infty &\leq \|\boldsymbol{x}\|_1 \leq n\,\|\boldsymbol{x}\|_\infty, && \forall \boldsymbol{x} \in \Re^n \\
\|\boldsymbol{x}\|_\infty &\leq \|\boldsymbol{x}\|_2 \leq \sqrt{n}\,\|\boldsymbol{x}\|_\infty, && \forall \boldsymbol{x} \in \Re^n
\end{aligned}$$

Other notable and useful inequalities are listed as follows:

(a) Holder's Inequality Holder's inequality relates the norm of the product of two vectors to the product of their norm. Let constants $p \in \Re$ and $q \in \Re$ satisfy the relation

$$\frac{1}{p} + \frac{1}{q} = 1$$

where $p > 1$, then

$$\left\|\boldsymbol{x}^T \boldsymbol{y}\right\|_1 \leq \|\boldsymbol{x}\|_p \|\boldsymbol{y}\|_q, \quad \forall \boldsymbol{x} \in \Re^n, \boldsymbol{y} \in \Re^n \tag{1.23}$$

(b) Cauchy-Schwarz Inequality The famous Cauchy-Schwarz inequality is a special form of Holder's inequality with $p = 2$ and $q = 2$. The inequality relates the dot product (or inner product) of two vectors to the product of their Euclidean norm. The Cauchy-Schwarz inequality has the following form:

$$\left\|\boldsymbol{x}^T \boldsymbol{y}\right\|_1 \leq \|\boldsymbol{x}\|_2 \|\boldsymbol{y}\|_2, \quad \forall \boldsymbol{x}, \boldsymbol{y} \in \Re^n. \tag{1.24}$$

(c) Minkowski's Inequality Minkowski's inequality relates the norm of the summation of two vectors to their individual norms. Let $p \in \Re$ and satisfy $p \geq 1$. Then

$$\|x + y\|_p \leq \|x\|_p + \|y\|_p, \quad \forall x \in \Re^n, y \in \Re^n \tag{1.25}$$

Note that this is also called "triangle inequality."

1.5.2 *Matrix and Matrix Norm*

A matrix is a collection of elements (numbers) arranged in rows and columns. For example, a matrix A with m rows and n column is represented as

$$A = \begin{bmatrix} a_{11} & \dots & a_{1n} \\ \vdots & \ddots & \vdots \\ a_{m1} & \dots & a_{mn} \end{bmatrix}$$

where a_{ij} $\forall i = 1, \dots, m$ and $\forall j = 1, \dots, n$ are matrix elements.

Matrices are divided into various categories based on the number of rows and columns and the nature of elements, such as (i) if $m = 1$, the matrix is called a *"row vector"*; (ii) if $n = 1$, the matrix is called a *"column vector"*; (iii) if $m = n$, the matrix is said to be a *"square matrix"*; (iv) for square matrices ($m = n$), if $a_{ij} \neq 0$ $\forall i = j$ and $a_{ij} = 0$ $\forall i \neq j$, the matrix A is known as a *"diagonal matrix"*; if $a_{ij} = 0$ $\forall i > j$, the matrix A is known as an *"upper triangular matrix"*; and if $a_{ij} = 0$ $\forall i < j$, the matrix A is known as a *"lower triangular matrix."* Note that a matrix can also be interpreted as a collection of rows or column vectors stacked together.

Eigenvalue and Eigenvector One of the key properties of matrices is their ability to act as linear operators, performing magnification (stretching/shrinking) and rotation operations on vectors. When a matrix multiplies a vector, the resulting vector undergoes stretching/shrinking and rotation, resulting in another vector. As shown in Fig. 1.7, the vector x is multiplied by the matrix A, resulting in a new vector $y = Ax$, which is stretched and rotated by angle $(\theta_y - \theta_x)$. For example, multiplication of vector $x = \begin{bmatrix} 1 \\ 0 \end{bmatrix}$ with matrix $A = \begin{bmatrix} 0 & 1 \\ 2 & 1 \end{bmatrix}$ leads to a new vector $y = Ax = \begin{bmatrix} 0 \\ 2 \end{bmatrix}$, which is stretched by two times and rotated by 90° compared to vector x.

An eigenvalue is a special scalar associated with a square matrix. If, upon multiplication, the direction of the vector remains unchanged, the scalar by which the vector is scaled is called an eigenvalue, and the nonzero vector that maintains

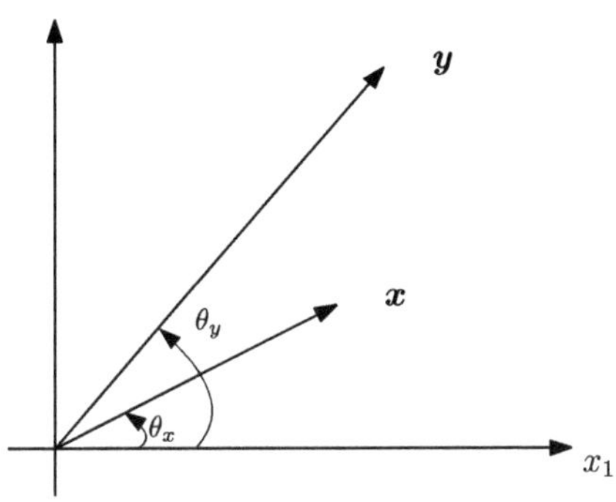

Fig. 1.7 Representation of vector rotation by matrix multiplication

Table 1.3 Various forms of matrix A

Terminology	Definition		Properties of eigenvalues
Symmetric matrix	$A^T = A$		$\lambda_i \in \Re \; \forall i$ are real
Skew-symmetric matrix	$A^T = -A$		$\lambda_i \; \forall i$ are zero or imaginary
Positive definite matrix	$x^T A x > 0$	$\forall x \neq 0$	$\lambda_i > 0, \quad \forall i$
Positive semi-definite matrix	$x^T A x \geq 0$	$\forall x \neq 0$	$\lambda_i \geq 0, \quad \forall i$
Negative definite matrix	$x^T A x < 0$	$\forall x \neq 0$	$\lambda_i < 0, \quad \forall i$
Negative semi-definite matrix	$x^T A x \leq 0$	$\forall x \neq 0$	$\lambda_i \leq 0, \quad \forall i$

its direction is referred to as an eigenvector. Mathematically, for a square matrix A, the relationship can be expressed as

$$Ax = \lambda x$$
$$(A - \lambda I)x = 0$$

If the nontrivial solution of $\|A - \lambda I\| = 0$ exist, then λ is the eigenvalue of matrix A, and x is the corresponding eigenvector

In a simple interpretation, eigenvalues indicate the extent to which a matrix will stretch or shrink a vector, while the corresponding eigenvectors represent the directions that remain unchanged by this matrix operation. Note that for real matrices, eigenvalues can also appear in complex pairs. For complex eigenvalues, the stretching operation is not quite intuitive.

Other important types of matrices are known as symmetric and skew-symmetric matrices, positive definite matrices, positive semi-definite matrices, negative definite matrices, and negative semi-definite matrices. These matrices are extensively used in the control system design of nonlinear systems. These matrices can also be distinguished by the nature of their eigenvalues. Considering a matrix A and λ_i as its ith eigenvalue, the definition and properties of these matrices are tabulated in Table 1.3.

It can be noted that the inequalities in the second column of Table 1.3 are the definition, even though the inequalities in the third column are often misunderstood as the definition, as both hold good in the "if and only if" sense.

Next, some important results related to matrix analysis are summarized below.

Theorem 1.1 *For any matrix $\boldsymbol{A} \in \Re^{m \times n}$, both $\boldsymbol{A}\boldsymbol{A}^T \in \Re^{m \times m}$ and $\boldsymbol{A}^T\boldsymbol{A} \in \Re^{n \times n}$ are square, symmetric and positive semi-definite.*

The proof can be outlined as follows:

1. *Square property: The product $\boldsymbol{A}\boldsymbol{A}^T$ has dimensions $m \times m$ and $\boldsymbol{A}^T\boldsymbol{A}$ has dimensions $n \times n$. Since $\boldsymbol{A}\boldsymbol{A}^T \in \Re^{m \times m}$ and $\boldsymbol{A}^T\boldsymbol{A} \in \Re^{n \times n}$, they are square matrices.*
2. *Symmetric property:*

$$(\boldsymbol{A}\boldsymbol{A}^T)^T = (\boldsymbol{A}^T)^T\boldsymbol{A}^T = \boldsymbol{A}\boldsymbol{A}^T$$

$$(\boldsymbol{A}^T\boldsymbol{A})^T = \boldsymbol{A}^T(\boldsymbol{A})^T = \boldsymbol{A}^T\boldsymbol{A}$$

   Hence, both are symmetric matrices.
3. *Positive semi-definiteness: For any vector $\boldsymbol{x} \in \Re^m$, consider the quadratic form $\boldsymbol{x}^T(\boldsymbol{A}\boldsymbol{A}^T)\boldsymbol{x}$, which can be analyzed as*

$$\boldsymbol{x}^T(\boldsymbol{A}\boldsymbol{A}^T)\boldsymbol{x} = (\boldsymbol{A}^T\boldsymbol{x})^T(\boldsymbol{A}^T\boldsymbol{x}) = \|\boldsymbol{A}^T\boldsymbol{x}\|^2 \geq 0.$$

   Similarly, for any vector $\boldsymbol{y} \in \Re^n$, consider the quadratic form $\boldsymbol{y}^T(\boldsymbol{A}^T\boldsymbol{A})\boldsymbol{y}$, which can be analyzed as

$$\boldsymbol{y}^T(\boldsymbol{A}^T\boldsymbol{A})\boldsymbol{y} = (\boldsymbol{A}\boldsymbol{y})^T(\boldsymbol{A}\boldsymbol{y}) = \|\boldsymbol{A}\boldsymbol{y}\|^2 \geq 0.$$

   This shows that both matrices are positive semi-definite.

Theorem 1.2 *For a symmetric positive semi-definite matrix $\boldsymbol{A}$, following inequality holds*

$$\lambda_{\min}(\boldsymbol{A})\|\boldsymbol{x}\|_2 \leq \boldsymbol{x}^T\boldsymbol{A}\boldsymbol{x} \leq \lambda_{\max}(\boldsymbol{A})\|\boldsymbol{x}\|_2$$

A pictorial representation of this inequality is shown in Fig. 1.8. The lower and upper bounds on $\boldsymbol{x}^T\boldsymbol{A}\boldsymbol{x}$ are shown in this picture, where $\boldsymbol{x} \in \Re^2$.

Matrix Norm Revising the fact that matrix $\boldsymbol{A}$ is a linear operator that transforms a vector $\boldsymbol{x}$ to another vector $\boldsymbol{y}$ in the sense $\boldsymbol{y} = \boldsymbol{A}\boldsymbol{x}$, the concept of a matrix norm is introduced to indicate the amount of magnification (stretching/shrinking) the matrix can do to a unit vector. Matrix norms are essential tools in control systems for analyzing and predicting system behavior, showing insights into stability, sensitivity, error, robustness, and convergence. Therefore, matrix norms play a critical role in designing effective control systems.

Fig. 1.8 Representation of Theorem 1.2

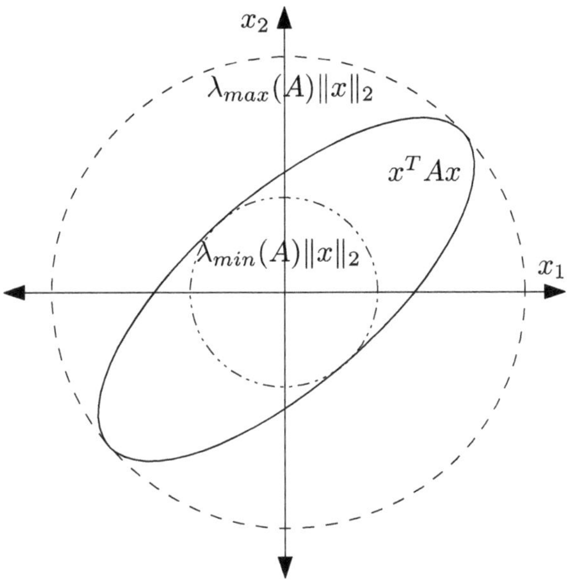

The p-norm of a matrix $\boldsymbol{A} \in \Re^{m\times n}$ is defined as

$$\|\boldsymbol{A}\|_p = \sup_{\boldsymbol{x}\neq 0} \frac{\|\boldsymbol{Ax}\|_p}{\|\boldsymbol{x}\|_p}$$
$$= \sup_{\|\boldsymbol{x}\|_p=1} \left(\|\boldsymbol{Ax}\|_p\right)$$

where $p \in \Re$. For matrices $\boldsymbol{A}$ and $\boldsymbol{B}$ with appropriate dimensions, the norm of a matrix has the following properties:

1. $\|\boldsymbol{A}\|_p > 0$ and $\|\boldsymbol{A}\|_p = 0$ only if $\boldsymbol{A} = 0$
2. $\|\alpha \boldsymbol{A}\|_p = |\alpha|\|\boldsymbol{A}\|_p$
3. $\|\boldsymbol{A} + \boldsymbol{B}\|_p \leq \|\boldsymbol{A}\|_p + \|\boldsymbol{B}\|_p$
4. $\|\boldsymbol{AB}\|_p \leq \|\boldsymbol{A}\|_p \|\boldsymbol{B}\|_p$

where α is a constant.

The commonly used first, second, and infinity matrix norms can be defined as

$$\|\boldsymbol{A}\|_1 = \max_j \sum_{i=1}^{m} |a_{ij}| \quad \text{(largest of the absolute column sums)}$$

$$\|\boldsymbol{A}\|_2 = \max_i \sqrt{\lambda_i \left(\boldsymbol{A}^T \boldsymbol{A}\right)}$$

$$\|\boldsymbol{A}\|_\infty = \max_i \sum_{j=1}^{m} |a_{ij}| \quad \text{(largest of the absolute row sums)}$$

Here, $\lambda_i\left(A^T A\right)$ represents the ith eigenvalue of matrix $A^T A$, and a_{ij} denotes the matrix element on the ith row and the jth column.

A special type of norm, known as *Frobenius norm*, is frequently used in neural network and adaptive control literature. The Frobenius norm of a matrix A is defined as

$$\|A_{m\times n}\|_F = \sqrt{\sum_{i=1}^{m}\sum_{j=1}^{n} a_{ij}^2} = \sqrt{\operatorname{tr}\left(A^T A\right)}$$

where tr(·) is the trace of the matrix in its argument, defined as the sum of diagonal elements of the matrix.

Example 1.11 Let a matrix A be defined as

$$A = \begin{bmatrix} 1 & -2 \\ 3 & 7 \end{bmatrix}$$

In that case, various norms of this matrix are calculated as follows:

(*i*) $\|A\|_1$ is the largest of the absolute column sum, which is $(|-2| + |7|)$. Hence, $\|A\|_1 = 9$.
(*ii*) To calculate $\|A\|_2$, first calculate $A^T A$.

$$A^T A = \begin{bmatrix} 10 & 19 \\ 19 & 53 \end{bmatrix}$$

The eigenvalues of the matrix $A^T A$ are 2.81 and 60.19. The square roots of the eigenvalues are 1.67 and 7.76. Hence, $\|A\|_2 = 7.76$.
(*iii*) $\|A\|_\infty$ is the largest of absolute row sum. Hence, $\|A\|_\infty = |3| + |7| = 10$.
(*iv*) To calculate $\|A\|_F$, first calculate $\operatorname{tr}\left(A^T A\right)$, which is $10 + 53 = 63$. Hence, $\|A\|_F = \sqrt{64} = 7.94$.

1.5.3 Function and Function Norm

A function defines the relationship between an independent variable and a dependent variable. For example, in the equation $y = f(x) : \Re^n \to \Re^m$, the function f specifies how changes in the independent variable $x \in \Re^n$ correspond to changes in the dependent variable $y \in \Re^m$.

Definition 1.1 A function $\boldsymbol{f}(\boldsymbol{x}) : \Re^n \to \Re^m$ is *continuous* at a point $\boldsymbol{x}$ if $\boldsymbol{f}(\boldsymbol{x} + \Delta\boldsymbol{x}) \to \boldsymbol{f}(\boldsymbol{x})$ as $\Delta\boldsymbol{x} \to \boldsymbol{0}$. In other words, a function $\boldsymbol{f} : \Re^n \to \Re^m$ is continuous in $\Re^n$ if given $\epsilon > 0$, there exists $\delta > 0$ such that

$$\|\boldsymbol{x} - \boldsymbol{y}\| < \delta \Longrightarrow \|\boldsymbol{f}(\boldsymbol{x}) - \boldsymbol{f}(\boldsymbol{y})\| < \epsilon$$

Definition 1.2 A function $f_i(x_j) : \Re \to \Re \; \forall \; 1 \leq i \leq m, 1 \leq j \leq n$ is said to be *differentiable* at a point x_j if the limit

$$\frac{\partial f_i(x_j)}{\partial x_j} \triangleq \lim_{\delta x_j \to 0} \frac{f_i(x_j + \delta x_j) - f_i(x_j)}{\delta x_j}$$

exists.

Definition 1.3 A function $\boldsymbol{f}(\boldsymbol{x}) : \Re^n \to \Re^m$ is *continuously differentiable* at point $\boldsymbol{x}$ if the partial derivatives $\partial f_i / \partial x_j$ exist and are continuous at $\boldsymbol{x}$ for $1 \leq i \leq m, 1 \leq j \leq n$.

Definition 1.4 A function $\boldsymbol{f}(\boldsymbol{x}) : \Re^n \to \Re^m$ is said to be *continuously differentiable of order k*, or C^k, if

$$D^a \boldsymbol{f} \triangleq \frac{\partial^{a_1}}{\partial x_1^{a_1}} \frac{\partial^{a_2}}{\partial x_2^{a_2}} \cdots \frac{\partial^{a_n}}{\partial x_n^{a_n}} \boldsymbol{f}$$

exists and is continuous for all points $(x_1, x_2, \ldots, x_n)$ in $\Re^n$ and all nonnegative integers $a_1, a_2, \ldots, a_n$ satisfying $\sum_{i=1}^{n} a_i \leq k$

Definition 1.5 If $f(\boldsymbol{x}) : \Re^n \to \Re$ is continuously differentiable, then the *gradient* of $f(\boldsymbol{x})$ is given by

$$\frac{\partial f(\boldsymbol{x})}{\partial \boldsymbol{x}} \triangleq \begin{bmatrix} \frac{\partial f}{\partial x_1} & \cdots & \frac{\partial f}{\partial x_n} \end{bmatrix}^T.$$

Definition 1.6 If $\boldsymbol{f}(\boldsymbol{x}) : \Re^n \to \Re^m$ is continuously differentiable, then the *Jacobian* of $\boldsymbol{f}(\boldsymbol{x})$ is given by

$$\frac{\partial \boldsymbol{f}(\boldsymbol{x})}{\partial \boldsymbol{x}} \triangleq \begin{bmatrix} \frac{\partial f_1}{\partial x_1} & \cdots & \frac{\partial f_1}{\partial x_n} \\ \vdots & \ddots & \vdots \\ \frac{\partial f_m}{\partial x_1} & \cdots & \frac{\partial f_m}{\partial x_n} \end{bmatrix}.$$

Function Norm Function norm is useful in control theory, as it helps in analyzing the function behavior, particularly in assessing properties like *stability* and *convergence*. Function norms are also applied in error estimation and optimization.

The $\mathcal{L}_p$ norm of a continuous function $\boldsymbol{f}(t) \in \Re^n$, $t \geq t_0 \in \Re$, is defined as

$$\|\boldsymbol{f}\|_p = \left(\int_{t_0}^{\infty} \sum_{i=1}^{n} |f_i(t)|^p dt \right)^{1/p}, \qquad \text{for } p \in [1, \infty)$$

$$\|\boldsymbol{f}\|_\infty = \sup_{t \in [t_0, \infty)} \max_{1 \leq i \leq n} |f_i(t)|$$

where t_0 is the initial time and the notation *sup* denotes the *supremum* for all $t \geq t_0$ that leads to the largest value of the argument.

Using the definition of $\mathcal{L}_p$ norm, the commonly used $\mathcal{L}_1$, $\mathcal{L}_2$, and $\mathcal{L}_\infty$ norms for a function $\boldsymbol{f}(t) \in \Re^n$ can be defined as

$$\|\boldsymbol{f}\|_1 = \int_{t_0}^{\infty} \sum_{i=1}^{n} |f_i(t)|\, dt$$

$$\|\boldsymbol{f}\|_2 = \sqrt{\int_{t_0}^{\infty} \sum_{i=1}^{n} |f_i(t)|^2\, dt}$$

$$\|\boldsymbol{f}\|_\infty = \sup_{t \geq t_0} \max_{1 \leq i \leq n} |f_i(t)|$$

These norms are highly useful in nonlinear system analysis and control system design. For instance, the $\mathcal{L}_2$ norm is used to analyze the energy of a signal, while the $\mathcal{L}_\infty$ norm is used to analyze the maximum magnitude of a signal.

Example 1.12 The $\mathcal{L}_1$, $\mathcal{L}_2$, and $\mathcal{L}_\infty$ norms of the function $\boldsymbol{f}(t)$

$$\boldsymbol{f}(t) = \begin{bmatrix} 2e^{-2t} \\ -e^{-t} \sin t \end{bmatrix}$$

for $t \geq 0$ can be calculated as

$$\begin{aligned} \|\boldsymbol{f}\|_1 &= \int_0^{\infty} \left(2e^{-2t} + e^{-t} \sin t\right) dt \\ &= \int_0^{\infty} 2e^{-2t}\, dt + \int_0^{\infty} e^{-t} \sin t\, dt \\ &= -2\left[\frac{e^{-2t}}{2}\right]_0^{\infty} + \frac{1}{(1)^2 + (1)^2} \end{aligned}$$

(continued)

Example 1.12 (continued)

$$
\begin{aligned}
&= 1 + \frac{1}{2} \\
&= 1.5 \\
\|f\|_2 &= \sqrt{\int_0^\infty \left(4e^{-4t} + e^{-2t}\sin^2 t\right) dt} \\
&= \sqrt{\int_0^\infty 4e^{-4t}\, dt + \int_0^\infty e^{-2t}\sin^2 t\, dt} \\
&= \sqrt{-4\left[\frac{e^{-4t}}{4}\right]_0^\infty + \frac{1^2}{2(2^2+4)}} \\
&= \sqrt{1 + \frac{1}{16}} \\
&= \frac{\sqrt{17}}{4} \\
&= 1.03 \\
\|f\|_\infty &= \sup_{t\geq 0}\left\{\max\left|2e^{-2t}\right|, \max\left|-e^{-t}\sin t\right|\right\} \\
&= \sup_{t\geq 0}\{2, 1\} \\
&= 2
\end{aligned}
$$

The component of the function $f(t)$ and corresponding function norms are shown in Fig. 1.9. It can be seen that the $\|f\|_\infty$ is the absolute maximum bound of function $f(t)$.

Note that in some cases, certain norms of a function may not exist, while others do. For example, consider the function $f(t) = \cos(t)$. The integral of $\cos(t)$ over the interval $t \in [0, \infty)$ is undefined because $\cos(t)$ does not decay to zero as $t \to \infty$, and it oscillates indefinitely. Consequently, both the $\mathscr{L}_1$ (absolute integrability) and $\mathscr{L}_2$ (square integrability) norms of this function do not exist over $[0, \infty)$. However, the $\mathscr{L}_\infty$ norm, which measures the maximum absolute value of the function over time, does exist. In fact, $\|f\|_\infty = 1$, as $\cos(t)$ oscillates between -1 and 1.

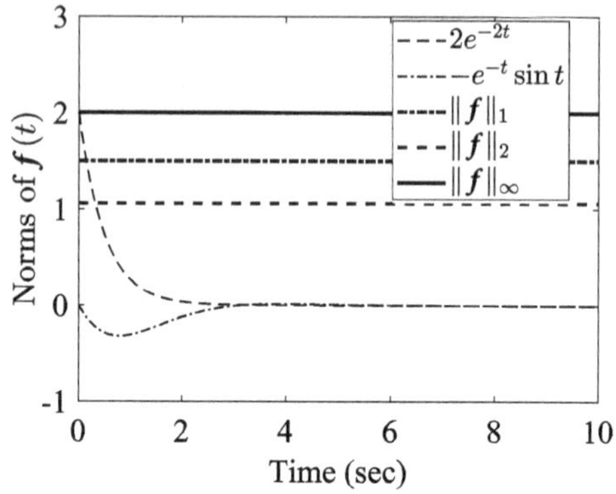

Fig. 1.9 Representation function $f(t)$ and corresponding function norms

1.6 Summary

This chapter provides an overview of nonlinear systems, emphasizing their significance in control design, as well as their classification and some distinctive characteristics. A concise comparison between nonlinear and linear systems is included to enhance clarity. An outline of linearization and equilibrium points is also incorporated in this chapter. Several examples are included to illustrate these concepts clearly. The chapter also introduces key mathematical preliminaries, such as vectors, matrices, functions, and their norms, which are frequently utilized throughout the book. Overall, this chapter establishes a good foundation for the rest of the book, offering essential background knowledge and setting the stage for subsequent chapters. While it briefly discusses several key concepts, readers seeking a more comprehensive exploration of these topics are encouraged to refer to additional sources given at the end of this chapter.

1.7 Exercise Problems

1. Linearize the following systems around origin (0, 0) and represent them in linear state space form (assume a, b, c, d are constants):

   (a) $\ddot{x} = a\dot{x}x^2 + \frac{bx}{(1-\dot{x}^2)} + x^3 u$

   (b) $\dddot{x} = a\ddot{x} + \frac{u}{x}$

   (c) $\ddddot{x} = a\ddot{x} + b\dot{x} + x^2 + cu$

   (d) $\begin{bmatrix} \ddot{x}_1 \\ \ddot{x}_2 \end{bmatrix} = \begin{bmatrix} a\dot{x}_1\dot{x}_2^2 + \frac{bx_2}{(1-\dot{x}_1^2)} + x_1\dot{x}_1 u_1 + \left(1+\dot{x}_2^2\right) u_2 \\ a\dot{x}_1 x_2^2 + bx_2\left(1-\dot{x}_1^2\right) + \left(1+x_2^4\right) u_1 + u_2 \end{bmatrix}$

2. Determine the equilibrium points and linearize the systems around them:
   (a) $\dot{x} = -x^2 + 3x$
   (b) $\dot{x} = -x + x^3$
   (c) $\dot{x} = x - x^3$

3. Determine the equilibrium points and linearize the systems around them:
   (a) $\ddot{x} = -\sin x$
   (b) $\ddot{x} = -x + \left(1 - x^2 - \dot{x}^2\right)\dot{x}$
   (c) $\begin{bmatrix} \dot{x}_1 \\ \dot{x}_2 \end{bmatrix} = \begin{bmatrix} x_2 \\ -\frac{g}{l}\sin x_1 + \omega^2 \sin x_1 \cos x_1 \end{bmatrix}$

4. Show that the Van der Pol oscillator described by the dynamics

$$\ddot{x} - \mu(1 - x^2)\dot{x} + x = 0$$

   has a stable limit cycle.

5. Determine the analytical solution of the following system:

$$\dot{x} = |x|x^2$$

   with a general initial condition x_0.
   (a) Let $x_0 = 1$. Does this solution have a finite escape time?
   (b) Repeat part a with $x_0 = -1$ and comment on the behavior.

6. Find the first, second, and infinite norms of the following vectors and validate the Holder and Cauchy-Schwartz inequalities:
   (a) $\boldsymbol{x} = [2, 1, 1, 3]^T$
   (b) $\boldsymbol{y} = [1, 1, 1, 1]^T$

7. Find the first, second, and infinite norms of the matrices defined as
   (a) $\boldsymbol{A} = \begin{bmatrix} 1 & 2 & 1 \\ 3 & 4 & 3 \\ -1 & 0 & -1 \end{bmatrix}$
   (b) $\boldsymbol{B} = \begin{bmatrix} 1 & 1 & 1 \\ 0 & 2 & 2 \\ 0 & 0 & 3 \end{bmatrix}$
   (c) $\boldsymbol{C} = \begin{bmatrix} 0 & 6 & -5 \\ 1 & 0 & 2 \\ 3 & 2 & 4 \end{bmatrix}$
   (d) $\boldsymbol{D} = \begin{bmatrix} 1 & 5 & 3 \\ 3 & -3 & 5 \\ 7 & 1 & 2 \end{bmatrix}$

8. Find out $\mathscr{L}_1$, $\mathscr{L}_2$, and $\mathscr{L}_\infty$ norms of the following functions for $t \geq 0$:
   (a) $f(t) = \exp^{-t}$
   (b) Impulse function
   (c) $f(t) = \begin{cases} \sin(t), & t \neq 4 \\ 2 & t = 4 \end{cases}, \quad \text{where} \quad t \leq \infty$
   (d) $f(t) = 3$

References

M.R. Akella, Adaptive control—a departure from the certainty-equivalence paradigm. J. Astronaut. Sci. **52**, 75–91 (2004)

A.M. Annaswamy, A.L. Fradkov, A historical perspective of adaptive control and learning. Annu. Rev. Control. **52**, 18–41 (2021)

I. Barkana, Simple adaptive control—a stable direct model reference adaptive control methodology—brief survey. Int. J. Adapt. Control Signal Process. **28**(7–8), 567–603 (2014)

S. Ge, C. Hang, T. Lee, T. Zhang, *Stable Adaptive Neural Network Control*, ser. The International Series on Asian Studies in Computer and Information Science (Springer, New York, 2013)

H.K. Khalil, *Nonlinear Systems* (Prentice-Hall, Englewood Cliffs, 2002)

M. Krstic, I. Kanellakopoulos, V. Petar, *Nonlinear and Adaptive Control Design* (Wiley, New York, 1995)

H. Márquez, *Nonlinear Control Systems: Analysis and Design* (Wiley, New York, 2003)

N. Nguyen, *Model-Reference Adaptive Control: A Primer*, ser. Advanced Textbooks in Control and Signal Processing (Springer International Publishing, New York, 2018)

J.-J.E. Slotine, W. Li, et al., *Applied Nonlinear Control*, vol. 199(1) (Prentice Hall, Englewood Cliffs, 1991)

B.L. Stevens, F.L. Lewis, E.N. Johnson, *Aircraft Control and Simulation: Dynamics, Controls Design, and Autonomous Systems* (Wiley, New York, 2015)

Chapter 2
Lyapunov Stability Theory

The development of Lyapunov theory marks a significant milestone in the field of nonlinear control design and analysis. Before the emergence of Lyapunov theory, analyzing the stability of nonlinear systems was challenging due to the difficulty in solving nonlinear differential equations. While some methods, such as the phase plane analysis, existed for a limited class of nonlinear systems, there was no uniform method to determine the stability of nonlinear systems. This highlights the necessity for a systematic approach, such as the Lyapunov theory, which not only provides insights into system stability (both in the neighborhood of equilibrium points and limit cycles) but also offers a foundation for designing effective control strategies.

As discussed in the previous chapter, the stability of a system is typically assessed relative to an equilibrium point (stability of limit cycles is beyond the scope of this book). Without any disturbance, the state of the system, starting from the equilibrium point, will remain at the same point indefinitely. A system is considered stable around an equilibrium point if, following a disturbance, the state of the system returns to the equilibrium point. If the state of the system remains at a new point after a disturbance, the equilibrium point is neutrally stable. However, if a slight disturbance causes the system to deviate from the equilibrium state, the equilibrium point is considered unstable. A representation of stable, unstable, and neutrally stable equilibrium points is shown in Fig. 2.1, where x_0 is an equilibrium point, x_0' represents a new position of the state after a disturbance, and the direction of the arrows shows the convergence or divergence of the state.

The above definition of stability suffices for a linear system, as a stable linear system is always globally exponentially stable. However, for a nonlinear system, stability depends on how the system state behaves around an equilibrium point for a given initial condition. Moreover, the definitions of stability differ for autonomous and nonautonomous nonlinear systems. It should be noted that the formal definitions of autonomous and nonautonomous nonlinear systems were provided in the previous chapter. The definitions of stability for both types of nonlinear systems are presented in the following sections.

K. Sachan et al., *Intelligent Adaptive Control with System Constraints*,
https://doi.org/10.1007/978-981-95-8220-4_2

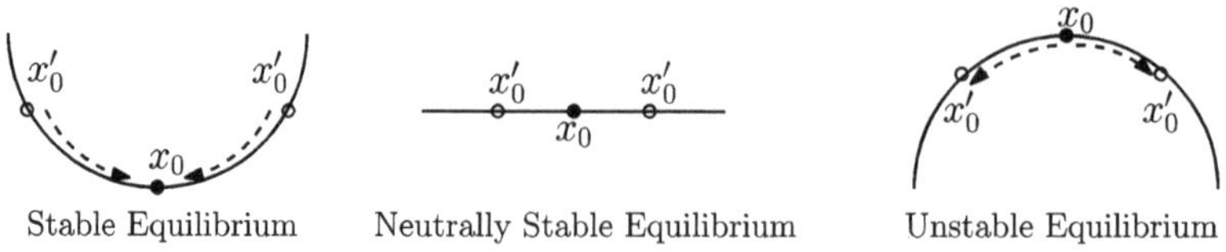

Fig. 2.1 Representation of stable, neutrally stable, and unstable equilibrium points

2.1 Stability of Autonomous Nonlinear Systems

2.1.1 Notions of Stability

The stability of a system around an equilibrium point can be analyzed by examining the behavior of its state with respect to initial conditions. If the state remains bounded without converging to the equilibrium point, the system is considered stable (if it stays within a bounded region) or unstable if it becomes unbounded. In contrast, if the state converges to the equilibrium point, the system is classified as asymptotically stable. The convergence to the equilibrium point can occur in various ways, which become special cases of asymptotic stability. For example, if this convergence to the equilibrium point occurs at an exponential rate, the system is referred to as exponentially stable.

The possible behavior of the state $\boldsymbol{x}(t) = [x_1, x_2]^T$ around an equilibrium point is illustrated in Fig. 2.2, where the origin is the equilibrium point. The cases in which the state trajectory does not converge to the equilibrium are shown in sub-figures (i) and (ii) of Fig. 2.2, whereas the cases in which the state trajectory does converge to the equilibrium point are shown in sub-figures (iii) and (iv).

The formal definitions of stability of nonlinear autonomous systems are given as follows.

Definition 2.1 (Equilibrium Point) A state $\boldsymbol{x}^*$ is an equilibrium point of the system if $\boldsymbol{x}(t) = \boldsymbol{x}^*$ at some t, and it remains there for all future time.

Example 2.1 The nonlinear system

$$\dot{x} = 3x - x^3$$

where system has three equilibrium points $\boldsymbol{x}^* = \{0, \pm\sqrt{3}\}$. The trajectory will converge to the equilibrium points based on the nature of the equilibrium points. Since trajectories are converging to $\sqrt{3}$ and $-\sqrt{3}$, both equilibrium points are stable, whereas $x^* = 0$ is an unstable equilibrium point. It can be seen in Fig. 2.3 that once trajectories $\boldsymbol{x}(t)$ reach equilibrium points $\boldsymbol{x}^*$, they remain there for all future time.

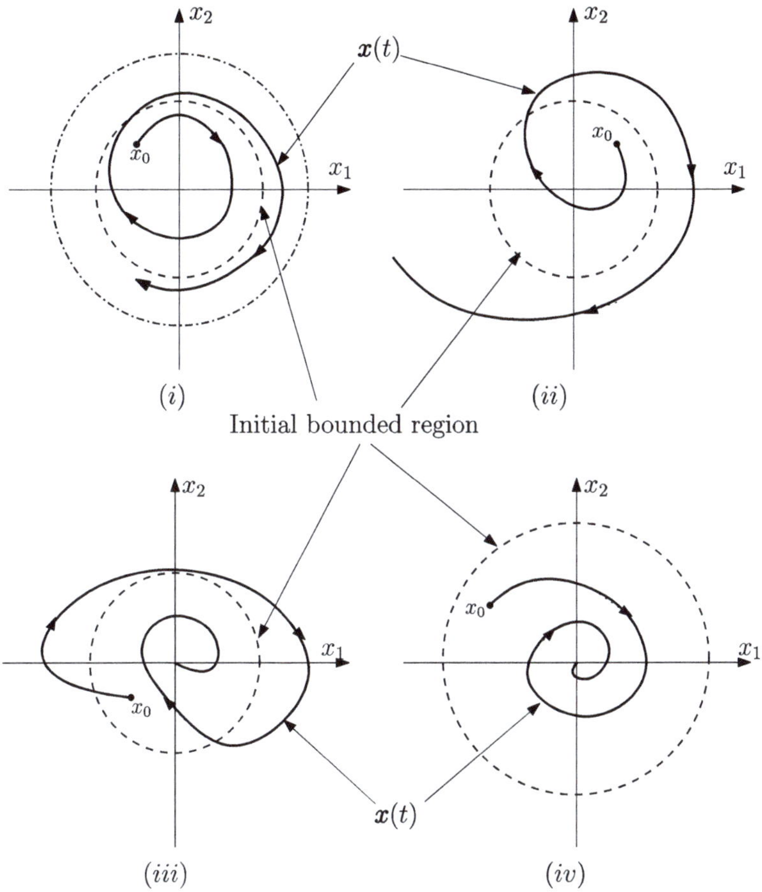

Fig. 2.2 Intuitive behavior of state around an equilibrium point for a typical second-order nonlinear system

Fig. 2.3 Equilibrium points and their characteristics

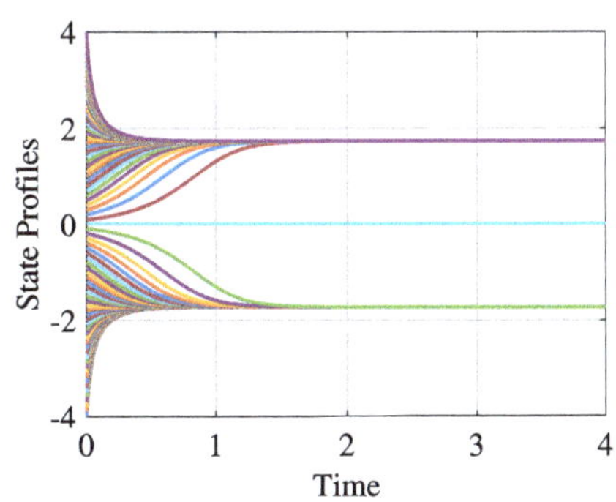

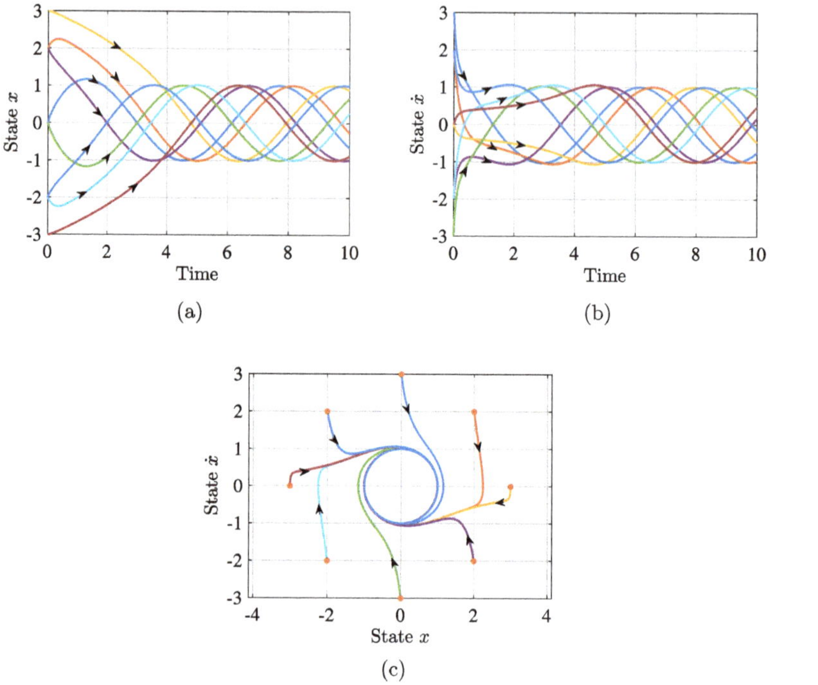

Fig. 2.4 (**a**) Time history of state x of nonlinear system. (**b**) Time history of state $\dot{x}$ of nonlinear system. (**c**) Phase plot of stable nonlinear system

Definition 2.2 (Stable Equilibrium Point) The equilibrium point $\boldsymbol{x}^*$ of a system starting at an initial condition $\boldsymbol{x}(t_0) = \boldsymbol{x}_0$ is said to be stable if, for any real number $R > 0$, there exists some other real number $r(R) > 0$ such that

$$\text{if } \left\|\boldsymbol{x}_0 - \boldsymbol{x}^*\right\| < r(R) \text{ then } \left\|\boldsymbol{x}(t) - \boldsymbol{x}^*\right\| < R, \ \forall t \geq t_0 \tag{2.1}$$

It is worth noting that this is the weakest notion of stability. Moreover, if the above condition is not satisfied, then the equilibrium point is said to be an *unstable equilibrium point*.

Example 2.2 Let the dynamics of a nonlinear system described by

$$\ddot{x} = -x + \left(1 - x^2 - \dot{x}^2\right)\dot{x}$$

The time histories of the system states are shown in Fig. 2.4a and b. One can observe that the system trajectories remain bounded for different initial

(continued)

Example 2.2 (continued)
conditions. The phase plot of both states is shown in Fig. 2.4c, where the stability property of the system is evident. Hence, it is a stable system.

Definition 2.3 (Convergent Equilibrium Point) The equilibrium point $\boldsymbol{x}^*$ of a system starting at an initial condition $\boldsymbol{x}(t_0) = \boldsymbol{x}_0$ is said to be convergent if there exists some real number $r > 0$ such that

$$\left\| \boldsymbol{x}_0 - \boldsymbol{x}^* \right\| < r \Rightarrow \lim_{t \to \infty} \left\| \boldsymbol{x}(t) - \boldsymbol{x}^* \right\| = 0 \tag{2.2}$$

Definition 2.4 (Asymptotically Stable Equilibrium Point) If an equilibrium point $\boldsymbol{x}^*$ is both stable and convergent, then it is said to be asymptotically stable.

Example 2.3 Let the dynamics of a nonlinear system be described by

$$\ddot{x} = -0.4x - \sin \dot{x}$$

The time histories of the system states are shown in Fig. 2.5a and b, where system trajectories are bounded and converge to the equilibrium point as $t \to \infty$ for different initial conditions. The phase plot of both states is shown in Fig. 2.5c, where the asymptotic stability property of the system is obvious. Hence, it is an asymptotically stable system.

It is worth emphasizing that asymptotic stability is generally the more desirable property in most practical applications. However, a key limitation of asymptotic stability is that it does not provide any information regarding the rate at which system trajectories approach the equilibrium point. To address this limitation, a stronger concept, known as exponential stability, is introduced. Exponential stability not only ensures convergence to the equilibrium but also characterizes the rate at which the state of the system converges.

Definition 2.5 (Exponentially Stable Equilibrium Point) The equilibrium point $\boldsymbol{x}^*$ of a system starting at an initial condition $\boldsymbol{x}(t_0) = \boldsymbol{x}_0$ is said to be (locally) exponentially stable if there exist two strictly positive real constants α and β, such that

$$\left\| \boldsymbol{x}(t) - \boldsymbol{x}^* \right\| \leq \alpha \left\| \boldsymbol{x}_0 - \boldsymbol{x}^* \right\| e^{-\beta(t-t_0)}, \quad \forall t \geq t_0 \tag{2.3}$$

where $\|\boldsymbol{x}_0 - \boldsymbol{x}^*\| < r$. The above is said to be *globally exponentially stable* if it holds for any $\boldsymbol{x}(t) \in \Re^n$

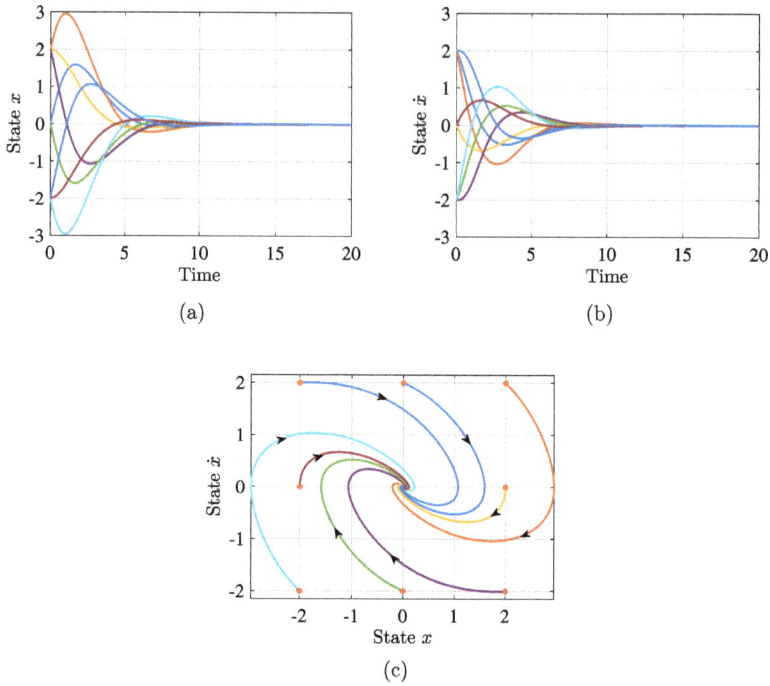

Fig. 2.5 (**a**) Time history of state x of nonlinear system. (**b**) Time history of state $\dot{x}$ of nonlinear system. (**c**) Phase plot of stable nonlinear system

Example 2.4 The dynamics of a nonlinear system described by

$$\dot{x} = -0.3x - x^3$$

is an exponentially stable system. This is because the time histories of the system for different initial conditions, shown in Fig. 2.6, are bounded and they converge to the equilibrium point at an exponential rate. Note that a detailed discussion of the convergence rate and associated stability analysis will be covered in the latter part of this section.

Example 2.5 The dynamics of a system be described by

$$\ddot{x} = -\dot{x} - x$$

(continued)

Example 2.5 (continued)
is an exponentially stable system. The time histories of the system are depicted in Fig. 2.7a and b, illustrating that the system trajectories remain bounded and converge to the equilibrium point as $t \to \infty$ for various initial conditions. The phase plot of both states, as shown in Fig. 2.7c, clearly demonstrates the exponential stability of the system. It is important to note that all stable linear systems exhibit global exponential stability.

Definition 2.6 (Unstable Equilibrium Point) The equilibrium point $\boldsymbol{x}^*$ of a system starting at an initial condition $\boldsymbol{x}(t_0) = \boldsymbol{x}_0$ is said to be unstable if it is not stable (even in the sense of weakest notion of stability given in (2.1)).

Example 2.6 The dynamics of a nonlinear system described by

$$\ddot{x} = -x\dot{x} + x^2$$

is an unstable stable system. This is because, as shown in Fig. 2.8a and b, the time histories of the system diverge if initial conditions are not at the equilibrium point and remain at the equilibrium point if the initial condition is at the equilibrium point.

In two dimensions, where $\boldsymbol{x}(t) = [x_1, x_2]^T$, various types of stability around the equilibrium point are illustrated in Fig. 2.9, with the origin representing the equilibrium point. Among the three types of stability, a stable equilibrium point is the weakest form, where system states are merely bounded. In control design problems, asymptotic stability is preferable because it ensures that the state of the system will be bounded and convergent. The exponential stability is the strongest

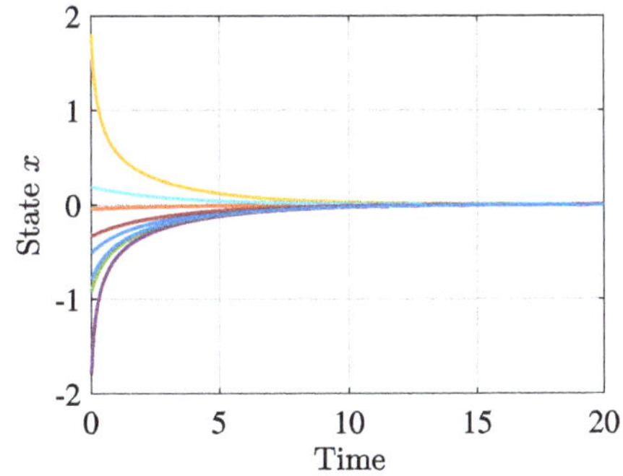

Fig. 2.6 Time history of state x of nonlinear system

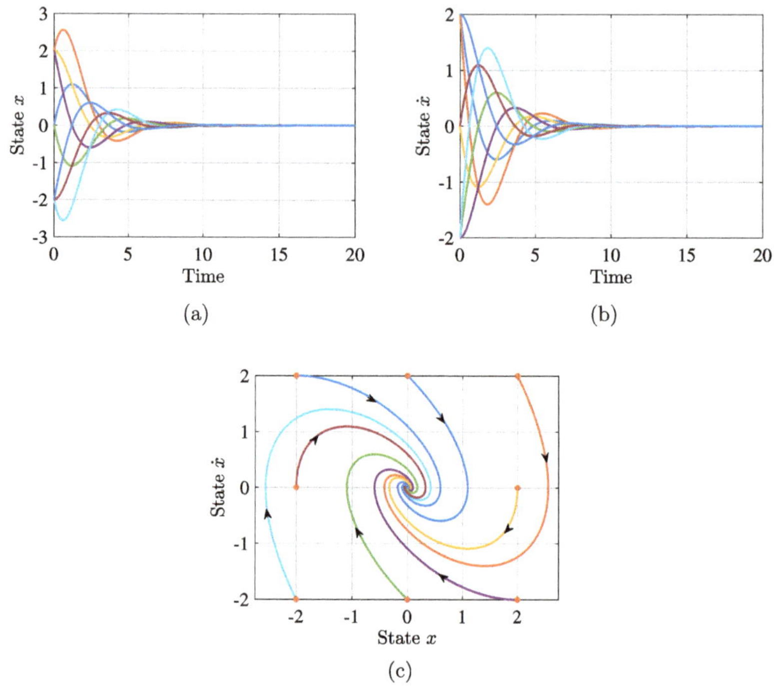

Fig. 2.7 (**a**) Time history of state x of nonlinear system. (**b**) Time history of state $\dot{x}$ of nonlinear system. (**c**) Phase plot of stable nonlinear system

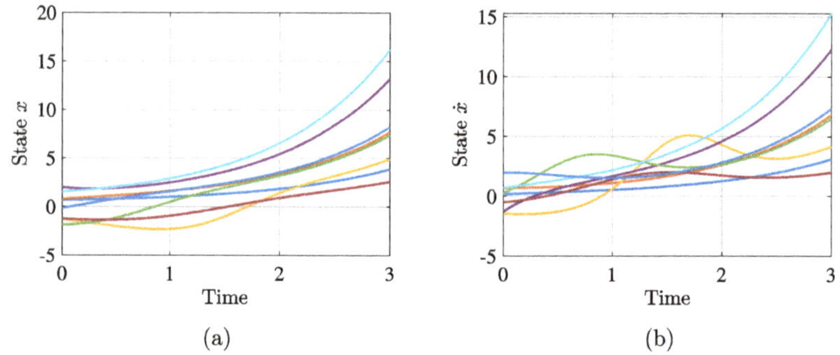

Fig. 2.8 (**a**) Time history of state x of nonlinear system. (**b**) Time history of state $\dot{x}$ of nonlinear system

form of stability as it characterizes the rate at which the state of the system converges to the equilibrium.

After understanding the notion of stability, the following subsection introduces Lyapunov stability theory to draw conclusions about the stability of nonlinear systems.

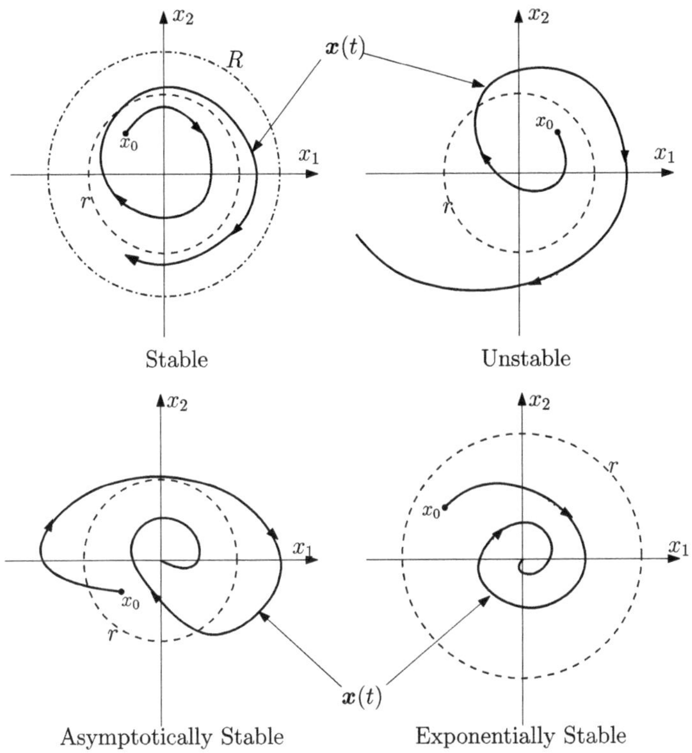

Fig. 2.9 Various types of stability around the equilibrium point (origin) for a typical second-order nonlinear system

2.1.2 Lyapunov Stability Theorems

The properties of positive definite functions and their variants are used in the Lyapunov stability theorems. The formal definitions of these functions are given below.

Definition 2.7 (Positive Definite Function) A function $V(\boldsymbol{x}) \in \Re$, $\boldsymbol{x} \in \Re^n$ is said to be a global *positive definite function* (PDF) if it satisfies the following conditions:

(i) $V(\boldsymbol{x}) = 0$ if $\boldsymbol{x} = 0$
(ii) $V(\boldsymbol{x}) > 0$ if $\boldsymbol{x} \neq 0$

Note that if the above condition is satisfied only in a neighborhood of the origin, then the function is referred to as a *locally* positive definite function.

Definition 2.8 (Positive Semi-definite Function) A function $V(\boldsymbol{x}) \in \Re, \boldsymbol{x} \in \Re^n$ is said to be a global *positive semi-definite function* (PSDF) if it satisfies the following conditions:

(i) $V(\boldsymbol{x}) = 0$ if $\boldsymbol{x} = 0$
(ii) $V(\boldsymbol{x}) \geq 0$ if $\boldsymbol{x} \neq 0$

Definition 2.9 (Negative Definite Function) A function $V(\boldsymbol{x}) \in \Re, \boldsymbol{x} \in \Re^n$ is said to be a global *negative definite function* (NDF) if it satisfies the following conditions:

(i) $V(\boldsymbol{x}) = 0$ if $\boldsymbol{x} = 0$
(ii) $V(\boldsymbol{x}) < 0$ if $\boldsymbol{x} \neq 0$

Definition 2.10 (Negative Semi-definite Function) A function $V(\boldsymbol{x}) \in \Re, \boldsymbol{x} \in \Re^n$ is said to be a global negative semi-definite function (NSDF) if it satisfies the following conditions:

(i) $V(\boldsymbol{x}) = 0$ if $\boldsymbol{x} = 0$
(ii) $V(\boldsymbol{x}) \leq 0$ if $\boldsymbol{x} \neq 0$

These functions are tabulated in Table 2.1 and illustrated in Fig. 2.10, where they exhibit distinct properties. A positive definite function always remains positive and only reaches zero at the origin. In contrast, a positive semi-definite function can reach zero at multiple points. The concept is analogous to negative definite functions, where they are always negative.

Lyapunov theorems for stable, asymptotically stable (AS), global asymptotically stable (GAS), exponentially stable (ES), and global exponentially stable (GES) equilibrium points, considering $\boldsymbol{x}^* = \boldsymbol{0}$ as the equilibrium point (where $\boldsymbol{0} \in \Re^n$ is a vector with elements as zero), can be stated as follows.

Theorem 2.1 *Let $\boldsymbol{x}^* = \boldsymbol{0}$ be an equilibrium point of a nonlinear autonomous system $\dot{\boldsymbol{x}} = \boldsymbol{f}(\boldsymbol{x})$, where $\boldsymbol{x} \in \Re^n, \boldsymbol{f} \in \Re^n$. Let $V(\boldsymbol{x}) \in \Re$ be a continuously*

Table 2.1 Definition of positive/negative definite functions

Positive definite function	$V(\boldsymbol{x}) = 0$ if $\boldsymbol{x} = 0$	$V(\boldsymbol{x}) > 0$ if $\boldsymbol{x} \neq 0$
Positive semi-definite function	$V(\boldsymbol{x}) = 0$ if $\boldsymbol{x} = 0$	$V(\boldsymbol{x}) \geq 0$ if $\boldsymbol{x} \neq 0$
Negative definite function	$V(\boldsymbol{x}) = 0$ if $\boldsymbol{x} = 0$	$V(\boldsymbol{x}) < 0$ if $\boldsymbol{x} \neq 0$
Negative semi-definite function	$V(\boldsymbol{x}) = 0$ if $\boldsymbol{x} = 0$	$V(\boldsymbol{x}) \leq 0$ if $\boldsymbol{x} \neq 0$

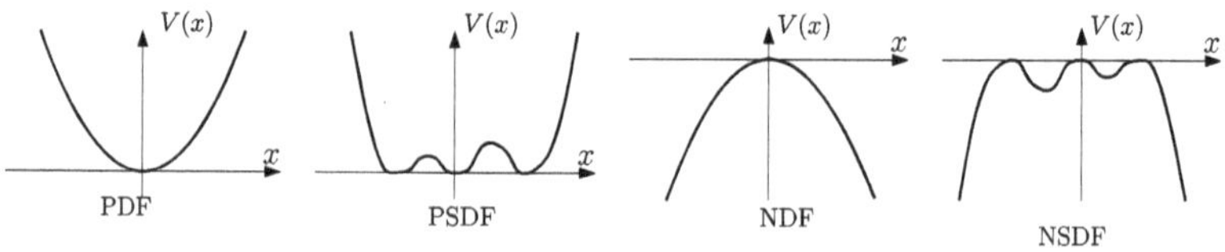

Fig. 2.10 Illustration of different types of positive/negative definite functions used in Lyapunov theory

Table 2.2 Sufficient conditions required for different types of stability of the equilibrium point $\boldsymbol{x}^* = \boldsymbol{0}$

Nature of equilibrium point	Nature of $V(\boldsymbol{x})$	Nature of $\dot{V}(\boldsymbol{x})$	RU condition
Stable	$V(\boldsymbol{x})$ is PDF	$\dot{V}(\boldsymbol{x})$ is NSDF	NA
Asymptotically stable	$V(\boldsymbol{x})$ is PDF	$\dot{V}(\boldsymbol{x})$ is NDF	NA
Globally asymptotically stable	$V(\boldsymbol{x})$ is PDF	$\dot{V}(\boldsymbol{x})$ is NDF	$V(\boldsymbol{x})$ is RU
Exponentially stable	$V(\boldsymbol{x})$ is PDF, $k_1\|\boldsymbol{x}\|^p \leq V(\boldsymbol{x}) \leq k_2\|\boldsymbol{x}\|^p$	$\dot{V}(\boldsymbol{x})$ is NDF, $\dot{V}(\boldsymbol{x}) < -k_3\|\boldsymbol{x}\|^p$	NA
Globally exponentially stable	$V(\boldsymbol{x})$ is PDF, $k_1\|\boldsymbol{x}\|^p \leq V(\boldsymbol{x}) \leq k_2\|\boldsymbol{x}\|^p$	$\dot{V}(\boldsymbol{x})$ is NDF, $\dot{V}(\boldsymbol{x}) < -k_3\|\boldsymbol{x}\|^p$	$V(\boldsymbol{x})$ is RU

*differentiable function. Then, the sufficient conditions required for different types of stability of the equilibrium point $\boldsymbol{x}^* = \boldsymbol{0}$ are presented in Table 2.2, where the acronym "NA" refers to not applicable and "RU" refers to V is **r**adially **u**nbounded, which means $V(\boldsymbol{x}) \to \infty$ as $\|\boldsymbol{x}\| \to \infty$ and constants $k_1 > 0, k_2 > 0, k_3 > 0$, and $p > 0$ are positive in nature.*

It is important to note that the Lyapunov theorems provide *sufficient conditions for stability*. If the conditions mentioned above are not satisfied for the chosen Lyapunov function $V(\boldsymbol{x})$, it does not necessarily mean that the system is unstable. Therefore, the careful selection of $V(\boldsymbol{x})$ is crucial for accurately determining the stability of the system.

Example 2.7 Consider the system defined by

$$\dot{x} = -x^2 - x \tag{2.4}$$

The following Lyapunov candidate is proposed:

$$V(x) = \frac{1}{2}x^2 \tag{2.5}$$

The derivative of V with time is obtained as

$$\dot{V}(x) = -x^3 - x^2 \tag{2.6}$$

The negative definiteness of $\dot{V}(x)$ cannot be concluded in this case, as it is not always ≤ 0. Therefore, no conclusion on stability and instability can be drawn with the above choice of Lyapunov candidate.

2.1.3 Stability of Nonzero Equilibrium Point

The various notions of stability introduced so far pertain to the stability of equilibrium points. In general, a given dynamical system may possess more than one isolated equilibrium point. It is common, particularly in the definitions and proofs of stability theorems, to assume that the equilibrium point under consideration is located at the origin, $\boldsymbol{x}^* = \mathbf{0}$. Therefore, to ensure consistency with the theoretical framework, if the equilibrium point is not located at the origin, a change of variables can be applied to reformulate the system such that the new equilibrium point is shifted to the origin at $\boldsymbol{x} = \mathbf{0}$.

To illustrate this, consider the system given by

$$\dot{\boldsymbol{x}} = \boldsymbol{f}(\boldsymbol{x}) \tag{2.7}$$

with the equilibrium point $\boldsymbol{x}^*$. Define a new variable $\boldsymbol{y}$ as

$$\boldsymbol{y} = \boldsymbol{x} - \boldsymbol{x}^* \tag{2.8}$$

Hence, the new system is obtained as

$$\dot{\boldsymbol{y}} = \boldsymbol{f}(\boldsymbol{x}) \tag{2.9}$$

$$= \boldsymbol{f}(\boldsymbol{y} + \boldsymbol{x}^*) = \boldsymbol{g}(\boldsymbol{y}) \tag{2.10}$$

The equilibrium point of the new system $\dot{\boldsymbol{y}} = \boldsymbol{g}(\boldsymbol{y})$ can be calculated by setting $\dot{\boldsymbol{y}}^* = \mathbf{0}$, i.e., $\mathbf{0} = \boldsymbol{g}(\boldsymbol{y}^*)$. At $\boldsymbol{y}^* = \mathbf{0}$, $\boldsymbol{g}(\mathbf{0}) = \boldsymbol{f}(\mathbf{0} + \boldsymbol{x}^*) = \boldsymbol{f}(\boldsymbol{x}^*) = \mathbf{0}$. Thus, the new system $\dot{\boldsymbol{y}} = \boldsymbol{g}(\boldsymbol{y})$ has the equilibrium point $\boldsymbol{y}^*$ at origin, i.e., $\boldsymbol{y}^* = \mathbf{0}$.

According to this, therefore, analyzing the stability of an equilibrium point $\boldsymbol{x}^*$ for the system $\dot{\boldsymbol{x}} = \boldsymbol{f}(\boldsymbol{x})$ is equivalent to analyzing the stability of the origin for the transformed system $\dot{\boldsymbol{y}} = \boldsymbol{g}(\boldsymbol{y})$.

Example 2.8 Consider the same nonlinear system:

$$\dot{x} = 3x - x^3$$

previously used to illustrate the stability of equilibrium points. This system has three equilibrium points, $\boldsymbol{x}^* = \{0, \pm\sqrt{3}\}$. As shown in Fig. 2.11, the trajectories converge to $x^* = \sqrt{3}$ and $x^* = -\sqrt{3}$, indicating that both of these equilibrium points are stable. In contrast, the equilibrium point at $x^* = 0$ is unstable, as nearby trajectories diverge from it.

In this example, the stability of these equilibrium points is analyzed using Lyapunov stability theory.

(continued)

Example 2.8 (continued)
Stability Analysis of the Equilibrium Point at $x^* = 0$ In this case, no variable transformation is required for the stability analysis. Selecting the Lyapunov function as

$$V(x) = \frac{1}{2}x^2$$

The time derivative of the Lyapunov function is given by

$$\dot{V}(x) = x\dot{x} = x(3x - x^3) = 3x^2 - x^4$$

From this expression, it can be observed that $\dot{V}(x) > 0$ in the region $-\sqrt{3} < |x| < \sqrt{3}$, indicating that the origin is an unstable equilibrium point. Therefore, the system is unstable at $x^* = 0$.

Stability Analysis of the Equilibrium Point at $x^* = \sqrt{3}$ Since the equilibrium point is not at the origin, the variable can be transformed as $y = x - \sqrt{3}$. Substituting this into the original system, the transformed system becomes

$$\dot{y} = -y^3 - 3\sqrt{3}y^2 - 6y$$

To analyze the stability of this system, the following Lyapunov function is chosen:

$$V(y) = \frac{1}{2}y^2$$

The time derivative of the Lyapunov function is given by

$$\dot{V}(y) = y\dot{y} = y(-y^3 - 3\sqrt{3}y^2 - 6y) = -y^4 - 3\sqrt{3}y^3 - 6y^2$$

Since $x^* = 0$ is also an equilibrium point, the stability of $x^* = \sqrt{3}$ is examined within the domain $(0, \infty)$. Under the coordinate transformation $y = x - \sqrt{3}$, this corresponds to analyzing stability in the region $(-\sqrt{3}, \infty)$. Analysis of the expression for $\dot{V}(y)$ shows that $\dot{V}(y) < 0$ for all $y \in (-\sqrt{3}, \infty)$ with $y \neq 0$ and that $\dot{V}(0) = 0$. Therefore, the equilibrium at $x = \sqrt{3}$ (i.e., $y = 0$) is asymptotically stable. Similarly, asymptotic stability can be shown for the equilibrium point at $x^* = -\sqrt{3}$.

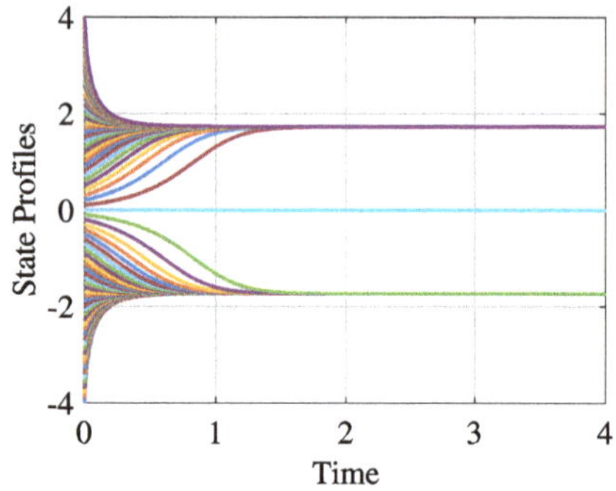

Fig. 2.11 Equilibrium points and their characteristics

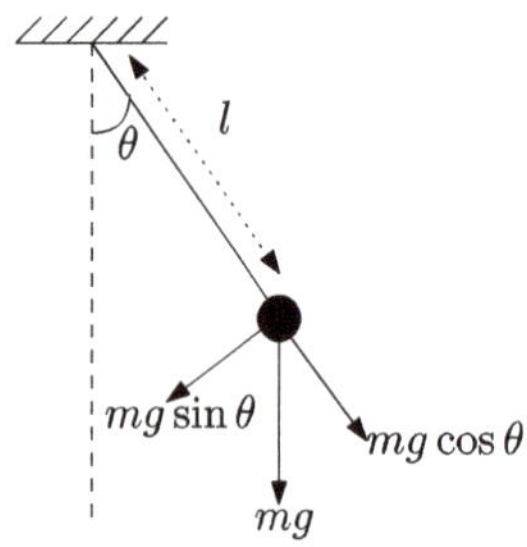

Fig. 2.12 Representation of a simple pendulum and acting forces

Many times, Lyapunov analysis leads only to a conclusion of stability, rather than asymptotic stability, which is often the desired result, as it ensures that the state trajectory converges to the equilibrium point. In fact, in many applications, one may infer asymptotic stability from the behavior of the system through observation, even when Lyapunov analysis only leads to stability. In such cases, it is advisable to carry out further analysis. To illustrate this, consider the following example.

Example 2.9 To illustrate this, consider the simple pendulum without friction with rod length l and mass m, as shown in Fig. 2.12.

The torque τ due to gravity can be written as

$$\tau = -mgl\sin\theta$$

where g is the acceleration due to gravity. Using Newton's second law for rotation, torque $\tau = ml^2\ddot{\theta}$. Hence, the following can be obtained:

$$ml^2\ddot{\theta} = -mgl\sin\theta$$

(continued)

Example 2.9 (continued)
Dividing both sides by mL^2, the pendulum dynamics is given by

$$\ddot{\theta} + \frac{g}{l}\sin\theta = 0$$

By defining $x_1 \triangleq \theta$ and $x_2 \triangleq \dot{\theta}$, the following dynamics in state-space form is obtained:

$$\begin{bmatrix} \dot{x}_1 \\ \dot{x}_2 \end{bmatrix} = \begin{bmatrix} x_2 \\ -\frac{g}{l}\sin x_1 \end{bmatrix} \tag{2.11}$$

To determine the stability of this system, consider the Lyapunov function as

$$\begin{aligned} V(x_1, x_2) &= \text{Kinetic Energy} + \text{Potential Energy} \\ &= \frac{mv^2}{2} + mgh = \frac{m(\dot{x}^2 + \dot{y}^2)}{2} + mgh \end{aligned}$$

Here, $x = l\sin x_1$, $y = -l\cos x_1$, and $h = l - l\cos x_1$. Substituting these parameters, the following is obtained:

$$\begin{aligned} V(x_1, x_2) &= \frac{ml^2\left[(\cos x_1)^2 + (\sin x_1)^2\right]x_2^2}{2} + mgl(1 - \cos x_1) \\ &= \frac{ml^2x_2^2}{2} + mgl(1 - \cos x_1) \end{aligned} \tag{2.12}$$

The time derivative of the Lyapunov function is obtained as

$$\begin{aligned} \dot{V}(x_1, x_2) &= ml^2x_2\dot{x}_2 + mgl\sin x_1\dot{x}_1 \\ &= -mglx_2\sin x_1 + mglx_2\sin x_1 \\ &\leq 0 \end{aligned}$$

Hence, the system is stable. This is compatible with intuition as well. This is because the pendulum is frictionless, and hence, it will keep oscillating around the equilibrium point.

Example 2.10 Next, let us consider the dynamics of a pendulum with friction. Hence, a new damping term ($k\dot{\theta}$) is added to the torque, where k is the damping constant. Thus, the torque τ due to gravity can be written as

$$\tau = -mgl \sin\theta - k\dot{\theta}$$

By comparing with torque $ml^2\ddot{\theta}$ and defining $x_1 \triangleq \theta$ and $x_2 \triangleq \dot{\theta}$, the following state-space form dynamics is obtained:

$$\begin{bmatrix} \dot{x}_1 \\ \dot{x}_2 \end{bmatrix} = \begin{bmatrix} x_2 \\ -\frac{g}{l}\sin x_1 - \frac{k}{ml^2}x_2 \end{bmatrix} \tag{2.13}$$

To determine the stability of this system, again consider the same Lyapunov function as in (2.12). However, the time derivative of the Lyapunov function is obtained as

$$\begin{aligned} \dot{V}(x_1, x_2) &= ml^2 x_2 \dot{x}_2 + mgl \sin x_1 \dot{x}_1 \\ &= -mglx_2 \sin x_1 - kx_2^2 + mglx_2 \sin x_1 \\ &= -kx_2^2 \\ &\leq 0 \end{aligned} \tag{2.14}$$

Since $\dot{V}(x_1, x_2) \leq 0$, the system can be proven only stable (not asymptotically stable) with the choice of Lyapunov function. However, intuitively, it is obvious that in the presence of friction, the pendulum will slowly converge to the equilibrium point. Therefore, this result is very restrictive, and on the other hand, $\dot{V}(\boldsymbol{x}) < 0$ is very difficult to prove.

2.1.4 *The Invariance Principle*

The stability analysis concluded in the above example highlights a critical limitation, as asymptotic stability is often the desired property in practical applications. This limitation primarily arises when the derivative of the Lyapunov function is negative semi-definite, which prevents the conclusion of asymptotic stability. This issue was extensively studied and mitigated using the *invariance principle*. In this approach, the concept of an invariant set, a generalization of the equilibrium point, is used to facilitate a more comprehensive stability analysis.

Definition 2.11 (Invariant Set) A set M is said to be an *invariant set* with respect to the system $\dot{\boldsymbol{x}} = \boldsymbol{f}(\boldsymbol{x})$ if

$$\boldsymbol{x}_0 \in M \Rightarrow \boldsymbol{x}(t) \in M, \ \forall t > 0$$

The example of the invariant set can be given as follows:

1. Since the state trajectory of a dynamical system that starts from an equilibrium point remains at the equilibrium point, any equilibrium point is an invariant set. Similarly, a limit cycle is also an invariant set.
2. Any trajectory of an autonomous system forms an invariant set. Moreover, the entire space $\Re^n$ is itself an invariant set.
3. In the context of Lyapunov analysis, if the function $V(\boldsymbol{x}) \leq l \ \forall \boldsymbol{x} \in \Re^n$ is continuously differentiable (not necessarily positive definite) and satisfies $\dot{V}(\boldsymbol{x}) < 0$ along the solutions of $\dot{\boldsymbol{x}} = \boldsymbol{f}(\boldsymbol{x})$, which means that if a trajectory crosses $V(\boldsymbol{x}) = c$, it can never come out again, then the set Ω_m defined by

$$\Omega_m = \{\boldsymbol{x} \in \Re^n : V(\boldsymbol{x}) \leq l\}$$

is an invariant set.

With the understanding of invariant sets, the following theorems can be stated as follows.

Theorem 2.2 *The equilibrium point* $\boldsymbol{x}^* = \boldsymbol{0}$ *of the autonomous system* $\dot{\boldsymbol{x}} = \boldsymbol{f}(\boldsymbol{x})$ *is asymptotically stable if there exists a function* $V(\boldsymbol{x})$ *satisfying*

(i) $V(\boldsymbol{x})$ *is positive definite* $\forall \boldsymbol{x} \in D$*, where we assume that* $0 \in D$.
(ii) $\dot{V}(\boldsymbol{x})$ *is negative semi-definite in a bounded region* $\Re \subset D$.
(iii) $\dot{V}(\boldsymbol{x})$ *does not vanish along any trajectory in* $\Re$*, other than the null solution* $\boldsymbol{x} = \boldsymbol{0}$.

Moreover, if the above condition holds good for $\Re^n$ *and* $V(\boldsymbol{x})$ *is radially unbounded, then* $\boldsymbol{x} = \boldsymbol{0}$ *is globally asymptotically stable.*

Theorem 2.3 (LaSalle's Theorem) *Let* $V : D \to \Re$ *be a continuously differentiable (not necessarily positive definite) function and*

(i) $M \subset D$ *be a compact set, which is invariant with respect to the solutions of* $\dot{\boldsymbol{x}} = \boldsymbol{f}(\boldsymbol{x})$.
(ii) $\dot{V} \leq 0$ *in* M.
(iii) $E = \{\boldsymbol{x} : \boldsymbol{x} \in M \text{ and } \dot{V}(\boldsymbol{x}) = 0\}$*, that is,* E *is the set of all points of* M *such that* $\dot{V} = 0$.
(iv) N *is the largest invariant set in* E.

Then, every solution starting in M *approaches* N *as* $t \to \infty$.

Note that LaSalle's theorem is a powerful tool for analyzing the stability of autonomous nonlinear systems because

1. In LaSalle's theorem, the function $V(\boldsymbol{x})$ is not required to be positive definite. It only needs to be continuously differentiable.
2. LaSalle's theorem applies not only to equilibrium points but also to more general dynamical behaviors such as limit cycles.

Example 2.11 Consider the same example of a pendulum with friction, where $V(x_1, x_2)$ in (2.12) and $\dot{V}(x_1, x_2)$ in (2.14) were obtained as

$$V(x_1, x_2) = \frac{ml^2 x_2^2}{2} + mgl(1 - \cos x_1)$$
$$\dot{V}(x_1, x_2) = -kx_2^2 \leq 0$$

Now, with the help of LaSalle's theorem, let us examine the condition:

$$\dot{V}(x_1, x_2) = 0 \; \forall t$$

This implies

$$-kx_2^2 = 0$$
$$x_2 = 0 \; \forall t$$

However, since $x_2 = 0, \; \forall t, \dot{x}_2 = 0$, and from the $\dot{x}_2$ equation of pendulum dynamics, $-\frac{g}{l}\sin x_1 - \frac{k}{ml^2}x_2 = 0$. This leads to $\sin x_1 = 0$, restricting x_1 to the interval $x_1 \in (-\pi, \pi)$ with the only solution $x_1 = 0$. Hence, $\dot{V}(x_1, x_2) = 0$ happens only at $x_1 = x_2 = 0$. Therefore, the equilibrium point $(x_1, x_2) = (0, 0)$ is the largest invariant set. Thus, it can be concluded from LaSalle's theorem that all state trajectories will converge to the equilibrium point as $t \to \infty$. This means that the equilibrium point is asymptotically stable.

2.1.5 Lyapunov Theorems for Linear Time Invariant Systems

As demonstrated in the previous chapter, nonlinear autonomous systems can be linearized around their equilibrium points, resulting in linear time-invariant (LTI) systems. By analyzing these linearized systems, one can infer the local stability behavior of the corresponding nonlinear systems. In this context, the following theorems can be stated.

Theorem 2.4 (Lyapunov's Indirect Theorem) *Let the linearized system about $\boldsymbol{x} = 0$ be*

$$\Delta\dot{\boldsymbol{x}} = \boldsymbol{A}\,\Delta\boldsymbol{x}$$

where $\boldsymbol{A} \in \Re^{n\times n}$ is the system matrix.

The theorem says that if the real part of all eigenvalues $Re(\lambda_i)$ $(i = 1, \ldots, n)$ of the matrix $\boldsymbol{A}$ satisfy

$$Re(\lambda_i) < 0 \quad \textit{(i.e., the linearized system is exponentially stable)}$$

then, for the nonlinear system, the origin is locally exponentially stable.

Theorem 2.5 (Lyapunov's Instability Theorem) *Consider the autonomous dynamical system and assume $\boldsymbol{x} = \boldsymbol{0}$ is an equilibrium point. Let $V(\boldsymbol{x}) : D \to \Re$ have the following properties:*

1. *$V(\boldsymbol{0}) = 0$.*
2. *There exist $\boldsymbol{x_0} \in \Re^n$, arbitrarily close to $\boldsymbol{x} = \boldsymbol{0}$, such that $V(\boldsymbol{x_0}) > 0$.*
3. *$\dot{V} > 0\ \forall \boldsymbol{x} \in U$, where the set U is defined as follows:*

$$U = \{\boldsymbol{x} \in D : \|\boldsymbol{x}\| \le \varepsilon \text{ and } V(\boldsymbol{x}) > 0\}$$

Under these conditions, $\boldsymbol{x} = \boldsymbol{0}$ is unstable.

Example 2.12 Determine the nature of stability of the following system around the equilibrium point $\boldsymbol{x} = [x_1, x_2]^T = [0, 0]^T$:

$$\dot{x}_1 = x_2 + x_1\left(4 - x_1^2 - x_2^2\right) \tag{2.15}$$

$$\dot{x}_2 = -x_1 + x_2\left(4 - x_1^2 - x_2^2\right) \tag{2.16}$$

To determine the nature of stability, consider the Lyapunov function as $V(x_1, x_2) = \frac{1}{2}(x_1^2 + x_2^2)$. Then, the time derivative of the Lyapunov function is calculated as

$$\begin{aligned}\dot{V}(x_1, x_2) &= x_1\dot{x}_1 + x_2\dot{x}_2 \\ &= (x_1^2 + x_2^2)\left(4 - (x_1^2 + x_2^2)\right)\end{aligned} \tag{2.17}$$

(continued)

Example 2.12 (continued)
Define

$$U = \left\{ \boldsymbol{x} \in \Re^2 : \|\boldsymbol{x}\| \leq \varepsilon, \quad 0 < \varepsilon < 2 \right\}$$

Then, $V(x_1, x_2) > 0$ and $\dot{V}(x_1, x_2) > 0$ in U. Hence, $\boldsymbol{x} = [0, 0]^T$ is unstable.

2.2 Stability of Nonautonomous Nonlinear Systems

For autonomous systems, stability is always defined with respect to an equilibrium point. However, in nonautonomous systems where time is explicitly present, the same definition of equilibrium point cannot be used for stability analysis. Instead, the definition of an equilibrium point for nonautonomous systems is given as follows.

Definition 2.12 (Equilibrium Point) For a nonautonomous nonlinear system $\dot{\boldsymbol{x}} = \boldsymbol{f}(\boldsymbol{x}, t)$ with initial time t_0, the equilibrium point $\boldsymbol{x}^*$ is defined as $\boldsymbol{f}(\boldsymbol{x}^*, t) = 0, \ \forall t \geq t_0$.

It is important to note that if a nonautonomous nonlinear system satisfies the above definition, then an equilibrium point will exist. However, for many types of nonautonomous nonlinear systems, an equilibrium point may not exist. The existence or nonexistence of equilibrium points primarily depends on the manner in which time t explicitly appears in the system equations. Therefore, the type of stability is different in both cases. This notion is clearly illustrated in the following examples.

Example 2.13 (Existence of Equilibrium Point) Consider the nonautonomous nonlinear system as

$$\dot{x} = t\,x^2 - x$$

where x and t represent the system state and time, respectively. This system has an equilibrium point at $x = 0$. Since the equilibrium point is defined for these systems, similar to the autonomous nonlinear systems, stability, asymptotic stability, and exponential stability can be defined for this nonautonomous system.

Example 2.14 (Nonexistence of Equilibrium Point) Consider the following nonautonomous nonlinear system:

$$\dot{x} = x^2 - t$$

where x and t represent the system state and time, respectively. This system does not have any equilibrium point as $x = 0$ leads to $\dot{x} = -t$. Since there is no equilibrium point for these types of systems, the concept of *uniform boundedness* is used to define stability.

2.2.1 Notions of Stability

In this section, we extend the definitions of stability from autonomous systems to nonautonomous systems, explicitly considering the initial time t_0. Let us consider a region $R > 0$ where the system must operate. If we can find another region r, possibly as a function of R, and initial time t_0, such that if the system starts in $r(R, t_0)$, it remains in R, then the system is stable. The formal definition of stability can be written as follows.

Definition 2.13 (Stability) The equilibrium point $\boldsymbol{x}^* = \mathbf{0}$ is stable at t_0 if for any $R > 0$, there exist a positive scalar $r(R, t_0) > 0$ such that

$$\|\boldsymbol{x}(t_0)\| < r(R, t_0) \Rightarrow \|\boldsymbol{x}(t)\| \leq R, \quad \forall t \geq t_0 \tag{2.18}$$

Otherwise, the equilibrium point $\boldsymbol{x}^* = \mathbf{0}$ is *unstable*.

Example 2.15 Consider the nonautonomous nonlinear system defined by

$$\dot{x} = -x \sin t$$

This system possesses an equilibrium point at $x = 0$. Since $\sin t$ is bounded for all $t \geq 0$, with $|\sin t| \leq 1$, the rate of change $\dot{x}$ oscillates around zero. Consequently, depending on the initial condition, the state trajectory will exhibit oscillatory behavior over time. However, since the state trajectory remains bounded, the equilibrium point at $x = 0$ is classified as stable. The time response of the system, depicted in Fig. 2.13, confirms the stable nature of the equilibrium point.

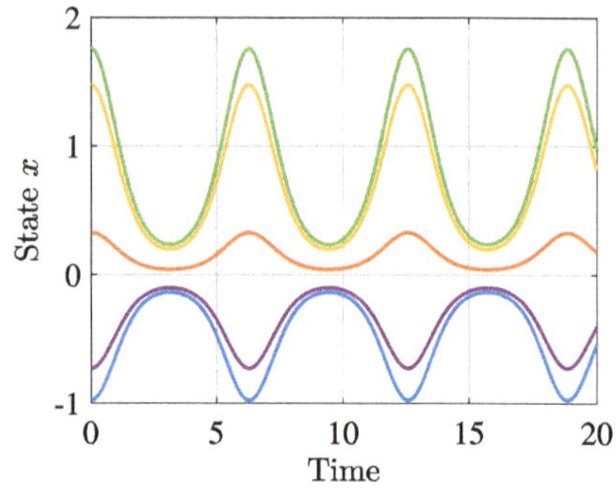

Fig. 2.13 Time history of state x of nonlinear system

The definition of stability can be further categorized based on the dependency of r on R and t_0. Similar to the definition of stability, the definitions of asymptotic and exponential stability can be extended to nonautonomous nonlinear systems, taking initial time into account.

Definition 2.14 (Asymptotic Stability) The equilibrium point $\boldsymbol{x}^* = \mathbf{0}$ is asymptotically stable at t_0 if it is "*stable*" and there exist a positive scalar $r(t_0)$ such that

$$\|\boldsymbol{x}(t_0)\| < r(t_0) \Rightarrow \|\boldsymbol{x}(t)\| \to 0, \quad \text{as} \quad t \to \infty \tag{2.19}$$

Here, r can be considered as a region of attraction.

Definition 2.15 (Global Asymptotic Stability) The equilibrium point $\boldsymbol{x}^* = \mathbf{0}$ is globally asymptotically stable at t_0 if it is "*stable*" and $\forall \boldsymbol{x}(t_0)$:

$$\|\boldsymbol{x}(t)\| \to 0, \quad \text{as} \quad t \to \infty \tag{2.20}$$

Example 2.16 Consider the nonautonomous nonlinear system:

$$\dot{x} = -0.3x - x^3 \sin t$$

This system has an equilibrium point at $x = 0$. The rate of change $\dot{x}$ comprises two terms: a linear damping term $-0.3x$ and a nonlinear oscillatory term $-x^3 \sin t$. The nonlinear term exhibits bounded oscillations due to the fact that $|\sin t| \leq 1$, while the linear term continuously drives the system toward the origin. As the state x approaches zero, the influence of the nonlinear term $-x^3 \sin t$ diminishes significantly due to the cubic dependence on x.

Hence, the state of the system converges to zero over time, indicating that the equilibrium point at $x = 0$ is asymptotically stable. The time history of the system, as illustrated in Fig. 2.14, further affirms the asymptotically stable behavior of the equilibrium point.

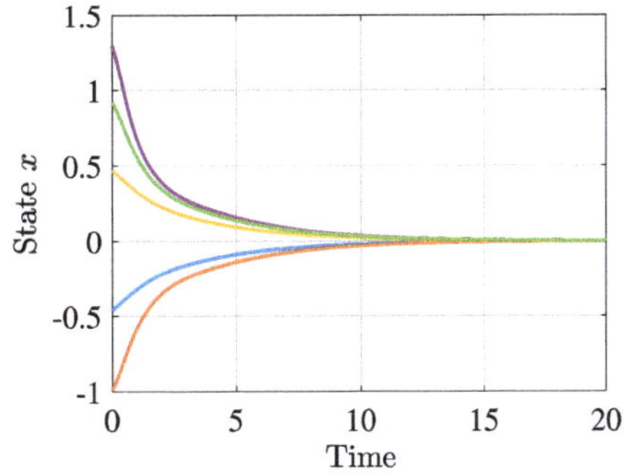

Fig. 2.14 Time history of state x of nonlinear system

Definition 2.16 (Exponential Stability) The equilibrium point $\boldsymbol{x}^* = \mathbf{0}$ is exponentially stable at t_0 if it is "*asymptotic stable*" and there exist two positive scalar α and λ such that for sufficiently small $\boldsymbol{x}(t_0)$, i.e.

$$\|\boldsymbol{x}(t_0)\| < r(t_0) \Rightarrow \|\boldsymbol{x}(t)\| \leq \alpha \|\boldsymbol{x}(t_0)\| e^{-\lambda(t-t_0)}, \quad \forall t \geq t_0 \tag{2.21}$$

Next, the definition of stability is given based on the boundedness of the response.

Definition 2.17 (Uniformly Bounded Stability) The equilibrium point $\boldsymbol{x}^* = \mathbf{0}$ is said to be uniformly bounded if, for any $R > 0$, there exists some $r(R) > 0$ *independent of the initial time* t_0 such that

$$\|\boldsymbol{x}(t_0)\| < r(R) \Rightarrow \|\boldsymbol{x}(t)\| \leq R, \forall t \geq t_0 \tag{2.22}$$

Definition 2.18 (Uniformly Ultimately Bounded Stability) The equilibrium point $\boldsymbol{x}^* = \mathbf{0}$ is said to be uniformly ultimately bounded if, for any $R > 0$, there exists some $r > 0$ independent of R and initial time t_0 such that

$$\|\boldsymbol{x}(t_0)\| < r \Rightarrow \|\boldsymbol{x}(t)\| \leq R, \forall t \geq t_0 + T \tag{2.23}$$

where $T = T(r)$ is some time interval after the initial time t_0.

Example 2.17 Consider the nonautonomous nonlinear system as

$$\dot{x} = -x + 2\ \sin t$$

The solution of the system can be obtained as

$$x = (x_0 + 1)e^{-t} + \sin t - \cos t$$

(continued)

Example 2.17 (continued)
Assuming by definition $\|x_0\| < r$ and since $e^{-t} \leq 1$ and $|\sin t - \cos t| \leq \sqrt{2}$, the *uniform bound* of the system is obtained as

$$\|x\| \leq \|x_0 + 1\| + \sqrt{2}$$

Furthermore, as $t \to \infty$, the *ultimate bound* of the system is obtained as

$$\|x\| \leq \sqrt{2}$$

Since the bound is independent of r, this is the ultimate bound for the system. The time history of the system is shown in Fig. 2.15a and b for different initial conditions. In Fig. 2.15a, the uniform bound and ultimate bound are shown. It can be noticed from Fig. 2.15b that the ultimate bound is the same for all initial conditions, as it is independent of initial conditions, but the uniform bound is different for all initial conditions.

After understanding the notion of stability for nonautonomous nonlinear systems, the following subsection introduces Lyapunov stability theory to assess the stability of these systems.

2.2.2 *Lyapunov Stability Theorems*

As mentioned in the previous section, the properties of positive definite functions and their variants are used in the analysis of Lyapunov stability theorems. Hence,

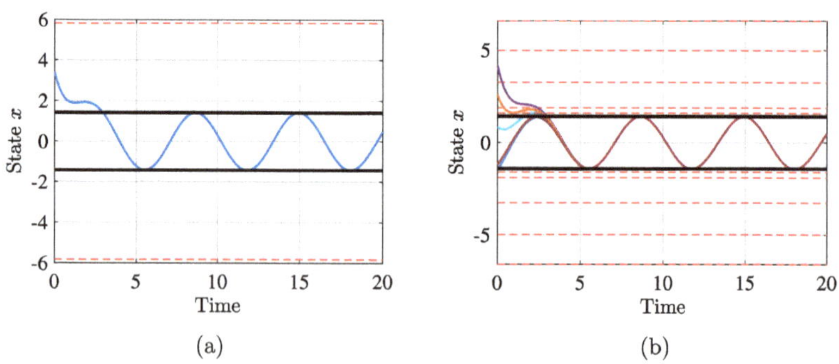

Fig. 2.15 (**a**) Time history of state x of nonlinear system. (**b**) Time history of state x of nonlinear system for multiple initial conditions

Table 2.3 Definition of positive/negative definite functions for nonautonomous nonlinear systems

Nature of function $V(\boldsymbol{x}, t)$	Properties of function $V_0(\boldsymbol{x})$	Conditions
Locally positive definite	Locally PDF	$V(\boldsymbol{x}, t) \geq V_0(\boldsymbol{x}),\ \forall t > t_0$
Locally positive semi-definite	Locally PSDF	$V(\boldsymbol{x}, t) \geq V_0(\boldsymbol{x}),\ \forall t > t_0$
Locally negative definite	Locally NDF	$V(\boldsymbol{x}, t) \leq V_0(\boldsymbol{x}),\ \forall t > t_0$
Locally negative semi-definite	Locally NSDF	$V(\boldsymbol{x}, t) \leq V_0(\boldsymbol{x}),\ \forall t > t_0$

for nonautonomous systems, these definitions have to be extended by considering time explicitly. Moreover, Lyapunov theory for nonautonomous systems requires the understanding of the *decrescent function* to prove the stability of systems. These functions are defined as follows.

Definition 2.19 Let $V(\boldsymbol{x}, t)$ be a scalar time-varying function with $V(\mathbf{0}, t) = 0$, and there exists a time-invariant $V_0(\boldsymbol{x})$. Then, based on the nature of $V_0(\boldsymbol{x})$, locally $V(\boldsymbol{x}, t)$ is defined in Table 2.3 in the neighborhood of the origin $\Re^n$. Globally, positive definite functions are defined if the neighborhood is $\Re^n$.

Definition 2.20 Let $V(\boldsymbol{x}, t)$ be a scalar time-varying function with $V(\mathbf{0}, t) = 0$, and there exists a time-invariant positive definite function $V_l(\boldsymbol{x})$ such that $\forall t \geq 0$, $V(\boldsymbol{x}, t) < V_l(\boldsymbol{x})$. Then, function $V(\boldsymbol{x}, t)$ is decrescent in nature.

In other words, a function is positive definite if it is lower bounded by a time-invariant positive definite function and decrescent if it is upper bounded by a time-invariant positive definite function.

Example 2.18 Consider the following nonautonomous nonlinear functions with states x_1 and x_2:

1. Let $V(x_1, x_2, t) = x_1^2 + (1 + x_2^2)e^{-t}$. This function is neither PDF nor decrescent in nature. To show that, one can notice that as $x_1 \to 0$ and $x_2 \to 0$, $V(0, t) \neq 0$. Hence, the nonautonomous function $V(x_1, x_2, t)$ is not a positive definite function. Furthermore, since $V(0, 0) \neq 0$, $V(x_1, x_2, t)$ cannot be bounded by any time-invariant PDF. Therefore, $V(x_1, x_2, t)$ is not decrescent in nature. Variation of $V(x_1, x_2, t)$ with respect to x_1, x_2 and time t is shown in Fig. 2.16.

   Since the function has three arguments, the function is plotted with respect to x_1 and x_2 axes at different time stamps. The different time instances t_1, t_2, and t_3 of time t show how the function is shifting from $t = 0$ to $t = \infty$.
2. Let $V(x_1, x_2, t) = (x_1^2 + x_2^2)e^{-t}$. This function is not PDF but decrescent in nature. To show that, one can notice that as $x_1 \to 0$ and $x_2 \to 0$, $V(0, t) = 0$. However, as $t \to \infty$, $V(0, \infty) = 0$. Therefore, $V(x_1, x_2, t)$ is

(continued)

Example 2.18 (continued)

not a PDF but a PSDF. Furthermore, since $V(x_1, x_2, t)$ can be bounded by $(x_1^2 + x_2^2)$, it is decrescent in nature. Variation of $V(x_1, x_2, t)$ with respect to x_1, x_2, and time t is presented in Fig. 2.17, where the upper bound of V is shown due to the decrescent nature. The different time instances t_1, t_2, and t_3 of time t show how the function is shifting from $t = 0$ to $t = \infty$.

3. Let $V(x_1, x_2, t) = x_1^2 + (1+t)x_2^2$. This function is PDF but not decrescent in nature. To show that, one can notice that as $x_1 \rightarrow 0$ & $x_2 \rightarrow 0$, $V(0, t) = 0$, and $V(x_1, x_2, t) > x_1^2 + x_2^2$, it is PDF in nature. Furthermore, as $t \rightarrow \infty$, $V(x_1, x_2, t) \rightarrow \infty$. Hence, it cannot be bounded by any time-invariant PDF. Thus, the function is not decrescent in nature. Variation of $V(x_1, x_2, t)$ with respect to x_1, x_2, and time t is presented in Fig. 2.18, where the lower bound of V is shown due to the PDF nature. The different time instances t_1, t_2, and t_3 of time t show how the function is shifting from $t = 0$ to $t = \infty$.
4. $V(x_1, x_2, t) = x_1^2 + (1+e^{-t})x_2^2$. This function is both PDF and decrescent in nature. To show that, one can notice that as $x_1 \rightarrow 0$ and $x_2 \rightarrow 0$, $V(0, t) = 0$ and $V(x_1, x_2, t) > x_1^2 + x_2^2$. Therefore, $V(x_1, x_2, t)$ is PDF in nature. Further, as $t \rightarrow \infty$, $V(x_1, x_2, t) \rightarrow 0$ and $V(x_1, x_2, t) < x_1^2 + 2x_2^2$. Thus, the function is decrescent in nature. Variation of $V(x_1, x_2, t)$ with respect to x_1, x_2, and time t is presented in Fig. 2.19a–c, where the upper bound of V is shown due to the decrescent nature and lower bound is shown due to the PDF nature. The different time instances t_1, t_2, and t_3 of time t show how the function is shifting from $t = 0$ to $t = \infty$.

In upcoming sections, it will be shown that both positive definiteness and decrescent properties are necessary for the asymptotic stability of the system.

Next, Lyapunov theorems for stable, uniform asymptotically stable (UAS), and global uniform asymptotically stable systems, considering $\boldsymbol{x}^* = \boldsymbol{0}$ as the equilibrium point, can be stated as follows.

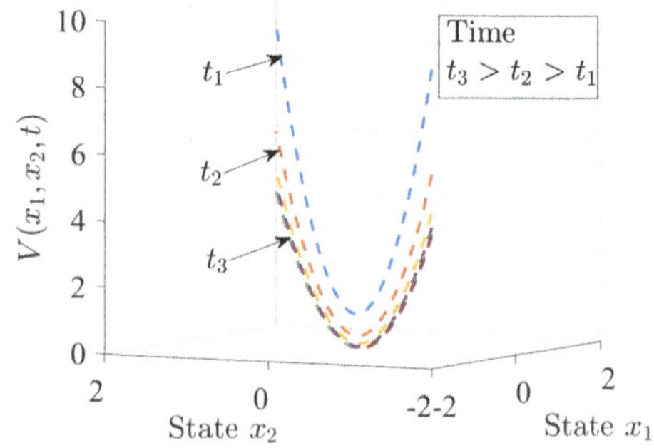

Fig. 2.16 Plot of nonautonomous function $V(x_1, x_2, t)$

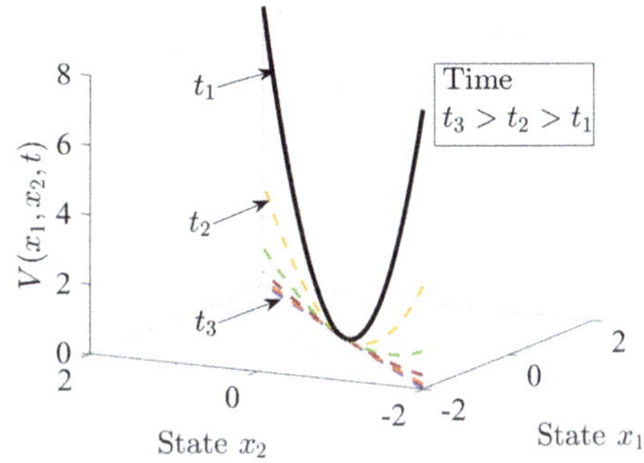

Fig. 2.17 Plot of nonautonomous function $V(x_1, x_2, t)$ and upper bound in solid

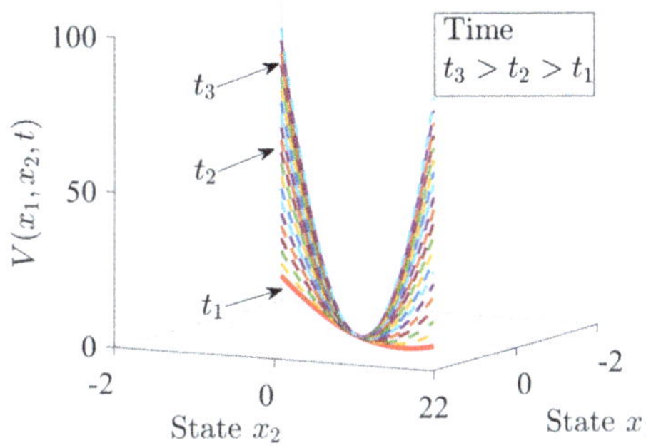

Fig. 2.18 Plot of nonautonomous function $V(x_1, x_2, t)$ and lower bound in solid

Theorem 2.6 *Let $\boldsymbol{x}^* = \mathbf{0}$ be an equilibrium point of a nonlinear nonautonomous system $\dot{\boldsymbol{x}} = \boldsymbol{f}(\boldsymbol{x}, t)$, where $\boldsymbol{x} \in \Re^n$, $\boldsymbol{f} \in \Re^n$. Let $V(\boldsymbol{x}, t) \in \Re$ be a continuously differentiable function. Then, the sufficient conditions required for different types of stability of the equilibrium point $\boldsymbol{x}^* = \mathbf{0}$ are presented in Table 2.4, where the acronym "NA" refers to not applicable and "RU" refers to* ***r****adially* ***u****nbounded.*

When the equilibrium point $\boldsymbol{x}^* = \mathbf{0}$ is not defined for nonautonomous nonlinear systems, then uniformly bounded stability in the sense of Lyapunov is summarized in the following theorem.

Theorem 2.7 *Let $V(\boldsymbol{x}, t) \in \Re$ be a continuously differentiable function. Then, the solution of the nonautonomous nonlinear systems is said to be uniformly bounded if*

(i) V is PDF and decrescent.
(ii) $\dot{V}$ is NDF and $\|\boldsymbol{x}\| \geq R$ for $t \in [0, \infty)$.

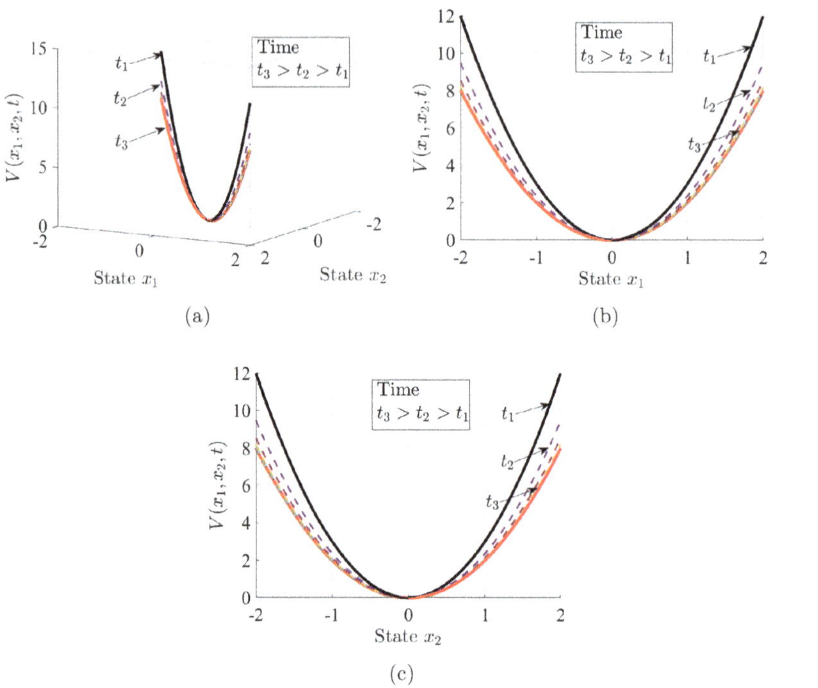

Fig. 2.19 (**a**) Plot of nonautonomous function $V(x_1, x_2, t)$ with upper and lower bounds in solid. (**b**) Plot of $V(x_1, x_2, t)$ with x_1. (**c**) Plot of $V(x_1, x_2, t)$ with x_2

Example 2.19 Consider the nonlinear nonautonomous system defined as

$$\dot{x}_1 = -x_1 - e^{-2t}x_2$$
$$\dot{x}_2 = x_1 - x_2$$

To determine the nature of stability of the system at the equilibrium point (0, 0), the following Lyapunov function is selected:

$$V(x_1, x_2, t) = x_1^2 + \left(1 + e^{-2t}\right) x_2^2$$

The chosen Lyapunov function is a valid Lyapunov function, as it is positive definite as it is lower bounded by a time-invariant positive definite function $(x_1^2 + x_2^2)$ and also decrescent as it is upper bounded by a time-invariant positive definite function $(x_1^2 + 2x_2^2)$.

(continued)

Example 2.19 (continued)
The time derivative of the Lyapunov function is

$$\begin{aligned}\dot{V}(x_1, x_2, t) &= -2\left[x_1^2 - x_1x_2 + x_2^2\left(1 + e^{-2t}\right)\right] \\ &\leq -2\left(x_1^2 - x_1x_2 + x_2^2\right) \\ &\leq -2\left(x_1^2 - x_1x_2 + 3x_2^2\right) \\ &\leq -2\left(\left(x_1 - \frac{1}{2}x_2\right)^2 + \frac{11}{4}x_2^2\right)\end{aligned}$$

which means that $\dot{V}(x_1, x_2, t)$ is negative definite. Thus, the equilibrium point is globally asymptotically stable.

2.2.3 Barbalat's Lemma

Similar to autonomous nonlinear systems, for a nonautonomous system, oftentimes it is difficult to achieve negative definiteness for $\dot{V}(\boldsymbol{x}, t)$. Therefore, the following lemma, commonly known as *Barbalat's lemma*, is used to conclude the asymptotic stability of the system.

Lemma 2.1 (Barbalat's Lemma) *If the differentiable function $f(t)$ has a finite limit as $t \to \infty$, $\dot{f}(t)$ is uniformly continuous, then $\dot{f}(t) \to 0$ as $t \to \infty$.*

Usually, the following variant of Barbalat's lemma is used in the Lyapunov-based stability analysis of nonautonomous nonlinear systems.

Lemma 2.2 (Lyapunov-like Lemma Slotine et al. 1991) *If a scalar function $V(\boldsymbol{x}, t)$ satisfies the following conditions*

- *$V(\boldsymbol{x}, t)$ is lower bounded.*
- *$\dot{V}(\boldsymbol{x}, t)$ is negative semi-definite.*
- *$\dot{V}(\boldsymbol{x}, t)$ is uniformly continuous in time.*

then $\dot{V}(\boldsymbol{x}, t) \to 0$ as $t \to \infty$. Note that the sufficient condition for a function to be uniformly continuous is that its derivative is both upper and lower bounded.

Table 2.4 Sufficient conditions required for different types of stability of the equilibrium point $\boldsymbol{x}^* = \mathbf{0}$

Nature of equilibrium point	Nature of $V(\boldsymbol{x}, t)$	Nature of $\dot{V}(\boldsymbol{x}, t)$	RU condition	Decrescent condition
Stable	$V(\boldsymbol{x}, t)$ is PDF	$\dot{V}(\boldsymbol{x}, t)$ is NSDF	NA	NA
Uniformly stable	$V(\boldsymbol{x}, t)$ is PDF	$\dot{V}(\boldsymbol{x}, t)$ is NSDF	NA	$V(\boldsymbol{x}, t)$ is decrescent
Uniformly asymptotically stable	$V(\boldsymbol{x}, t)$ is PDF	$\dot{V}(\boldsymbol{x}, t)$ is NDF	NA	$V(\boldsymbol{x}, t)$ is decrescent
Globally uniformly asymptotically stable (GUAS)	$V(\boldsymbol{x}, t)$ is PDF	$\dot{V}(\boldsymbol{x}, t)$ is NDF	$V(\boldsymbol{x}, t)$ is RU	$V(\boldsymbol{x}, t)$ is decrescent

Example 2.20 Consider the nonautonomous nonlinear system given by

$$\dot{e} = w\phi(t) - e$$
$$\dot{w} = -e\phi(t)$$

where e and w are the system states and $\phi(t)$ is the continuous bounded function.

To establish the stability of the system, consider the Lyapunov function as

$$V(e, w) = \frac{e^2 + w^2}{2}$$

The Lyapunov stability analysis can be carried out as follows:

$$\dot{V}(e, w, t) = e\,(w\phi(t) - e) + w(-e\phi(t))$$
$$\dot{V}(e, w, t) = -e^2 + ew\phi(t) - we\phi(t)$$
$$\dot{V}(e, w) = -e^2$$

Since $\dot{V}(e, w) \leq 0$, the error e and w are bounded. However, the convergence of error e cannot be concluded. Hence, Barbalat's lemma is used to check the convergence of error. The derivative of $\dot{V}(e, w)$ is calculated as

$$\ddot{V}(e, w, t) = -2e\,(w\phi(t) - e)$$

Since $\phi(t)$ is bounded by construction and e and w are previously shown to be bounded through Lyapunov analysis, $\ddot{V}(e, w)$ is bounded. This means that $\dot{V}(e, w)$ is uniformly continuous. Hence, using Barbalat's lemma, $\dot{V}(e, w) \to 0$, i.e., $e \to 0$.

2.3 Summary

This chapter provides a comprehensive overview of various types of stability for both autonomous and nonautonomous nonlinear systems. It introduces the Lyapunov stability theory for both system types, focusing on analyzing the behavior of nonlinear systems around equilibrium points. To enhance clarity, Lyapunov theorems for different types of stability are presented in a tabular format, aiding readers in understanding the concepts more effectively. Given the differences in stability analysis between autonomous and nonautonomous systems, several examples are included for better illustration. Additionally, key theorems and lemmas related to the stability analysis of nonlinear systems are discussed. However, this chapter only provides a concise summary of stability theory for nonlinear systems, limited to the scope of this book. For a more in-depth understanding of stability analyses of nonlinear systems, readers are encouraged to refer to sources such as Márquez (2003), Slotine et al. (1991), Khalil (2002), Narendra and Annaswamy (2012), Lavretsky and Wise (2012), Nguyen (2018), and Ge et al. (2013).

2.4 Exercise Problems

1. Determine the stability of the following systems around the equilibrium point:
   (a) $\dot{x} = -x^2 + 3x$
   (b) $\dot{x} = -x + x^3$
   (c) $\dot{x} = x - x^3$
   (d) $\dot{x} = -x\left(\frac{-1}{2} + \sin^2 x\right)$
2. Determine the stability of the following systems around the equilibrium point:
   (a) $\ddot{x} = -\sin x$
   (b) $\ddot{x} = -x + \left(1 - x^2 - \dot{x}^2\right)\dot{x}$
   (c) $\ddot{x} - 0.7(1 - x^2)\dot{x} + x = 0$
   (d) $\ddot{x} + \dot{x} + \frac{x}{1+\sin^2 x} = 0$
3. Find the equilibrium points and determine the stability of the following systems around them:
   (a) $\begin{bmatrix} \dot{x}_1 \\ \dot{x}_2 \end{bmatrix} = \begin{bmatrix} x_2 \\ -2\sin x_1 + \sin x_1 \cos x_1 \end{bmatrix}$
   (b) $\begin{bmatrix} \dot{x}_1 \\ \dot{x}_2 \end{bmatrix} = \begin{bmatrix} x_2 \\ -x_2 - x_1 - (x_1 + x_2)^2 x_2 \end{bmatrix}$
   (c) $\begin{bmatrix} \dot{x}_1 \\ \dot{x}_2 \end{bmatrix} = \begin{bmatrix} -x_1^3 - x_1 x_2^2 + x_1 \\ -x_2^3 - x_1^2 x_2 + x_2 \end{bmatrix}$
4. Show that if a function $V(t)$ satisfies the inequality

$$\dot{V}(t) \leq -\alpha V(t), \quad \alpha \in \mathbb{R}$$

then

$$V(t) \leq e^{-\alpha t} V(0)$$

where $V(0)$ is the initial condition of $V(t)$ at $t = 0$.

5. Analyze the stability of the following nonlinear system:

(a) $\begin{bmatrix} \dot{x}_1 \\ \dot{x}_2 \end{bmatrix} = \begin{bmatrix} x_2 + x_1\left(1 - x_1^2 - x_2^2\right) \\ -x_1 + x_2\left(1 - x_1^2 - x_2^2\right) \end{bmatrix}$

(b) $\begin{bmatrix} \dot{x}_1 \\ \dot{x}_2 \end{bmatrix} = \begin{bmatrix} x_2 \\ -x_2 - 5x_1 - (x_1 + x_2)^2 x_2 \end{bmatrix}$

(c) $\begin{bmatrix} \dot{x}_1 \\ \dot{x}_2 \end{bmatrix} = \begin{bmatrix} x_2 \\ -\frac{x_1}{k^2 - x_1^2} - x_2 \end{bmatrix}$, where k is a positive constant.

6. Find out the positive definite and decrescent properties of the following functions:

(a) $V(x_1, x_2, t) = (2 + \sin^2 t)(x_1^2 + x_2^2)$
(b) $V(x_1, x_2, t) = x_1^2 + (1 + e^{-t})x_2^2$
(c) $V(x_1, x_2, t) = \frac{(x_1^2 + x_2^2)(t^2 + 1)}{x_2^2 + 1}$
(d) $V(x_1, x_2, t) = (2x_1^2 + x_2^2)(1 + \sin^2 t))$

7. Determine which of the following functions can be selected as a Lyapunov function:

(a) $V(x_1, x_2, t) = \frac{(x_1^2 + x_2^2)(t^2 + 1)}{t^2 + 5}$
(b) $V(x_1, x_2, t) = (2x_1^2 + x_2^2)e^{-3t}$
(c) $V(x_1, x_2, t) = (2x_1^2 + x_2^2)\cos^2 t)$
(d) $V(x_1, x_2, t) = \frac{(x_1^2 + x_2^2)(e^{-t} + 1)}{x_1^2 + 2}$

8. Determine the stability of the following nonautonomous system:

(a) $\dot{x} = \frac{-x}{(1+t)^2}$
(b) $\dot{x} = \frac{-x}{(1+t)}$
(c) $\begin{bmatrix} \dot{x}_1 \\ \dot{x}_2 \end{bmatrix} = \begin{bmatrix} -x_1 - x_1 x_2^2 \cos^2 t \\ -x_2 - x_1^2 x_2 \sin^2 t \end{bmatrix}$
(d) $\begin{bmatrix} \dot{x}_1 \\ \dot{x}_2 \end{bmatrix} = \begin{bmatrix} -2x_1 - e^{-t}x_2 \\ x_1 - 2x_2 \end{bmatrix}$
(e) $\begin{bmatrix} \dot{x}_1 \\ \dot{x}_2 \end{bmatrix} = \begin{bmatrix} -x_2 + x_1^2 \\ -x_1 - 2x_2 \end{bmatrix}$

9. Determine the stability and ultimate bound of the following nonautonomous system:

(a) $\dot{x} = -x^3 + \cos t$
(b) $\dot{x} = (\frac{\cos t}{2} - 1)x - \sin^2 t$
(c) $\dot{x} = -(2\cos^2 t + 1)x + \sin t$
(d) $\dot{x} = -(e^{-2t} + 1)x + \cos t$

References

S. Ge, C. Hang, T. Lee, T. Zhang, *Stable Adaptive Neural Network Control*. The International Series on Asian Studies in Computer and Information Science Series (Springer US, New York, 2013)

H. K. Khalil, *Nonlinear Systems* (Prentice-Hall, Englewood Cliffs, 2002).

E. Lavretsky, K. Wise, *Robust and Adaptive Control: With Aerospace Applications*. Advanced Textbooks in Control and Signal Processing Series (Springer London, London, 2012)

H. Márquez, *Nonlinear Control Systems: Analysis and Design* (Wiley, Hoboken, 2003).

K. Narendra, A. Annaswamy, *Stable Adaptive Systems*. Dover Books on Electrical Engineering Series (Dover Publications, New York, 2012)

N. Nguyen, *Model-Reference Adaptive Control: A Primer*. Advanced Textbooks in Control and Signal Processing Series (Springer International Publishing, Cham, 2018)

J.-J. E. Slotine, W. Li et al., *Applied Nonlinear Control*, vol. 199, no. 1 (Prentice Hall, Englewood Cliffs, 1991)

Chapter 3
Nonlinear Control Design Philosophies

This chapter focuses on the design of control techniques for various classes of nonlinear systems. Stability theory and mathematical tools previously discussed serve as the framework to ensure stability of the closed-loop system. The motivation behind delving into nonlinear control emerges from the need to control systems whose behavior cannot be adequately modeled by linear system dynamics. Nonlinear control is a field that addresses the challenges posed by nonlinear systems, aiming to achieve desired performance objectives. However, it is worth reiterating that nonlinear systems, by definition, are not necessarily linear systems, and hence, the nonlinear control design techniques are equally valid for linear systems as well.

Several nonlinear control methods are introduced to design controllers for different classes of nonlinear systems. Some of these methods include feedback linearization-based dynamic inversion (DI), backstepping control, and sliding mode control (SMC). Each of these approaches comes with advantages and limitations, making them suitable and/or advantageous for specific scenarios or system characteristics. A concise summary of each control design approach is presented in the subsequent sections.

3.1 Dynamic Inversion Control

The dynamic inversion (DI) design technique is a readily available and easy-to-understand nonlinear control design technique that exploits the stability theory of linear systems to design an output tracking controller for nonlinear systems. The technique is essentially based on the philosophy of feedback linearization. In the feedback linearization philosophy, an appropriate coordinate transformation is first applied to reformulate the system into a form that resembles a linear system in the transformed coordinates. Subsequently, any conventional linear control design method can be used to design the controller in the transformed coordinates. Such

K. Sachan et al., *Intelligent Adaptive Control with System Constraints*,
https://doi.org/10.1007/978-981-95-8220-4_3

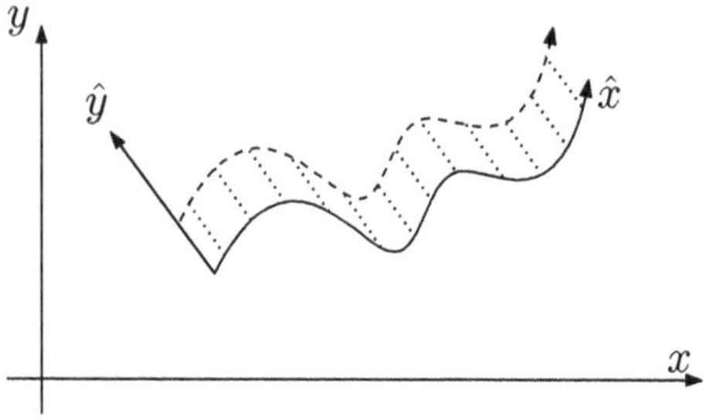

Fig. 3.1 Philosophy of dynamic inversion control

a nonlinear coordinate transformation is indeed feasible, provided that certain mathematical conditions are met. Hence, any linear control design technique, mostly the standard proportional-integral-derivative (PID) controller, can be used in the transformed linear system dynamics, ultimately yielding a nonlinear state feedback controller when converted back to the original coordinates. It is important to note that feedback linearization may be implemented either via input-state linearization or input-output linearization. However, the stringent conditions required for input-state linearization are rarely satisfied in practice. Consequently, instead of taking such a rigorous approach, a simplified, easy-to-understand approach is presented here, which essentially embeds both input-output linearization and PID control design. An interested reader can find more details about the feedback linearization technique from numerous textbooks, such as Slotine et al. (1991); Isidori (2013).

Intuitively, to illustrate the elegance of the DI control design philosophy, consider a reference signal represented by the dotted line in Fig. 3.1, plotted in the two-dimensional xy frame of reference. A conventional linear controller designed to track this reference signal typically requires linearization at multiple operating points, resulting in a gain-scheduled controller that is only locally stable. However, if the xy frame can be transformed into a new coordinate frame $\hat{x}\hat{y}$, such that at each point the distance between $\hat{x}$ and $\hat{y}$ remains constant, the reference signal in this transformed coordinate system becomes a step signal. In this case, a linear controller can be designed in the $\hat{x}\hat{y}$ frame to track the step signal. This controller can then be transformed back into the original coordinates, resulting in a nonlinear state feedback controller that effectively achieves the desired tracking performance.

In the input-output feedback linearization (DI control design) approach, the core idea involves inverting the system model by constructing an appropriate control law such that the tracking error dynamics behave like a stable linear system. As a result, the tracking error asymptotically converges to zero, thereby ensuring perfect tracking performance. The DI control design facilitates the effective application of well-established linear control strategies to regulate the behavior of nonlinear systems. The DI control design has the following advantages: (i) It is a simple design with a systematic design procedure. (ii) It leads to a "closed-form solution" for the controller, which is preferable for both online and practical implementation. (iii) DI controller leads to global exponential stability of the closed-loop output error dynamics.

Due to these benefits, the DI control design has recently been found to be applicable in various domains, including the control of aerospace systems, robotics, and mechanical systems. Without going through the standard involved way found in standard textbooks (see Slotine et al. 1991), an attempt is made here to follow an alternate, simplified approach which is both easy to understand and easy to implement. To elucidate the input-output feedback linearization control design procedure, consider the following control-affine square nonlinear system:

$$\dot{\boldsymbol{x}} = \boldsymbol{f}(\boldsymbol{x}) + \boldsymbol{G}(\boldsymbol{x})\,\boldsymbol{u} \tag{3.1}$$

$$\boldsymbol{y} = \boldsymbol{h}(\boldsymbol{x}) \tag{3.2}$$

where variables $\boldsymbol{x} \in \Re^n$, $\boldsymbol{u} \in \Re^m$, and $\boldsymbol{y} \in \Re^m$ represent the state, input, and output vectors of the system, respectively. Vectors $\boldsymbol{f}(\boldsymbol{x})$ and $\boldsymbol{h}(\boldsymbol{x})$ are nonlinear functions of $\boldsymbol{x}$, and $\boldsymbol{G}(\boldsymbol{x}) \in \Re^{n \times m}$ is the control input gain matrix.

The aim of the control design is to design a stable control law that guarantees accurate tracking of the desired output command, expressed as $\boldsymbol{y} \to \boldsymbol{y}^*$ as $t \to \infty$, where $\boldsymbol{y}^*$ denotes the desired command to be tracked. It is assumed that the desired command is continuous and smooth (i.e., the derivative is also continuous).

The output dynamics of the system can be derived as follows:

$$\dot{\boldsymbol{y}} = \left(\frac{\partial \boldsymbol{h}}{\partial \boldsymbol{x}}\right)\dot{\boldsymbol{x}} \tag{3.3}$$

$$= \left(\frac{\partial \boldsymbol{h}}{\partial \boldsymbol{x}}\right)[\boldsymbol{f}(\boldsymbol{x}) + \boldsymbol{G}(\boldsymbol{x})\boldsymbol{u}] \tag{3.4}$$

Here, $\left(\frac{\partial \boldsymbol{h}}{\partial \boldsymbol{x}}\right)$ represents the Jacobian matrix:

$$\left(\frac{\partial \boldsymbol{h}}{\partial \boldsymbol{x}}\right) \triangleq \begin{bmatrix} \frac{\partial h_1}{\partial x_1} & \cdots & \frac{\partial h_1}{\partial x_n} \\ \vdots & \ddots & \vdots \\ \frac{\partial h_m}{\partial x_1} & \cdots & \frac{\partial h_m}{\partial x_n} \end{bmatrix}$$

Defining $\boldsymbol{f}_y(\boldsymbol{x}) \triangleq \left(\frac{\partial \boldsymbol{h}}{\partial \boldsymbol{x}}\right)\boldsymbol{f}(\boldsymbol{x})$ and $\boldsymbol{G}_y(\boldsymbol{x}) \triangleq \left(\frac{\partial \boldsymbol{h}}{\partial \boldsymbol{x}}\right)\boldsymbol{G}(\boldsymbol{x})$, the output dynamics can be rewritten as

$$\dot{\boldsymbol{y}} = \boldsymbol{f}_y(\boldsymbol{x}) + \boldsymbol{G}_y(\boldsymbol{x})\boldsymbol{u} \tag{3.5}$$

It can be observed that, upon differentiating the output, the control input $\boldsymbol{u}$ appears explicitly in the output dynamics. This explicit appearance of $\boldsymbol{u}$ is essential for DI control design, as it enables direct manipulation of the system output through the control input. Consequently, the output selection becomes a critical factor in the DI control design process. This characteristic is known as the "relative degree" of the system, defined as the number of times the output must be differentiated until the control variable appears explicitly in the dynamics. Ideally, for effective DI control design, the relative degree should match the order of the system. The significance of output selection and relative degree in DI design will be further illustrated through examples.

Now, starting with the definition of the error as $\boldsymbol{e} \triangleq \boldsymbol{y}^* - \boldsymbol{y}$, the closed-loop error dynamics can be defined as

$$\dot{\boldsymbol{e}} + \boldsymbol{K}\boldsymbol{e} = \boldsymbol{0} \tag{3.6}$$

where $\boldsymbol{0} = [0, \ldots, 0]^T \in \Re^n$ denotes the vector whose all entries are zero and $\boldsymbol{K}$ is a positive definite matrix, usually selected as a diagonal matrix with positive diagonal elements.[1] This equation has a stable solution as $\boldsymbol{e} = \exp(-\boldsymbol{K}t)\,\boldsymbol{e}_0$, where $\boldsymbol{e}_0$ is the initial condition of $\boldsymbol{e}$ and "exp" is the exponential function. It can be seen from the solution that $\boldsymbol{e}(t) \to 0$ as $t \to \infty$.

Upon conducting further algebraic manipulations, the resulting equation takes the following form:

$$\dot{\boldsymbol{e}} + \boldsymbol{K}\,\boldsymbol{e} = \boldsymbol{0} \tag{3.7}$$

$$\dot{\boldsymbol{y}}^* - \dot{\boldsymbol{y}} + \boldsymbol{K}(\boldsymbol{y}^* - \boldsymbol{y}) = \boldsymbol{0} \tag{3.8}$$

$$\dot{\boldsymbol{y}}^* - \boldsymbol{f}_y(\boldsymbol{x}) - \boldsymbol{G}_y(\boldsymbol{x})\,\boldsymbol{u} + \boldsymbol{K}(\boldsymbol{y}^* - \boldsymbol{y}) = \boldsymbol{0} \tag{3.9}$$

The control input $\boldsymbol{u}$ can be extracted through the following expression:

$$\boldsymbol{u} = \left[\boldsymbol{G}_y(\boldsymbol{x})\right]^{-1}\left(\dot{\boldsymbol{y}}^* - \boldsymbol{f}_y(\boldsymbol{x}) + \boldsymbol{K}(\boldsymbol{y}^* - \boldsymbol{y})\right) \tag{3.10}$$

It is crucial to emphasize that for a valid solution, the inverse $\left[\boldsymbol{G}_y(\boldsymbol{x})\right]^{-1}$ must exist at all times during the control application. This analysis synthesized a closed-loop control system that ensures the output of the system tracks the desired command. The control architecture of DI control is presented in Fig. 3.2, where DI control $\boldsymbol{u}$ is expressed in (3.10).

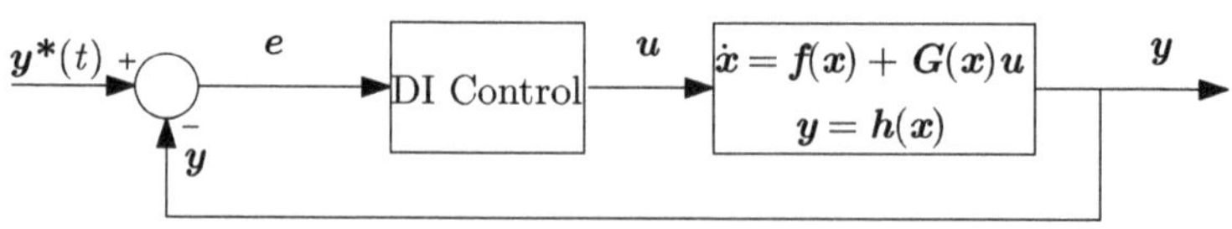

Fig. 3.2 Dynamic inversion control architecture

[1] Eigenvalues of a diagonal matrix are the diagonal elements.

Example 3.1 The DI control design steps are illustrated by considering the following nonlinear system given by

$$\dot{x} = x + x^2 + (1 + x^2)u \tag{3.11}$$

The control objective is to design a stabilizing controller, i.e., $y \to y^*$, where the desired output y^* is assumed to be 0. The first step in DI control design is to choose the system output, which is selected as $y = x$. Hence, the system error is defined as $e = x$. The next step involves the selection of stable error dynamics, which is selected as $\dot{e} + ke = 0$, where $k > 0$ is a constant gain. By substituting $e = x$, the resultant error dynamics is rewritten as

$$\dot{x} + kx = 0$$

Next, using the expression of $\dot{x}$ from (3.11), one can write

$$x + x^2 + (1 + x^2)u + kx = 0 \tag{3.12}$$

From (3.12), the expression for the control input u can be extracted as

$$u = -\frac{1}{(1 + x^2)}\left(x + x^2 + kx\right) \tag{3.13}$$

It can be noted that the selection of error dynamics is dictated by the relative degree of the problem. For example, based on the selection of system output, if the relative degree of the system is two, then a second-order error dynamics needs to be enforced, which is represented by the following equation:

$$\ddot{\boldsymbol{e}} + \boldsymbol{K}_v \dot{\boldsymbol{e}} + \boldsymbol{K}_p \boldsymbol{e} = \boldsymbol{0} \tag{3.14}$$

where positive definite diagonal controller gain matrices $\boldsymbol{K}_v$ and $\boldsymbol{K}_p$ are usually selected as $\boldsymbol{K}_v = \mathrm{diag}(2\zeta_i \omega_{n_i})$ and $\boldsymbol{K}_p = \mathrm{diag}(\omega_{n_i}^2)$, respectively, where $\zeta_i > 0$, $\omega_{n_i} > 0$, and index $i = 1, \ldots, p$, where p depends on the size of output vector that needs to be controlled.

Example 3.2 The importance of the selection of system output and the concept of relative degree in DI design are demonstrated using the following nonlinear system given by

$$\dot{x_1} = 2x_1x_2 + 3x_1^2 + 2 + 5\left(1 + x_1^2x_2^2\right)u$$

$$\dot{x_2} = -3x_1 \tag{3.15}$$

The control objective is to design a stabilizing DI controller, i.e., $x_1 \to 0$ and $x_2 \to 0$ as $t \to \infty$.

As mentioned earlier, a careful output selection is crucial in DI control design. For instance, if output y is selected as $y = x_1$, which is tempting as $x_1 \to 0$ is one of the objectives. However, as $x_1 \to 0$, $\dot{x_2} \to 0$. That means x_2 will not change after that. Hence, driving $x_2 \to 0$ after that is impossible. Hence, y is selected as

$$y = x_2 \tag{3.16}$$

Note that $x_2 \to 0$ and $\dot{x_2} \to 0$ happen when a second-order dynamics is enforced. Hence, consider the following stable error dynamics as

$$\ddot{e} + 2\zeta\omega_n\dot{e} + \omega_n^2 e = 0 \tag{3.17}$$

where ζ and ω_n are selected as $0 < \zeta < 1$ and $\omega_n > 0$, respectively. Since the desired signal is zero, i.e., $y^* = \dot{y}^* = \ddot{y}^* = 0, \forall t$, the dynamics can be rewritten as

$$\ddot{y} + 2\zeta\omega_n\dot{y} + \omega_n^2 y = 0 \tag{3.18}$$

The time derivative of output $y = x_2$ leads to

$$\dot{y} = -3x_1$$

$$\ddot{y} = -3\left[2x_1x_2 + 3x_1^2 + 2 + 5\left(1 + x_1^2x_2^2\right)u\right]$$

Upon substitution of $\ddot{y}$, $\dot{y}$, and y in (3.18), the following is obtained:

$$-3\left[2x_1x_2 + 3x_1^2 + 2 + 5\left(1 + x_1^2x_2^2\right)u\right] - 6\zeta\omega_n x_1 + \omega_n^2 x_2 = 0$$

(continued)

Example 3.2 (continued)
The control expression can be extracted as

$$u = \frac{1}{15\left(1 + x_1^2 x_2^2\right)}\left[-6x_1x_2 - 9x_1^2 - 6 - 6\zeta\omega_n x_1 + \omega_n^2 x_2\right] \quad (3.19)$$

Here, as $\left(1 + x_1^2 x_2^2\right) \neq 0$, the controller is well defined.

Note that only a well-defined relative degree in DI control does not guarantee that the controller will perform satisfactorily. A scenario may appear in the DI control design where, despite a well-defined relative degree through output selection, the system may lose controllability; for example, the coefficient of the control input may vanish under certain conditions. In such cases, one has to be careful about the domain of applicability of the controller. This point is explained in the following example.

Example 3.3 Consider the following nonlinear system given by

$$\begin{aligned} \dot{x}_1 &= x_2 \\ \dot{x}_2 &= -3x_1x_2 + u \end{aligned} \quad (3.20)$$

The objective is to design a stabilizing DI controller, i.e., $(x_1,\ x_2) \rightarrow (0,\ 0)$ as $t \rightarrow \infty$.

To illustrate the concept, select the system output as $y = x_1^2$. Then, the time derivatives of the output can be calculated as

$$\dot{y} = 2x_1x_2$$

$$\ddot{y} = 2x_2^2 + 2x_1\left(-3x_1x_2 + u\right) \quad (3.21)$$

$$= 2x_2^2 - 6x_1^2x_2 + 2x_1u \quad (3.22)$$

where it can be noted that when $x_1 \rightarrow 0$, the term $2x_1u \rightarrow 0$. Hence, at $x_1 = 0$, i.e., on the entire x_2 axis, the system loses controllability and the control is not effective.

Now consider the output of the system as $y = x_1$. The time derivatives are calculated as $\dot{y} = x_2$ and $\ddot{y} = -3x_1x_2 + u$, which leads to a stabilizing controller. Therefore, a careful output selection is necessary in DI control design.

The DI control design has certain other generic limitations as well that require special attention (a few are pointed out here through examples). The major limitations and their possible solutions are described here:

- **Inverse of the matrix $G_y(x)$:** The inverse of the matrix $G_y(x)$ must always exist for the effective performance of the DI controller. In scenarios where the control gain matrix is singular, a possible solution is to stop the control update near singularity and hope for the best that it will come out of it. However, this strategy results in degraded tracking performance wherever the controller update is stopped.
- **Dimension of control input gain matrix $G_y(x)$:** The dimension of the matrix $G_y(x) \in \Re^{m \times m}$ is square in the context of the DI control design procedure. This implies that the input dimension is considered equal to the output dimension. However, this assumption may not hold true for certain applications. If the dimension of the control input exceeds that of the output, an optimization approach can be followed, along with tracking performance. This approach is known as optimal dynamic inversion (Padhi and Kothari 2007; Padhi and Balakrishnan 2007). Conversely, when the dimension of the control input is less than that of the output, the control design objectives cannot be adequately fulfilled, resulting in degraded control performance.
- **Stability of internal dynamics:** If the total "relative degree" of the problem is lower than the degree of the system, the controller cannot control all the states. The dynamics of these uncontrolled states are known as "internal dynamics" of the system. In this scenario, the internal dynamics may exhibit instability and necessitate additional investigation. In such scenarios, the first step involves examining the stability of the "zero dynamics" of the system. The asymptotic stability of these "zero dynamics" is sufficient for ensuring local stability of the internal dynamics. Note that a nonlinear system is classified as a "minimum phase" if its zero dynamics are stable. An alternative method, known as the output redefinition technique, is also presented in the literature to stabilize the internal dynamics of the system (Gopalswamy and Hedrick 1993; Wallner and Well 2002).
- **Sensitive to modeling inaccuracies:** The design of the DI controller requires an accurate model of the system. Therefore, modeling uncertainties of the system may degrade the performance of DI control and, in severe cases, may lead to system instability. To address this limitation, various adaptive control methods have been documented in the literature (Slotine et al. 1991; Padhi et al. 2007). These methods serve as effective strategies to cope with and mitigate the impact of modeling uncertainties on the DI control system.
- **Cancellation of useful system nonlinearities:** The formulation of the DI controller leads to the cancellation of useful system nonlinearities. This happens because the DI control uses the inverse dynamics of the system in the formulation of the control law, aiming to neutralize nonlinearities, occasionally leading to the cancellation of even useful nonlinearities (Lee et al. 2009). This is illustrated in (3.24) of Example 3.4.

3.2 Lyapunov Stability Theory-Based Control Design

As the name suggests, Lyapunov stability theory-based controllers are constructed using Lyapunov stability theory. The advantages of such controllers are that they can be designed to preserve certain beneficial nonlinear terms of the system dynamics, although this is not always guaranteed. Additionally, these controllers can be designed to be robust to model uncertainties. A brief outline of the control design is given below.

Consider a nonlinear system given by $\dot{\boldsymbol{x}} = \boldsymbol{f}(\boldsymbol{x}, \boldsymbol{u})$, where $\boldsymbol{x}$ and $\boldsymbol{u}$ represent states and control input of the system, respectively. The objective of the control design is to design a feedback control law $\boldsymbol{u} = \boldsymbol{\phi}(\boldsymbol{x})$ such that closed-loop system $\dot{\boldsymbol{x}} = \boldsymbol{f}(\boldsymbol{x}, \boldsymbol{\phi}(\boldsymbol{x}))$ is asymptotically stable, where $\boldsymbol{\phi}(\boldsymbol{x})$ is a nonlinear function.

The controller design steps for this system are listed as follows: (i) Select a positive definite Lyapunov function $V(\boldsymbol{x})$, and (ii) select state feedback control law $\boldsymbol{u} = \boldsymbol{\phi}(\boldsymbol{x})$ such that $\dot{V}(\boldsymbol{x}) \leq -V_1(\boldsymbol{x})$, where $V_1(\boldsymbol{x})$ is a positive definite function.

The following examples will give a better overview of the concept.

Example 3.4 Design a stabilizing controller for the following system given by

$$\dot{x} = x + x^2 - x^3 + u$$

To design the controller for this system, select the Lyapunov function as

$$\begin{aligned} V(x) &= \frac{1}{2}x^2 \\ \dot{V} &= x\dot{x} \\ \dot{V} &= x\left(x + x^2 - x^3 + u\right) \end{aligned}$$

To make the $\dot{V}$ negative definite, the feedback controller is constructed as

$$u = -2x - x^2 \tag{3.23}$$

Upon substitution of the control law, the time derivative of the Lyapunov function is modified as

$$\begin{aligned} \dot{V} &= x\left(-x - x^3\right) \\ \dot{V} &= -(x^2 + x^4) \\ &\leq -V_1(x) \end{aligned}$$

(continued)

Example 3.4 (continued)
where $V_1(x) = x^2 + x^4$ is a positive definite function. Since $V(x)$ is positive definite and $\dot{V}$ is negative definite, the closed-loop system will be asymptotically stable. By substituting the control law into the system dynamics, the closed-loop system is obtained as

$$\dot{x} = -x - x^3$$

which has the beneficial nonlinearity x^3. To compare with the DI controller, one can select output $y = x$ and closed-loop stable error dynamics as $\dot{e}+ke = 0$, where $k > 0$. Since $e = x$, DI controller for this system is obtained as

$$\begin{aligned} x + x^2 - x^3 + u_{DI} + kx &= 0 \\ u_{DI} &= -(k+1)x - x^2 + x^3 \end{aligned} \tag{3.24}$$

which leads to a closed-loop stable system as

$$\dot{x} = -kx$$

Thus, by comparing (3.23) and (3.24), it can be seen that the Lyapunov-based control design preserves the beneficial nonlinearities of the system.

Example 3.5 To demonstrate the limitation of the control design, consider the same system as used in the previous example with unknown constants α and β:

$$\dot{x} = \alpha x + \beta x^2 - x^3 + u$$

Using the same controller ($u = -2x - x^2$) designed in the previous example, the closed-loop system is obtained as

$$\dot{x} = (\alpha - 2)x + (\beta - 1)x^2 - x^3$$

One can observe here that the stability of the system depends on the values of α and β. Hence, the global stability of the system is reduced to local stability in the presence of uncertainties. Note that the robustness of the system can be improved by either selecting a more stability-enforcing control law or selecting a different Lyapunov function. However, selecting a Lyapunov function that leads to the desired performance may be difficult to construct.

3.3 Backstepping Control

Backstepping control is a recursive stabilizing-based method that utilizes Lyapunov functions to design controllers for strict feedback or lower triangular systems. This method is also known as the integrator backstepping method, as control acts on the system through an integrator. The major advantages of backstepping control include the capability to handle uncertainties and disturbances, and it guarantees the global asymptotic stability of the closed-loop system. This section gives an introduction to basic backstepping control. The reader can refer to Krstic et al. (1995); Márquez (2003) for a deeper understanding of backstepping control and the variants.

Without loss of generality, an asymptotically stabilizing backstepping controller design is presented for a system with the following system dynamics:

$$\begin{bmatrix} \dot{\boldsymbol{x}} \\ \dot{\zeta} \end{bmatrix} = \begin{bmatrix} \boldsymbol{f}(\boldsymbol{x}) + \boldsymbol{G}(\boldsymbol{x})\zeta \\ u \end{bmatrix} \tag{3.25}$$

where $\boldsymbol{x} \in \Re^n$, $\zeta \in \Re$, and $[\boldsymbol{x}, \zeta]^T \in \Re^{n+1}$ are the states of the system and $u \in \Re$ is the control input. It is assumed in the control design that functions $\boldsymbol{f}$ and $\boldsymbol{G}$ are continuous and smooth, and $\boldsymbol{f}(\mathbf{0}) = \mathbf{0}$. Note that ζ act as a control input for the subsystem $\dot{\boldsymbol{x}} = \boldsymbol{f}(\boldsymbol{x}) + \boldsymbol{G}(\boldsymbol{x})\zeta$. Hence, the objective is to design a state feedback asymptotically stabilizing controller such that both $\boldsymbol{x} \to \mathbf{0}$ and $\zeta \to 0$.

The representation of the system (3.25), where the control input u acts on the system through the integrator, is shown in Fig. 3.3.

Since ζ acts as an input to the system $\dot{\boldsymbol{x}} = \boldsymbol{f}(\boldsymbol{x}) + \boldsymbol{G}(\boldsymbol{x})\zeta$, let us assume that there exists a state feedback control law $\zeta = \phi(\boldsymbol{x})$, $\phi(0) = 0$ such that the time derivative of the Lyapunov function $V_1(\boldsymbol{x})$ satisfy

$$\dot{V}_1(\boldsymbol{x}) = \frac{\partial V_1}{\partial \boldsymbol{x}}[\boldsymbol{f}(\boldsymbol{x}) + \boldsymbol{G}(\boldsymbol{x})\phi(\boldsymbol{x})] \tag{3.26}$$

$$\leq -V_a(\boldsymbol{x}) \tag{3.27}$$

where $V_a(\boldsymbol{x})$ is positive definite function and hence $-V_a(\boldsymbol{x}) \leq 0\ \forall \boldsymbol{x}$. Note that in practice, one has to design such a $\phi(\boldsymbol{x})$ first.

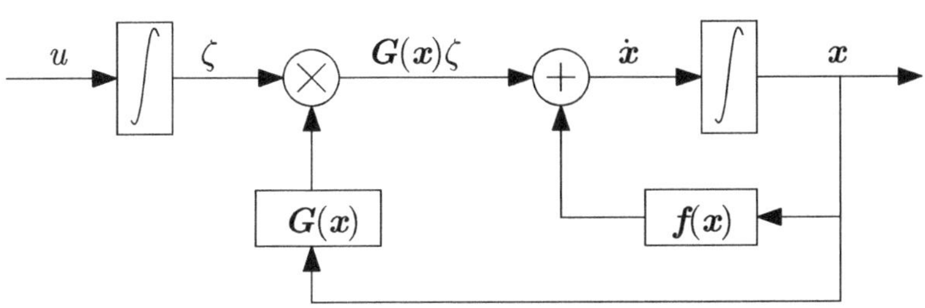

Fig. 3.3 Pictorial representation of system shown in (3.25)

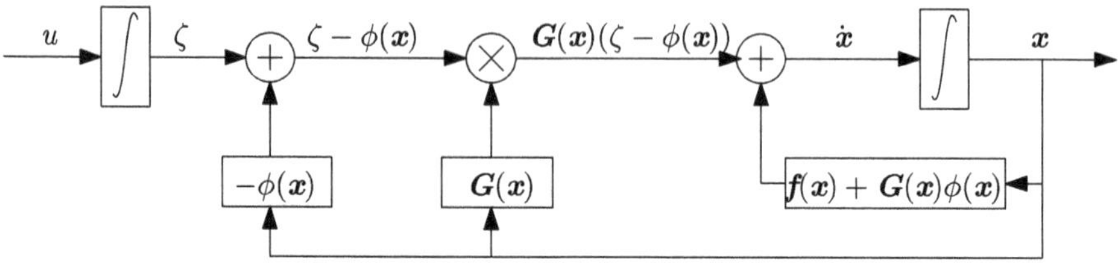

Fig. 3.4 Modified system (3.25) after adding and subtracting $\boldsymbol{G}(\boldsymbol{x})\phi(\boldsymbol{x})$

Upon careful analysis of this control law, one can observe that when $\boldsymbol{x} = \mathbf{0}$, $\zeta = \phi(\mathbf{0}) = 0$ and $\dot{\boldsymbol{x}} = \boldsymbol{f}(\mathbf{0}) = \mathbf{0}$. However, when $\zeta \to 0$, $\dot{\boldsymbol{x}} = \boldsymbol{f}(\boldsymbol{x})$. Thus, $\dot{\boldsymbol{x}} = \boldsymbol{f}(\boldsymbol{x})$ need not be $\mathbf{0}$. This means $\boldsymbol{x}(t)$ need not remain at zero. This is one of the core problems of the existing design structure, and mathematical manipulation is required to ensure that $\boldsymbol{x}(t) \to 0$ happens first.

To achieve this goal, first add and subtract $\boldsymbol{G}(\boldsymbol{x})\phi(\boldsymbol{x})$ as shown in Fig. 3.4:

$$\begin{aligned}\dot{\boldsymbol{x}} &= \boldsymbol{f}(\boldsymbol{x}) + \boldsymbol{G}(\boldsymbol{x})\zeta - \boldsymbol{G}(\boldsymbol{x})\phi(\boldsymbol{x}) + \boldsymbol{G}(\boldsymbol{x})\phi(\boldsymbol{x}) \\ &= \boldsymbol{f}(\boldsymbol{x}) + \boldsymbol{G}(\boldsymbol{x})\phi(\boldsymbol{x}) + \boldsymbol{G}(\boldsymbol{x})\left(\zeta - \phi(\boldsymbol{x})\right)\end{aligned}$$

Next, define a new error variable z as $z \triangleq \zeta - \phi(\boldsymbol{x})$. The system dynamics can now be written as

$$\dot{\boldsymbol{x}} = \boldsymbol{f}(\boldsymbol{x}) + \boldsymbol{G}(\boldsymbol{x})\phi(\boldsymbol{x}) + \boldsymbol{G}(\boldsymbol{x})z$$

One can notice that when $z \to 0$, $\dot{\boldsymbol{x}} = \boldsymbol{f}(\boldsymbol{x}) + \boldsymbol{G}(\boldsymbol{x})\phi(\boldsymbol{x})$, which is asymptotically stable (i.e., $\boldsymbol{x} \to 0$) as $\zeta = \phi(\boldsymbol{x})$ is designed to ensure this. Continuing with the dynamics of variable z, the following is obtained:

$$\begin{aligned}\dot{z} &= \dot{\zeta} - \dot{\phi} \\ &= u - \dot{\phi}\end{aligned}$$

Since $\phi(\boldsymbol{x})$ is stepped back by differentiation, this is known as backstepping, shown in Fig. 3.5.

Next, by defining a new "pseudo-control" variable $v = u - \dot{\phi}$, new transformed system can be written as

$$\begin{bmatrix}\dot{\boldsymbol{x}} \\ \dot{z}\end{bmatrix} = \begin{bmatrix}\boldsymbol{f}(\boldsymbol{x}) + \boldsymbol{G}(\boldsymbol{x})\phi(\boldsymbol{x}) + \boldsymbol{g}(\boldsymbol{x})z \\ v\end{bmatrix} \tag{3.28}$$

The revised block diagram for the transformed system is shown in Fig. 3.6, which is equivalent to the original system (3.25).

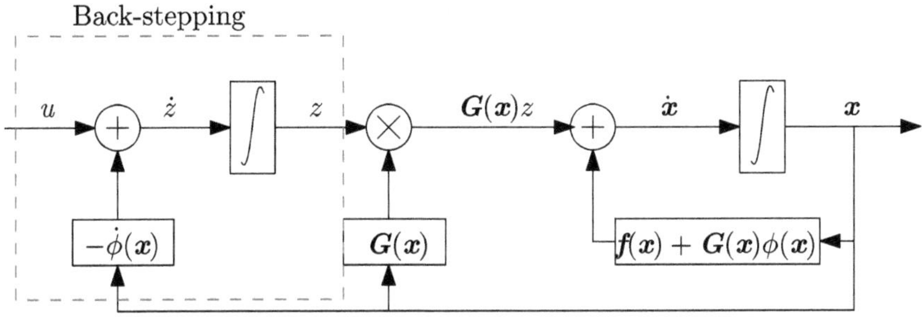

Fig. 3.5 Representation of backstepping of term $\phi(x)$

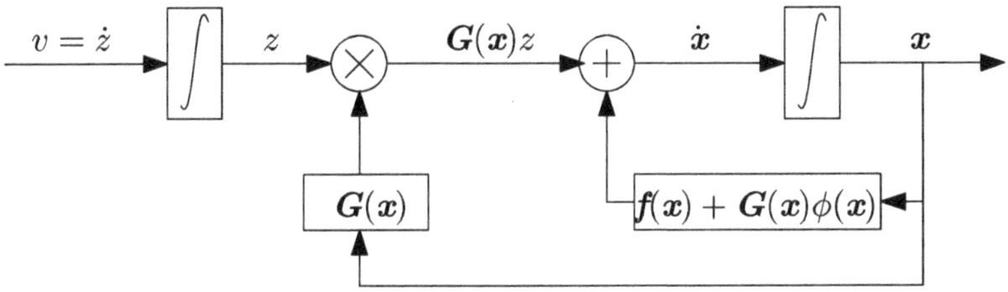

Fig. 3.6 Transformed system after change of variables

Next, to design a stabilizing controller for the transformed system, the following Lyapunov function is selected:

$$V(\boldsymbol{x}, z) = V_1(x) + \frac{1}{2}z^2 \tag{3.29}$$

where $V_1(x)$ is a positive definite function. The time derivative of the Lyapunov function is obtained as

$$\dot{V} = \left(\frac{\partial V_1}{\partial \boldsymbol{x}}\right)^T [\boldsymbol{f}(\boldsymbol{x}) + \boldsymbol{G}(\boldsymbol{x})\phi(\boldsymbol{x}) + \boldsymbol{G}(\boldsymbol{x})z] + zv \tag{3.30}$$

since $\frac{\partial V_1}{\partial \boldsymbol{x}}[\boldsymbol{f}(\boldsymbol{x}) + \boldsymbol{G}(\boldsymbol{x})\phi(\boldsymbol{x})] \leq -V_a(\boldsymbol{x})$, the following is obtained:

$$\dot{V} \leq -V_a(\boldsymbol{x}) + \left(\frac{\partial V_1}{\partial \boldsymbol{x}}\right)^T \boldsymbol{G}(\boldsymbol{x})z + vz \tag{3.31}$$

Define the control

$$v = -\left(\frac{\partial V_1}{\partial \boldsymbol{x}}\right)^T \boldsymbol{G}(\boldsymbol{x}) - kz \tag{3.32}$$

where k is a positive constant gain. The resultant expression is obtained as

$$\dot{V} \leq -V_a(\boldsymbol{x}) - kz^2 < 0 \quad \text{(NDF)} \tag{3.33}$$

It is obvious from (3.33) that the overall system is asymptotically stable. Finally, the control law is extracted from (3.32) as

$$\begin{aligned} u &= v + \dot{\phi} \\ &= \dot{\phi} - \left(\frac{\partial V_1}{\partial \boldsymbol{x}}\right)^T \boldsymbol{G}(\boldsymbol{x}) - kz \\ &= \dot{\phi} - \left(\frac{\partial V_1}{\partial \boldsymbol{x}}\right)^T \boldsymbol{G}(\boldsymbol{x}) - k\left(\zeta - \phi(\boldsymbol{x})\right) \end{aligned} \tag{3.34}$$

where $\dot{\phi} = \left(\frac{\partial \phi}{\partial \boldsymbol{x}}\right)^T [\boldsymbol{f}(\boldsymbol{x}) + \boldsymbol{G}(\boldsymbol{x})\zeta]$. Since $\zeta = \phi(\boldsymbol{x})$, $\dot{\phi}$ can be written as $\dot{\phi} = \left(\frac{\partial \phi}{\partial \boldsymbol{x}}\right)^T [\boldsymbol{f}(\boldsymbol{x}) + \boldsymbol{G}(\boldsymbol{x})\phi(\boldsymbol{x})]$.

As one can see, the backstepping technique is based on the "sufficiency condition" of the Lyapunov stability theory. Hence, it often leads to a robust controller. This is because of the fact that it relies on the negative definiteness of a Lyapunov function derivative, which is bounded away from zero (the margin to the left from zero is a reflection of its robustness). Note that the backstepping procedure can also be applied recursively to higher-dimensional systems and to systems having more general system dynamics. An interested reader can see Márquez (2003) for details.

The steps involved in the design can be summarized using the following example.

Example 3.6 Consider the following system:

$$\dot{x}_1 = x_1^2 - x_1^3 + x_2 \tag{3.35}$$

$$\dot{x}_2 = u \tag{3.36}$$

Here, as explained earlier, $x_2 = \phi(x_1)$ is the control for subsystem $\dot{x}_1 = x_1^2 - x_1^3 + x_2$, and hence, the first step is to determine $\phi(x_1)$. To design $x_2 = \phi(x_1)$ for the subsystem (3.35), consider the Lyapunov function:

$$V_1(x_1) = \frac{1}{2}x_1^2 \tag{3.37}$$

The time derivative of this Lyapunov function is

$$\dot{V}_1(x_1) = x_1\dot{x}_1 = x_1\left(x_1^2 - x_1^3 + x_2\right) \tag{3.38}$$

(continued)

Example 3.6 (continued)
To make $\dot{V}_1(x_1) \leq 0$, x_2 can be selected as $x_2 = -x_1^2 - x_1$, leading to

$$\dot{V}_1(x_1) = -\left(x_1^2 + x_1^4\right) \tag{3.39}$$

which is clearly a negative definite function for (3.35). Thus, a valid selection of x_2 is $x_2 \triangleq \phi(x_1) = -x_1^2 - x_1$.

Next, by adding and subtracting $\phi(x_1)$ and defining the error z as $z \triangleq x_2 - \phi(x_1)$, the modified system can be written as

$$\dot{x}_1 = x_1^2 - x_1^3 + \phi(x_1) + z \tag{3.40}$$

$$\dot{z} = u - \dot{\phi}(x_1) \tag{3.41}$$

where by definition $v = \dot{z}$. To design the pseudo-control v, consider the following Lyapunov function:

$$V(x_1, z) = V_1(x_1) + \frac{1}{2}z^2 \tag{3.42}$$

The time derivative of the Lyapunov function is

$$\dot{V}(x_1, z) = \left(\frac{\partial V_1}{\partial x_1}\right)\left[x_1^2 - x_1^3 + \phi(x_1) + z\right] + zv \tag{3.43}$$

$$= \dot{V}_1(x_1) + \left(\frac{\partial V_1}{\partial x_1} + v\right)z \tag{3.44}$$

Note $\dot{V}_1(x_1)$ is already NDF for subsystem (3.35). Hence, selecting $v = -\left(\frac{\partial V_1}{\partial x_1}\right) - kz$, where $k > 0$, $\dot{V}(x_1, z) = \dot{V}_1(x_1) - kz^2$, which is NDF for the system.

Finally, the control input of the system u can be obtained as

$$u = v + \dot{\phi} \tag{3.45}$$

$$= -\left(\frac{\partial V_1}{\partial x_1}\right) - kz + \left(\frac{\partial \phi}{\partial x_1}\right)\dot{x}_1 \tag{3.46}$$

$$= -x_1 - k\left(x_2 - \phi(x_1)\right) + (-2x_1 - 1)\left(x_1^2 - x_1^3 + x_2\right) \tag{3.47}$$

which leads to

$$u = -x_1 - k\left(x_2 + x_1^2 + x_1\right) + (-2x_1 - 1)\left(x_1^2 - x_1^3 + x_2\right) \tag{3.48}$$

It is worth mentioning that the backstepping design has certain limitations. It can only be applied to dynamic systems in strict feedback or in triangular form. Additionally, its design is often complex and tedious, as it requires the calculation of intermediate stability terms and their derivatives, a challenge known as the "explosion of complexity" problem (Yu and Lin 2017). To address this issue, the *dynamic surface control* design has been introduced in the literature. An interested reader can refer to Yip and Hedrick (1998) for more details.

3.4 Sliding Mode Control

Sliding mode control (SMC) is a robust nonlinear control technique that belongs to the class of *variable structure control* methods, wherein the controller continuously enforces the dynamic behavior of the system to achieve specific control objectives (Bartolini 2008). A unique advantage of sliding mode control lies in its ability to drive the system to the sliding surface in finite time, while maintaining robustness against model uncertainties and external disturbances. Hence, it requires less system information for controller design compared to conventional control approaches.

Sliding mode control design comprises two main stages: (i) the design of a sliding hyperplane, such that the system trajectory slides along this hyperplane (also referred to as the sliding surface) toward the origin, known as the sliding stage, and (ii) the enforcement of the system trajectory to reach this hyperplane, referred to as the reaching stage. Both stages of the control process are illustrated in Figs. 3.7 and 3.8. Figure 3.7 depicts the temporal evolution of the system trajectories, wherein, after the reaching stage, the states slide along the sliding surface. In Fig. 3.8, a phase portrait plot represents the state evolution during both stages. Notably, chattering is observed in Fig. 3.8 as the state slides along the surface. This phenomenon is a well-known limitation of sliding mode control and

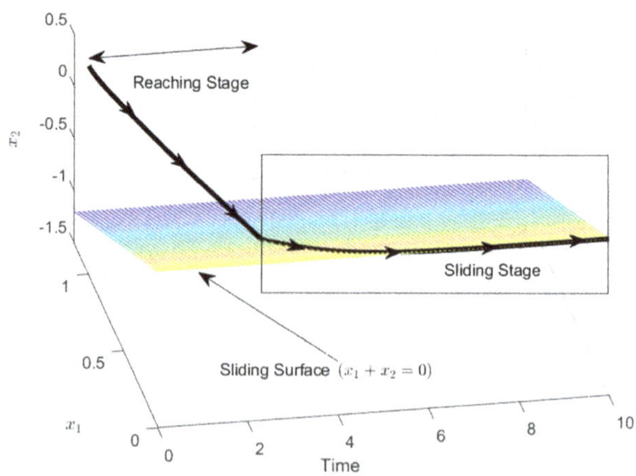

Fig. 3.7 Representation of sliding mode control design stages with time

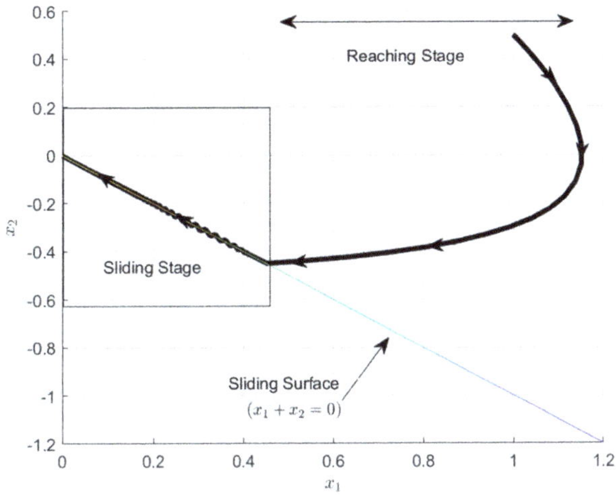

Fig. 3.8 Representation of sliding mode control design stages in phase portrait plot

arises due to high-frequency switching. It occurs because exact convergence to the sliding surface is impeded by factors such as time discretization and control signal quantization (Perruquetti and Barbot 2002). Strategies to mitigate this issue are presented in the latter part of this section.

Since the underlying philosophy of sliding mode control comprises two distinct stages, the control law in SMC can be mathematically expressed as follows:

$$\boldsymbol{u} = \boldsymbol{u}_c + \boldsymbol{u}_e \tag{3.49}$$

where $\boldsymbol{u}_c$ and $\boldsymbol{u}_e$ denote the *corrective control* vector and the *equivalent control* vector, respectively. The corrective control $\boldsymbol{u}_c$ ensures that the system trajectories reach the sliding surface, whereas the equivalent control $\boldsymbol{u}_e$ enforces the sliding motion along the surface.

A common selection of the sliding variable $\boldsymbol{s}$ is

$$\boldsymbol{s} = \dot{\boldsymbol{e}} + \boldsymbol{\lambda}\boldsymbol{e} \tag{3.50}$$

where $\boldsymbol{s}$ is a vector of the sliding variables $\boldsymbol{s} = [s_1, s_2, \ldots, s_n]$, $\dot{\boldsymbol{e}}$ is the derivative of the error vector $\dot{\boldsymbol{e}} = [\dot{e}_1, \dot{e}_2, \ldots, \dot{e}_n]$, $\boldsymbol{e}$ is the error vector $\boldsymbol{e} = [e_1, e_2, \ldots, e_n]$, and $\boldsymbol{\lambda}$ is a diagonal positive definite matrix $\boldsymbol{\lambda} = \text{diag}(\lambda_1, \lambda_2, \ldots, \lambda_n)$ with positive elements. The reason for selecting the sliding variable $\boldsymbol{s}$ as in (3.50) is that $\boldsymbol{s} = \boldsymbol{0}$ will convert the dynamics of the system into a stable first-order linear dynamics. Hence, it is aimed to drive the system to the surface $\boldsymbol{s} = \boldsymbol{0}$ and maintain it on that surface.

By setting $\boldsymbol{s} = \boldsymbol{0}$ in (3.50), one can write

$$\dot{\boldsymbol{e}}(t) + \boldsymbol{\lambda}\boldsymbol{e}(t) = 0 \tag{3.51}$$

Since this is a stable linear differential equation, one could conclude that the error will asymptotically converge to zero. Note that the diagonal elements of $\boldsymbol{\lambda}$ control the convergence speed of elements of the error vector $\boldsymbol{e}(t)$ on the sliding surface.

The following example illustrates the SMC design.

Example 3.7 Consider the following system:

$$\dot{x}_1 = x_2 \tag{3.52}$$

$$\dot{x}_2 = x_1^2 - x_1^3 + x_2 + u \tag{3.53}$$

The objective is to design a sliding mode control to drive the system states to zero.

Note that if $x_2 = -x_1$, (3.52) becomes $\dot{x}_1 = -x_1$, which is a stable linear dynamics. This motivates us to start by designing the sliding surface as

$$s = x_1 + x_2 \tag{3.54}$$

Using (3.53), the time derivative of s is

$$\dot{s} = \dot{x_1} + \dot{x_2} \tag{3.55}$$

$$= x_2 + x_1^2 - x_1^3 + x_2 + u \tag{3.56}$$

Next, to ensure $s \to 0$, consider a Lyapunov function as

$$V(s) = \frac{1}{2}s^2 \tag{3.57}$$

Using (3.56), the time derivative of $V(s)$ is

$$\dot{V}(s) = s\dot{s} \tag{3.58}$$

$$= s(x_2 + x_1^2 - x_1^3 + x_2 + u) \tag{3.59}$$

However, to ensure asymptotic stability, $\dot{V}(s)$ can be selected as

$$\dot{V}(s) = -ks\ \text{sgn}(s) \tag{3.60}$$

where k is a positive gain and "sgn" is signum function.

In sliding mode control, the *signum function* is fundamental due to its ability to generate discontinuous control that ensures *finite-time convergence* of the system states to the sliding surface, thereby enhancing *robustness*

(continued)

Example 3.7 (continued)
against model uncertainties and external disturbances. Although smooth alternatives, such as the quadratic function in (3.60), can provide asymptotic stability, they do not guarantee finite-time convergence. The signum function is preferred in conventional SMC for its simplicity, robustness, and guaranteed convergence. A comparison between the selection of signum and quadratic functions in (3.60) and their impact on control performance is tabulated in Table 3.1.
The finite-time convergence property is explained in detail as follows:

Finite-Time Convergence The chosen time derivative of $V(s)$ can be rewritten as

$$\dot{V} = -ks\,\mathrm{sgn}(s) = -k|s| \tag{3.61}$$

where, due to the signum function, if $s > 0$, then $\dot{V} = -k$, and if $s < 0$, then $\dot{V} = -k$. In both cases, the system state moves at a constant rate, leading to the explicit convergence time. This is illustrated in Fig. 3.9.

From (3.57), the following can be obtained:

$$\dot{V} = -k\sqrt{2V} \tag{3.62}$$

which can be rearranged as

$$\frac{dV}{\sqrt{V}} = -k\sqrt{2}\,dt \tag{3.63}$$

Integrating both sides from $V(0) = \frac{1}{2}s_0^2$ to $V(t_r) = 0$, the following is obtained:

$$\int_{V(0)}^{0} \frac{dV}{\sqrt{V}} = -k\sqrt{2}\int_{0}^{t_r} dt \tag{3.64}$$

Evaluating the integrals, the following is yielded after mathematical analysis:

$$2\sqrt{V(0)} = k\sqrt{2}t_r \quad \Rightarrow \quad t_r = \frac{|s_0|}{k} \tag{3.65}$$

Thus, the system reaches the sliding surface $s = 0$ in finite time t_r.

(continued)

Example 3.7 (continued)

By comparing (3.59) and (3.60), the control expression can be obtained as

$$u = -k\,\mathrm{sgn}(s) - (2x_2 + x_1^2 - x_1^3) \tag{3.66}$$

$$= u_c + u_e \tag{3.67}$$

where corrective control $u_c = -k\mathrm{sgn}(s)$ ensures that the system trajectories reach the sliding surface, whereas the equivalent control $u_e = -(2x_2 + x_1^2 - x_1^3)$ enforces the sliding motion along the surface.

By selecting the initial conditions as $x_1(0) = 1$ and $x_2(0) = 0.5$ and choosing the control gain $k = 0.5$, the simulation results are presented in Figs. 3.10, 3.11, 3.12, 3.13, 3.14, and 3.15. The time responses of the state variables are shown in Fig. 3.10, where both x_1 and x_2 asymptotically approach zero, thereby satisfying the control design objectives.

However, based solely on the time histories, it is difficult to distinguish between the reaching phase and the sliding phase of the sliding mode control behavior. To illustrate this, a phase portrait is shown in Fig. 3.11. Moreover, a three-dimensional plot including the sliding surface is provided in Fig. 3.12, offering clearer insights into the state evolution and convergence trajectory.

Figures 3.13 and 3.14 illustrate the convergence to the sliding surface, with the finite-time convergence validated using the theoretical expression $t_r = \frac{|s(0)|}{k} = \frac{1.5}{0.5} = 3$, as derived in (3.65).

Finally, the control input trajectories are depicted in Fig. 3.15, where the control successfully drives the system to meet the design requirements. Nonetheless, the well-known issue of *chattering* caused by the high-frequency switching is clearly observed. In practical systems, such oscillations may excite unmodeled high-frequency dynamics or lead to excessive wear of mechanical actuators.

Mitigation of Chattering The main drawback of using the signum function is chattering, which is high-frequency oscillations due to discontinuities. In practice, there are several ways to mitigate chattering, such as (i) continuous approximation of the signum function, (ii) implementation of higher-order SMC, (iii) adjustment of the switching surface, and (iv) boundary layer approximation around the sliding surface (Slotine et al. 1991; Utkin et al. 1999).

(continued)

Example 3.7 (continued)
Considering *boundary layer approximation* in this example, the signum function sign(s) is approximated as

$$\text{sat}\left(\frac{s}{\phi}\right) = \begin{cases} 1 & s > \phi \\ \dfrac{s}{\phi} & |s| \leq \phi \\ -1 & s < -\phi \end{cases} \tag{3.68}$$

where the parameter ϕ defines the width of the boundary layer around the sliding surface, within which the control law transitions smoothly.

By selecting $\phi = 0.1$, the corresponding simulation results are shown in Figs. 3.16, 3.17, and 3.18. It is evident that this smooth approximation effectively mitigates the chattering phenomenon. However, as a consequence of the continuous control action, the system no longer exhibits strict finite-time convergence, and the convergence time cannot be determined precisely.

Although this section presents the fundamental concepts of sliding mode control in a simplified manner for clarity and instructional purposes, the theoretical framework can be extended to more general nonlinear systems. For a comprehensive and rigorous treatment of sliding mode control theory and its application to complex dynamical systems, the reader is referred to Slotine et al. (1991); Utkin et al. (1999).

3.5 Summary

This chapter presents a concise overview of commonly used nonlinear control design techniques for nonlinear systems, namely, dynamic inversion control, back-stepping control, and sliding mode control. It explores the fundamental principles behind each method, including outlining their strengths and a few limitations. An interested reader is encouraged to see the references for more in-depth details.

Table 3.1 Comparison between signum and quadratic function selection in (3.60)

Criterion	Signum function	Quadratic function
Convergence time	Finite	Asymptotic
Robustness	Strong	Usually weak
Chattering	Present	Absent
Implementation	Requires approximation	Smooth control

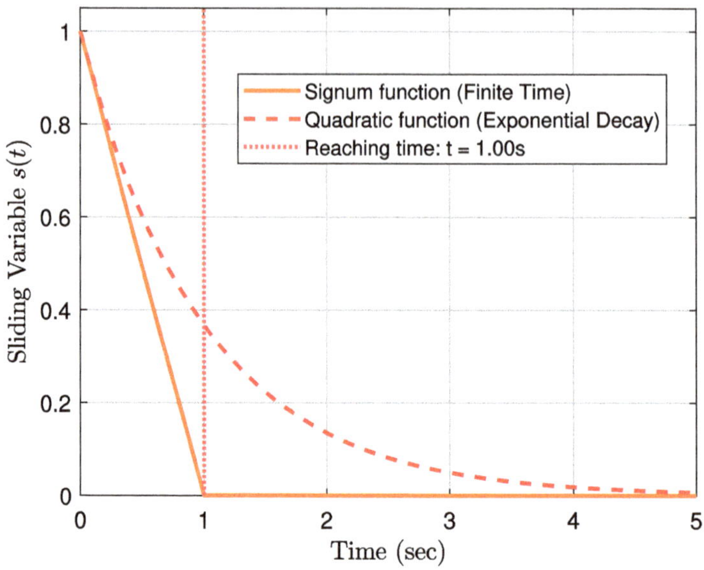

Fig. 3.9 Effect on convergence by selecting signum and quadratic functions in (3.60) for gain $k = 1$

Fig. 3.10 Time histories of states $x_1(t)$ and $x_2(t)$ showing convergence to the origin

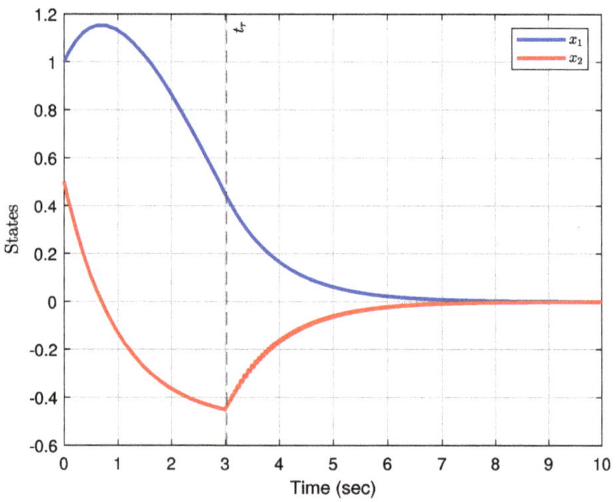

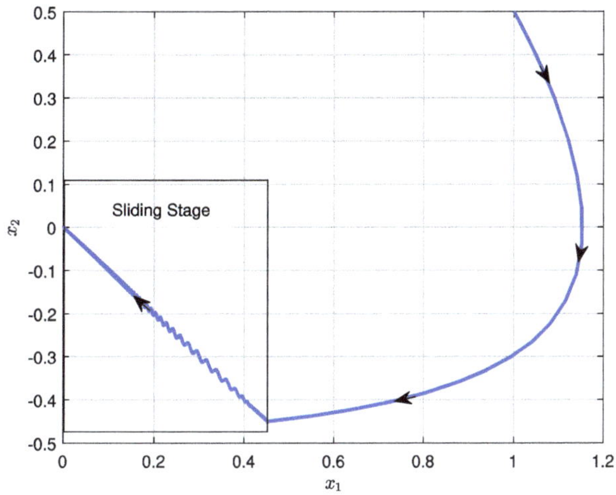

Fig. 3.11 Phase portrait illustrating the trajectory of the system states and the transition between reaching and sliding phases

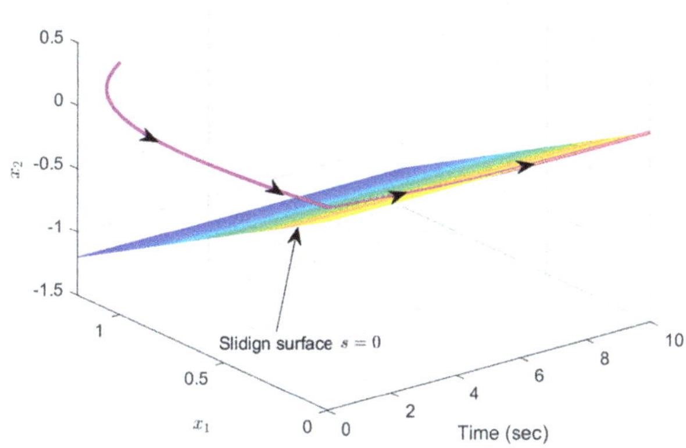

Fig. 3.12 Three-dimensional plot showing state evolution with the sliding surface

3.6 Exercise Problems

Design different types of nonlinear controllers, namely, DI, backstepping, SMC, and a generic Lyapunov-based controller, for the following systems and determine the stability of the closed-loop systems based on the output selection. Additionally, how do these controllers compare in terms of ease of implementation and robustness to modeling uncertainties, and what are their limitations? Assume the problem is a stabilization problem, i.e., $x \to 0$ as $t \to \infty$.

1. $\dot{x} = -x + x^3 + u, \quad y = x.$
2. $\dot{x} = -x - x^3 + u, \quad y = x.$
3. $\dot{x} = -x^2 + (x - 3)\,u, \quad y = x.$

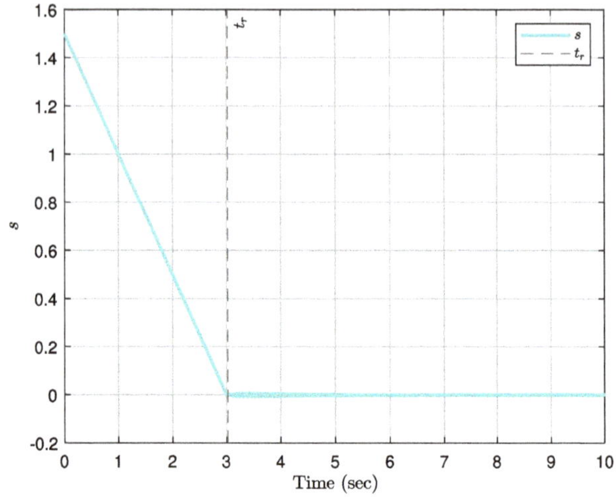

Fig. 3.13 Time evolution of the sliding variable $s(t)$ demonstrating convergence toward the sliding surface $s = 0$

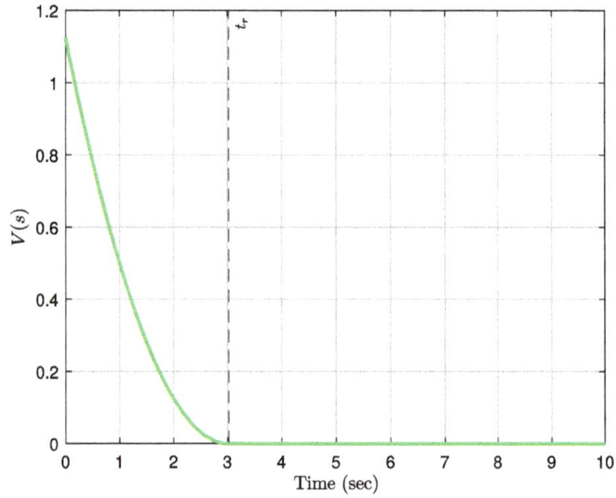

Fig. 3.14 Evolution of Lyapunov function with time

4. $\begin{bmatrix} \dot{x}_1 \\ \dot{x}_2 \\ \dot{x}_3 \end{bmatrix} = \begin{bmatrix} x_1 + x_2 \\ x_3 \\ u \end{bmatrix}, \quad y = x_1 \text{ and } y = x_2.$
5. $\begin{bmatrix} \dot{x}_1 \\ \dot{x}_2 \\ \dot{x}_3 \end{bmatrix} = \begin{bmatrix} -x_1 + x_2 \\ x_3 \\ u \end{bmatrix}, \quad y = x_1 \text{ and } y = x_2.$
6. $\ddot{x} = -\sin x + u, \quad y = x.$
7. $\ddot{x} = \left(1 - x^2 - \dot{x}^2\right)\dot{x} + u, \quad y = x.$
8. $\ddot{x} - 0.7(1 - x^2)\dot{x} - x^3 + (1 + \dot{x}^2)u = 0, \quad y = x.$
9. $\ddot{x} + u + \frac{x}{1+\sin^2 x} = 0, \quad y = x.$

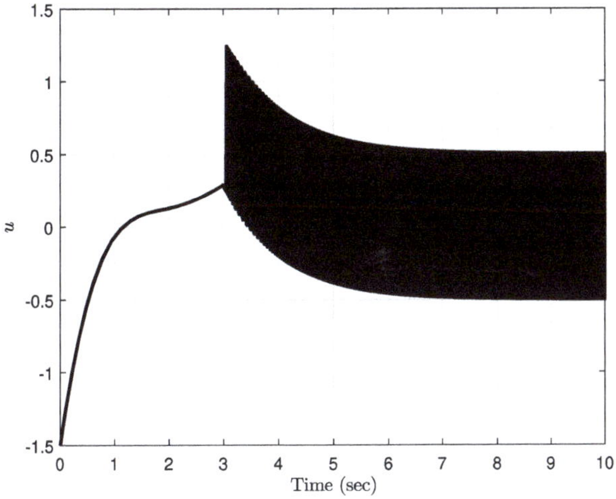

Fig. 3.15 Demanded sliding mode control input $u(t)$ time history and the presence of chattering

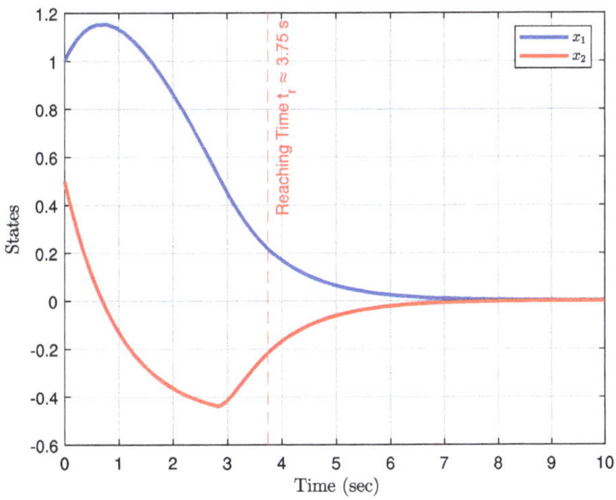

Fig. 3.16 Time histories of states $x_1(t)$ and $x_2(t)$ showing convergence to the origin

10. $\begin{bmatrix} \dot{x}_1 \\ \dot{x}_2 \end{bmatrix} = \begin{bmatrix} x_2 + u \\ -2\sin x_1 \cos x_2 + x_1 \end{bmatrix}$
11. $\begin{bmatrix} \dot{x}_1 \\ \dot{x}_2 \end{bmatrix} = \begin{bmatrix} x_2 + u \\ -x_2 - x_1 - (x_1 + x_2)^2 x_2 + u \end{bmatrix}$
12. $\begin{bmatrix} \dot{x}_1 \\ \dot{x}_2 \end{bmatrix} = \begin{bmatrix} -x_1^3 - x_1 x_2^2 + x_1^2 u \\ -x_2^3 - x_1^2 x_2 + x_2 + u \end{bmatrix}$

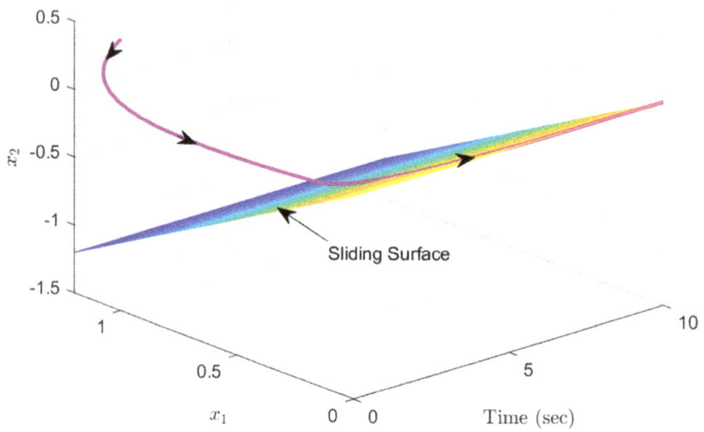

Fig. 3.17 Three-dimensional plot showing state evolution with the sliding surface

Fig. 3.18 Demanded sliding mode control input $u(t)$ time history and the removal of chattering

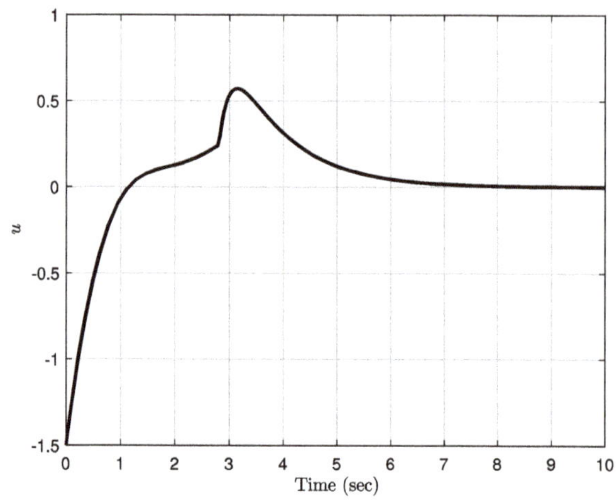

References

G. Bartolini, *Modern Sliding Mode Control Theory: New Perspectives and Applications*, ser. Lecture Notes in Control and information Sciences Modern Sliding Mode Control Theory (Springer, Berlin, 2008)

S. Gopalswamy, J.K. Hedrick, Tracking nonlinear non-minimum phase systems using sliding control. Int. J. Control. **57**(5), 1141–1158 (1993)

A. Isidori, *Nonlinear Control Systems*, ser. Communications and Control Engineering (Springer, London, 2013)

M. Krstic, I. Kanellakopoulos, V. Petar, *Nonlinear and Adaptive Control Design* (Wiley, New York, 1995)

D. Lee, H. Jin Kim, S. Sastry, Feedback linearization vs. adaptive sliding mode control for a quadrotor helicopter. Int. J. Control. Autom. Syst. **7**(3), 419–428 (2009)

H. Márquez, *Nonlinear Control Systems: Analysis and Design* (Wiley, New York, 2003)

R. Padhi, M. Kothari, An optimal dynamic inversion-based neuro-adaptive approach for treatment of chronic myelogenous leukemia. Comput. Methods Prog. Biomed. **87**(3), 208–224 (2007)

R. Padhi, S. Balakrishnan, Optimal dynamic inversion control design for a class of nonlinear distributed parameter systems with continuous and discrete actuators. IET Control Theory Appl. **1**(6), 1662–1671 (2007)

R. Padhi, N. Unnikrishnan, S.N. Balakrishnan, Model-following neuro-adaptive control design for non-square, non-affine nonlinear systems. IET Control Theory Appl. **1**(6), 1650–1661 (2007)

W. Perruquetti, J.-P. Barbot, *Sliding Mode Control in Engineering* (CRC Press, New York, 2002)

J.-J.E. Slotine, W. Li, et al., *Applied Nonlinear Control*, vol. 199(1) (Prentice Hall, Englewood Cliffs, NJ, 1991)

V.I. Utkin, J. Guldner, J. Shi, *Sliding Mode Control in Electromechanical Systems*, 1st ed. (CRC Press, Boca Raton, FL, 1999)

E. Wallner, K. Well, *Attitude Control of a Reentry Vehicle with Internal Dynamics*

P.P. Yip, J.K. Hedrick, Adaptive dynamic surface control: a simplified algorithm for adaptive backstepping control of nonlinear systems. Int. J. Control. **71**(5), 959–979 (1998)

X. Yu, Y. Lin, Adaptive backstepping quantized control for a class of nonlinear systems. IEEE Trans. Autom. Control **62**(2), 981–985 (2017)

Chapter 4
Constrained Control Design Philosophies

System constraints, mainly in the form of state and control constraints, are inherently present in nearly all real-world applications. These constraints are particularly prevalent in mechanical, aerospace, and electrical systems, often manifesting as saturation limits, physical boundaries, or design specifications (Karason and Annaswamy, 1994; Zhang et al., 2018; Chen et al., 2011; Tee et al., 2009, 2011). For instance, in aerospace systems, parameters such as the angle of attack and sideslip angle of a hypersonic vehicle must remain within prescribed bounds to maintain stable flight conditions (Rollins et al., 2013). Similarly, mechanical systems, such as robotic manipulators, frequently operate in physically constrained environments that impose strict design requirements (Huang et al., 2006). Naturally, incorporating such constraints into the control design process can enhance both the performance and applicability of the system.

The conventional control design methods discussed in the previous chapter typically do not explicitly consider these constraints during the design phase. Instead, they focus on analyzing the steady-state response of a system, specifically addressing the characteristics of closed-loop error convergence. Consequently, the classical controllers may result in performance disparities when constraints are present. This necessitates the development of a controller that incorporates system constraints in the design process and makes the necessary adjustments to adhere to imposed constraints, while ensuring the desired closed-loop steady-state error response. However, note that it may take a longer time to achieve the objectives for obvious reasons, which is very much acceptable.

4.1 Description of System Constraints

System constraints are primarily classified into different types for control design purposes.

K. Sachan et al., *Intelligent Adaptive Control with System Constraints*,
https://doi.org/10.1007/978-981-95-8220-4_4

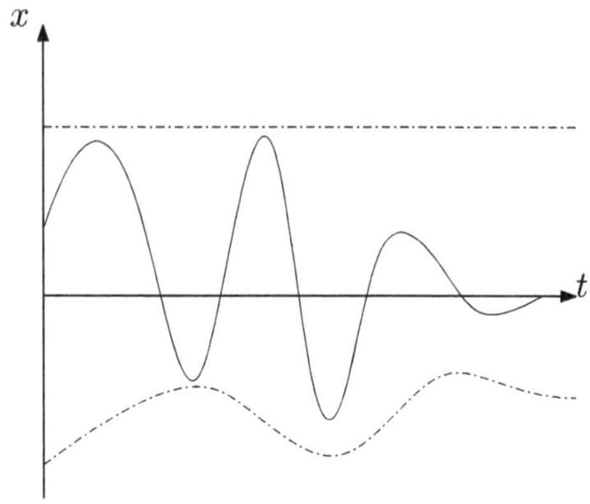

Fig. 4.1 Pictorial representation of state constraints

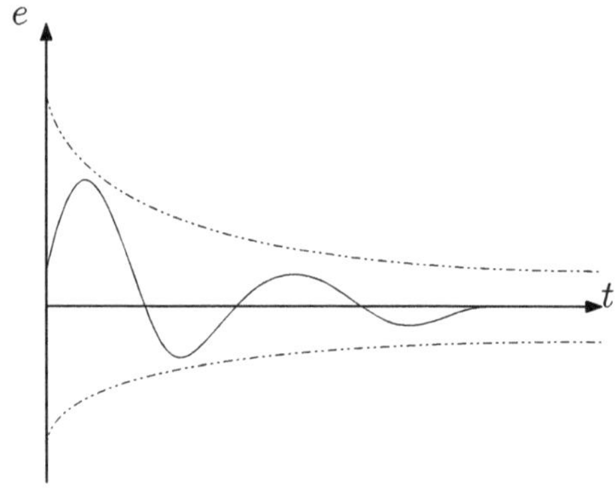

Fig. 4.2 Pictorial representation of error constraints

1. *State constraints:* As the name suggests, state constraints are boundaries imposed on the state of the system. Output constraints are a subset of state constraints applied to the output of the system. Figure 4.1 illustrates a typical representation of a state constraint, where the state x (solid line) is bounded by the corresponding user-imposed state constraints (dotted-dashed line). It is important to note that state constraints can be either constant or time-varying in nature, depending on the specific requirements of the application.
2. *Error constraints:* Error constraints are applied to the closed-loop error of the system state or outputs, with the objective of ensuring a desired transient response of the system. A visual representation of error constraints is depicted in Fig. 4.2, where the error e (solid line) is constrained by the corresponding user-imposed error limits (dotted-dashed line). Unlike state constraints, error constraints are imposed to exhibit converging behavior, ensuring that the closed-loop error diminishes over time, satisfying the key performance requirement.
3. *Input constraints:* In a real system, the control input is generally constrained by the limits of the actuator hardware. Consequently, input constraints are applied to the control input of the system as a saturation. A representative figure of an input constraint is illustrated in Fig. 4.3, where the control input u is restricted by the

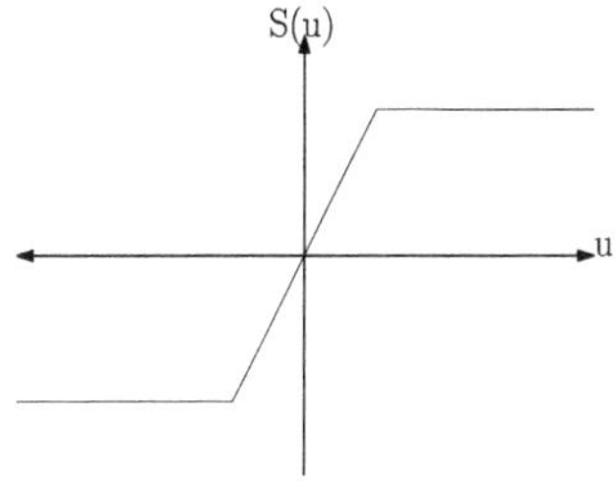

Fig. 4.3 Pictorial representation of input constraints

physical actuator limit denoted by $S(u)$. Note that in many practical applications, constraints on the control rate also matter. However, control constraints and control rate constraints are beyond the scope of this book.

4.2 Control Designs for Constrained Systems

Over the past few decades, significant research has been devoted to incorporating system constraints into the process of designing control laws. Optimal control theory is a valuable approach for attaining control solutions that adhere to constraints in linear and nonlinear systems (see references such as Mayne et al. (2000); Bemporad (1998); DeHaan and Guay (2005) and related literature). However, the implementation of an optimal control algorithm requires substantial computational resources. In response to this challenge, control designs based on Lyapunov stability theory, which is valid for nonlinear systems in general, have been developed as an alternative, allowing the realization of analytic closed-loop control solutions while accommodating constraints. Table 4.1 presents a summary of constraints and their respective control design techniques available in the literature.

Table 4.1 Analytic closed-loop control design methods for constrained systems

Constraint type	Notable control design method
State constraint	1. Barrier Lyapunov function (BLF) method (Tee and Ge, 2011; Zhong-Liang et al., 2016; Tang et al., 2016; Zhao et al., 2018)
	2. Dynamic inversion equivalent approach (Sachan and Padhi, 2018a,b, 2020a)
	3. Transformation approach (Sachan and Padhi, 2018c, 2019, 2020b)
Error constraint	1. Prescribed performance method (Bechlioulis and Rovithakis, 2008, 2009; Wang et al., 2016)
	2. Funnel control method (Ilchmann et al., 2006; Hopfe et al., 2010; Chowdhury and Khalil, 2019)
Input constraint	1. Nussbaum gain method (Ge et al., 2004; Wen et al., 2011; Habibi et al., 2019)
	2. Auxiliary system method (Li et al., 2014a,b; Zhou et al., 2017)

Since the emphasis of this chapter is on state-constrained control designs, the upcoming sections are dedicated to providing a detailed discussion on the synthesis of state-constrained control. For those interested in diving deeper into the subject of error and input-constrained control designs, it is recommended to refer to the literature mentioned in Table 4.1.

4.3 State-Constrained Control Designs

As outlined, the design of state-constrained controllers predominantly relies on three methods, namely, (i) the barrier Lyapunov function (BLF)-based control method, (ii) the dynamic inversion equivalent approach, and (iii) the transformation method. In the BLF method, initially imposed state constraints are converted into error constraints, after which a stable controller is designed using the Lyapunov stability theory. The second method is inspired by BLF and similar to dynamic inversion (DI) control, where the control law is derived from the error dynamics. However, in this approach, a nonlinear stable error dynamics is enforced, leading to a state-constrained solution. Consequently, this method is named as the *"dynamic inversion equivalent approach"* by the authors. In the transformation method-based control design approach, on the other hand, the imposed state constraints are converted into error constraints, and then, the error-constrained system is transformed into an equivalent unconstrained system. Subsequently, a stable controller is designed using conventional control techniques, indirectly stabilizing the constrained system. The following sections provide detailed discussions of these approaches.

4.4 BLF-Based Control Designs

BLF-based controllers are designed using barrier Lyapunov functions, which differ from the more commonly used quadratic Lyapunov functions (QLF). The key characteristic of the BLF, crucial in state-constrained control design, is that as its value approaches infinity, the value of its argument reaches the constraint boundary. To illustrate this point, consider the scalar representations of QLF (V_{QLF}) and symmetric BLF (V_{BLF}) in Fig. 4.4. Specifically, note that the value of BLF becomes infinite as its argument approaches the imposed constraint $|\zeta_i|$, i.e., as $e_i \to |\zeta_i|$, the function $V_{\text{BLF}} \to \infty$.
Formally, a BLF is defined as follows.

A barrier Lyapunov function can be described as a scalar function $V(\boldsymbol{x})$, defined with respect to the system $\dot{\boldsymbol{x}} = \boldsymbol{f}(\boldsymbol{x})$ on an open region D containing the origin that is continuous, positive definite, has continuous first-order partial derivatives at every point of D, has the property $V(\boldsymbol{x}) \to \infty$ as $\boldsymbol{x}$ approaches the boundary of D, and satisfies $V(\boldsymbol{x}(t)) \leq b \; \forall t \geq 0$ along the solution of $\dot{\boldsymbol{x}} = \boldsymbol{f}(\boldsymbol{x})$ for $\boldsymbol{x}(\mathbf{0}) \in D$ and some positive constant b.

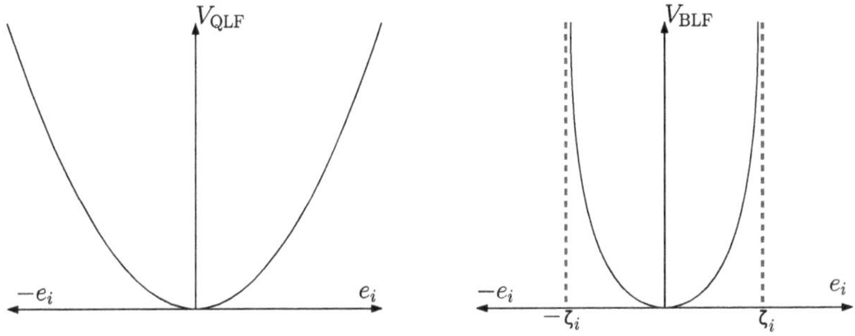

Fig. 4.4 Pictorial representation of QLF and BLF

Based on the nature of imposed constraints, the selection of BLF can be symmetric or asymmetric. A symmetric BLF can be represented as

$$V(e) = \frac{1}{2}\log\left(\frac{\zeta_e^2}{\zeta_e^2 - e^2}\right)$$

where "log" is the natural logarithm (with base "e," not to be confused with the error $e(t)$). The term ζ_e is the imposed symmetric bound on error e. The asymmetric BLF can be represented as

$$V(e) = \frac{1}{2}\gamma^*(e)\log\left(\frac{\overline{\zeta}_e^2}{\overline{\zeta}_e^2 - e^2}\right) + \frac{1}{2}\left(1 - \gamma^*(e)\right)\log\left(\frac{\underline{\zeta}_e^2}{\underline{\zeta}_e^2 - e^2}\right)$$

where "log" is the natural logarithm and terms $\overline{\zeta}_e$ and $\underline{\zeta}_e$ are imposed asymmetric upper and lower bounds on error e. The switching function $\gamma^*(e)$ switches based on the value of error e, defined as

$$\gamma^*(e) = \begin{cases} 1, & \text{if } 0 < e < \overline{\zeta}_e \\ 0, & \text{if } -\underline{\zeta}_e < e \leq 0 \end{cases}$$

Note that the asymmetric BLF can be converted into symmetric BLF by selecting imposed constraints as $-\underline{\zeta}_e = \overline{\zeta}_e$.

The primary challenge in BLF-based control design is the identification of an appropriate BLF for stability analysis. This complexity restricts the applicability of BLF. The BLF-based control design method is largely applicable to strict-feedback (or triangular) form of nonlinear systems. The control design procedure using BLF and QLF for a nonlinear system is identical, but the stability analysis is relatively difficult with BLF. The BLF-based control designs in this book are presented for nonlinear systems with *relative degrees of one and two*, defined in Sect. 1.2.2 of Chap. 1.

To explain the step-by-step process of BLF-based state-constrained control design, consider the examples provided below.

Example 4.1 Consider the nonlinear system given by

$$\dot{x} = 4x^2 + (1 + x^2)u, \tag{4.1}$$

where x is the state of the system and u is the control input. The control design aims to stabilize the closed-loop system while ensuring that the state of the system remains bounded by the user-imposed constraint ζ, i.e., $|x| < \zeta$.

Since the aim is to stabilize the system, the control design can start with the definition of system error, defined as

$$e = x - x^*, \tag{4.2}$$

where x^* is the desired trajectory, set to zero in the context of the stabilization problem (thus, $x^* = 0$). Therefore, the condition $|x| < \zeta$ is equivalent to $|e| < \zeta$.

The time derivative of e can be calculated as

$$\begin{aligned} \dot{e} &= \dot{x} - \dot{x}^* \\ &= \dot{x} \\ &= 4x^2 + (1 + x^2)u. \end{aligned} \tag{4.3}$$

Consider the following barrier Lyapunov function:

$$V = \frac{1}{2}\log\left(\frac{\zeta^2}{\zeta^2 - e^2}\right) \tag{4.4}$$

Note that this function is valid only if $|e(0)| < \zeta$. Thus, a restriction in this control design is that the initial condition of the system should be within this bound. If that is not the case, initially one can use QLF-based design and switch to the BLF-based design once the trajectory comes inside this bound. However, care should be taken during the switching process to ensure a smooth transition between controllers to prevent undesirable transients.

The time derivative of the BLF in (4.4) is obtained as

$$\dot{V} = \frac{e\dot{e}}{\zeta^2 - e^2} \tag{4.5}$$

(continued)

Example 4.1 (continued)

Substituting the values of e and $\dot{e}$ from (4.2) and (4.3), the time derivative of BLF in (4.5) is

$$\dot{V} = \frac{e\left[4x^2 + (1 + x^2)u\right]}{\zeta^2 - e^2} \tag{4.6}$$

From this expression, to ensure $\dot{V} < 0$ (NDF), the control input u can be selected as

$$u = \frac{-4x^2 - \frac{ke}{(\zeta^2 - e^2)}}{1 + x^2} \tag{4.7}$$

where $k > 0$ is a positive controller gain. Substituting (4.7) into (4.6), it is clear that

$$\dot{V} = -\frac{ke^2}{(\zeta^2 - e^2)^2} \tag{4.8}$$

which is clearly a negative definite function. Thus, the controller in (4.7) is a stabilizing controller, ensuring that the closed-loop system is asymptotically stable. It is important to gain insight into the term $(\zeta^2 - e^2)$ in the control input u, which becomes very small as the error approaches the constraint boundary, resulting in a large control effort. Interpreted differently, this term effectively acts as a dynamic gain, adjusting the control effort to prevent constraint violations. However, due to this characteristic, the control effort can sometimes become excessively large, as observed in the simulation results.

A further analysis resulted in more insight into this approach. Since $\dot{V} < 0, \forall t$, the value of BLF at any time will be less than V^0, where V^0 is the value of V at $t = 0$, i.e.

$$\frac{1}{2}\log\left(\frac{\zeta^2}{\zeta^2 - e^2}\right) < V^0 \tag{4.9}$$

Note that $V^0 = 0$ means the error is at equilibrium. Hence, for a nonzero error, V^0 should be positive, i.e., $V^0 > 0$. After simplification, one can arrive at

$$|e| < \zeta\sqrt{1 - \exp(-2V^0)} \tag{4.10}$$

(continued)

Example 4.1 (continued)
where "exp" is the natural exponential function. By analyzing (4.10), since $\exp(-2V^0) \in (0, 1)$, the term $\sqrt{1 - \exp(-2V^0)} < 1$. Hence, it can be concluded that the closed-loop error e will always be bound by the user-imposed constraint ζ. Moreover, $e = x$; hence, $|x| < \zeta, \forall t$.

For a pictorial representation, selecting the state constraint as $\zeta = 0.18$ in this example, the state and control histories are plotted in Figs. 4.5 and 4.6 for different initial conditions.

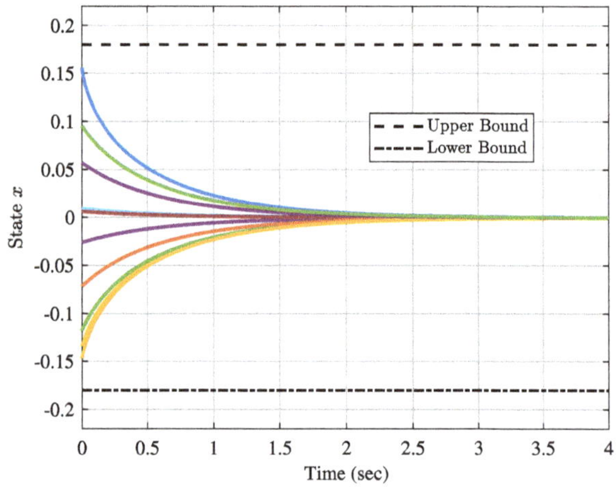

Fig. 4.5 State histories obtained using control (4.7)

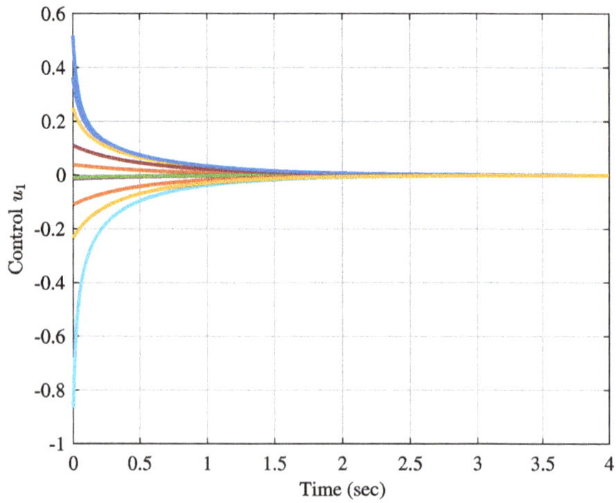

Fig. 4.6 Control histories (4.7)

Example 4.2 Consider a second-order nonlinear system given by

$$\begin{aligned}\dot{x}_1 &= x_2\\ \dot{x}_2 &= 4x_1x_2 + x_2^2 + u\end{aligned} \tag{4.11}$$

where x_1 and x_2 represent the state of the system and u is the control input. The goal of the control design is to stabilize the closed-loop system, i.e., $(x_1, x_2) \to (0, 0)$, while ensuring that the states x_1 and x_2 remain bounded by user-defined constraints ζ_1 and ζ_2, respectively, i.e., $|x_1| < \zeta_1$ and $|x_2| < \zeta_2$.

To design the controller, first define the closed-loop error as $e = x_1$. Thus, the time derivatives of the error are obtained as $\dot{e} = x_2$. Next, the BLF is chosen as

$$V = \frac{1}{2}\log\left(\frac{\zeta_1^2}{\zeta_1^2 - e^2}\right) + \frac{1}{2}\log\left(\frac{\zeta_2^2}{\zeta_2^2 - \dot{e}^2}\right) \tag{4.12}$$

Note that this function is valid only if $|e(0)| = |x_1(0)| < \zeta_1$ and $|\dot{e}(0)| = |x_2(0)| < \zeta_2$, and we assume that this condition is satisfied.

Next, substituting the values of e and $\dot{e}$ as $e = x_1$ and $\dot{e} = x_2$, the Lyapunov function can be rewritten as

$$V = \frac{1}{2}\log\left(\frac{\zeta_1^2}{\zeta_1^2 - x_1^2}\right) + \frac{1}{2}\log\left(\frac{\zeta_2^2}{\zeta_2^2 - x_2^2}\right)$$

The time derivative of BLF is obtained as

$$\dot{V} = \frac{x_1\dot{x}_1}{\zeta_1^2 - x_1^2} + \frac{x_2\dot{x}_2}{\zeta_2^2 - x_2^2} \tag{4.13}$$

Substituting the values of $\dot{x}_1$ and $\dot{x}_2$ from (4.11) into (4.13), the time derivative of BLF becomes

$$\dot{V} = \frac{x_1x_2}{\zeta_1^2 - x_1^2} + \frac{x_2\left[4x_1x_2 + x_2^2 + u\right]}{\zeta_2^2 - x_2^2} \tag{4.14}$$

Closely examining the expression of $\dot{V}$ in (4.14), one can select the control input u expression as

$$u = -4x_1x_2 - x_2^2 - \frac{\left(\zeta_2^2 - x_2^2\right)x_1}{\zeta_1^2 - x_1^2} - k\frac{x_2}{\zeta_2^2 - x_2^2} \tag{4.15}$$

(continued)

Example 4.2 (continued)
where $k > 0$ is a positive control gain. Substituting the control expression in (4.14) leads to

$$\dot{V} = -k\frac{x_2^2}{\left(\zeta_2^2 - x_2^2\right)^2} \tag{4.16}$$

It is a function of x_2. Hence, it can only be a negative semi-definite function. Thus, the controller design in (4.16) is a stabilizing controller, and the closed-loop system is stable (however, not necessarily asymptotically stable).

The nature of the stability of the closed-loop system can be further analyzed using LaSalle's theorem. Analyzing the condition $\dot{V} = 0$ leads to $x_2 = 0$. By substituting the control expression, the closed-loop dynamics can be obtained as

$$\begin{aligned}\dot{x}_1 &= x_2\\ \dot{x}_2 &= -\frac{\left(\zeta_2^2 - x_2^2\right)x_1}{\zeta_1^2 - x_1^2} - k\frac{x_2}{\zeta_2^2 - x_2^2}\end{aligned} \tag{4.17}$$

Evaluating the closed-loop dynamics on the invariant set $x_2 = 0$ leads to (i) $\dot{x}_1 = 0$; i.e, x_1 is constant, (ii) $\frac{(\zeta_2^2 - x_2^2)x_1}{\zeta_1^2 - x_1^2} = 0$, i.e, $x_1 = 0$. Therefore, the largest invariant set for which $\dot{V} = 0$ is $x_1 = 0, x_2 = 0$. Hence, by LaSalle's theorem, all trajectories starting within the constrained domain asymptotically converge to the origin.

Considering $\zeta_1 = \zeta_2 = 0.18$ and gain $k = 0.05$, the simulation results for the system (4.11) are presented. The state histories are plotted in Fig. 4.7a, b for various initial conditions. As theoretically predicted, the controller ensures the asymptotic convergence of system state, i.e., $x_1 \to 0$ and $x_2 \to 0$. The corresponding control histories are depicted in Fig. 4.7c.

4.5 Dynamic Inversion Equivalent Approach

Inspired by the term $(\zeta^2 - e^2)$ that appears in the BLF-based controller, this approach, using this term, constructs a nonlinear stable ordinary differential equation (ODE). Instead of deriving the controller through a barrier Lyapunov function, the control expression is obtained from this stable nonlinear ODE, which is similar to dynamic inversion control. However, the stability proof of such complex

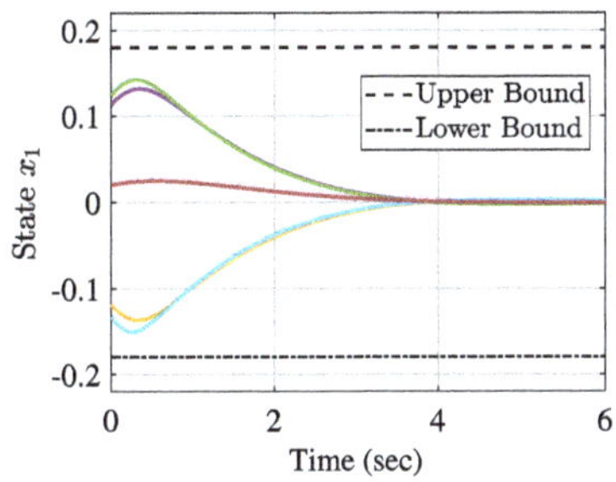

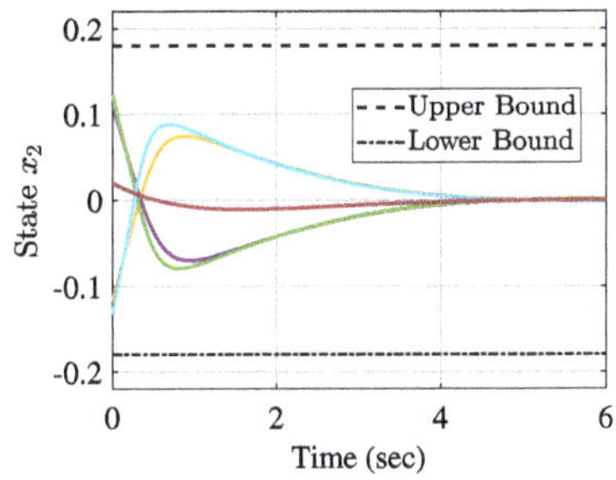

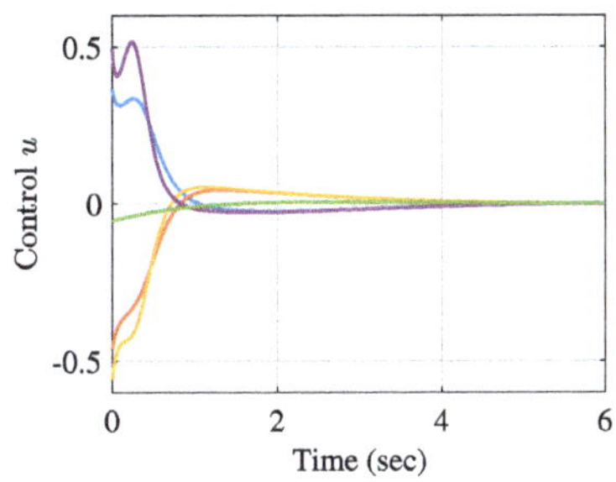

Fig. 4.7 State and control histories obtained for various initial conditions using BLF-based control (4.15)

nonlinear ODEs is an active area of research. As a result, this method is currently applicable only to the state-constrained nonlinear systems with a relative degree of one. The transformation-based approach, presented in the next section, is preferred when designing controllers for more complex nonlinear systems with state constraints.

The control design is illustrated considering the following nonlinear system:

$$\dot{\boldsymbol{x}} = \boldsymbol{f}(\boldsymbol{x}) + \boldsymbol{G}(\boldsymbol{x})\,\boldsymbol{u}$$
$$\boldsymbol{y} = \boldsymbol{x}$$

where $\boldsymbol{x} \in \Re^n$, $\boldsymbol{y} \in \Re^n$, and $\boldsymbol{u} \in \Re^n$ are the state, output, and control input vectors with the same dimensions, respectively. The components of vector $\boldsymbol{f}\,(\boldsymbol{x}) \in \Re^n$ and the matrix $\boldsymbol{G}\,(\boldsymbol{x}) \in \Re^{n\times n}$ are assumed to be continuous and smooth functions of the states, and matrix $\boldsymbol{G}\,(\boldsymbol{x})$ is assumed to be invertible.

The objective of the control design is to track y_i^*, i.e., $y_i \rightarrow y_i^*$ while ensuring that state remains bounded by user-imposed ζ_{x_i}, i.e., $|x_i| < \zeta_{x_i}, \forall i$. Since the desired signals are always bounded, it is assumed that $y_i^* < \zeta_{y_i^*}$. Note that the subscript "i" is used to represent the ith component of a vector.

To achieve the control design objectives, the ith component of error $\boldsymbol{e}$ is defined as

$$e_i = y_i - y_i^*, \qquad i = 1, 2, \ldots, n. \tag{4.18}$$

In order to derive the state feedback controller, the following BLF is used:

$$V = \frac{1}{2}\sum_{i=1}^{n} \log\left(\frac{\zeta_{e_i}^2}{\zeta_{e_i}^2 - e_i^2}\right), \tag{4.19}$$

where ζ_{e_i} is the limit on error e_i, defined as $\zeta_{e_i} = \zeta_{x_i} - \zeta_{y_i^*}$. Note that, for (4.19) to be a valid function, the error $e_i(0)$ is restricted in domain $|e_i(0)| < \zeta_{e_i}$.

The time derivative of the Lyapunov function (4.19) is given by

$$\dot{V} = \sum_{i=1}^{n} \frac{\dot{e}_i e_i}{\zeta_{e_i}^2 - e_i^2} \tag{4.20}$$

Equation (4.20) can also be written in a compact form as

$$\dot{V} = (\boldsymbol{B}_L \boldsymbol{e})^T \dot{\boldsymbol{e}} \tag{4.21}$$

where $\boldsymbol{e} = [e_1, e_2, \ldots, e_n]^T$ is an error vector and $[\boldsymbol{B}_L] \in \Re^{n\times n}$ is a diagonal matrix, defined as

$$\boldsymbol{B}_L = \text{diag}\left(\frac{1}{\zeta_{e_1}^2 - e_1^2}, \frac{1}{\zeta_{e_2}^2 - e_2^2}, \ldots, \frac{1}{\zeta_{e_n}^2 - e_n^2}\right). \tag{4.22}$$

To fulfill the requirement of Lyapunov stability theory, let the time derivative of the Lyapunov function be

$$\dot{V} = -\boldsymbol{e}^T \boldsymbol{K}_{blf} \boldsymbol{e} \tag{4.23}$$

where $\left[\boldsymbol{K}_{blf}\right] \in \Re^{n\times n}$ is a positive definite diagonal matrix expressed as

$$\boldsymbol{K}_{blf} = \text{diag}\left(\frac{k_1}{\left(\zeta_{e_1}^2 - e_1^2\right)^{\lambda_1}}, \ldots, \frac{k_n}{\left(\zeta_{e_n}^2 - e_n^2\right)^{\lambda_n}}\right), \tag{4.24}$$

where $k_i, \forall i$ are positive controller gains and $\lambda_i, \forall i$ are positive even integer. Equating (4.21) with (4.23) results in

$$(\boldsymbol{B}_L \boldsymbol{e})^T \dot{\boldsymbol{e}} = -\boldsymbol{e}^T \boldsymbol{K}_{blf} \boldsymbol{e} \tag{4.25}$$

Since $\boldsymbol{B}_L^T = \boldsymbol{B}_L$, (4.25) can be rewritten as

$$\boldsymbol{e}^T \boldsymbol{B}_L \dot{\boldsymbol{e}} = -\boldsymbol{e}^T \boldsymbol{K}_{blf} \boldsymbol{e}, \tag{4.26}$$

which yields the error dynamics as

$$\dot{\boldsymbol{e}} + \left[\boldsymbol{B}_L^{-1} \boldsymbol{K}_{blf}\right] \boldsymbol{e} = \boldsymbol{0}, \tag{4.27}$$

where $\left[\boldsymbol{B}_L^{-1} \boldsymbol{K}_{blf}\right] \in R^{n\times n}$ is given by

$$\left[\boldsymbol{B}_L^{-1} \boldsymbol{K}_{blf}\right] = \text{diag}\left(\frac{k_i}{\left(\zeta_{e_i}^2 - e_i^2\right)^{\lambda_i - 1}}\right), \forall i. \tag{4.28}$$

Note that the error dynamics in (4.27) is nonlinear and can be expressed into individual ith components as

$$\dot{e}_i = -\frac{k_i e_i}{\left(\zeta_{e_i}^2 - e_i^2\right)^{\lambda_i - 1}}, \qquad i = 1, 2, \ldots, n. \tag{4.29}$$

Since $\dot{V} < 0$ using this error dynamics, $V(t) < V_0, \forall t \geqslant 0$, where V_0 is the value of Lyapunov function at time $t = 0$. Since each individual element of $V(t)$ is positive, the following relation holds:

$$\frac{1}{2}\log\left(\frac{\zeta_{e_i}^2}{\zeta_{e_i}^2 - e_i^2}\right) < V_0, \forall i \tag{4.30}$$

which can be simplified as

$$\zeta_{e_i}^2 \exp(-2V_0) < \left(\zeta_{e_i}^2 - e_i^2\right)$$
$$|e_i| < \sqrt{\zeta_{e_i}^2 \left(1 - \exp(-2V_0)\right)}$$

where "exp" represents the exponential function. Since $V_0 \in [0, \infty)$, the following can be concluded:

$$|e_i| < \zeta_{e_i}, \forall i. \tag{4.31}$$

Since $|y_i^*| \leqslant \zeta_{y_i^*}$ and $|e_i| < \zeta_{e_i}$, from the relation $x_i = e_i + y_i^*$, the constraint on x_i is $|x_i| < \zeta_{y_i^*} + \zeta_{e_i}$. As $\zeta_{x_i} = \zeta_{y_i^*} + \zeta_{e_i}$, it can be concluded that $|x_i| < \zeta_{x_i}, \forall i$.

It can be observed that the error dynamics in (4.29) is similar to the stable error dynamics used in dynamic inversion control design, with the only difference being the term $\left(\zeta_{e_i}^2 - e_i^2\right)$. Consequently, one can begin designing the controller directly using this error dynamics (4.27), bypassing the Lyapunov selection steps, and can achieve a stable closed-loop system. Due to the similarity in the structure of the error dynamics, this method is referred to as the dynamic inversion equivalent approach.

As mentioned, the controller can be designed directly using the following nonlinear error dynamics:

$$\dot{\boldsymbol{e}} + \left[\boldsymbol{B}_L^{-1}\boldsymbol{K}_{blf}\right]\boldsymbol{e} = \boldsymbol{0}$$

Upon substitution of system dynamics

$$\boldsymbol{f}(\boldsymbol{x}) + \boldsymbol{G}(\boldsymbol{x})\boldsymbol{u} - \dot{\boldsymbol{y}}^* + \left[\boldsymbol{B}_L^{-1}\boldsymbol{K}_{blf}\right]\boldsymbol{e} = \boldsymbol{0} \tag{4.32}$$

The control $\boldsymbol{u}$ can be extracted as

$$\boldsymbol{u} = [\boldsymbol{G}(\boldsymbol{x})]^{-1}\left(-\left[\boldsymbol{B}_L^{-1}\boldsymbol{K}_{blf}\right]\boldsymbol{x} - \boldsymbol{f}(\boldsymbol{x}) + \dot{\boldsymbol{y}}^*\right) \tag{4.33}$$

It is worth noting that, by substituting $\lambda_i = 1, \ \forall i$, the controller transforms into the dynamic inversion (DI) controller.

Example 4.3 Consider the following nonlinear system:

$$\dot{x_1} = x_1 + x_2^2 + \left(1 + x_2^2\right)u_1 + u_2$$

$$\dot{x_2} = x_2 + x_3^2 + \left(1 + x_3^2\right)u_2 + u_3$$

$$\dot{x_3} = x_3 + x_1^2 + \left(1 + x_1^2\right)u_3 + u_1$$

where x_1, x_2, and x_3 are the states and u_1, u_2, and u_3 are the control inputs, respectively. Assuming $y_i^* = 0$, a stable controller is designed with state constraints defined by $|x_i| < k_{x_i}$, where $k_{x_i} = 0.5$ for $i = 1, 2, 3$. Simulation results are obtained using the controller with unity gains. The simulations are conducted with the state-constrained controller (designed using the dynamic inversion equivalent approach) for randomly selected initial conditions within the range $(-0.5, 0.5)$. The results, shown in Fig. 4.8a–d, demonstrate that the proposed state-constrained controller successfully prevents constraint violations.

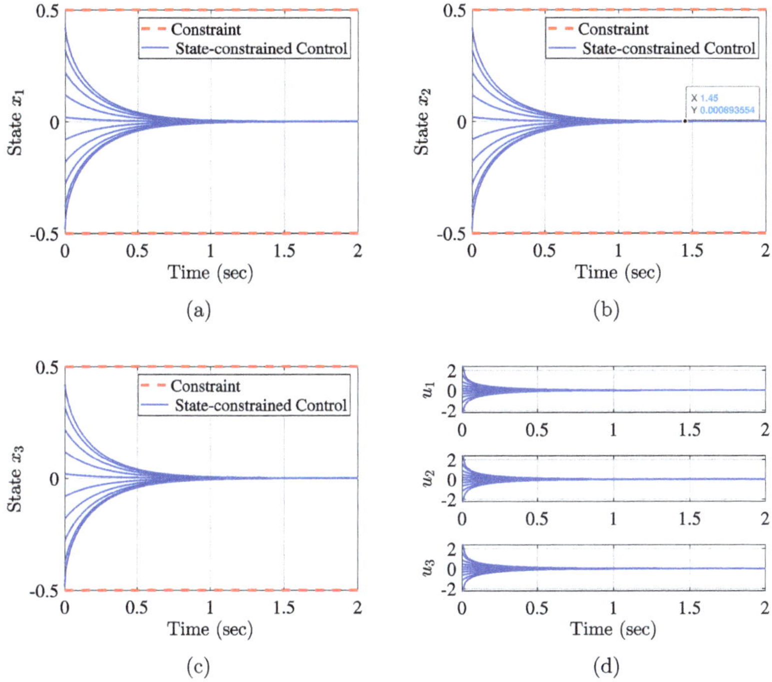

Fig. 4.8 State and control histories obtained for various initial conditions using the state-constrained controller

4.6 Transformation-Based Approach

The transformation-based state-constrained control design involves a three-step process: (i) *Conversion of imposed state constraints*: Initially imposed state constraints are transformed into error constraints; (ii) *transformation of the error-constrained system*: The error-constrained system is then converted into an equivalent unconstrained system; and (iii) *controller design for the unconstrained system*: A controller is designed for the unconstrained system, which indirectly stabilizes the constrained system. It is crucial to note that defining the transformation is a critical aspect of this design process.

To illustrate the control design procedure using the transformation method, let us consider the same second-order system used in Example 4.2.

Example 4.4 Consider the second-order system given in (4.11):

$$\begin{aligned}\dot{x}_1 &= x_2 \\ \dot{x}_2 &= 4x_1x_2 + x_2^2 + u\end{aligned} \tag{4.34}$$

where x_1 and x_2 represent the state of the system and u is the control input. The objective of the control design is to stabilize the closed-loop system while ensuring that the states x_1 and x_2 remain bounded by the user-imposed constraints ζ_1 and ζ_2, i.e., $|x_1| < \zeta_1$ and $|x_2| < \zeta_2$.

Given that this is a stabilization problem, the conversion of constraints to error constraints in the first step of control design is straightforward, i.e., $|e| < \zeta_1$ and $|\dot{e}| < \zeta_2$.

In the second step, consider the following transformation:

$$s = k_1 \frac{e}{\zeta_1^2 - e^2} + k_2 \frac{\dot{e}}{\zeta_2^2 - \dot{e}^2} \tag{4.35}$$

where s is the transformation variable and $k_1 > 0, k_2 > 0$ are constant positive gain tuning parameters. It can be observed that although the variables e and $\dot{e}$ are constrained, the transformation variable s is unconstrained. Further details about the properties of this transformation are provided in Chap. 8.

Since $e = x_1$ and $\dot{e} = x_2$, transformation in variable s can be expressed as

$$s = k_1 \frac{x_1}{\zeta_1^2 - x_1^2} + k_2 \frac{x_2}{\zeta_2^2 - x_2^2} \tag{4.36}$$

In the third step, the controller will be designed for the dynamics of the transformation variable. To design the controller, the following QLF is chosen:

$$V = \frac{1}{2}s^2 \tag{4.37}$$

This function is valid only if $|x_1(0)| < \zeta_1$ and $|x_2(0)| < \zeta_2$. Thus, similar to BLF-based methods, this method also has the same assumption on initial condition selection.

The time derivative of QLF in (4.37) is obtained as

$$\dot{V} = s\dot{s} \tag{4.38}$$

(continued)

Example 4.4 (continued)
The time derivative of s from (4.36) is calculated as

$$\dot{s} = k_1 \frac{\left(\zeta_1^2 + x_1^2\right) x_2}{\left(\zeta_1^2 - x_1^2\right)^2} + k_2 \frac{\left(\zeta_2^2 + x_2^2\right) \dot{x}_2}{\left(\zeta_2^2 - x_2^2\right)^2} \tag{4.39}$$

Substituting the values of s and $\dot{s}$ into (4.38), the time derivative of QLF becomes

$$\dot{V} = \frac{k_1 \left(\zeta_1^2 + x_1^2\right) s x_2}{\left(\zeta_1^2 - x_1^2\right)^2} + \frac{k_2 \left(\zeta_2^2 + x_2^2\right) s \dot{x}_2}{\left(\zeta_2^2 - x_2^2\right)^2} \tag{4.40}$$

By substituting $\dot{x}_2$ from (4.34), the following is obtained:

$$\dot{V} = s \left[k_1 \frac{\left(\zeta_1^2 + x_1^2\right) x_2}{\left(\zeta_1^2 - x_1^2\right)^2} + \frac{k_2 \left(\zeta_2^2 + x_2^2\right) \left(4x_1 x_2 + x_2^2 + u\right)}{\left(\zeta_2^2 - x_2^2\right)^2} \right] \tag{4.41}$$

To make the $\dot{V}$ negative definite,the control input u is selected as

$$u = -4x_1 x_2 - x_2^2 - \left[k_3 s + \frac{\left(\zeta_2^2 - x_2^2\right)^2}{\left(\zeta_2^2 + x_2^2\right)} \frac{k_1 x_2 \left(\zeta_1^2 + x_1^2\right)}{k_2 \left(\zeta_1^2 - x_1^2\right)^2} \right], \tag{4.42}$$

The following is obtained:

$$\dot{V} = -k_2 k_3 \frac{\left(\zeta_2^2 + x_2^2\right)}{\left(\zeta_2^2 - x_2^2\right)^2} s^2 \tag{4.43}$$

which is a negative definite function. Thus, the controller design in (4.43) is a stabilizing controller, and the closed-loop system is asymptotically stable, i.e., $s \to 0$ as $t \to \infty$. By analyzing the transformation, it can be concluded that as $s \to 0$, $x_1 \to 0$ and $x_2 \to 0$ simultaneously.

Considering state constraints $\zeta_1 = \zeta_2 = 0.18$ and gains $k_1 = k_2 = 1, k_3 = 0.05$, the simulation results for the system (4.34) are presented. The state histories are plotted in Fig. 4.9a, b for various initial conditions. As theoretically predicted, the controller ensures both $x_1 \to 0$ and $x_2 \to 0$. The corresponding control histories are shown in Fig. 4.9c.

A comparison between the results of BLF-based control (with gain $k = 0.05$) and transformation-based control is illustrated in Fig. 4.10a–c. The state histories are presented in Fig. 4.10a, b for various initial conditions. It can

(continued)

Example 4.4 (continued)
be observed that both the BLF-based and transformation-based controllers show comparable performance in terms of state regulation. However, the transformation-based approach requires a higher initial control effort. This increased demand arises due to the more strict modulation of the state trajectories inherent in the transformation-based method. Nevertheless, this effect can be mitigated through tuning of the control gains. The corresponding comparison of the control input histories is depicted in Fig. 4.10c.

4.7 Summary

This chapter presents an overview of nonlinear control design techniques for systems associated with physical, operational, or user-imposed state constraints. It provides a concise discussion of system constraints and introduces foundational

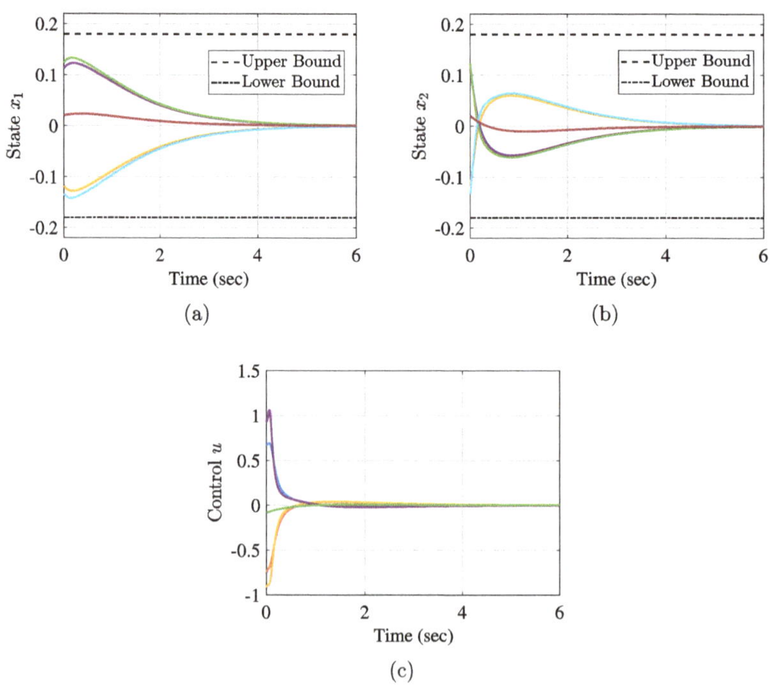

Fig. 4.9 State and control histories obtained for various initial conditions using the transformation-based state-constrained controller

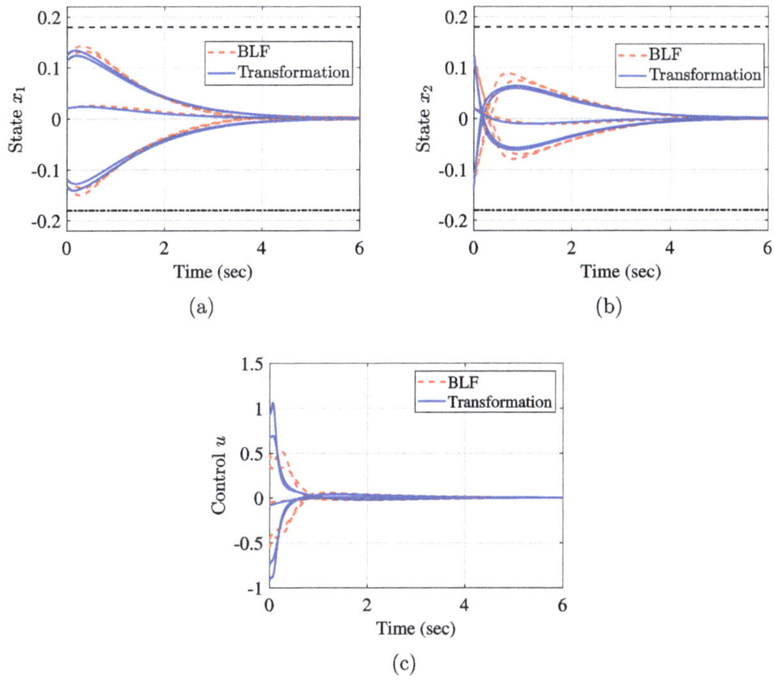

Fig. 4.10 State and control histories obtained for various initial conditions using the transformation-based controller and BLF-based controller

principles of control strategies specifically developed to address state constraints (which are scalable to output constraints). Advanced nonlinear control design methodologies, such as barrier Lyapunov function-based technique, dynamic inversion equivalent approach, and transformation-based strategy, are presented for different classes of nonlinear systems and examined for their ability to stabilize systems while ensuring the imposed state constraints.

4.8 Exercise Problems

1. Consider the BLF given by

$$V(e) = \frac{1}{p}\gamma^*(e)\log\left(\frac{\overline{\zeta}_e^p}{\overline{\zeta}_e^p - e^p}\right) + \frac{1}{p}\left(1 - \gamma^*(e)\right)\log\left(\frac{\underline{\zeta}_e^p}{\underline{\zeta}_e^p - e^p}\right) \tag{4.44}$$

where p is an even integer, "log" is the natural logarithm, and terms $\overline{\zeta}_e$ and $\underline{\zeta}_e$ are imposed upper and lower bounds on e. The switching function $\gamma^*(e)$ switches based on the value of e, defined as

$$\gamma^*(e) = \begin{cases} 1, & \text{if } 0 < e < \overline{\zeta}_e \\ 0, & \text{if } -\underline{\zeta}_e < e \leq 0 \end{cases}$$

Show the continuity of the time derivative of the BLF $V(e)$ around $e = 0$.

2. Show the steps involved in the calculation of time derivative $\dot{V}$ from V, where the time derivative of

$$V = \frac{1}{2}\log\left(\frac{\zeta^2}{\zeta^2 - e^2}\right)$$

is calculated as

$$\dot{V} = \frac{e\dot{e}}{\zeta^2 - e^2}$$

3. Consider the closed-loop error dynamics:

$$\dot{e} + \frac{ke}{\left(\zeta_e^2 - e^2\right)} + d = 0$$

where e is error, d is a constant unknown parameter, k is a positive constant, and ζ_e is a constant symmetric constraint imposed on e.
Show that for any bounded value of d, the error dynamics will remain stable without violation of the imposed constraint, i.e., $|e| < \zeta_e$.

4. Show the steps involved in the calculation of the time derivative $\dot{s}$ from s, where the time derivative of

$$s = \frac{x_1}{\zeta_1^2 - x_1^2} + \frac{x_2}{\zeta_2^2 - x_2^2}$$

is calculated as

$$\dot{s} = \frac{\left(\zeta_1^2 + x_1^2\right)x_2}{\left(\zeta_1^2 - x_1^2\right)^2} + \frac{\left(\zeta_2^2 + x_2^2\right)\dot{x}_2}{\left(\zeta_2^2 - x_2^2\right)^2}$$

5. Design a state-constrained controller using the dynamic inversion equivalent approach for the following nonlinear systems, assuming the imposed state constraint ζ is $\zeta = 0.5$. Additionally, add constant model uncertainties in the

systems and verify the robustness of the controller to modeling uncertainties. (Assume the problem is a stabilization problem.)

(a) $\dot{x} = -x + x^3 + u, \quad y = x.$
(b) $\dot{x} = -x - x^3 + u, \quad y = x.$
(c) $\dot{x} = -x^2 + \left(x^2 - 3\right) u, \quad y = x.$

6. Design and compare state-constrained controllers using BLF-based approach and transformation-based approach for the following nonlinear systems assuming imposed state constraints ζ_1 and ζ_2 are $\zeta_1 = 0.3$ and $\zeta_2 = 0.8$, respectively. (Assume the problem is a stabilization problem.)

(a) $\begin{bmatrix} \dot{x}_1 \\ \dot{x}_2 \end{bmatrix} = \begin{bmatrix} x_2 \\ -x_2 - x_1 - (x_1 + x_2)^2 x_2 + u \end{bmatrix}$
(b) $\ddot{x} = -\sin x + u, \quad y = x.$
(c) $\ddot{x} - 0.7(1 - x^2)\dot{x} - x^3 + (1 + \dot{x}^2)u = 0, \quad y = x.$

7. Design a state-constrained controller using a transformation-based approach for the following nonlinear systems, assuming imposed state constraints ζ_1, ζ_2, and ζ_3 are $\zeta_1 = 0.3$, $\zeta_2 = 0.6$, and $\zeta_3 = 0.8$, respectively. (Assume the problem is a stabilization problem.)

(a) $\begin{bmatrix} \dot{x}_1 \\ \dot{x}_2 \\ \dot{x}_3 \end{bmatrix} = \begin{bmatrix} x_1 + x_2 \\ x_3 \\ u \end{bmatrix}, \quad y = x_1.$

(b) $\begin{bmatrix} \dot{x}_1 \\ \dot{x}_2 \\ \dot{x}_3 \end{bmatrix} = \begin{bmatrix} -x_1 + x_2 \\ x_3 \\ u \end{bmatrix}, \quad y = x_1.$

References

C.P. Bechlioulis, G.A. Rovithakis, Robust adaptive control of feedback linearizable MIMO nonlinear systems with prescribed performance. IEEE Trans. Automat. Contr. **53**(9), 2090–2099 (2008)

C.P. Bechlioulis, G.A. Rovithakis, Adaptive control with guaranteed transient and steady state tracking error bounds for strict feedback systems. Automatica **45**(2), 532–538 (2009)

A. Bemporad, Reference governor for constrained nonlinear systems. IEEE Trans. Autom. Control **43**(3), 415–419 (1998)

M. Chen, S.S. Ge, B. Ren, Adaptive tracking control of uncertain MIMO nonlinear systems with input constraints. Automatica **47**(3), 452–465 (2011)

D. Chowdhury, H.K. Khalil, Funnel control for nonlinear systems with arbitrary relative degree using high-gain observers. Automatica **105**, 107–116 (2019)

D. DeHaan, M. Guay, Extremum-seeking control of state-constrained nonlinear systems. Automatica **41**(9), 1567–1574 (2005)

S.S. Ge, F. Hong, T.H. Lee, Adaptive neural control of nonlinear time-delay systems with unknown virtual control coefficients. IEEE Trans. Syst. Man Cybernet. Part B (Cybernet.) **34**(1), 499–516 (2004)

H. Habibi, H. Rahimi Nohooji, I. Howard, Backstepping Nussbaum gain dynamic surface control for a class of input and state constrained systems with actuator faults. Inf. Sci. **482**(2019), 27–46 (2019)

N. Hopfe, A. Ilchmann, E.P. Ryan, Funnel control with saturation: nonlinear SISO systems. IEEE Trans. Automat. Contr. **55**(9), 2177–2182 (2010)

L. Huang, S.S. Ge, T.H. Lee, Position/force control of uncertain constrained flexible joint robots. Mechatronics **16**(2), 111–120 (2006)

A. Ilchmann, E.P. Ryan, P. Townsend, Tracking control with prescribed transient behaviour for systems of known relative degree. Syst. Control Lett. **55**(5), 396–406 (2006)

S.P. Karason, A.M. Annaswamy, Adaptive control in the presence of input constraints. IEEE Trans. Autom. Control **39**(11), 2325–2330 (1994)

Y. Li, T. Li, X. Jing, Indirect adaptive fuzzy control for input and output constrained nonlinear systems using a barrier Lyapunov function. Int. J. Adapt. Control Signal Process. **28**(2), 184–199 (2014a)

Y. Li, S. Tong, T. Li, Adaptive fuzzy output-feedback control for output constrained nonlinear systems in the presence of input saturation. Fuzzy Sets Syst. **248**(20102012), 138–155 (2014b)

D.Q. Mayne, J.B. Rawlings, C.V. Rao, P.O.M. Scokaert, Constrained model predictive control: stability and optimality. Automatica **36**(6), 789–814 (2000)

E. Rollins, J. Valasek, J.A. Muse, M.A. Bolender, Nonlinear adaptive dynamic inversion applied to a generic hypersonic vehicle, in *AIAA Guidance, Navigation, and Control (GNC) Conference* (2013)

K. Sachan, R. Padhi, Barrier Lyapunov function based state-constrained control for a class of nonlinear systems. IFAC-PapersOnLine **51**(1), 7–12 (2018a)

K. Sachan, R. Padhi, State-constrained robust adaptive cruise control design for air-breathing hypersonic vehicles, in *2018 AIAA Guidance, Navigation, and Control Conference* (2018b)

K. Sachan, R. Padhi, Barrier Lyapunov function based output-constrained control of nonlinear Euler-Lagrange systems, in *2018 15th International Conference on Control, Automation, Robotics and Vision, ICARCV 2018* (2018c)

K. Sachan, R. Padhi, Output-constrained robust adaptive control for uncertain nonlinear MIMO systems with unknown control directions. IEEE Control Syst. Lett. **3**(4), 823–828 (2019)

K. Sachan, R. Padhi, Nonlinear robust neuro-adaptive flight control for hypersonic vehicles with state constraints. Control Eng. Pract. **102**, 104526 (2020a)

K. Sachan, R. Padhi, Safety-constrained robust adaptive control for a class of MIMO non-linear systems. IET Control Theory Appl. **14**(7), 982–990 (2020b)

Z.L. Tang, S.S. Ge, K.P. Tee, W. He, Adaptive neural control for an uncertain robotic manipulator with joint space constraints. Int. J. Control **89**(7), 1428–1446 (2016)

K.P. Tee, S.S. Ge, Control of nonlinear systems with partial state constraints using a barrier Lyapunov function. Int. J. Control **84**(12), 2008–2023 (2011)

K.P. Tee, S.S. Ge, E.H. Tay, Barrier Lyapunov Functions for the control of output-constrained nonlinear systems. Automatica **45**(4), 918–927 (2009)

K.P. Tee, B. Ren, S.S. Ge, Control of nonlinear systems with time-varying output constraints. Automatica **47**(11), 2511–2516 (2011)

W. Wang, D. Wang, Z. Peng, T. Li, Prescribed performance consensus of uncertain nonlinear strict-feedback systems with unknown control directions. IEEE Trans. Syst. Man, Cybern. Syst. **46**(9), 1279–1286 (2016)

C. Wen, J. Zhou, Z. Liu, H. Su, Robust adaptive control of uncertain nonlinear systems in the presence of input saturation and external disturbance. IEEE Trans. Automat. Contr. **56**(7), 1672–1678 (2011)

Z. Zhang, G. Duan, Y. Hu, Robust adaptive control for a class of semi-strict feedback systems with state and input constraints. Int. J. Robust Nonlinear Control **28**(9), 3189–3211 (2018)

K. Zhao, Y. Song, T. Ma, L. He, Prescribed performance control of uncertain euler-lagrange systems subject to full-state constraints. IEEE Trans. Neural Netw. Learn. Syst. **29**(8), 3478–3489 (2018)

T. Zhong-Liang, G.S. Sam, T.K. Peng, H. Wei, Robust adaptive neural tracking control for a class of perturbed uncertain nonlinear systems with state constraints. IEEE Trans. Syst. Man, Cybern. Syst. **PP**(99), 1–12 (2016)

Q. Zhou, L. Wang, C. Wu, H. Li, H. Du, Adaptive fuzzy control for nonstrict-feedback systems with input saturation and output constraint. IEEE Trans. Syst. Man, Cybern. Syst. **47**(1), 1–12 (2017)

Chapter 5
Adaptive Control Design Philosophies

5.1 Motivation

As the name suggests, adaptive control means control with adaptation. Here, adaptation is concerned with altering the controller parameters to achieve or sustain a certain performance level measured by control performance indices. It is one of the powerful solutions to handle the control of systems with uncertainties. In this type of control, the parameters of the controller are continuously adjusted to reach and/or attain the required performance in the presence of uncertainties in tracking or stabilization control problems (Krstic et al., 1995). Adaptive control has been acknowledged as one of the significant developments in control theory, as it introduced the learning ability to the feedback controllers (Krstic et al., 1995). It can be used to deal with different types of uncertainties that could arise during the operation of the system consistently. For instance, in aircraft, rockets, and missiles, some significant uncertainties might emerge, such as the change in the system mass or inertia. These uncertainties might degrade the performance of classical controllers. Adaptive control could handle such uncertain or unknown system parameters or dynamics to achieve or sustain the desired level of performance by the automatic adjustment of the controller parameters. Adaptive control has a plethora of applications in power systems, mechanical systems, aircraft systems, and process control (Narendra, 2012).

Adaptive control has two main parts: the adjustable controller and the algorithm of adaptation, as depicted in Fig. 5.1. The algorithm of adaptation adjusts the parameters of the controller based on the comparison between the current performance index and the desired performance index (Landau et al., 2011).

In comparison to the classical feedback controllers, the performance of adaptive controllers can be improved with time as their parameters are adapted online, while the performance of the conventional ones will converge to a certain limit as their gains are fixed, and they can not be tuned again (Landau et al., 2011). The main

K. Sachan et al., *Intelligent Adaptive Control with System Constraints*,
https://doi.org/10.1007/978-981-95-8220-4_5

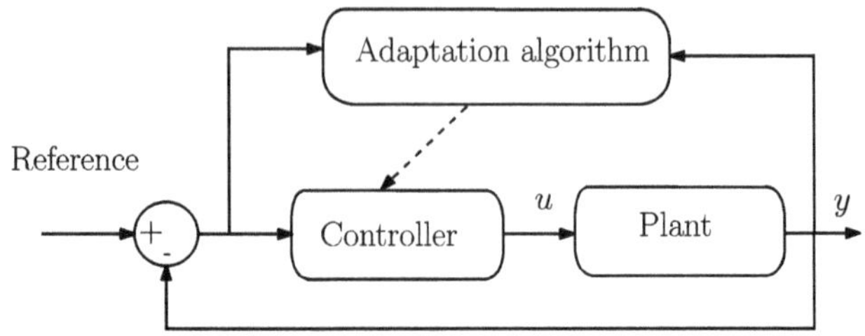

Fig. 5.1 Adaptive control

Table 5.1 Classical feedback controllers vs. adaptive controllers

Classical feedback control	Adaptive control
Controller gains are fixed after tuning	Controller gains are continuously adjusted
The design does not explicitly consider uncertainties	The design handles uncertainties
Performance is fixed	Performance can be improved with adaptation

differences between classical controllers and adaptive controllers are summarized in Table 5.1.

The question of "when" to use adaptive controllers and "when" to settle with fixed gain controllers depends solely on the control problem at hand. If the control problem incorporates unknown plant parameters or uncertain parameters that would significantly impact the control performance, then adaptive controllers might be considered. One should also remember that adaptive controllers contain adaptation algorithms, and thus, they would impose additional computational cost in the control task. A trade-off between performance and computational cost should be considered when deciding whether to select adaptive controllers or not.

5.2 Modeling Uncertainties

Uncertainties occur when the given system parameters are not exactly known or prone to change. Adaptive control methods deal with uncertainties, especially parametric uncertainties, and thus, it is useful to determine the type of uncertainty to decide if adaptive control would achieve its control goals or not. Various types of modeling uncertainties have been assumed to be associated with nonlinear systems, which are defined as follows.

Definition 5.1 *Matched uncertainty* is the type of uncertainty that enters the system through the control channel. For example, the following spring-mass-damper system has matched uncertainty:

$$\dot{x}_1 = x_2$$

$$\dot{x}_2 = -\frac{c}{m}x_2 - \frac{k}{m}x_1 + d + u$$

where d represents a matched uncertainty.

Definition 5.2 *Unmatched (mismatched) uncertainty* is a kind of uncertainty that does not go through the control channel. For example, the following spring-mass-damper system has unmatched uncertainty:

$$\dot{x}_1 = x_2 + d$$

$$\dot{x}_2 = -\frac{c}{m}x_2 - \frac{k}{m}x_1 + u$$

where d depicts an unmatched uncertainty in the system.

Definition 5.3 (Nguyen, 2018) *Structured uncertainty* is a source of uncertainty with uncertain parameters but known functional characteristics. It is also often referred to as parametric uncertainty. For example, a spring-mass-damper system with an uncertain spring constant is given as

$$m\ddot{x}(t) + c\dot{x}(t) + \bar{k}x(t) = u(t)$$

where $\bar{k}$ is an uncertain parameter and the function $x(t)$ associated with $\bar{k}$ is a known function.

Definition 5.4 (Nguyen, 2018) *Unstructured uncertainty* is a source of uncertainty for which neither parameters nor function characteristics are known. For example, a spring-mass-damper system with an uncertain spring characteristic is given as

$$m\ddot{x}(t) + c\dot{x}(t) + d\left(x(t), \bar{k}\right) = u(t)$$

where $d\left(x(t), \bar{k}\right)$ is an uncertain function.

Definition 5.5 (Nguyen, 2018) *Control input uncertainty* is a type of uncertainty that exists in the control input matrix of a class of MIMO affine-in-control systems of the form:

$$\dot{\boldsymbol{x}} = \boldsymbol{f}(\boldsymbol{x}) + \boldsymbol{G}(\boldsymbol{x})\boldsymbol{u}$$

where $\boldsymbol{G}(\boldsymbol{x})$ is unknown. This type of uncertainty is also known as the unknown control direction (UCD) of a system.

5.3 Overview of Neural Networks

Neural networks (NNs), inspired by biological neural systems, are one of the most prominent elements of AI methodologies. These networks are connected in parallel, processing neuron elements with interconnections through synaptic weights. NNs are composed of parallel connected layers, which are input, hidden, and output layers as follows:

- Input layer: This layer accepts raw input $\boldsymbol{x} \in \Re^m$ such as system states or tracking errors.
- Hidden layer/layers: In this layer, the nonlinear transformation of the input data is learned using nonlinear activation functions $\boldsymbol{\phi}(\boldsymbol{x}) \in \Re^n$ such as the Gaussian activation function, sigmoid, or hyperbolic tangent.
- Output layer: In this layer, the final output of the network is estimated $\hat{\boldsymbol{y}}(\boldsymbol{x}) \in \Re^l$.

5.3.1 *Function Approximation Using Neural Networks*

NNs can learn from their environment by using various forms of data from specific domains, such as numerical, visual, or vocal data, by adjusting their weights. After training, NNs can perform predictive tasks for the same/similar tasks. For instance, a NN trained on historical meteorological data can predict weather patterns for similar inputs. Key features of NNs, such as universal function approximation, parallel processing, nonlinearity, adaptability, and the availability of extensive software libraries, have facilitated their application to a wide range of domains (Haykin, 2009).

From a control system perspective, the function approximation capabilities of NNs are particularly valuable when addressing uncertain nonlinear control problems. In such cases, physics-based modeling often proves challenging due to the inherent complexity and the presence of modeling uncertainties. On the other hand, the widespread availability of data, made possible by observation and data acquisition systems, further enhances the utility of NNs in nonlinear control systems.

Theorem 5.1 (Universal Approximation Theorem (UAT)) *For an unknown continuous function $f \ : \ \in \Re^n \rightarrow \Re$ defined in a compact set $\Omega \in \Re$, it can be estimated using a single-hidden layer NN as $\hat{f}$ with an arbitrary accuracy $\epsilon > 0$ in the compact set $\Omega \in \Re^m$ as follows:*

$$\sup_{\boldsymbol{x}\in\Omega} \left|\hat{f}(\boldsymbol{x}) - f(\boldsymbol{x})\right| \leq \epsilon$$

$$\sup_{\boldsymbol{x}\in\Omega} \left|\hat{\boldsymbol{w}}^T \boldsymbol{\phi}(\boldsymbol{P}^T \boldsymbol{x} + \boldsymbol{b}) - f(\boldsymbol{x})\right| \leq \epsilon$$

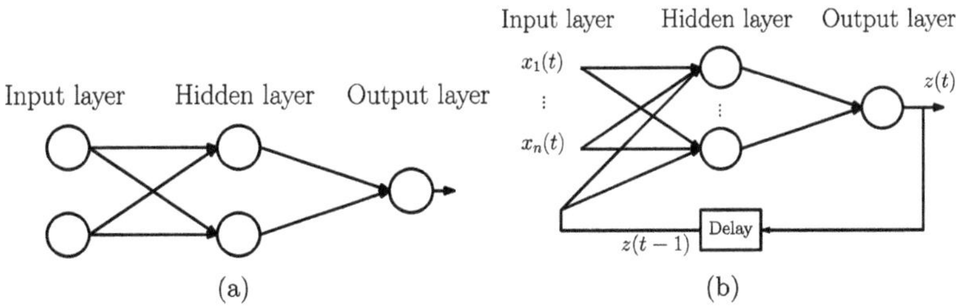

Fig. 5.2 Feedforward neural networks and recurrent neural networks. (**a**) FFNN. (**b**) RNN

where sup *is the supremum,* $\hat{\boldsymbol{w}} \in \Re^m$ *is an output weights vector,* $\boldsymbol{\phi}(\cdot) \in \Re^m$ *is an activation function,* $\boldsymbol{P} \in \Re^{n\times m}$ *is an input weight vector,* $\boldsymbol{x} \in \Re^n$ *is an input vector, and* $\boldsymbol{b} \in \Re^m$ *is a bias vector.*

UAT is a convenient base for using NNs in the estimation of unknown nonlinear functions, such as uncertainties and unmodeled dynamics in nonlinear systems.

5.3.2 *Types of Neural Networks*

NNs can be classically categorized according to how they process the data flow in the model, as feed-forward neural networks (FFNNs) and recurrent neural networks (RNNs). In the FFNNs, the model processes the data from input to the hidden layers and then to the output layers, while in RNNs, the model processes the data in the same manner, but there is always feedback or recurrency involved, as shown in Fig. 5.2a, b. While RNN might not be used in this book, it is added for completeness of the information.

NNs can be used for regression, i.e., mapping of independent variables to dependent variables, and classification when categorizing input variables into distinct classes. FFNNs are typically used in tasks that are nonsequential in nature, while RNNs are used for sequential or temporal tasks such as time series analysis and natural language processing (Tsantekidis et al., 2022). Noticeably, in adaptive control, FFNNs can be used to estimate the unknown dynamics of a given system, as in the case of indirect adaptive control, or can be utilized as direct adaptive controllers to estimate the control law directly (Ge et al., 2013).

5.4 Uncertainty Estimation

Adaptive control usually involves online parameter estimation. To estimate the uncertainties in a nonlinear system, a linearly parameterized or linear in parameter estimator can be used in the following form:

$$\hat{d}(\boldsymbol{x}) = \hat{\boldsymbol{w}}^T \boldsymbol{\phi}(\boldsymbol{x}) \tag{5.1}$$

where $\hat{d}(\boldsymbol{x}) \in \Re$ is the estimation of the unknown uncertainty, $\hat{\boldsymbol{w}}$ is an adjustable weight vector, and $\boldsymbol{\phi}$ is a nonlinear basis function.

Among the different choices of estimators, NNs have great potential in estimating such nonlinearities due to their universal approximation capabilities (Hornik et al., 1989).

Consider a nonlinear system of the form:

$$\dot{\boldsymbol{x}} = f(\boldsymbol{x}) + u + d(\boldsymbol{x}) \tag{5.2}$$

Using (5.2), the uncertainty $d(\boldsymbol{x})$ can be written as

$$d(\boldsymbol{x}) = \dot{\boldsymbol{x}} - f(\boldsymbol{x}) - u \tag{5.3}$$

The estimation error $\epsilon(\boldsymbol{x})$ is expressed as the difference between the real and approximated uncertainty:

$$\begin{aligned} \epsilon(\boldsymbol{x}) &= d(\boldsymbol{x}) - \hat{d}(\boldsymbol{x}) \\ &= \dot{\boldsymbol{x}} - f(\boldsymbol{x}) - u - \hat{d}(\boldsymbol{x}) \end{aligned} \tag{5.4}$$

Let us define a scalar objective function j used to minimize the approximation error $\epsilon(x)$ as

$$\begin{aligned} j &= \frac{1}{2}\epsilon^2(\boldsymbol{x}) \\ &= \frac{1}{2}\left(\dot{\boldsymbol{x}} - f(\boldsymbol{x}) - u - \hat{d}(\boldsymbol{x})\right)^2 \\ &= \frac{1}{2}\left(\dot{\boldsymbol{x}} - f(\boldsymbol{x}) - u - \hat{\boldsymbol{w}}^T \boldsymbol{\phi}(\boldsymbol{x})\right)^2 \end{aligned} \tag{5.5}$$

Gradient descent (GD) is usually utilized to update the weight vector $\hat{\boldsymbol{w}}$ as follows:

$$\begin{aligned} \dot{\hat{\boldsymbol{w}}} &= -\gamma \frac{\partial j}{\partial \hat{\boldsymbol{w}}} \\ &= -\gamma \frac{\partial j}{\partial \hat{d}(\boldsymbol{x})} \frac{\partial \hat{d}(\boldsymbol{x})}{\partial \hat{\boldsymbol{w}}} \\ &= -\gamma \epsilon(\boldsymbol{x}) \boldsymbol{\phi}(x) \end{aligned} \tag{5.6}$$

where $\gamma > 0$ is the learning rate.

Example 5.1 Consider the following scalar system with unknown uncertainty $d(x) = 0.4\sin(0.5x(t)) + 0.4$ (assumed to be unknown):

$$\dot{x} = -x^2 + d(x) \tag{5.7}$$

A NN is used to approximate the uncertainty as $\hat{d}(x) = \hat{\boldsymbol{w}}^T\boldsymbol{\phi}(x)$. $\boldsymbol{\phi}(x)$ is defined as $\phi_i(x) = \exp(-\frac{\|(x-c_i)\|}{2\sigma^2})$, $\boldsymbol{\phi}(x) = \left[\frac{\phi_1(x)}{\sum_{i=1}^{5}\phi_i(x)}, \ldots, \frac{\phi_5(x)}{\sum_{i=1}^{5}\phi_i(x)}\right]^T$, and "exp" is the exponential function $[c_1, c_2, c_3, c_4, c_5] = [-2, -1, 0, 1, 2]$ and $\sigma = 1$.

The approximation error $\epsilon(x)$ is defined as

$$\begin{aligned}\epsilon(x) &= d(x) - \hat{d}(x)\\ &= \dot{x} + x^2 - \hat{\boldsymbol{w}}^T\boldsymbol{\phi}(x)\end{aligned} \tag{5.8}$$

The objective function j is defined as

$$j = \frac{1}{2}\epsilon^2(x)$$

The weights $\hat{\boldsymbol{w}}$ are adapted using GD as

$$\begin{aligned}\dot{\hat{\boldsymbol{w}}} &= -\gamma\frac{\partial j}{\partial \hat{\boldsymbol{w}}}\\ &= -\gamma\frac{\partial j}{\partial \epsilon(x)}\frac{\partial \epsilon(x)}{\partial \hat{d}(x)}\frac{\partial \hat{d}(x)}{\partial \hat{\boldsymbol{w}}}\\ &= -\gamma\epsilon(x)\boldsymbol{\phi}(x)\end{aligned} \tag{5.9}$$

Using (5.9), the update law of NN weights $\hat{\boldsymbol{w}}$ is given as

$$\dot{\hat{\boldsymbol{w}}} = -\gamma\boldsymbol{\phi}(x)\epsilon(x) \tag{5.10}$$

The results of estimating $d(x)$ online using NN are depicted in Fig. 5.3a. The estimation of $\hat{d}(x)$ converges to its true value $d(x)$ in almost 3 seconds. This is the time taken by NN to learn the nonlinear disturbance function.

Evolution of NN weight norm $\|\hat{\boldsymbol{w}}\|$ is also depicted in Fig. 5.3b. $\|\hat{\boldsymbol{w}}\|$ also converges to its final value almost 1 within 4 seconds.

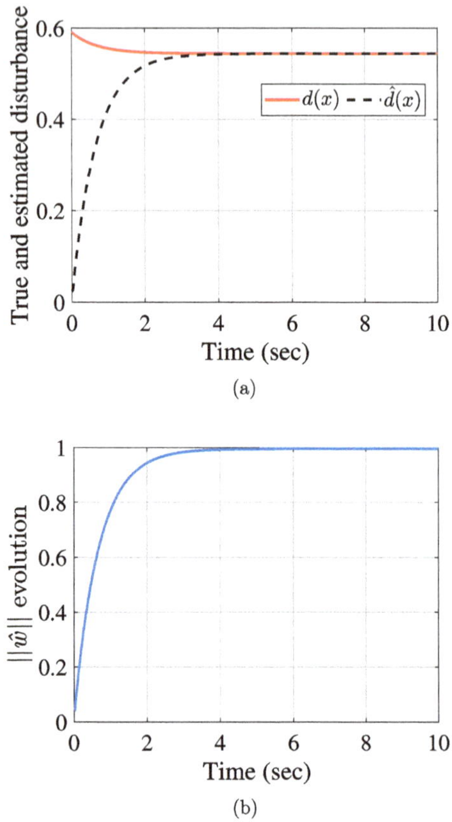

Fig. 5.3 (**a**) d estimation using NN. (**b**) Evolution of $\|\hat{\boldsymbol{w}}\|$

5.5 Direct and Indirect Adaptive Control

Adaptive controllers can be classified as either direct, where the parameters of the controller are directly adjusted based on the differences between the actual outputs of the system and the desired outputs, or indirect , where the plant is identified first, and then, the design of the controller and its adaptation are carried out using the identified plant (Landau et al., 2011).

In direct adaptive control, as shown in Fig. 5.4, an adaptive controller is directly used to control a plant/process with no system identification required. It also contains an adaptation algorithm that is usually used to adapt the controller parameters online, utilizing a certain objective function.

Consider a dynamic system of the following form:

$$
\begin{aligned}
\dot{x}_1(t) &= x_2(t) \\
\dot{x}_2(t) &= f(x_1(t), x_2(t), \boldsymbol{\theta}) + u(t) \\
y(t) &= x_2(t)
\end{aligned}
$$

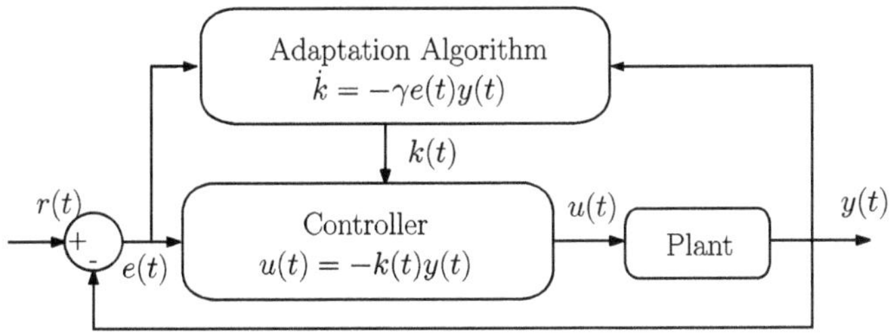

Fig. 5.4 Direct adaptive control

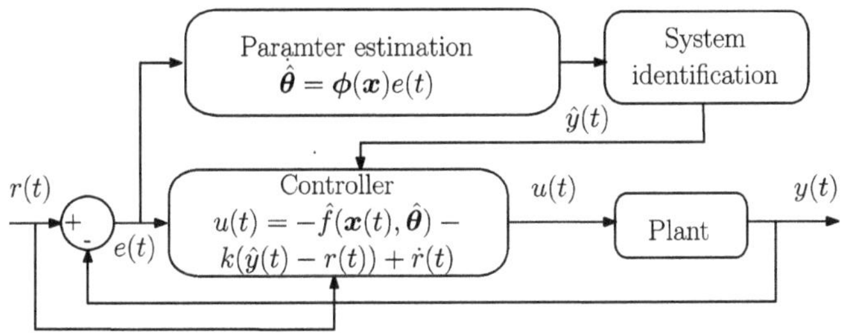

Fig. 5.5 Indirect adaptive control

The system is controlled to track a certain reference $r(t)$ in the presence of parametric uncertainties $\boldsymbol{\theta} \in \Re^p$. A direct adaptive tracking controller of the above system takes the following form:

$$u(t) = -k(t)y(t) \tag{5.11}$$

where $k(t)$ is a controller parameter adapted online.

The tracking error is represented as $e(t) = y(t) - r(t)$. The controller gain $k(t)$ is adapted as follows:

$$\dot{k}(t) = -\gamma e(t)y(t) \tag{5.12}$$

where γ is the adaptation gain.

One of the main advantages of direct adaptive controllers is their simple implementation algorithm; however, they might suffer from transient issues.

In the indirect adaptive control scheme, parameter estimation and system identification, i.e., model of the system dynamics, is required to update the control signal as shown in Fig. 5.5.

Consider the following system:

$$\dot{x}_1(t) = x_2(t)$$
$$\dot{x}_2(t) = f(x_1(t), x_2(t), \boldsymbol{\theta}) + u(t)$$

$$y(t) = x_2(t)$$

The controller aims to track a reference signal $r(t)$. Parameter estimation is the first step in the indirect adaptive control as follows:

$$\dot{\hat{\boldsymbol{\theta}}} = \boldsymbol{\phi}(\boldsymbol{x})e(t) \tag{5.13}$$

where $\boldsymbol{\phi}(\boldsymbol{x})$ is a regressor vector of the states.

The estimated parameters can be used for system identification as follows:

$$\begin{aligned} \dot{\hat{x}}_2(t) &= \hat{f}(x_1(t), x_2(t), \hat{\boldsymbol{\theta}}) + u(t) \\ \hat{y}(t) &= \hat{x}_2(t) \end{aligned} \tag{5.14}$$

where $\hat{f}(x_1(t), x_2(t), \hat{\boldsymbol{\theta}}) = \hat{\boldsymbol{\theta}}^T \boldsymbol{\phi}(\boldsymbol{x})$. An indirect adaptive controller might be designed as follows:

$$u(t) = -\hat{f}(x_1(t), x_2(t), \hat{\boldsymbol{\theta}}) - k(\hat{y}(t) - r(t)) + \dot{r}(t) \tag{5.15}$$

where $k > 0$ is the feedback gain and $\hat{y}(t)$ is an estimation of the actual plant output. The control law in (5.15) yields the following closed-loop error dynamics:

$$e(t) = x_2(t) - r(t) \tag{5.16}$$

Taking the time derivative of $e(t)$ leads to

$$\dot{e}(t) = \dot{x}_2(t) - \dot{r}(t) \tag{5.17}$$

Using the control law in (5.15) $\dot{x}_2$ becomes

$$\dot{x}_2(t) = f(x_1(t), x_2(t), \boldsymbol{\theta}) - \hat{f}(x_1(t), x_2(t), \hat{\boldsymbol{\theta}}) - k(\hat{y}(t) - r(t)) + \dot{r} \tag{5.18}$$

Using (5.18), $\dot{e}(t)$ can be written as

$$\dot{e}(t) = f(x_1(t), x_2(t), \boldsymbol{\theta}) - \hat{f}(x_1(t), x_2(t), \hat{\boldsymbol{\theta}}) - k(\hat{y}(t) - r(t)) \tag{5.19}$$

When the parameter estimation is accurate then $\hat{f} \approx f$ and $\hat{y}(t) \approx y(t)$ then $\dot{e}(t)$ becomes

$$\dot{e}(t) = -k(e(t)) \tag{5.20}$$

This result ensures exponential tracking of $e(t)$ toward a neighborhood of zero.

Table 5.2 Comparison between direct and indirect adaptive control

Direct adaptive control	Indirect adaptive control
Does not require an accurate model of the plant	Requires an accurate parameterized model of the plant
Does not require parameter estimation	Needs parameter estimation
Stability depends on the control law design and adaptation process	Stability analysis depends on the convergence of the parameters estimation
Performance depends on the parameter estimation quality	Performance depends on the quality of the adaptation of the control law

In comparison to direct adaptive controllers, indirect adaptive controllers have better robustness, but they require more computational effort when implemented in hardware.

Comparison between direct and indirect adaptive controllers is depicted in Table 5.2.

5.6 Basic Adaptive Control Design

Adaptive controllers are utilized to handle model uncertainties by applying an adaptation mechanism to update their parameters online. The basic concept of adaptive control lies in utilizing an adaptation of the controller parameters to achieve the control objective by considering the uncertainties in the controlled system.

Example 5.2 Consider a nonlinear system with the following dynamics:

$$\ddot{x}(t) = u(t) - a\dot{x}(t) - x(t) \tag{5.21}$$

Assuming $x(t) = x_1(t)$ and $\dot{x}(t) = x_2(t)$, the above system can be rewritten as

$$\begin{aligned} \dot{x}_1(t) &= x_2(t) \\ \dot{x}_2(t) &= u(t) - ax_2(t) - x_1(t) \end{aligned} \tag{5.22}$$

Let a be an unknown parameter, and the velocity of the system is to be controlled to a specific trajectory $x_{2d}(t) = \sin t$.

The tracking error is defined as $e(t) = x_2(t) - x_{2d}(t)$, and its derivative is $\dot{e}(t) = \dot{x}_2(t) - \dot{x}_{2d}(t)$. In the case when a is known, a dynamic inversion-based

(continued)

Example 5.2 (continued)
control law of the following form could achieve the tracking performance:

$$u(t) = \dot{x}_{2d}(t) - ke(t) + ax_2(t) + x_1(t) \tag{5.23}$$

where k is a control gain selected to achieve the desired performance specifications. Substituting the above control law in (5.21) leads to

$$\dot{e}(t) + ke(t) = 0 \tag{5.24}$$

which achieves asymptotic tracking, i.e., $e(t) \to 0$ as $t \to \infty$. However, since a is unknown, the above control law could not be directly used. Nevertheless, an estimate of the unknown parameter a is used to find an approximate control law, and thus, the approximate control law becomes

$$\hat{u}(t) = \dot{x}_{2d}(t) - ke(t) + \hat{a}x_1(t) + x_2(t) \tag{5.25}$$

where $\hat{a}$ is the estimation of the unknown parameter a.

Plugging in the above control law into the system dynamics (5.21) leads to

$$\dot{e}(t) = -ke(t) + (\hat{a} - a)x_2(t) \tag{5.26}$$

The parameter estimation error is defined as $\tilde{a} = \hat{a} - a$, and since a is constant, the time derivative of the estimation error is expressed as $\dot{\tilde{a}} = \dot{\hat{a}}$.

Choosing a Lyapunov function candidate V as

$$V = \frac{1}{2}e^2(t) + \frac{1}{2\gamma}\tilde{a}^2 \tag{5.27}$$

The time derivative of the Lyapunov function is written as

$$\dot{V} = e(t)\dot{e}(t) + \frac{1}{\gamma}\tilde{a}\dot{\tilde{a}} \tag{5.28}$$

Choosing the adaptation law of the parameter estimation $\hat{a}$ as

$$\dot{\hat{a}} = -\gamma e(t)x_2(t) \tag{5.29}$$

Using the above adaptation law and considering (5.26), (5.28) can be rewritten as

$$\dot{V} = e(t)(-ke(t) + (\hat{a} - a)x_2(t)) - \frac{1}{\gamma}(\tilde{a}\gamma e(t)x_2(t))$$

(continued)

Example 5.2 (continued)

$$= -k^2 e(t) + e(t)(\hat{a} - a)x_2(t) - \frac{1}{\gamma}(\tilde{a}\gamma e(t)x_2(t))$$

$$= -k^2 e(t) + e(t)\tilde{a}x_2(t) - \frac{1}{\gamma}(\tilde{a}\gamma e(t)x_2(t))$$

$$= -k^2 e(t) + (e(t)\tilde{a}x_2(t)) - (e(t)\tilde{a}x_2(t))$$

$$= -ke^2(t) \quad (5.30)$$

The initial conditions are $[x_1(0), x_2(0), x_3(0)] = [1, 1, 0]$. $\gamma = 15$ and gain $k = 3$. Velocity tracking results, as shown in Fig. 5.6a, demonstrate that the velocity tracks the desired velocity accurately after almost 2.35 seconds. On the other hand, the estimation of the parameter a, as shown in Fig. 5.6b, takes almost 4 seconds to converge to its true value (2.5). The control signal starts from -1 and moves in a sinusoidal shape to 4 as depicted in Fig. 5.6c.

5.7 Model Reference Adaptive Control (MRAC)

The core concept in model reference adaptive control (MRAC) is the model matching, where the controller parameters are adjusted such that the output of a controlled system matches the output of a desired reference model (Nguyen, 2018).

Consider a single-input-single-output (SISO) dynamic system represented as

$$\dot{\boldsymbol{x}}(t) = \boldsymbol{f}(\boldsymbol{x}(t)) + \boldsymbol{G}(\boldsymbol{x}(t))u(t) \quad (5.31)$$
$$y(t) = h(\boldsymbol{x}(t))$$

where $\boldsymbol{x} \in \Re^n$ is a vector the system states, $\boldsymbol{f}(\cdot) \in \Re^n$ is a nonlinear function of state, $\boldsymbol{G}(\cdot) \in \Re^n$ is the control gain matrix, $\boldsymbol{u} \in \Re$ is the control input, $y \in \Re$ is the system output, and $h \in \Re$ is an output function. Consider a reference model describing the desired system behavior be given as

$$\dot{\boldsymbol{x}}_m(t) = \boldsymbol{A}_m\boldsymbol{x}_m(t) + \boldsymbol{b}_m r(t) \quad (5.32)$$
$$y_m(t) = \boldsymbol{c}_m\boldsymbol{x}_m(t)$$

where $\boldsymbol{A}_m \in \Re^{n\times n}$ is a Hurwitz matrix representing the desired system matrix, $\boldsymbol{x}_m \in \Re^n$ are the reference states, $\boldsymbol{b}_m \in \Re^n$ are the input gain vector, $r \in \Re$ is the reference command, $y_m \in \Re$ is the reference model output, and $\boldsymbol{c}_m \in \Re^{1\times n}$ is the

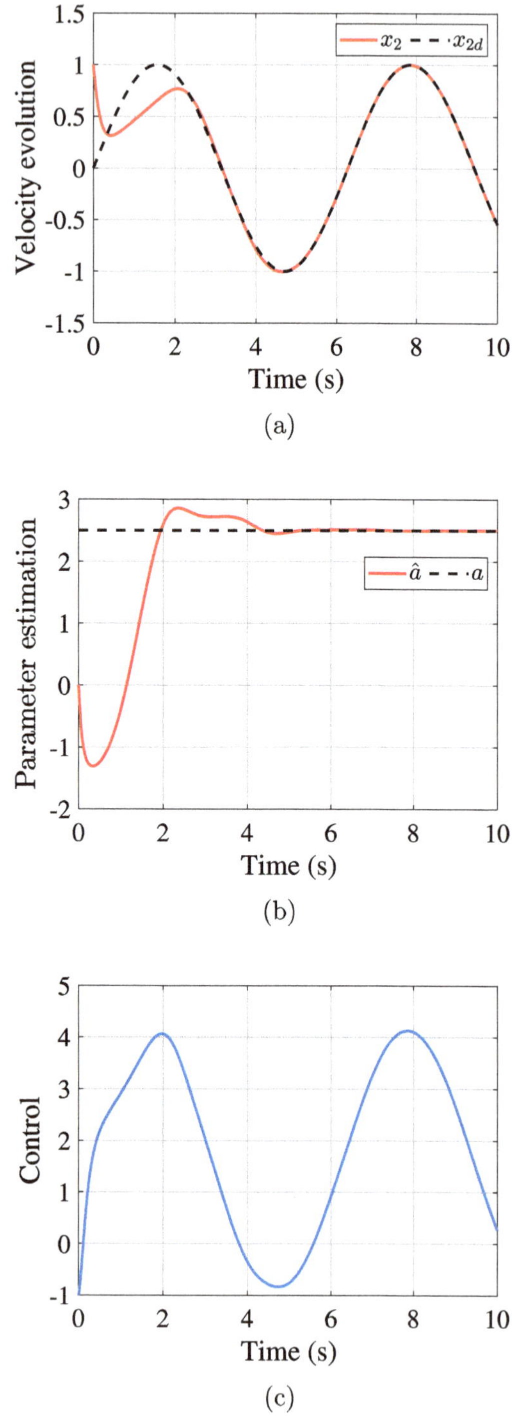

Fig. 5.6 (**a**) State history. (**b**) a estimation. (**c**) Control histories—Example 5.2

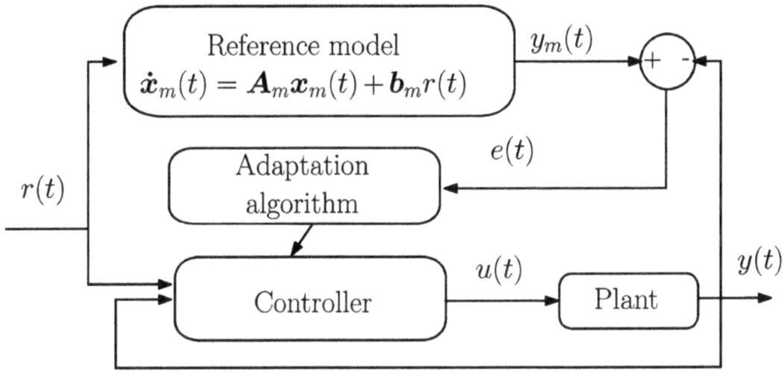

Fig. 5.7 MRAC

reference model output vector. The reference model is stable if and only if all the eigenvalues of the matrix $\boldsymbol{A}_m$ have strictly negative real poles.

The tracking error $e(t)$ is given as

$$e(t) = y_m(t) - y(t) \tag{5.33}$$

The objective of the MRAC is to reduce the tracking error so that the output y matches the reference model output y_m, and thus, the tracking error e converges to zero by adapting the controller parameters. MRAC can be used to control systems with structured and unstructured uncertainties (Nguyen, 2018). MRAC can be classified as direct and indirect MRAC. Using direct or indirect MRAC depends on the system to be controlled, available computational capabilities, and the required adaptation speed. For systems with imprecise models, direct MRAC is more suitable. For systems with high variation in parameters, indirect MRAC can be more accurate and robust. The MRAC block diagram is illustrated in Fig. 5.7.

5.7.1 Direct MRAC

Let us consider a scalar dynamic system represented as

$$\dot{x}(t) = ax(t) + b(u(t) + f(x(t))) \tag{5.34}$$

where $a < 0$ and $b > 0$ are unknown parameters and f is a smooth unknown uncertainty function.

A model reference is designed as

$$\dot{x}_m(t) = a_m x_m(t) + b_m r(t) \tag{5.35}$$

where $a_m < 0, b_m > 0$, and r is a continuous bounded reference command function. The uncertainty $f(x)$ can be estimated using a linearly parameterized estimator such as NNs as follows:

$$f(x) = \boldsymbol{w}^T \boldsymbol{\phi}(x) + \epsilon \tag{5.36}$$

where $\boldsymbol{w}^T$ is an ideal constant weight vector and $\boldsymbol{\phi}$ is a nonlinear bounded activation function of the NN.

An ideal controller can be designed to eliminate the uncertainty effect and force x to follow x_m as follows:

$$u(t) = k_1 x(t) + k_2 r(t) - \boldsymbol{w}^T \boldsymbol{\phi}(x) - \epsilon \tag{5.37}$$

where k_1 and k_2 are ideal constant gains.

Since k_1, k_2, and $\boldsymbol{w}$ are ideal unknown values, an approximate control law can be used as follows:

$$\hat{u}(t) = \hat{k}_1(t)x(t) + \hat{k}_2(t)r(t) - \hat{\boldsymbol{w}}^T \boldsymbol{\phi}(x) \tag{5.38}$$

where $\hat{k}_1(t)$, $\hat{k}_2(t)$, and $\hat{\boldsymbol{w}}$ are estimates of their ideal values.

The tracking error is defined as

$$e(t) = x_m(t) - x(t) \tag{5.39}$$

The parameter estimation errors are defined as

$$\tilde{k}_1 = k_1 - \hat{k}_1 \tag{5.40}$$

$$\tilde{k}_2 = k_2 - \hat{k}_2 \tag{5.41}$$

$$\tilde{\boldsymbol{w}} = \boldsymbol{w} - \hat{\boldsymbol{w}} \tag{5.42}$$

To derive the adaptation laws of the parameters k_1, k_2, and $\hat{\boldsymbol{w}}$, Lyapunov stability analysis can be utilized. Consider a positive definite function defined as follows:

$$V(e, \tilde{k}_1, \tilde{k}_2, \tilde{\boldsymbol{w}}) = e^2 + |b| \left(\frac{\tilde{k}_1^2}{\gamma_1} + \frac{\tilde{k}_2^2}{\gamma_2} + \tilde{\boldsymbol{w}}^T \Gamma \tilde{\boldsymbol{w}} \right) \tag{5.43}$$

where γ_1 and γ_1 are positive adaptation gains and Γ is a positive definite adaptation matrix of the weight vector.

The time derivative of the above function is negative semi-definite when the following holds:

$$-ex + \frac{\dot{\hat{k}}_1}{\gamma_2} = 0 \tag{5.44}$$

$$-er + \frac{\dot{\hat{k}}_2}{\gamma_2} = 0 \tag{5.45}$$

$$-e\boldsymbol{\phi}(x) + \Gamma^{-1}\dot{\hat{\boldsymbol{w}}} = 0 \tag{5.46}$$

Based on the above, the following adaptation laws can be found:

$$\dot{\hat{k}}_1 = -\gamma_1 e(t)x(t) \tag{5.47}$$

$$\dot{\hat{k}}_2 = -\gamma_2 e(t)r(t) \tag{5.48}$$

$$\dot{\hat{\boldsymbol{w}}} = -\Gamma e(t)\boldsymbol{\phi}(x) \tag{5.49}$$

Example 5.3 Let us consider the system described by the following:

$$\dot{x}(t) = ax(t) + b(u(t) + f(x(t)))$$

where $a < 0$ and $b > 0$ are unknown constants and the uncertainty $f(x)$ is unknown with true value of $f(x) = \tanh(x)$, where tanh is the hyperbolic tangent function. The reference model is given as $\dot{x}_m(t) = x_m(t) + r(t)$ and $r(t) = \sin(t)$.

A direct MRAC controller of the following form is utilized to control the above system with the control law:

$$\hat{u}(t) = \hat{k}_1(t)(t)x + \hat{k}_2(t)r(t) - \hat{\boldsymbol{w}}^T\boldsymbol{\phi}(x) \tag{5.50}$$

where $\hat{k}_1$ is the feedback control gain and $\hat{k}_2$ is the feedforward control gain.

The tracking error is given by

$$e(t) = x_m(t) - x(t) \tag{5.51}$$

The adaptation laws of the gains k_1 and k_2 are selected as

$$\dot{\hat{k}}_1 = -\gamma_1 e(t)x(t) \tag{5.52}$$

$$\dot{\hat{k}}_2 = -\gamma_2 e(t)r(t) \tag{5.53}$$

$$\dot{\hat{\boldsymbol{w}}} = -\Gamma_3 e(t)\boldsymbol{\phi}(x) \tag{5.54}$$

where $\gamma_1 = 10$, $\gamma_2 = 10$, and $\Gamma_3 = \text{diag}(2, \ldots, 2)$ are the adaptation gains.

The tracking results, as shown in Fig. 5.8a, depict that the state $x(t)$ accurately follows the reference model state $x_m(t)$ after the initial learning.

(continued)

Example 5.3 (continued)
The adaptive gains $\hat{k}_1$ and $\hat{k}_2$, as demonstrated in Fig. 5.8b, also quickly converge to approximately -3 and 1.7, respectively. The control signal, as shown in Fig. 5.8c, varies between -2 and 2 with a sinusoidal shape.

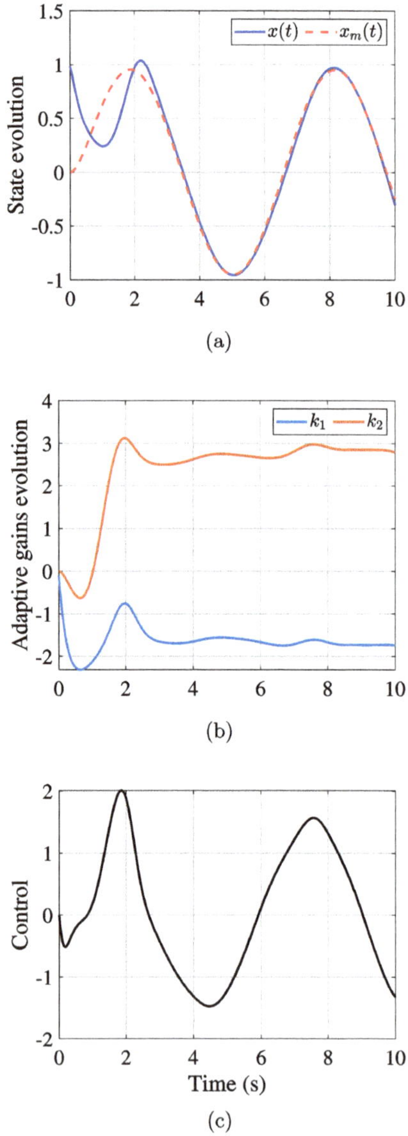

Fig. 5.8 (**a**) State history. (**b**) Adaptive gains. (**c**) Control signal—Example 5.3

5.7.2 Indirect MRAC

Indirect MRAC is a class of indirect adaptive controllers that concerns forcing a system state to a certain trajectory defined by a reference model. In indirect MRAC, plant modeling is required.

Consider a class of a dynamic system represented as

$$\dot{x}(t) = ax(t) + b(u(t) + f(x)) \tag{5.55}$$

where $a < 0$ and $b > 0$ are unknown parameters.

A reference model is defined as

$$\dot{x}_m(t) = a_m x_m(t) + b_m r(t) \tag{5.56}$$

where x_m is the state of the reference model and r is the reference input signal.

The estimates of the parameters a and b are expressed as $\hat{a}$ and $\hat{b}$. The estimation of the unknown uncertainty is expressed as $\hat{f}(x) = \hat{\boldsymbol{w}}^T \boldsymbol{\phi}(x)$.

The tracking error $e(t)$ is defined as $e(t) = x_m(t) - x(t)$. The adaptation laws of the parameters are

$$\dot{\hat{a}} = -\gamma_1 e(t) x(t) \tag{5.57}$$

$$\dot{\hat{b}} = -\gamma_2 e(t) u(t) \tag{5.58}$$

$$\dot{\hat{\boldsymbol{w}}} = -\Gamma_3 e(t) \boldsymbol{\phi}(x) \tag{5.59}$$

where γ_1 and γ_2 are adaptation gains.

An indirect MRAC can be designed as

$$\hat{u}(t) = k_1(t)x(t) + k_2(t)r(t) - \hat{\boldsymbol{w}}^T \boldsymbol{\phi}(x) \tag{5.60}$$

The gains k_1 and k_2 are calculated as

$$k_1(t) = \frac{a_m - \hat{a}(t)}{\hat{b}(t)} \tag{5.61}$$

$$k_2(t) = \frac{b_m}{\hat{b}(t)} \tag{5.62}$$

Example 5.4 A dynamic system is expressed as

$$\dot{x}(t) = \theta_1 x(t) + \theta_2 (u(t) + f(x)) \tag{5.63}$$

(continued)

Example 5.4 (continued)
where θ_1 and θ_2 are unknown parameters with true values of 1 and 1, respectively, and $f(x)$ is unknown certainty with true value of $f(x) = 0.1\cos x$.

The following reference model is used:

$$\dot{x}_m(t) = a_m x_m(t) + b_m r(t) \tag{5.64}$$

where $a_m = -2$, $b_m = 1$, and $r(t) = \sin(t)$.

The parameters θ_1 and θ_2 are adapted online using the following adaptation laws:

$$\dot{\hat{\theta}}_1 = -\gamma_1 e(t) x(t) \tag{5.65}$$

$$\dot{\hat{\theta}}_2 = -\gamma_2 e(t) u(t) \tag{5.66}$$

$$\dot{\hat{\boldsymbol{w}}} = -\Gamma_3 e(t) \boldsymbol{\phi}(x) \tag{5.67}$$

where $\gamma_1 = 0.05$, $\gamma_2 = 0.015$, and $\Gamma_3 = \text{diag}(0.1, \ldots, 0.1)$ are adaptation gains and $e(t) = x(t) - x_m(t)$ is the tracking error. The controller is designed as

$$u(t) = k_1(t)x(t) + k_2(t)r(t) - \hat{\boldsymbol{w}}^T \boldsymbol{\phi}(x) \tag{5.68}$$

where the gains k_1 and k_2 are computed from the estimation of the unknown parameters and the reference model parameters as

$$k_1 = \frac{(a_m - \hat{\theta}_1)}{\hat{\theta}_2} \tag{5.69}$$

$$k_2 = \frac{b_m}{\hat{\theta}_2} \tag{5.70}$$

The tracking results as depicted in Fig. 5.9a, b demonstrate that the state x tracks the reference model state x_m with an initial tracking error of -0.5, which gradually approaches zero with time. The control signal, as depicted in Fig. 5.9c, takes a sinusoidal shape bounded between -0.6 and 0.6.

The estimations of the parameters θ_1 and θ_2 converge to their true values ($\theta_1 = 1$, $\theta_2 = 1$) with the progress of online adaptation as shown in Fig. 5.10a, b. The values of the gains k_1 and k_2, as shown in Fig. 5.10c, oscillate largely at the beginning of the learning. This happens due to oscillation in the parameter estimation of θ_1 and θ_2. However, these gains converge to $k_1 = -3$ and $k_2 = 1$ after some learning takes place. The ideal values of the gains can be calculated in case when θ_1 and θ_2 are known as

(continued)

Example 5.4 (continued)
$k_1 = \frac{a_m - \theta_1}{\theta_2} = \frac{(-2-1)}{1} = -3$ and $k_2 = \frac{b_m}{\theta_2} = \frac{1}{1} = 1$, but since θ_1 and θ_2 are unknown, they are estimated using $\hat{\theta}_1$ and $\hat{\theta}_2$. It is noted that in the indirect MRAC framework, achieving precise online parameters estimation and accurate model matching is not a trivial problem due to several reasons, such as

1. Parameter estimation equations are nonlinear even when the plant is linear. For instance, when estimating parameters such as θ_1, an adaptation law such as $\dot{\hat{\theta}}_1 = -\gamma e(t)x(t)$, when it is extended, will become as $\dot{\hat{\theta}}_1 = -\gamma(x(t) - x_m(t))x(t) = -\gamma(x^2(t) - x(t)x_m(t))$; thus, it is nonlinear in the state $x(t)$ since $x^2(t)$ exist.
2. Persistent excitation (PE) is missing in the indirect MRAC since $x(t)$ and $u(t)$ are not always persistent excited.
3. Parameter adaptation and control are coupled in indirect MRAC, and hence, parameter adaptation errors affect control performance.
4. Division by estimated parameters can lead to instability, particularly when these parameters are close to zero. For instance, to calculate the controller gains, one needs to use these equations $k_1 = \frac{a_m - \hat{a}(t)}{\hat{b}(t)}$ $k_2 = \frac{b_m}{\hat{b}(t)}$, which contain $\hat{b}(t)$ in the denominator.

Readers seeking a deeper understanding of MRAC are referred to references (Nguyen, 2018; Stepanyan and Krishnakumar, 2019; Astrom and Wittenmark, 2008) for detailed explanations and theoretical insights.

In the subsequent section, the concept of L_1 adaptive control is discussed, highlighting its capability to deliver improved transient performance while maintaining robustness.

5.8 L_1 Adaptive Control

The L_1 adaptive controller[1] is employed where the system requires an improved transient response, fast adaptation, and robustness. Evolved from MRAC, the L_1 adaptive control framework (Hovakimyan and Cao, 2010) incorporates adaptive gain logic and a low-pass filter in controller command computation. This design

[1] The simulation results have been generated by Dr. Amit Kumar Tripathi, a graduate student at the Indian Institute of Science, Bangalore, India. The authors express their sincere gratitude to him for his valuable contribution in providing the relevant results.

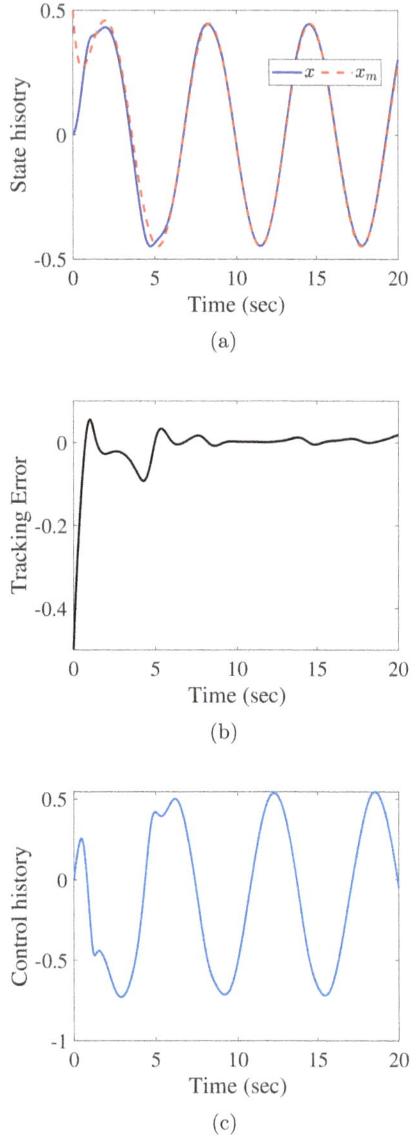

Fig. 5.9 (**a**) State tracking evolution. (**b**) Tracking error evolution. (**c**) Control history

effectively mitigates high-frequency signal components in the controller output, enhancing system performance. The L_1 adaptive control architecture is shown in Fig. 5.11, where the equations and variables are defined in subsequent paragraphs.

Consider the following class of systems as described in (Cao and Hovakimyan, 2008):

$$\dot{\boldsymbol{x}}(t) = \boldsymbol{A}\boldsymbol{x}(t) + \boldsymbol{b}\big(u(t) + \boldsymbol{\theta}^T \boldsymbol{x}(t)\big), \quad \boldsymbol{x}(\boldsymbol{0}) = \boldsymbol{x}_0 \tag{5.71}$$

Fig. 5.10 (**a**) θ_1 estimation. (**b**) θ_2 estimation. (**c**) Gain estimation—Example 5.4

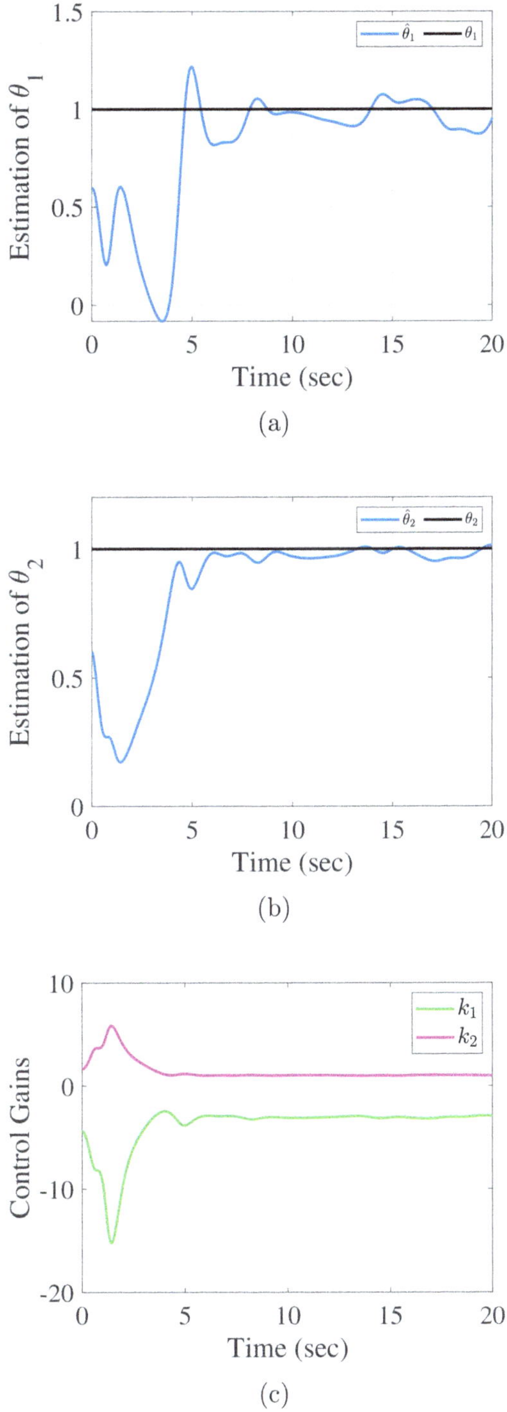

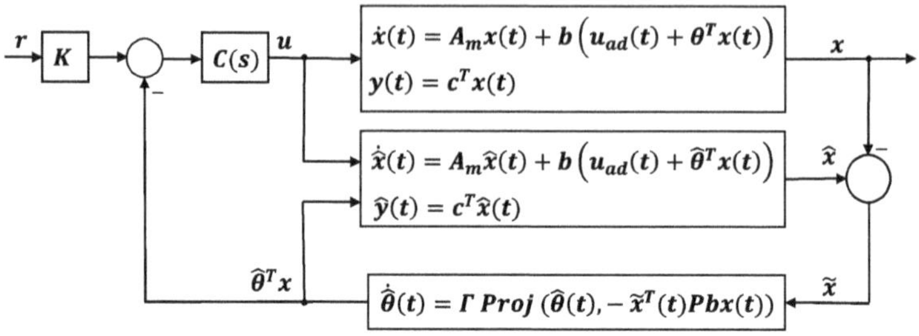

Fig. 5.11 Closed-loop L_1 adaptive control architecture

$$y(t) = \boldsymbol{c}^T \boldsymbol{x}(t) \tag{5.72}$$

where $\boldsymbol{x}(t) \in \Re^n$ is the system state vector and $u(t) \in \Re$ is the control input. The vectors $\boldsymbol{b} \in \Re^n$ and $\boldsymbol{c} \in \Re^n$ are known control input and output vectors, respectively. System matrix $\boldsymbol{A} \in \Re^{n\times n}$ is a known matrix with the condition that the system should be controllable. The parameter uncertainty $\boldsymbol{\theta} \in \Re^n$ is an unknown bounded parameter, and $y(t)$ is a regulated output. Note that the uncertainties selected here are matched uncertainties, chosen to demonstrate the design concept. Nevertheless, this adaptive technique is also effective in the presence of unmatched uncertainties.

Here, a two-component control structure is employed to adapt to the parameter uncertainty in order to stabilize the closed-loop system. The first component, $u_m(t)$, represents the nominal full-state feedback control, and the second component is the adaptive control term, as shown in (5.74) and (5.73), respectively. The controller is designed to ensure accurate tracking and robustness with parameter uncertainty compensation, which results in fast adaptation.

The controller is first decomposed as

$$u(t) = u_m(t) + u_{ad}(t) \tag{5.73}$$

with

$$u_m(t) = -\boldsymbol{k}_m \boldsymbol{x}(t) \tag{5.74}$$

where $\boldsymbol{k}_m \in \Re^{1\times n}$ is a full-state feedback gain vector. It is selected in such a way that the closed-loop system matrix $\boldsymbol{A}_m$, given in (5.75), becomes a Hurwitz matrix:

$$\boldsymbol{A}_m = \boldsymbol{A} - \boldsymbol{b}\boldsymbol{k}_m \tag{5.75}$$

It is well known that if all the eigenvalues of $\boldsymbol{A}_m$ remain in the left half of the complex plane, then the matrix $\boldsymbol{A}_m$ is called the Hurwitz matrix (Khalil, 2002) or stability matrix. Like the concept of dynamic inversion (see Sect. 3.1 of Chap. 3), the fundamental concept, $\dot{\boldsymbol{x}} = \boldsymbol{A}\boldsymbol{x}$, is asymptotically stable if $\boldsymbol{A}$ is a Hurwitz and is utilized while designing the L_1 adaptive controller.

In general, all practical systems are nonlinear in nature. However, since a linearized model around an operating point represents the system quite close to the nonlinear model, the linearized model is exploited for the control design.

Using (5.73), (5.74), and (5.75), in (5.71), the linearized state and output equation with a constant adaptive gain component is given by

$$\dot{\boldsymbol{x}}(t) = \boldsymbol{A}_m \boldsymbol{x}(t) + \boldsymbol{b}\big(u_{ad}(t) + \boldsymbol{\theta}^T \boldsymbol{x}(t)\big), \quad \boldsymbol{x}(0) = \boldsymbol{x}_0 \tag{5.76}$$

$$y(t) = \boldsymbol{c}^T \boldsymbol{x}(t) \tag{5.77}$$

where $\boldsymbol{A}_m$ is the Hurwitz matrix, $\boldsymbol{b}$ is the control input gain vector, and $\boldsymbol{\theta}$ is the parameter uncertainty in the current problem formulation. Next, a state predictor model is used to approximate the uncertainties, which is described as follows:

$$\dot{\hat{\boldsymbol{x}}}(t) = \boldsymbol{A}_m \hat{\boldsymbol{x}}(t) + \boldsymbol{b}\big(u_{ad}(t) + \hat{\boldsymbol{\theta}}^T \boldsymbol{x}(t)\big), \quad \hat{\boldsymbol{x}}(0) = \boldsymbol{x}_0 \tag{5.78}$$

$$\hat{y}(t) = \boldsymbol{c}^T \hat{\boldsymbol{x}}(t) \tag{5.79}$$

where $\hat{\boldsymbol{x}}(t) \in \Re^n$ is the predictor of the state $\boldsymbol{x}(t)$ and $\hat{\boldsymbol{\theta}}(t)$ is the estimate of unknown constant parameter uncertainty $\boldsymbol{\theta}$. Next, the following projection-type adaptive law is used to compute the estimate $\hat{\boldsymbol{\theta}}(t)$, which is given as

$$\dot{\hat{\boldsymbol{\theta}}}(t) = \Gamma \, \mathrm{Proj}\Big(\hat{\boldsymbol{\theta}}(t), -\tilde{\boldsymbol{x}}^T \boldsymbol{P} \boldsymbol{b}\, \boldsymbol{x}(t)\Big), \quad \hat{\boldsymbol{\theta}}(0) = \hat{\boldsymbol{\theta}}_0 \tag{5.80}$$

where $\tilde{\boldsymbol{x}}(t) = \hat{\boldsymbol{x}}(t) - \boldsymbol{x}(t)$, Γ is the gain in adaptation, and $\boldsymbol{P} = \boldsymbol{P}^T$ is a positive definite matrix. Note that $\mathrm{Proj}\Big(\hat{\boldsymbol{\theta}}(t),\ -\tilde{\boldsymbol{x}}^T \boldsymbol{P} \boldsymbol{b}\, \boldsymbol{x}(t)\Big)$ is a "projection operator" that ensures that the parameter estimates $\hat{\boldsymbol{\theta}}(t)$ remain bounded, while preserving the direction of the adaptation law as much as possible. The projection operator is computed as

$$\mathrm{Proj}(\hat{\boldsymbol{\theta}}, \boldsymbol{\varphi}) = \begin{cases} \boldsymbol{\varphi}, & \text{if } f(\hat{\theta}) < 0 \\ \boldsymbol{\varphi} - \frac{\nabla f}{|\nabla f|}\big(\frac{\nabla f}{|\nabla f|}, \boldsymbol{\varphi}\big) f(\theta), & \text{if } f(\hat{\theta}) \geq 0 \text{ and } \nabla f^T \boldsymbol{\varphi} > 0 \\ \boldsymbol{\varphi}, & \text{if } f(\hat{\theta}) \geq 0 \text{ and } \nabla f^T \boldsymbol{\varphi} \leq 0 \end{cases}$$

where $\hat{\boldsymbol{\theta}}(t)$ is the current parameter estimate vector at time t, $\theta_{\max}$ is the upper bound on the norm of the parameter vector, and $\boldsymbol{\varphi}(t) = -\tilde{\boldsymbol{x}}^T \boldsymbol{P} \boldsymbol{b}\, \boldsymbol{x}(t)$ is the adaptation direction. Here, $f(\hat{\theta})$ is defined as

$$f(\hat{\theta}) = \frac{(\epsilon_\theta + 1)\hat{\theta}^T \hat{\theta} - \theta_{max}^2}{\epsilon_\theta \theta_{max}^2}, \epsilon_\theta > 0, \; f(\hat{\theta}) \leq c, \;\; 0 \leq c \leq 1$$

where ϵ_θ is the projection tolerance bound. The detailed description of the projection operator can be found in Hovakimyan and Cao (2010), Kulkarni et al. (2010).

The value of $\boldsymbol{P}$ is obtained from the solution of Lyapunov equation $\boldsymbol{A}_m^T\boldsymbol{P} + \boldsymbol{P}\boldsymbol{A}_m = -\boldsymbol{Q}$, given in Slotine and Li (1991), where $\boldsymbol{Q}$ is a positive definite matrix. Such a selection of P leads to stable estimation. Note that since $-\tilde{\boldsymbol{x}}^T\boldsymbol{P}\boldsymbol{b}\boldsymbol{x}(t)$ is the adaptation direction, a properly chosen $\boldsymbol{P}$ results in faster convergence.

The nominal controller term in adaptive control can be designed using any standard state feedback control design technique. However, the nominal controller assumes perfect cancellation of uncertainties, which is practically not possible and hence leads to poor reference command tracking. Therefore, the adaptive control law is designed to approximate parameter uncertainties, which results in proper tracking performance.

Adaptive control: The adaptive component of control u_{ad} in the Laplace domain is defined as

$$u_{ad}(s) = -C(s)(\hat{\eta}(s) - k_g r(s)) \tag{5.81}$$

$$\hat{\eta}(s) \triangleq \mathcal{L}\{\hat{\boldsymbol{\theta}}^T\boldsymbol{x}(t)\} \tag{5.82}$$

where $r(s)$ and $\hat{\eta}(s)$ are Laplace transform ($\mathcal{L}$) of $r(t)$ and $\hat{\boldsymbol{\theta}}^T\boldsymbol{x}(t)$, respectively. The reference input gain k_g is defined as $k_g = -(\boldsymbol{c}^T\boldsymbol{A}_m^{-1}\boldsymbol{b})^{-1}$, and $C(s)$ represents a low-pass filter $C(s) = \frac{\omega_c}{s+\omega_c}$ with bounded-input-bounded-output stability. It is a strictly proper function with DC gain $C(0) = 1$ and cutoff frequency ω_c rad/s.

The low-pass filter is added to remove high-frequency components of uncertainty from the control command. The cut-off frequency range of the low-pass filter is defined by a parameter λ, which is calculated from L_1 Norm. The expression for calculating λ is given by

$$\lambda \triangleq \left[L\,\|G(s)\|_{L_1}\right] < 1 \tag{5.83}$$

where $\|G(s)\|_{L_1}$ is the L_1 norm of $G(s)$ and L is the maximum bound on parameter uncertainty, defined as $L \triangleq \max\|\theta\|_1$. The term $G(s)$ is defined as $G(s) \triangleq H(s)(1 - C(s))$, where $H(s) \triangleq (sI - A_m)^{-1}b$.

Note that in (5.83), the value of L is fixed, and hence, to satisfy the inequality, the $\|G(s)\|_{L_1}$ value should be selected appropriately. Since $G(s)$ depends on the low-pass filter $C(s)$, the bandwidth can be adjusted by changing the cut-off frequency to satisfy the inequality.

Example 5.5 To illustrate the concept of L_1 adaptive control for the class of system described in (5.71), the short-period mode dynamics of an aircraft

(continued)

Example 5.5 (continued)
is considered (Tripathi and Patel, 2019). This is chosen because the short-period mode is a high-frequency mode in aircraft dynamics. The states of the system in the short-period mode are angle of attack α and pitch rate q, $x = [\alpha, q]$, and control input is elevator deflection δ_e, $u = \delta_e$. The linear system matrix of short-period mode dynamics is defined as $\boldsymbol{A} = [-0.1, 1; -4, -3]$, and the input and output vectors are given as $\boldsymbol{b} = [0.7, -0.3]^T$ and $\boldsymbol{c} = [0, 1]^T$, respectively. The parameter uncertainty vector is represented as $\boldsymbol{\theta} = [1.4, -1.45]^T$. The initial state vector is selected as $\boldsymbol{x}(0) = [0, 0]^T$. The reference pitch rate command 2°/s is set to be tracked under parameter uncertainty. The adaptive gain used for the simulation is $\Gamma = 10000$.

In this example, the max limits on parameter uncertainty are taken as $[-2, 2]^T$. Therefore, L can be computed as $L = 4$. The L_1 norm computation for short-period mode is carried out in Fig. 5.12, where the low-pass filter used for the simulation is $C(s) = \frac{(3w_c{}^2 s + w_c{}^3)}{(s + w_c{}^3)}$. It can be noticed from Fig. 5.12 that the allowed range of filter bandwidth for which L_1 norm is below 1 is at cut-off frequency $w_c \geq 4.65\ rad/s$. It can be shown further that by using proper tuning of the low-pass filter cut-off frequency, the tracking performance of the controller can be improved significantly.

The command tracking result of reference command tracking for reduced order mode is shown in Fig. 5.13, where the desired tracking is achieved with the application of L_1 adaptive control. The zoomed view of the pitch rate response is shown in Fig. 5.14, where it can be noticed that minor oscillations are seen in the response in the vicinity of the reference pitch rate. The L_1 adaptive control elevator deflection command is shown in Fig. 5.15.

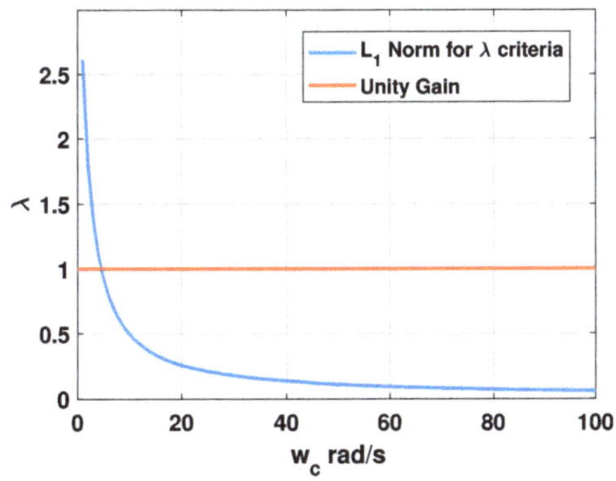

Fig. 5.12 L_1 Norm computation for a short-period mode

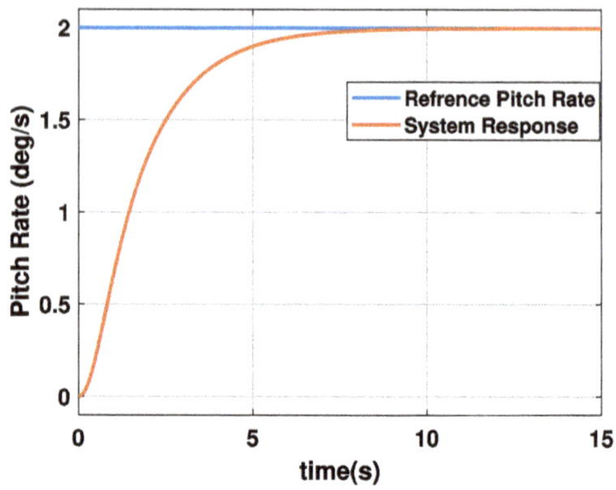

Fig. 5.13 Reference pitch rate q tracking in short-period mode

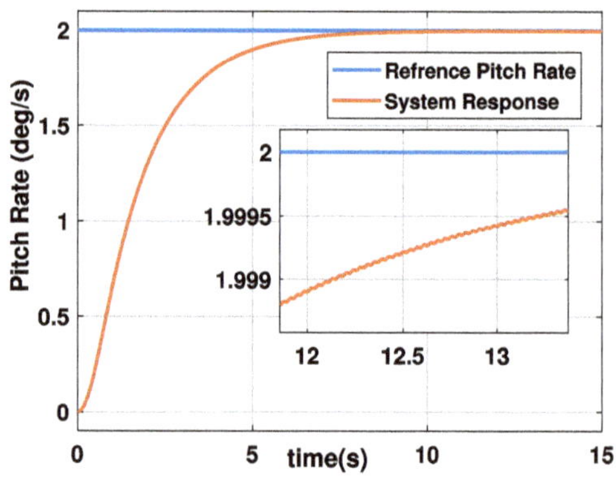

Fig. 5.14 Reference pitch rate q tracking in short-period mode (zoomed view)

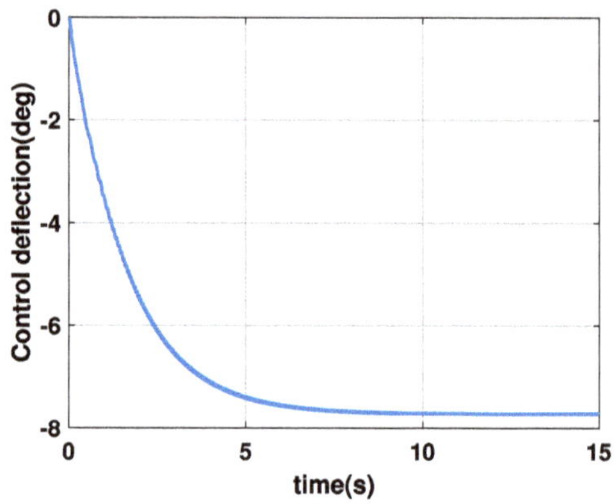

Fig. 5.15 L_1 adaptive control command

5.9 Summary

This chapter briefly discusses adaptive control philosophy by introducing the different uncertainty types and their potential estimation approach. Also, it provides insight into direct and indirect adaptive control techniques while focusing on model reference adaptive control and L_1 adaptive control as the main techniques for direct and indirect adaptive control. While a brief discussion on different types of adaptive control techniques is presented in this chapter, readers seeking a more thorough understanding can see additional sources (Narendra and Annaswamy, 2012; Slotine and Li, 1991; Lavretsky and Wise, 2012; Nguyen, 2018; Landau, 1979).

It is noteworthy that all adaptive control designs presented in this chapter are applicable primarily to specific classes of linear and nonlinear systems, such as control-affine or single-input-single-output (SISO) systems. This limitation underscores the necessity for a more general, universally applicable adaptive control design methodology that can be used for a broader range of nonlinear systems. In light of this, the subsequent chapter presents a detailed development of a neuro-adaptive model-following control strategy, which is applicable to the generic class of nonlinear systems.

5.10 Exercise Problems

1. Determine the type of uncertainty in the following systems:

   (a) $\begin{bmatrix} \dot{x}_1 \\ \dot{x}_2 \end{bmatrix} = \begin{bmatrix} x_2 \\ a x_1 x_2 + u \end{bmatrix}$

   (b) $\begin{bmatrix} \dot{x}_1 \\ \dot{x}_2 \end{bmatrix} = \begin{bmatrix} x_2 \\ x_1 x_2 + f(x_1, x_2) + u \end{bmatrix}$

   (c) $\begin{bmatrix} \dot{x}_1 \\ \dot{x}_2 \end{bmatrix} = \begin{bmatrix} x_2 \\ a x_2 + x_1 + f(x_1, x_2) + u \end{bmatrix}$

2. For the following system, design a direct MRAC controller.

$$\dot{x}_1 = x_2$$
$$\dot{x}_2 = a x_1 x_2 + u$$

   where $a > 0$ is an unknown parameter. Its true value is 1. The reference model is given as $\dot{x}_m = -x_m + r(t)$; use firstly $r(t) = 1$ and secondly $r(t) = \sin(t)$.

3. For the following system, design an indirect MRAC controller.

$$\dot{x}_1 = x_2$$
$$\dot{x}_2 = a x_1^2 x_2 + u$$

where $a > 0$ is an unknown parameter. Its true value is 1. The reference model is given as $\dot{x}_m = -x_m + r(t)$; use $r(t) = 1$ and $r(t) = \sin(t)$. Plot the results and comment on the performance.

4. For the following system, design an indirect MRAC controller.

$$\dot{x}_1 = x_2$$

$$\dot{x}_2 = x_1^2 x_2 + u + d(x)$$

where $d(x)$ is an unknown uncertainty defined as $d(x) = 0.1\sin(x_1(t))$. The reference model is given as $\dot{x}_m = -x_m + r(t)$; use $r(t) = 1$ and $r(t) = \sin(t)$.

5. Design an L_1 adaptive controller for the following DC motor system:

$$\dot{\boldsymbol{x}} = \begin{bmatrix} 0 & 1 \\ 0 & -\frac{b}{j} \end{bmatrix} \boldsymbol{x} + \begin{bmatrix} 0 \\ \frac{b}{j} \end{bmatrix} u + \begin{bmatrix} 0 \\ \frac{1}{j} \end{bmatrix} d(t)$$

$$y = x_1$$

where b, j, and k are unknown constants with true values of $b = 0.1$ (N.m.s/rad), $j = 0.01$ (kg.m^2), and $k = 0.01$ (N.m/V) and $d(t)$ (N.m) is an unknown external disturbance with true value given as $d(t) = 0.05\sin(2t)$. The goal of the control is to track the rotor position according to the following desired dynamics:

$$\dot{\boldsymbol{x}}_m = \begin{bmatrix} 0 & 1 \\ -\omega_n^2 & -2\zeta\omega_n \end{bmatrix} \boldsymbol{x}_m + \begin{bmatrix} 0 \\ \omega_n^2 \end{bmatrix} r(t)$$

where $\omega_n = 5$ (rad/sec) is the natural frequency and $\zeta = 0.7$ is the damping ratio and $r(t) = \sin(t)$ (rad).

References

K.J. Astrom, B. Wittenmark, *Adaptive Control* (Dover Publications, Garden City, 2008)

C. Cao, N. Hovakimyan, Design and analysis of a novel L_1 adaptive control architecture with guaranteed transient performance. IEEE Trans. Autom. Control **53**(2), 586–591 (2008)

S. Ge, C. Hang, T. Lee, T. Zhang, *Stable Adaptive Neural Network Control*, ser. The International Series on Asian Studies in Computer and Information Science (Springer US, New York, 2013)

S.S. Haykin, *engNeural Networks and Learning Machines*, 3rd edn. (Prentice Hall, New York, 2009)

K. Hornik, M. Stinchcombe, H. White, Multilayer feedforward networks are universal approximators. Neural Netw. **2**(5), 359–366 (1989)

N. Hovakimyan, C. Cao, *L_1 Adaptive Control Theory: Guaranteed Robustness with Fast Adaptation*, vol. 21 (SIAM, Philadelphia, 2010)

H.K. Khalil, *Nonlinear Systems* (Prentice-Hall, Hoboken, 2002)

V.V. Kulkarni, A. Paranjape, K.R. Ghusinga, N. Hovakimyan, Synthesis of robust tunable oscillators using mitogen activated protein kinase cascades. Syst. Synthetic Biol. **4**(4), 331–341 (2010)

M. Krstic, I. Kanellakopoulos, V. Petar, *Nonlinear and Adaptive Control Design* (Wiley, New York, 1995)

Y.D. Landau, *Adaptive Control: The Model Reference Approach* (Marcel Dekker, Inc., New York City, 1979)

I.D. Landau, R. Lozano, M. M'Saad, A. Karimi, *Introduction to Adaptive Control* (Springer, London, 2011), pp. 1–33

E. Lavretsky, K. Wise, *Robust and Adaptive Control: With Aerospace Applications*, ser. Advanced Textbooks in Control and Signal Processing (Springer, London, 2012)

K.S. Narendra, *Applications of Adaptive Control* (Elsevier, Amsterdam, 2012)

K. Narendra, A. Annaswamy, *Stable Adaptive Systems*, ser. Dover Books on Electrical Engineering (Dover Publications, Garden City, 2012)

N. Nguyen, *Model-Reference Adaptive Control: A Primer*, ser. Advanced Textbooks in Control and Signal Processing (Springer International Publishing, Cham, 2018)

J.-J.E. Slotine, W. Li, et al., *Applied Nonlinear Control*, vol. 199, no. 1 (Prentice Hall, Englewood Cliffs, 1991)

V. Stepanyan, K.S. Krishnakumar, Input constrained M-MRAC for multirotors operating in an urban environment, in *AIAA Scitech 2019 Forum*, no. January (American Institute of Aeronautics and Astronautics, Reston, 2019), pp. 1–22

A.K. Tripathi, V.V. Patel, Autonomous landing for small uavs using L_1 adaptive control tracking applications, in *2019 Fifth Indian Control Conference (ICC)* (IEEE, Piscataway, 2019), pp. 472–477

A. Tsantekidis, N. Passalis, A. Tefas, Chapter 5 - recurrent neural networks, in *Deep Learning for Robot Perception and Cognition*, ed. by A. Iosifidis, A. Tefas (Academic Press, San Diego, 2022), pp. 101–115

Chapter 6
Neuro-adaptive Control Design

The neuro-adaptive controller is a model-following adaptive control design framework that falls under the category of indirect adaptive control. The approach involves designing an additional control input online, which, when combined with a nominal controller, ensures overall system stability while enhancing the performance of the plant. A key advantage of the neuro-adaptive technique lies in its universality; i.e., this method can be seamlessly integrated not only with dynamic inversion but also with any of the control design strategies discussed in the preceding Chap. 3. The neuro-adaptive controller serves as a powerful and versatile tool, applicable to a broad class of nonlinear systems, and is highly beneficial for addressing diverse control challenges across various engineering applications.

This chapter provides a comprehensive exposition of neuro-adaptive control design methodologies. It begins with a basic design procedure for a simple nonlinear system with structured model uncertainties. Following this, the concept of gradient learning is introduced, which enables not only the estimation of uncertainties but also their rate of change. Incorporating this gradient information leads to smoother transient responses and improved neural network approximation performance. Subsequently, the chapter presents an extended neuro-adaptive control approach that addresses both structured and unstructured model uncertainties for Euler-Lagrange systems. In the latter sections, the neuro-adaptive control technique is generalized for application to a wide class of nonlinear systems. It is worth emphasizing here that the neuro-adaptive approach discussed here constitutes a generic and systematic control design framework. Accordingly, various design variants are introduced throughout the chapter, allowing the readers to select appropriate strategies based on specific application requirements.

K. Sachan et al., *Intelligent Adaptive Control with System Constraints*,
https://doi.org/10.1007/978-981-95-8220-4_6

6.1 Overview of Neuro-adaptive Control Design

A neuro-adaptive controller is synthesized to ensure that the actual system follows the desired reference model, assuming that a suitable controller has already been designed for the reference model (therefore also referred to as model-following adaptive control design). The desired controller can be developed using standard nonlinear control techniques discussed in the previous Chap. 3 for a nominal model of a system. However, since the desired system model may not accurately represent the actual system (due to modeling uncertainties), the desired controller may not perform as expected in reality. To address this problem, the neuro-adaptive control design approach follows two steps, namely, (i) capturing the unknown function resulting from the model uncertainties using neural networks (which exploits the universal function approximation property of neural networks) and (ii) designing a controller that drives the state of the actual plant to track the state trajectories of the reference model.

It can be emphasized that, in many practical applications, the basis functions of the neural network in neuro-adaptive control design are derived directly from the system dynamics. Consequently, the weights of the neural network are updated in accordance with the underlying system behavior. Since the physics of the system dynamics is incorporated into the learning process, this adaptation mechanism is consistent with the philosophy of physics-informed neural network (PINN), wherein the neural network weights are updated in accordance with the governing system dynamics. Therefore, when the basis functions are extracted from the system dynamics, the neuro-adaptive control design can also be regarded as a PINN-based neuro-adaptive control design. Nonetheless, it is important to note that the neuro-adaptive control framework is inherently more general and can also address completely unknown forms of uncertainty by employing generic basis functions.

Before delving into the detailed mathematics of the design, it is essential to first clarify the basic terminology used in the design for better clarity. The design involves three types of system models: (i) *Actual system*: This represents the real plant model, which includes unknown uncertainties. (ii) *Desired closed-loop system*: This refers to the "nominal mathematical model" of the system (which contains only known terms in the system dynamics with known parameters) and also a control design based on this desired nominal model. (iii) *Approximated system*: This system is constructed to design a weight update rule to update the weights of neural networks (Stone-Weierstrass theorem Haykin 2009), which is utilized to approximate the unknown uncertainties of the actual system.

The overall design philosophy is illustrated in Fig. 6.1, where the reference signal represents the target that the actual system should track. The desired controller is responsible for achieving the objective of the nominal system. The neural network, using the universal approximation capability, approximates the function resulting from model uncertainties, enabling the approximated system to become closer to the actual system. A controller is then designed to minimize the error between the approximated system and the desired system, ultimately reducing the error between

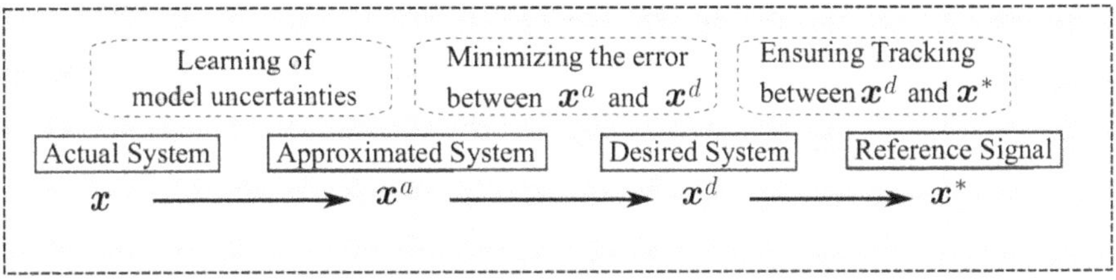

Fig. 6.1 Philosophy of neuro-adaptive control design

the two to zero. To illustrate this concept, consider the nominal nonlinear system as

$$\dot{x} = 2\sin x + u \tag{6.1}$$

and the actual nonlinear system as

$$\dot{x} = 2.1\sin x + u = 2\sin x + u + 0.1\sin x \tag{6.2}$$

where x is the state and u is the nominal control input. Here, the term $0.1\sin x$ represents the modeling uncertainties, and the approximated system is designed to approximate these modeling uncertainties as a function using a neural network.

In certain applications, if only the parameters are uncertain (e.g., the scalar coefficient 0.1 is unknown), but the functional form of the uncertainty is known (e.g., $\sin x$ is known), then the associated uncertainty is referred to as *structured uncertainty*. In such cases, the basis function of a neural network can be chosen as $\sin x$, allowing the unknown parameter to be learned without approximation error. This approach is essentially a PINN-based neuro-adaptive control design. Conversely, if the functional form of the uncertainty is also unknown (e.g., the entire term $0.1\sin x$ is unknown), the uncertainty is classified as *unstructured uncertainty*. In this scenario, generic basis functions, such as Gaussian basis functions, are typically employed in the neural network to approximate the uncertainty. More information about model uncertainties is provided in Chap. 5.

In practical scenarios, most applications involve unstructured uncertainties, as accurately modeling the entire plant is often challenging. However, if the plant is well understood and the modeling is highly accurate, it is reasonable to assume structured uncertainties for the purpose of control design. Control design and analysis in the presence of unstructured uncertainties is significantly more complex compared to the case of structured uncertainties. Therefore, this chapter begins with the development of a neuro-adaptive control design for systems with structured uncertainties using a simple nonlinear model to establish a fundamental understanding. Subsequently, the design methodology is extended to handle more complex nonlinear systems with unstructured uncertainties.

6.2 Basic Controller Design Procedure

To understand the design process, consider a simple nonlinear system with relative degree one represented as

$$\dot{\boldsymbol{x}} = \boldsymbol{f}(\boldsymbol{x}) + \boldsymbol{G}(\boldsymbol{x})\boldsymbol{u} + \boldsymbol{d}(\boldsymbol{x}) \tag{6.3}$$

where $\boldsymbol{x} \in \Re^n$ and $\boldsymbol{u} \in \Re^n$ are the states and control input, respectively, $\boldsymbol{G}(\boldsymbol{x}) \in \Re^{n\times n}$ is the control effectiveness matrix, and $\boldsymbol{d}(\boldsymbol{x})$ represents the disturbance function resulting from bounded structured model uncertainties of the actual system. The vector $\boldsymbol{f}(\boldsymbol{x}) \in \Re^n$ is assumed to be known, and the matrix $\boldsymbol{G}(\boldsymbol{x}) \in \Re^{n\times n}$ is assumed to be known and non-singular.

The primary objective is to design an adaptive state feedback controller for the system described in (6.3) to ensure accurate tracking of a given reference signal, denoted as $\boldsymbol{x}^*(t)$.

The control design strategy is illustrated in Fig. 6.1. It should be noted that the desired controller design step can be omitted from the overall design, and a controller can be directly designed to minimize the error between the approximated system and the reference signal. However, for applications requiring smooth responses, it is often preferable to include the desired controller design step. In this approach, an approximated system is first constructed to learn the model uncertainties using neural networks. This step ensures that the states of the actual system track the states of the approximated system with a desired accuracy, i.e., $\boldsymbol{x} \to \boldsymbol{x}^a$. Subsequently, an asymptotically stable tracking controller is formulated to minimize the error between the states of the approximated system and the reference signal, driving it to zero, i.e., $\boldsymbol{x}^a \to \boldsymbol{x}^*$. It should be noted that the desired control design step is not included in the control design.

The approximated dynamical system is constructed as

$$\dot{\boldsymbol{x}}^a = \boldsymbol{f}(\boldsymbol{x}) + \boldsymbol{G}(\boldsymbol{x})\boldsymbol{u} + \hat{\boldsymbol{d}}(\boldsymbol{x}) + \boldsymbol{K}\boldsymbol{e} \tag{6.4}$$

with the same initial condition as the actual system, where $\boldsymbol{e} = (\boldsymbol{x} - \boldsymbol{x}^a)$, $\hat{\boldsymbol{d}}(\boldsymbol{x})$ is the approximation of the unknown disturbance function $\boldsymbol{d}(\boldsymbol{x})$, and $\boldsymbol{K}$ is a positive definite matrix (usually selected as a diagonal matrix with positive elements).

After outlining the dynamics of both the actual and approximated systems, the subsequent subsection introduces the first step of the adaptive control design process, i.e., $\boldsymbol{x} \to \boldsymbol{x}^a$.

6.2.1 Approximation of Structured Model Uncertainties

This step involves approximating the model uncertainties of the system using a neural network. To develop a weight update rule for adjusting the neural network weights, the error between the actual and approximated systems is defined as follows:

$$e = x - x^a \tag{6.5}$$

The time differentiation of (6.5) gives

$$\dot{e} = \dot{x} - \dot{x}^a \tag{6.6}$$

Substituting (6.3) and (6.4) into (6.6) leads to

$$\dot{e} = d(x) - \hat{d}(x) - Ke \tag{6.7}$$

Since the matrix K is usually selected as $\mathbf{diag}(k_i) > 0, \forall i$, to ensure that it is a positive definite diagonal matrix, Eq. (6.7) can be analyzed in terms of its ith component. Here, i^{th} components of e, $d(x)$, and $\hat{d}(x)$ are represented by e_i, $d_i(x)$, and $\hat{d}_i(x)$, respectively. Therefore, the ith component of the error dynamics in Eq. (6.7) is given by

$$\dot{e}_i = d_i(x) - \hat{d}_i(x) - k_i e_i \tag{6.8}$$

where $d_i(x),\ i = 1, \ldots, n$ is the ith component of $d(x)$ which is the uncertainty in the ith state equation. Since each element of $d(x)$ is represented by a separate neural network, utilizing linear-in-weight architecture of neural networks, the structured uncertainties $d_i(x)$ (assuming approximation error $\epsilon_i = 0$) and the approximation of uncertainties $\hat{d}_i(x)$ can be expressed as the outputs of the corresponding neural networks, i.e.

$$d_i(x) = w_i^T \phi(x) \tag{6.9}$$

$$\hat{d}_i(x) = \hat{w}_i^T \phi(x) \tag{6.10}$$

where w_i and $\hat{w}_i$ are the optimal and approximated weights and $\phi(x)$ is the known basis function vector. Note that w_i is a vector with constant elements.

By substituting (6.9) and (6.10) into (6.8), the following can be obtained:

$$\dot{e}_i = w_i^T \phi(x) - \hat{w}_i^T \phi(x) - k_i e_i \tag{6.11}$$

$$\dot{e}_i = \left[w_i^T - \hat{w}_i^T \right] \phi(x) - k_i e_i \tag{6.12}$$

Defining $\tilde{w}_i \triangleq w_i - \hat{w}_i$, where $\tilde{w}_i$ is weight update error, the following is obtained:

$$\dot{e}_i = \tilde{w}_i^T \phi(x) - k_i e_i \tag{6.13}$$

For ensuring stability of the closed-loop error dynamics described in Eq. (6.13), the following quadratic Lyapunov function is selected:

$$V_i = \frac{\beta_i e_i^2}{2} + \frac{\tilde{\boldsymbol{w}}_i^T \tilde{\boldsymbol{w}}_i}{2\gamma_i} \tag{6.14}$$

where β_i is a positive constant and $\gamma_i > 0$ is known as the learning rate constant.

The time derivative of the Lyapunov function in (6.14) is

$$\dot{V}_i = \beta_i e_i \dot{e}_i + \frac{\tilde{\boldsymbol{w}}_i^T \dot{\tilde{\boldsymbol{w}}}_i}{\gamma_i} \tag{6.15}$$

Using error dynamics (6.13), the time derivative of the Lyapunov function in (6.15) is modified as

$$\begin{aligned} \dot{V}_i &= \beta_i e_i \left(\tilde{\boldsymbol{w}}_i^T \boldsymbol{\phi}(\boldsymbol{x}) - k_i e_i \right) - \frac{\tilde{\boldsymbol{w}}_i^T \dot{\hat{\boldsymbol{w}}}_i}{\gamma_i} \\ &= \tilde{\boldsymbol{w}}_i^T \left[\beta_i e_i \boldsymbol{\phi}(\boldsymbol{x}) - \frac{\dot{\hat{\boldsymbol{w}}}_i}{\gamma_i} \right] - k_i \beta_i e_i^2 \end{aligned} \tag{6.16}$$

Select the weight update rule as

$$\dot{\hat{\boldsymbol{w}}}_i = \gamma_i \beta_i e_i \boldsymbol{\phi}(\boldsymbol{x}) \tag{6.17}$$

The resultant $\dot{V}_i$ expression in (6.16) can be written as

$$\begin{aligned} \dot{V}_i &= -k_i \beta_i e_i^2 \\ &\leq 0 \end{aligned} \tag{6.18}$$

where $\dot{V}_i$ is negative semi-definite. This indicates that the closed-loop system is stable but not necessarily asymptotically stable. To demonstrate the asymptotic stability of the system, the second derivative of the Lyapunov function with respect to time is computed as

$$\begin{aligned} \ddot{V}_i &= -2k_i \beta_i e_i \dot{e}_i \\ &= -2k_i \beta_i e_i \left(\tilde{\boldsymbol{w}}_i^T \boldsymbol{\phi}(\boldsymbol{x}) - k_i e_i \right) \end{aligned} \tag{6.19}$$

Since e_i, $\tilde{\boldsymbol{w}}_i$, and $\boldsymbol{\phi}(\boldsymbol{x})$ are bounded, $\ddot{V}_i$ is bounded. Therefore, using the Barbalat lemma, $\dot{V}_i \rightarrow 0$. Hence, $e_i \rightarrow 0$. Thus, without compromising the closed-loop stability of the system, the neural network will learn the model uncertainties using this learning rule.

6.2.2 Tracking Control Design with Improved Model

After capturing model uncertainties, the next step is to drive the error between the approximated and reference signal to zero. To achieve $\boldsymbol{x}^a \rightarrow \boldsymbol{x}^*$, consider the following closed-loop nonlinear stable error dynamics:

$$\dot{\boldsymbol{e}}^a + \boldsymbol{K}^a \boldsymbol{e}^a = 0 \tag{6.20}$$

where $\boldsymbol{e}^a$ is the error, defined as $\boldsymbol{e}^a = \boldsymbol{x}^a - \boldsymbol{x}^*$, and $\boldsymbol{K}^a$ is a diagonal positive definite gain matrix.

By substituting $\boldsymbol{e}^a = \boldsymbol{x}^a - \boldsymbol{x}^*$, Eq. (6.20) is modified as

$$\left(\dot{\boldsymbol{x}}^a - \dot{\boldsymbol{x}}^*\right) + \boldsymbol{K}^a(\boldsymbol{x}^a - \boldsymbol{x}^*) = \boldsymbol{0} \tag{6.21}$$

which can be further modified by substituting $\dot{\boldsymbol{x}}^a$ from (6.4)

$$\boldsymbol{f}(\boldsymbol{x}) + \boldsymbol{G}(\boldsymbol{x})\boldsymbol{u} + \hat{\boldsymbol{d}}(\boldsymbol{x}) + \boldsymbol{K}(\boldsymbol{x} - \boldsymbol{x}^a) - \dot{\boldsymbol{x}}^* + \boldsymbol{K}^a(\boldsymbol{x}^a - \boldsymbol{x}^*) = 0 \tag{6.22}$$

where $\hat{\boldsymbol{d}}(\boldsymbol{x})$ is already obtained as $\hat{d}_i(\boldsymbol{x}) = \hat{\boldsymbol{w}}_i^T \boldsymbol{\phi}(\boldsymbol{x})$ and updated using the learning rule constructed in (6.17). Thus, the adaptive control $\boldsymbol{u}$ can be extracted as

$$\boldsymbol{u} = -\left[\boldsymbol{G}(\boldsymbol{x})\right]^{-1}\left(\boldsymbol{f}(\boldsymbol{x}) + \hat{\boldsymbol{d}}(\boldsymbol{x}) + \boldsymbol{K}(\boldsymbol{x} - \boldsymbol{x}^a) + \boldsymbol{K}^a(\boldsymbol{x}^a - \boldsymbol{x}^*) - \dot{\boldsymbol{x}}^*\right) \tag{6.23}$$

This controller ensures that $\boldsymbol{x} \rightarrow \boldsymbol{x}^a \rightarrow \boldsymbol{x}^*$ as $t \rightarrow \infty$.

Example 6.1 Design a neuro-adaptive controller for the system given by

$$\dot{x} = (x + x^2) + (1 + x^2)u + d(x)$$

where x and u are the state and control input for the system, respectively, and $d(x)$ is the unknown function present in the system dynamics. The objective of the controller is to stabilize the plant.

The desired known system dynamics extracted from the actual system is defined by

$$\dot{x}_d = (x_d + x_d^2) + (1 + x_d^2)u_d$$

The stabilizing desired controller for the desired system can be designed using dynamic inversion control design. The desired control expression is given as

$$u_d = -\frac{(2x_d + x_d^2)}{(1 + x_d^2)}$$

(continued)

Example 6.1 (continued)
Following the adaptive control design steps, the following adaptive control expression is obtained:

$$u = -\frac{(x_d + x_d^2) + (1 + x_d^2)u_d - k(x_a - x_d) - (x + x^2) - \hat{d}(x) - k_a(x - x_a)}{(1 + x^2)}$$

where the term $\hat{d}(x)$ is calculated as $\hat{d}(x) = \hat{\boldsymbol{w}}^T \boldsymbol{\phi}(x)$ and weights are updated from the learning rule $\dot{\hat{w}} = \gamma \beta e_a \boldsymbol{\phi}(x)$.

To simulate the controller response, the unknown function $d(x)$ is assumed as $d(x) = \sin(\frac{\pi x}{2})$. The controller gains k, k_a, γ, and β are selected as $k = 2.5$, $k_a = 1$, $\gamma = 30$, and $\beta = 1$, respectively. The basis functions to approximate the unknown function are chosen as $\boldsymbol{\phi}(x) = \left[\frac{x}{x_0}, \frac{x^2}{x_0^2}, \frac{x^3}{x_0^3}\right]^T$ with $x_0 = -1.5$ as an initial condition.

The state and control trajectories of the system are illustrated in Fig. 6.2a and b. The results indicate the performance of the desired controller for both the desired and actual systems. While the desired controller performs well for the desired system, its performance degrades for the actual system due to model uncertainties. With the application of adaptive control, the desired tracking is achieved for the actual system. The approximation of the unknown functions and the evolution of weight updates are shown in Fig. 6.3a and b, which highlight the effectiveness of the constructed weight update rule.

6.3 Gradient Learning-Based Control Design

Gradient learning-based neuro-adaptive control design is a novel approach that aims to achieve faster and smoother adaptation of modeling uncertainties. The underlying design philosophy of this controller is similar to the approach outlined in the previous section. However, the key difference lies in the approximation of modeling uncertainties. In this method, the neural network is designed to learn not only the modeling uncertainties themselves but also their gradients. This learning mechanism enables fast and smooth approximation of modeling uncertainties and, ultimately, leads to better overall control performance.

To understand the control design approach, consider the following nonlinear system represented as

$$\dot{\boldsymbol{x}} = \boldsymbol{f}(\boldsymbol{x}) + \boldsymbol{G}(\boldsymbol{x})\boldsymbol{u} + \boldsymbol{d}(\boldsymbol{x}) \tag{6.24}$$

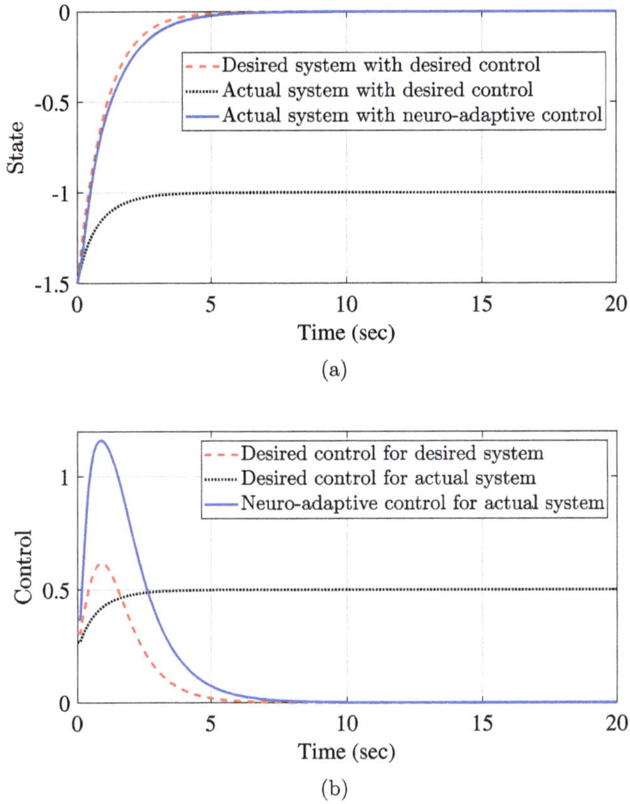

Fig. 6.2 (**a**) State histories. (**b**) Control histories

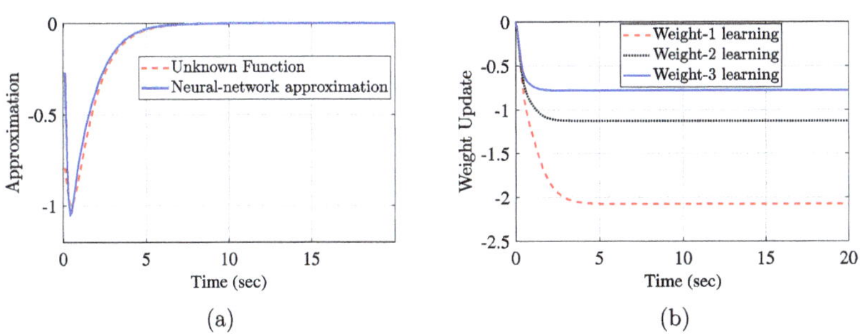

Fig. 6.3 (**a**) Unknown function and its approximation. (**b**) Weight update histories

where $\boldsymbol{x} \in \Re^n$ and $\boldsymbol{u} \in \Re^n$ are the states and control input, respectively, leading to the relative degree one system. The term $\boldsymbol{G}(\boldsymbol{x}) \in \Re^{n\times n}$ represents the control effectiveness matrix, and $\boldsymbol{d}(\boldsymbol{x})$ represents bounded unstructured model uncertainties of the actual system. The vector $\boldsymbol{f}(\boldsymbol{x}) \in \Re^n$ and matrix $\boldsymbol{G}(\boldsymbol{x}) \in \Re^{n\times n}$ are assumed to be known. Moreover, matrix $\boldsymbol{G}(\boldsymbol{x})$ is assumed to be invertible.

The control design strategy is similar to that discussed in the previous section. Hence, following the same design philosophy, the approximated system is constructed as

$$\dot{\boldsymbol{x}}^a = \boldsymbol{f}(\boldsymbol{x}) + \boldsymbol{G}(\boldsymbol{x})\boldsymbol{u} + \hat{\boldsymbol{d}}(\boldsymbol{x}) + \boldsymbol{K}\boldsymbol{e} \tag{6.25}$$

with the same initial condition as the actual system, where $\boldsymbol{e} = (\boldsymbol{x} - \boldsymbol{x}^a)$, $\hat{\boldsymbol{d}}(\boldsymbol{x})$ is the approximation of $\boldsymbol{d}(\boldsymbol{x})$, and $\boldsymbol{K}$ is a positive definite diagonal matrix.

To design the controller, the first step is to approximate modeling uncertainties, i.e., $\boldsymbol{x} \to \boldsymbol{x}^a$. Therefore, in the next subsection, this step is explained in detail, along with the discussion on gradient learning.

6.3.1 Approximation with Gradient Learning

This step involves approximating the model uncertainties of the system using a neural network. In the previous section, the weight update rule was a direct function of the basis function. This update rule can be improved by learning the gradient information of the uncertainty. This refinement enables the model to achieve faster and smoother approximation.

To develop the new weight update rule, the error between the actual and approximated systems is defined as follows:

$$\boldsymbol{e} = \boldsymbol{x} - \boldsymbol{x}^a \tag{6.26}$$

The time differentiation of (6.26) gives

$$\dot{\boldsymbol{e}} = \dot{\boldsymbol{x}} - \dot{\boldsymbol{x}}^a \tag{6.27}$$

Substituting (6.24) and (6.25) into (6.27) leads to

$$\dot{\boldsymbol{e}} = \boldsymbol{d}(\boldsymbol{x}) - \hat{\boldsymbol{d}}(\boldsymbol{x}) - \boldsymbol{K}\boldsymbol{e} \tag{6.28}$$

where the matrix $\boldsymbol{K} = \mathrm{diag}(k_i), \forall i$ is a positive definite diagonal matrix. The ith component of the error dynamics in Eq. (6.28) can be expressed as follows:

$$\dot{e}_i = d_i(\boldsymbol{x}) - \hat{d}_i(\boldsymbol{x}) - k_i e_i \tag{6.29}$$

The unstructured modeling uncertainties $d_i(\boldsymbol{x})$ and the approximation of uncertainties $\hat{d}_i(\boldsymbol{x})$ are defined as

$$d_i(\boldsymbol{x}) = \boldsymbol{w}_i^T \boldsymbol{\phi}(\boldsymbol{x}) + \epsilon_i \tag{6.30}$$

$$\hat{d}_i(\boldsymbol{x}) = \hat{\boldsymbol{w}}_i^T \boldsymbol{\phi}(\boldsymbol{x}) \tag{6.31}$$

where $\epsilon_i > 0$ is the approximation error; $\boldsymbol{w}_i$ and $\hat{\boldsymbol{w}}_i$ are the optimal and approximated weights, respectively; and $\boldsymbol{\phi}(\boldsymbol{x})$ is a known basis function vector. Note that $\boldsymbol{w}_i$ is a vector with constant elements.

By substituting (6.30) and (6.31) into (6.29), the following can be obtained:

$$\dot{e}_i = \tilde{\boldsymbol{w}}_i^T \boldsymbol{\phi}(\boldsymbol{x}) + \epsilon_i - k_i e_i \tag{6.32}$$

where weight error $\tilde{\boldsymbol{w}}_i$ is defined as $\tilde{\boldsymbol{w}}_i \triangleq \boldsymbol{w}_i - \hat{\boldsymbol{w}}_i$.

To this point, the control design steps were the same as the design explained in the previous section. However, the weight update rule is derived differently in the current design. The following quadratic Lyapunov function is selected for deriving the stable learning rule:

$$V_i = \frac{\beta_i e_i^2}{2} + \frac{\tilde{\boldsymbol{w}}_i^T \tilde{\boldsymbol{w}}_i}{2\gamma_i} + \left[\frac{\partial d_i(\boldsymbol{x})}{\partial \boldsymbol{x}} - \frac{\partial \hat{d}_i(\boldsymbol{x})}{\partial \boldsymbol{x}}\right]^T \frac{\Theta_i}{2} \left[\frac{\partial d_i(\boldsymbol{x})}{\partial \boldsymbol{x}} - \frac{\partial \hat{d}_i(\boldsymbol{x})}{\partial \boldsymbol{x}}\right] \tag{6.33}$$

where Θ_i and β_i are positive constants and $\gamma_i > 0$ is known as the learning rate constant. The terms $\frac{\partial d_i(\boldsymbol{x})}{\partial \boldsymbol{x}}$ and $\frac{\partial \hat{d}_i(\boldsymbol{x})}{\partial \boldsymbol{x}}$ are gradients of model uncertainties and approximated model uncertainties, respectively.

The time derivative of the Lyapunov function in (6.33) is

$$\dot{V}_i = \beta_i e_i \dot{e}_i + \frac{\tilde{\boldsymbol{w}}_i^T \dot{\tilde{\boldsymbol{w}}}_i}{\gamma_i} + \frac{d}{dt}\left[\tilde{\boldsymbol{w}}_i^T \left[\frac{\partial \boldsymbol{\phi}}{\partial \boldsymbol{x}}\right] \frac{\Theta_i}{2} \left[\frac{\partial \boldsymbol{\phi}}{\partial \boldsymbol{x}}\right]^T \tilde{\boldsymbol{w}}_i\right] \tag{6.34}$$

$$= \beta_i e_i \dot{e}_i + \frac{\tilde{\boldsymbol{w}}_i^T \dot{\tilde{\boldsymbol{w}}}_i}{\gamma_i} + \tilde{\boldsymbol{w}}_i^T \left[\frac{\partial \boldsymbol{\phi}}{\partial \boldsymbol{x}}\right] \Theta_i \left[\left[\frac{\partial \boldsymbol{\phi}}{\partial \boldsymbol{x}}\right]^T \dot{\tilde{\boldsymbol{w}}}_i + \frac{d}{dt}\left[\frac{\partial \boldsymbol{\phi}}{\partial \boldsymbol{x}}\right]^T \tilde{\boldsymbol{w}}_i\right] \tag{6.35}$$

where $\frac{\partial \boldsymbol{\phi}}{\partial \boldsymbol{x}}$ is a Jacobian matrix. Further, using error dynamics (6.32), the time derivative of the Lyapunov function is modified as

$$\dot{V}_i = \beta_i e_i \left(\tilde{\boldsymbol{w}}_i^T \boldsymbol{\phi}(\boldsymbol{x}) + \epsilon_i - k_i e_i\right) - \frac{\tilde{\boldsymbol{w}}_i^T \dot{\hat{\boldsymbol{w}}}_i}{\gamma_i}$$

$$-\tilde{\boldsymbol{w}}_i^T \left[\frac{\partial \boldsymbol{\phi}}{\partial \boldsymbol{x}}\right] \Theta_i \left[\frac{\partial \boldsymbol{\phi}}{\partial \boldsymbol{x}}\right]^T \dot{\hat{\boldsymbol{w}}}_i + \tilde{\boldsymbol{w}}_i^T \left[\frac{\partial \boldsymbol{\phi}}{\partial \boldsymbol{x}}\right] \Theta_i \frac{d}{dt}\left[\frac{\partial \boldsymbol{\phi}}{\partial \boldsymbol{x}}\right]^T \tilde{\boldsymbol{w}}_i$$

$$
\begin{aligned}
&= \tilde{\boldsymbol{w}}_i^T \left[\beta_i e_i \boldsymbol{\phi}(\boldsymbol{x}) - \frac{\dot{\hat{\boldsymbol{w}}}_i}{\gamma_i} - \left[\frac{\partial \boldsymbol{\phi}}{\partial \boldsymbol{x}} \right] \Theta_i \left[\frac{\partial \boldsymbol{\phi}}{\partial \boldsymbol{x}} \right]^T \dot{\hat{\boldsymbol{w}}}_i \right] \\
&\quad -k_i \beta_i e_i^2 + \beta_i \epsilon_i e_i + \tilde{\boldsymbol{w}}_i^T \left[\frac{\partial \boldsymbol{\phi}}{\partial \boldsymbol{x}} \right] \Theta_i \frac{d}{dt} \left[\frac{\partial \boldsymbol{\phi}}{\partial \boldsymbol{x}} \right]^T \tilde{\boldsymbol{w}}_i \\
&= \tilde{\boldsymbol{w}}_i^T \left[\beta_i e_i \boldsymbol{\phi}(\boldsymbol{x}) - \left(\frac{I_P}{\gamma_i} + \left[\frac{\partial \boldsymbol{\phi}}{\partial \boldsymbol{x}} \right] \Theta_i \left[\frac{\partial \boldsymbol{\phi}}{\partial \boldsymbol{x}} \right]^T \right) \dot{\hat{\boldsymbol{w}}}_i \right] \\
&\quad -k_i \beta_i e_i^2 + \beta_i \epsilon_i e_i + \tilde{\boldsymbol{w}}_i^T \left[\frac{\partial \boldsymbol{\phi}}{\partial \boldsymbol{x}} \right] \Theta_i \frac{d}{dt} \left[\frac{\partial \boldsymbol{\phi}}{\partial \boldsymbol{x}} \right]^T \tilde{\boldsymbol{w}}_i
\end{aligned}
\tag{6.36}
$$

where I_P is an identity matrix.

Select the weight update rule as

$$
\dot{\hat{\boldsymbol{w}}}_i = \left[\frac{I_P}{\gamma_i} + \left[\frac{\partial \boldsymbol{\phi}}{\partial \boldsymbol{x}} \right] \Theta_i \left[\frac{\partial \boldsymbol{\phi}}{\partial \boldsymbol{x}} \right]^T \right]^{-1} \beta_i e_i \boldsymbol{\phi}(\boldsymbol{x}) \tag{6.37}
$$

The resultant $\dot{V}_i$ expression can be written as

$$
\dot{V}_i = -k_i \beta_i e_i^2 + \beta_i \epsilon_i e_i + \tilde{\boldsymbol{w}}_i^T \left[\frac{\partial \boldsymbol{\phi}}{\partial \boldsymbol{x}} \right] \Theta_i \frac{d}{dt} \left[\frac{\partial \boldsymbol{\phi}}{\partial \boldsymbol{x}} \right]^T \tilde{\boldsymbol{w}}_i \tag{6.38}
$$

The condition for which $\dot{V}_i$ becomes negative definite is

$$
\left[k_i \beta_i e_i^2 - \beta_i \epsilon_i e_i - \tilde{\boldsymbol{w}}_i^T \left[\frac{\partial \boldsymbol{\phi}}{\partial \boldsymbol{x}} \right] \Theta_i \frac{d}{dt} \left[\frac{\partial \boldsymbol{\phi}}{\partial \boldsymbol{x}} \right]^T \tilde{\boldsymbol{w}}_i \right] > 0 \tag{6.39}
$$

This inequality indicates that the closed-loop error will be bounded by a small region. Furthermore, the term $\frac{d}{dt}\left[\frac{\partial \boldsymbol{\phi}}{\partial \boldsymbol{x}}\right] = \left[\frac{\partial^2 \boldsymbol{\phi}}{\partial \boldsymbol{x}^2}\right]\dot{\boldsymbol{x}}$ contains a double partial derivative of basis function $\boldsymbol{\phi}$. Since typical basis functions are commonly selected as smooth functions, the Taylor series expansion of $\boldsymbol{\phi}(\boldsymbol{x})$ about zero results in the cancellation of dominant (linear and constant) terms in the entries of the matrix $\left[\frac{\partial^2 \boldsymbol{\phi}}{\partial \boldsymbol{x}^2}\right]$. This reduces the region of error bound.

Finally, using the weight update rule in (6.37), model uncertainties of the system can be approximated as $\hat{d}_i(\boldsymbol{x}) = \hat{\boldsymbol{w}}_i^T \boldsymbol{\phi}(\boldsymbol{x})$. This completes the first step of the neuro-adaptive control design.

6.3.2 Asymptotically Stable Tracking Control Design

After capturing model uncertainties, the next step is to drive the error between the approximated and reference signal to zero. To achieve $\boldsymbol{x}^a \rightarrow \boldsymbol{x}^*$, consider the following closed-loop nonlinear stable error dynamics:

$$\dot{\boldsymbol{e}}^a + \boldsymbol{K}^a \boldsymbol{e}^a = 0 \tag{6.40}$$

where $\boldsymbol{e}^a$ is the error, defined as $\boldsymbol{e}^a = \boldsymbol{x}^a - \boldsymbol{x}^*$, and $\boldsymbol{K}^a$ is a diagonal positive definite gain matrix.

By substituting $\boldsymbol{e}^a = \boldsymbol{x}^a - \boldsymbol{x}^*$, (6.40) is modified as

$$\left(\dot{\boldsymbol{x}}^a - \dot{\boldsymbol{x}}^*\right) + \boldsymbol{K}^a(\boldsymbol{x}^a - \boldsymbol{x}^*) = 0 \tag{6.41}$$

which can be further modified by substituting $\dot{\boldsymbol{x}}^a$ from (6.25)

$$\boldsymbol{f}(\boldsymbol{x}) + \boldsymbol{G}(\boldsymbol{x})\boldsymbol{u} + \hat{\boldsymbol{d}}(\boldsymbol{x}) + \boldsymbol{K}(\boldsymbol{x} - \boldsymbol{x}^a) - \dot{\boldsymbol{x}}^* + \boldsymbol{K}^a(\boldsymbol{x}^a - \boldsymbol{x}^*) = 0 \tag{6.42}$$

where $\hat{\boldsymbol{d}}(\boldsymbol{x})$ is already obtained as $\hat{d}_i(\boldsymbol{x}) = \hat{\boldsymbol{w}}_i^T \boldsymbol{\phi}(\boldsymbol{x})$ and updated using the learning rule constructed in (6.37). Thus, the adaptive control $\boldsymbol{u}$ can be extracted as

$$\boldsymbol{u} = -\left[\boldsymbol{G}(\boldsymbol{x})\right]^{-1}\left(\boldsymbol{f}(\boldsymbol{x}) + \hat{\boldsymbol{d}}(\boldsymbol{x}) + \boldsymbol{K}(\boldsymbol{x} - \boldsymbol{x}^a) + \boldsymbol{K}^a(\boldsymbol{x}^a - \boldsymbol{x}^*) - \dot{\boldsymbol{x}}^*\right) \tag{6.43}$$

This controller ensures stable tracking, i.e., $\boldsymbol{x} \rightarrow \boldsymbol{x}^a \rightarrow \boldsymbol{x}^*$ as $t \rightarrow \infty$.

Example 6.2 Consider the previous example and design a neuro-adaptive controller with gradient learning. The dynamics of the system is given by

$$\dot{x} = (x + x^2) + (1 + x^2)u + d(x)$$

where x and u are the state and control input for the system, respectively, and $d(x)$ is the unknown function present in the system dynamics. The objective of the controller is to stabilize the plant.

The desired known system dynamics extracted from the actual system is defined by

$$\dot{x}_d = (x_d + x_d^2) + (1 + x_d^2)u_d$$

The stabilizing desired controller for the desired system can be designed using dynamic inversion control design. The desired control expression is given as

(continued)

Example 6.2 (continued)

$$u_d = -\frac{(2x_d + x_d^2)}{(1 + x_d^2)}$$

Following the adaptive control design steps, the following adaptive control expression is obtained:

$$u = -\frac{(x_d + x_d^2) + (1 + x_d^2)u_d - k(x_a - x_d) - (x + x^2) - \hat{d}(x) - k_a(x - x_a)}{(1 + x^2)}$$

where the term $\hat{d}(x)$ is calculated as $\hat{d}(x) = \hat{\boldsymbol{w}}^T \boldsymbol{\phi}(x)$ and weights are updated from the learning rule $\dot{\hat{w}} = \left[\frac{I_P}{\gamma} + \left[\frac{\partial \phi}{\partial x}\right] \Theta \left[\frac{\partial \phi}{\partial x}\right]^T\right]^{-1} \beta e_a \boldsymbol{\phi}(x)$.

To simulate the controller response, the unknown function $d(x)$ is assumed as $d(x) = \sin(\frac{\pi x}{2})$. The controller gains k, k_a, γ, Θ, and β are selected as $k = 2.5$, $k_a = 1$, $\gamma = 30$, $\Theta = 1$, and $\beta = 1$, respectively. The basis functions to approximate the unknown function are chosen as $\boldsymbol{\phi}(x) = \left[\frac{x}{x_0}, \frac{x^2}{x_0^2}, \frac{x^3}{x_0^3}\right]^T$ with $x_0 = -1.5$ as an initial condition. The gradient of the basis function is calculated as $\frac{\partial \boldsymbol{\phi}(x)}{\partial x} = \left[\frac{1}{x_0}, \frac{2x}{x_0^2}, \frac{3x^2}{x_0^3}\right]^T$.

The state and control histories of the system are illustrated in Fig. 6.4a and b, where it is apparent that the performance of the adaptive controller is better in the presence of modeling uncertainties. On the other hand, as expected, the desired controller works well only for the desired dynamics. The approximations of the unknown functions and the weight update histories are depicted in Fig. 6.5a and b. It is evident that the weight update is fast, and the approximation of uncertainties is better with the gradient-based learning rule.

6.4 Control of Euler-Lagrange Systems with Structured Uncertainties

In previous sections, the adaptive controller is designed for square-affine nonlinear systems. However, the neuro-adaptive controller design for non-square-affine systems is not as straightforward as it is for square-affine systems and requires a slightly different approach. In this section, a neuro-adaptive control design strategy is presented for EL systems with structured model uncertainties. EL systems belong to the class of non-square-affine nonlinear systems with dimensions $2m = n$,

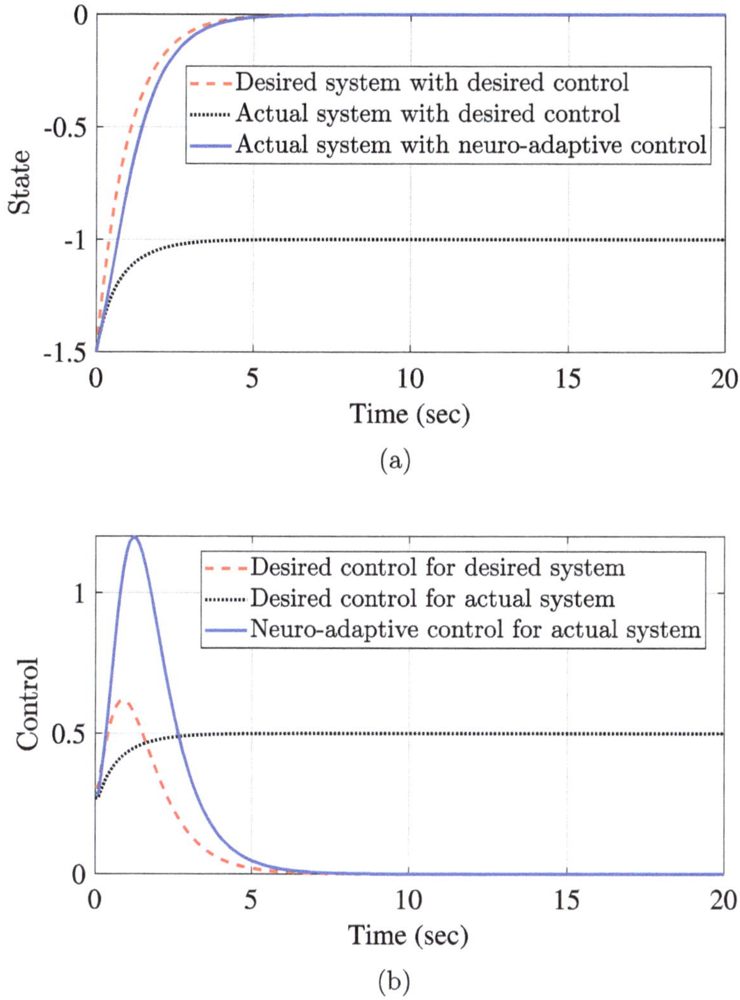

Fig. 6.4 (**a**) State histories. (**b**) Control histories

meaning that the dimension of the control input is half the dimension of the state. EL systems are an important category of nonlinear systems, as they result directly from Newton's laws of motion. The dynamics of many practical systems, including mechanical and electrical systems, can be modeled using EL systems.

The dynamics of EL systems is given by

$$\begin{aligned}\dot{\boldsymbol{x}}_1 &= \boldsymbol{x}_2 \\ \dot{\boldsymbol{x}}_2 &= \boldsymbol{f}(\boldsymbol{x}) + \boldsymbol{G}(\boldsymbol{x})\boldsymbol{u} + \boldsymbol{d}(\boldsymbol{x})\end{aligned} \tag{6.44}$$

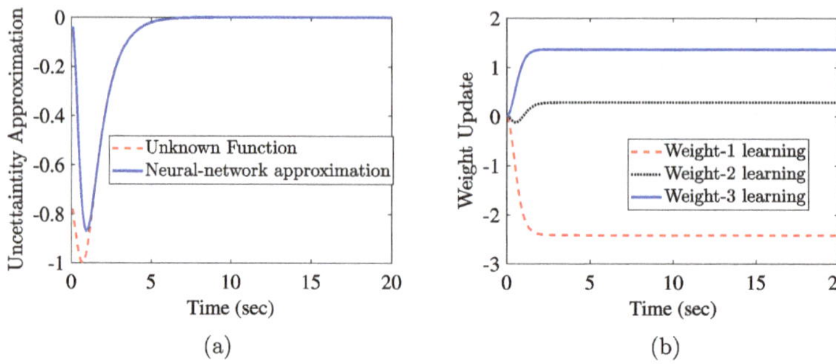

Fig. 6.5 (**a**) Unknown function and its approximation. (**b**) Weight update histories

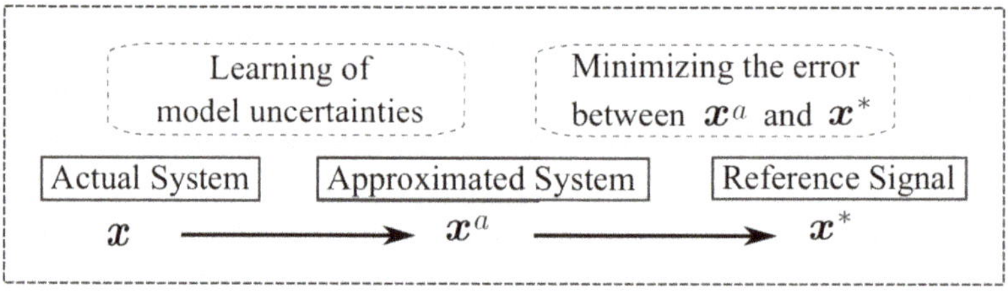

Fig. 6.6 Neuro-adaptive control design philosophy

where total states are defined as $\boldsymbol{x} \triangleq \left[\boldsymbol{x}_1^T, \boldsymbol{x}_2^T\right]^T$ and kinematic and dynamic states of the system are defined as $\boldsymbol{x}_1 \triangleq [x_{11}, x_{12}, \ldots, x_{1n}]^T \in \Re^n$ and $\boldsymbol{x}_2 \triangleq [x_{21}, x_{22}, \ldots, x_{2n}]^T \in \Re^n$, respectively. System output is selected as $\boldsymbol{x}_1 \in \Re^n$, and the control input to the system is denoted by $\boldsymbol{u} \in \Re^n$. Nonlinear terms $\boldsymbol{f}(\boldsymbol{x})$ and $\boldsymbol{G}(\boldsymbol{x})$ are assumed to be known, and the control effectiveness matrix $\boldsymbol{G}(\boldsymbol{x})$ is assumed to be invertible for all time. The system is assumed to be associated with structured model uncertainties $\boldsymbol{d}(\boldsymbol{x})$.

The overall neuro-adaptive control design philosophy in this section is similar to that in the previous sections; i.e., a neural network-based approximated model is constructed to approximate modeling uncertainties. Subsequently, a tracking controller is designed to ensure that the system tracks a reference signal accurately. However, due to the differences in the system structure, a new approximated model has been developed specifically to handle the uncertainties in EL systems. Figure 6.6 illustrates the philosophy of the neuro-adaptive control approach. The synthesis of this control design, along with a detailed explanation, is presented in the following subsections.

In the adaptive control formulation, the first step is to approximate the uncertainties of the actual system using an approximated system. Hence, a new approximated system is constructed as

$$\dot{\boldsymbol{x}}_1^a = \boldsymbol{x}_2^a$$

$$\dot{\boldsymbol{x}}_2^a = \boldsymbol{f}(\boldsymbol{x}) + \boldsymbol{G}(\boldsymbol{x})\boldsymbol{u} + \hat{\boldsymbol{d}}(\boldsymbol{x}) + \boldsymbol{K}_1\,\boldsymbol{e} + \boldsymbol{K}_2\dot{\boldsymbol{e}} \tag{6.45}$$

with the same initial condition as the actual system, where $\boldsymbol{x}_1^a$ and $\boldsymbol{x}_2^a$ are state variables of the approximated system. Error variables $\boldsymbol{e}$ and $\dot{\boldsymbol{e}}$ are defined as $\boldsymbol{e} = \boldsymbol{x}_1 - \boldsymbol{x}_1^a$ and $\dot{\boldsymbol{e}} = \boldsymbol{x}_2 - \boldsymbol{x}_2^a$, respectively. The matrices $\boldsymbol{K}_1$ and $\boldsymbol{K}_2$ are positive definite and diagonal in nature. The term $\hat{\boldsymbol{d}}(\boldsymbol{x})$ is the approximation of model uncertainties, obtained as $\hat{d}_i(\boldsymbol{x}) = \hat{\boldsymbol{w}}_i^T\boldsymbol{\phi}(\boldsymbol{x}), \forall i$.

6.4.1 Approximation of Structured Model Uncertainties

Using systems (6.44) and (6.45), the approximation error is defined as

$$\boldsymbol{e} = \boldsymbol{x}_1 - \boldsymbol{x}_1^a \tag{6.46}$$

The double time derivative of the error in (6.46) is

$$\ddot{\boldsymbol{e}} = \dot{\boldsymbol{x}}_2 - \dot{\boldsymbol{x}}_2^a \tag{6.47}$$

By substituting $\dot{\boldsymbol{x}}_2$ and $\dot{\boldsymbol{x}}_2^a$ from (6.44) and (6.45), the following is obtained:

$$\ddot{\boldsymbol{e}} = \boldsymbol{d}(\boldsymbol{x}) - \hat{\boldsymbol{d}}(\boldsymbol{x}) - \boldsymbol{K}_1\,\boldsymbol{e} - \boldsymbol{K}_2\,\dot{\boldsymbol{e}} \tag{6.48}$$

Since $\boldsymbol{K}_1$ and $\boldsymbol{K}_2$ are diagonal matrices, the ith component of closed-loop error dynamics (6.48) can be expressed as

$$\ddot{e}_i = d_i(\boldsymbol{x}) - \hat{d}_i(\boldsymbol{x}) - k_{1_i}e_i - k_{2_i}\dot{e}_i \tag{6.49}$$

Using the structures of $d_i(\boldsymbol{x}) = \boldsymbol{w}_i^T\boldsymbol{\phi}(\boldsymbol{x})$ and $\hat{d}_i(\boldsymbol{x}) = \hat{\boldsymbol{w}}_i^T\boldsymbol{\phi}(\boldsymbol{x})$, the following is obtained:

$$\ddot{e}_i = \tilde{\boldsymbol{w}}_i^T\boldsymbol{\phi}(\boldsymbol{x}) - k_{1_i}e_i - k_{2_i}\dot{e}_i \tag{6.50}$$

where $\tilde{\boldsymbol{w}}_i^T$ is defined as $\tilde{\boldsymbol{w}}_i^T \triangleq \boldsymbol{w}_i^T - \hat{\boldsymbol{w}}_i^T$. To analyze the stability of error dynamics (6.50), the following Lyapunov function is selected:

$$V_i = \frac{k_{1_i}k_{2_i}e_i^2}{2} + \frac{c_{2_i}\dot{e}_i^2}{2} + \frac{\tilde{\boldsymbol{w}}_i^T\tilde{\boldsymbol{w}}_i}{2\gamma_i} \tag{6.51}$$

where $\gamma_i > 0$ is the learning rate constant. The time derivative of (6.51) is obtained as

$$\dot{V}_i = k_{1_i}k_{2_i}e_i\dot{e}_i + k_{2_i}\dot{e}_i\ddot{e}_i - \frac{\tilde{\boldsymbol{w}}_i^T\dot{\hat{\boldsymbol{w}}}_i}{\gamma_i} \tag{6.52}$$

Using (6.50), the derivative of the Lyapunov function in (6.52) can be expressed as

$$\dot{V}_i = k_{1_i} k_{2_i} e_i \dot{e}_i - k_{1_i} k_{2_i} e_i \dot{e}_i - \left(k_{2_i} \dot{e}_i\right)^2 + \tilde{\boldsymbol{w}}_i^T \boldsymbol{\phi}(\boldsymbol{x}) k_{2_i} \dot{e}_i - \frac{\tilde{\boldsymbol{w}}_i^T \dot{\hat{\boldsymbol{w}}}_i}{\gamma_i} \tag{6.53}$$

which can be simplified as

$$\dot{V}_i = -\left(k_{2_i} \dot{e}_i\right)^2 + \tilde{\boldsymbol{w}}_i^T \left(\boldsymbol{\phi}(\boldsymbol{x}) k_{2_i} \dot{e}_i - \frac{\dot{\hat{\boldsymbol{w}}}_i}{\gamma_i}\right) \tag{6.54}$$

By selecting the weight learning rule as

$$\dot{\hat{\boldsymbol{w}}}_i = \gamma_i k_{2_i} \dot{e}_i \boldsymbol{\phi}(\boldsymbol{x}), \tag{6.55}$$

The time derivative of the Lyapunov function in (6.54) is modified as

$$\dot{V}_i = -\left(k_{2_i} \dot{e}_i\right)^2 \tag{6.56}$$

Since $\dot{V}_i \leq 0$, e_i, $\dot{e}_i$, and $\tilde{\boldsymbol{w}}_i$ are bounded. Furthermore, as basis functions $\boldsymbol{\phi}(\boldsymbol{x})$ are chosen to be bounded in nature, by analyzing (6.50), $\ddot{e}_i$ is also bounded. Therefore, all closed-loop signals are bounded and stable in the sense of Lyapunov.

The nature of $\dot{e}_i$ can be further analyzed by calculating the time derivative of $\dot{V}_i$ in (6.56) as

$$\ddot{V}_i = -2k_{2_i}^2 \dot{e}_i \ddot{e}_i \tag{6.57}$$

Since each term of $\ddot{V}_i$ is bounded, using Barbalat's lemma, $\dot{V}_i \rightarrow 0$ as $t \rightarrow \infty$. Thus, $\dot{e}_i \rightarrow 0$, i.e., $\boldsymbol{x}_2 \rightarrow \boldsymbol{x}_2^a$ as $t \rightarrow \infty$. Since $\dot{e}_i \rightarrow 0$, $\ddot{e}_i \rightarrow 0$ as $t \rightarrow \infty$. Hence, from (6.50), there will be a finite error, which is $e_i = \frac{\tilde{\boldsymbol{w}}_i^T \boldsymbol{\phi}(x)}{k_{1_i}}$. This error can be minimized with proper selection of gain k_{1_i}. Moreover, since $\boldsymbol{x}_2 \rightarrow \boldsymbol{x}_2^a$ is already proven and modeling uncertainties are present in $\dot{\boldsymbol{x}}_2$ and $\dot{\boldsymbol{x}}_2^a$ dynamics, $\tilde{\boldsymbol{w}}_i^T$ can be further minimized with accurate choice of basis functions.

After the approximation of model uncertainties, the next step is to design an asymptotically stable controller for the approximated system to track the reference signal, which is presented as follows.

6.4.2 Tracking Control Design

In order to design a tracking controller, the error between the approximated system in (6.45) and the reference output $\boldsymbol{x}^*$ is defined as

$$\boldsymbol{e}^a = \boldsymbol{x}_1^a - \boldsymbol{x}^* \tag{6.58}$$

The time derivatives of (6.58) can be calculated as

$$\dot{e}^a = x_2^a - \dot{x}^* \tag{6.59}$$

$$\ddot{e}^a = \dot{x}_2^a - \ddot{x}^* \tag{6.60}$$

By substituting $\dot{x}_2^a$ from (6.45), the error dynamics can be written as

$$\ddot{e}^a = f(x) + G(x)u + \hat{d}(x) + K_1\, e + K_2\, \dot{e} - \ddot{x}^* \tag{6.61}$$

To stabilize the error dynamics (6.61), a closed-form adaptive control law is constructed as

$$u = G(x)^{-1}\Big(- f(x) - K_1^a e^a - K_2^a \dot{e}^a - K_1 e - K_2 \dot{e} - \hat{d}(x) + \ddot{x}^* \Big)$$

where the gain matrices K_2^a and K_1^a are positive definite and diagonal in nature. By substituting the control expression in (6.61), the following closed-loop error dynamics is obtained:

$$\ddot{e}^a = -K_1^a e^a - K_2^a \dot{e}^a \tag{6.62}$$

which is an exponentially stable linear closed-loop dynamics. Hence, error $e^a \to \mathbf{0}$ and $\dot{e}^a \to \mathbf{0}$ exponentially. Thus, both steps of control design will ensure that the actual system output tracks the reference output with small approximation errors.

Example 6.3 The effectiveness of the control design is demonstrated using the Van der Pol system. This system is selected for several reasons: (i) It represents a nonlinear system with relative degree two, (ii) it is a non-square problem ($m = 1, n = 2$), (iii) the homogeneous form of the system has an unstable equilibrium at the origin, and (iv) the system exhibits limit cycle behavior. These properties make the Van der Pol system an ideal candidate for testing the efficacy of the control design.

Consider the dynamics of the Van der Pol system given by

$$\begin{aligned} \dot{x}_1 &= x_2 \\ \dot{x}_2 &= (1 - x_1^2)x_2 - x_1 + d(x_1, x_2) + (1 + x_1^2 + x_2^2)u \end{aligned} \tag{6.63}$$

where x_1 and x_2 are states, u is the control input for the system, and $d(x_1, x_2)$ is the unknown function present in the system dynamics. The objective of the controller is to stabilize the plant.

(continued)

Example 6.3 (continued)
The desired known system dynamics extracted from the actual system is defined by

$$\dot{x}_1^d = x_2^d$$
$$\dot{x}_2^d = (1-(x_1^d)^2)x_2^d - x_1^d + (1+(x_1^d)^2+(x_2^d)^2)u^d \tag{6.64}$$

The stabilizing desired controller for the desired system can be designed using dynamic inversion control design. The desired control expression is given as

$$u^d = -\frac{(1-(x_1^d)^2)x_2^d - x_1^d + 2\zeta\omega_n x_2^d + \omega_n^2 x_1^d}{(1+(x_1^d)^2+(x_2^d)^2)}$$

where $\omega_n = \frac{20}{2\pi}$ and $\zeta = 0.7$. Following the adaptive control design steps, the adaptive control expression is obtained as

$$u = \frac{1}{(1+x_1^2+x_2^2)}\begin{pmatrix} -(1-x_1^2)x_2 + x_1 - k_1(x_1 - x_1^a) - k_2(x_2 - x_2^a) \\ -k_1^a(x_1^a - x_1^d) - k_2^a(x_2^a - x_2^d) - \hat{d}(x_1, x_2) + \dot{x}_2^d \end{pmatrix}$$

where the term $\hat{d}(x_1, x_2)$ is calculated as $\hat{d}(x_1, x_2) = \hat{\boldsymbol{w}}^T \boldsymbol{\phi}(x_1, x_2)$ and weights are updated from the learning rule $\dot{\hat{w}} = \gamma k_2 \dot{e} \boldsymbol{\phi}(x_1, x_2)$.

To simulate the controller response, the unknown function $d(x_1, x_2)$ is assumed as $d(x) = 2\cos(x_2)$. The controller gains k_1, k_2, k_1^a, k_2^a, and γ are selected as $k_1 = 1$, $k_2 = 5$, $k_1^a = 1$, $k_2^a = 5$, and $\gamma = 10$, respectively. The basis functions to approximate the unknown function are chosen as $\boldsymbol{\phi}(x) = kron(\boldsymbol{c}_1, \boldsymbol{c}_2)$, where $\boldsymbol{c}_1 = [1, \sin(x_1), \cos(x_1)]^T$ and $\boldsymbol{c}_2 = [1, \sin(x_2), \cos(x_2)]^T$, where $kron(*, *)$ represents the "Kronecker product."

The position, velocity, and control histories of the system are illustrated in Figs. 6.7a and b, and 6.8a. Once again, the performance of the desired controllers is not inadequate when applied to the actual system in the presence of uncertainties. However, as demonstrated in theory, the adaptive controller ensures the desired tracking with a small finite error. The approximation of the unknown function is depicted in Fig. 6.8b, where the neural network is used to approximate the unknown function.

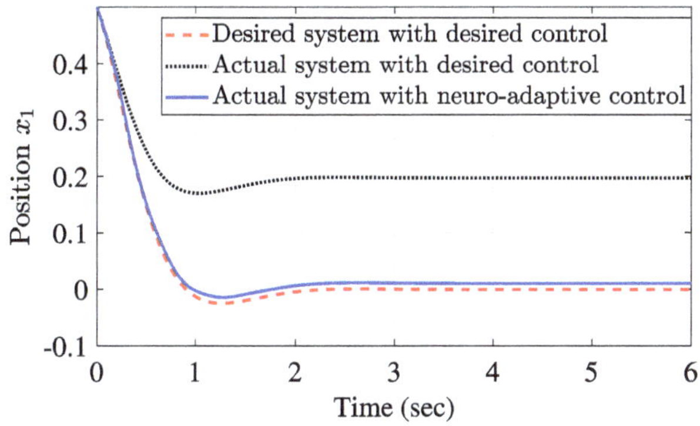

(a)

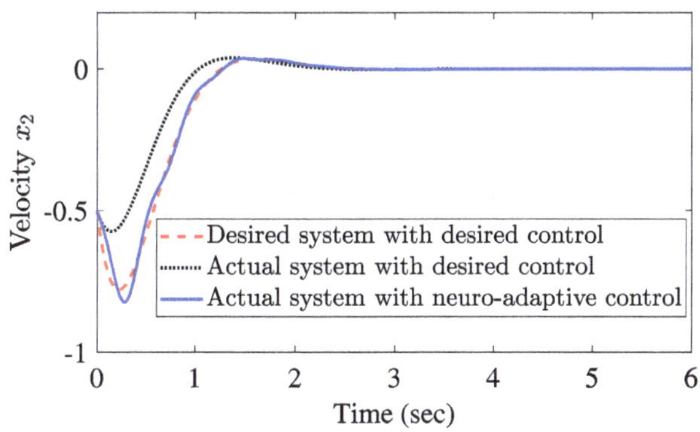

(b)

Fig. 6.7 (**a**) Position histories. (**b**) Velocity histories

6.5 Control of Euler-Lagrange Systems with Unstructured Uncertainties

In the previous section, the controller was designed considering structured uncertainties, where the form of the uncertainties was known. However, in many practical scenarios, the exact nature of the uncertainties is unknown. Therefore, in this section, a neuro-adaptive controller is developed for Euler-Lagrange (EL) systems, accounting for unstructured uncertainties.

The state-space model of EL systems is given by

$$\begin{aligned} \dot{\boldsymbol{x}}_1 &= \boldsymbol{x}_2 \\ \dot{\boldsymbol{x}}_2 &= \boldsymbol{f}(\boldsymbol{x}) + \boldsymbol{G}(\boldsymbol{x})\boldsymbol{u} + \boldsymbol{d}(\boldsymbol{x}) \end{aligned} \tag{6.65}$$

Fig. 6.8 (**a**) Control histories. (**b**) Unknown function and its approximation

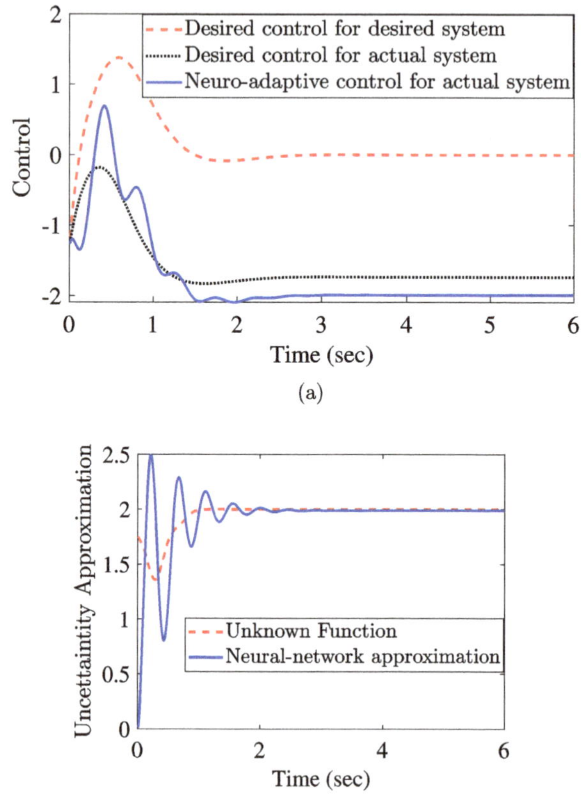

where states of the system are defined as $\boldsymbol{x} = \left[\boldsymbol{x}_1^T, \boldsymbol{x}_2^T\right]^T$, and kinematic and dynamic states of the system are defined as $\boldsymbol{x}_1 = [x_{11}, x_{12}, \ldots, x_{1n}]^T \in \Re^n$ and $\boldsymbol{x}_2 = [x_{21}, x_{22}, \ldots, x_{2n}]^T \in \Re^n$, respectively. The system output is selected as $\boldsymbol{x}_1 \in \Re^n$, and the control input to the system is denoted by $\boldsymbol{u} \in \Re^n$. Lumped unstructured modeling uncertainties of the system are represented by $\boldsymbol{d}(\boldsymbol{x})$. Nonlinear terms $\boldsymbol{f}(\boldsymbol{x})$ and $\boldsymbol{G}(\boldsymbol{x})$ are assumed to be known, and the control effectiveness matrix $\boldsymbol{G}(\boldsymbol{x})$ is assumed to be invertible for all time.

The objective is to design controller $\boldsymbol{u}$ in such a way that the output of the actual system tracks the reference signal, i.e., $\boldsymbol{x}_1 \rightarrow \boldsymbol{x}^*$ as $t \rightarrow \infty$.

The control design philosophy in this section is similar to that of the previous section. However, to enhance the robustness of the controller in the presence of unstructured uncertainties, a new error transformation is introduced in the control formulation. This modification necessitates a different construction of the approximated system compared to the previous section. The synthesis of the overall control design, incorporating these changes, is explained as follows.

6.5.1 Approximation of Unstructured Model Uncertainties

In this subsection, the weight update rule is developed using the approximated system to update the weights of the neural networks in order to approximate model uncertainties. To construct the approximated system, consider the kinematic and dynamic states of the approximated system as $\boldsymbol{x}_1^a$ and $\boldsymbol{x}_2^a$, respectively. The error between $\boldsymbol{x}_1$ and $\boldsymbol{x}_1^a$ can be defined as

$$\boldsymbol{e} = \boldsymbol{x}_1 - \boldsymbol{x}_1^a \tag{6.66}$$

The time derivatives of the approximation error (6.66) are obtained as

$$\dot{\boldsymbol{e}} = \boldsymbol{x}_2 - \boldsymbol{x}_2^a \tag{6.67}$$

$$\ddot{\boldsymbol{e}} = \dot{\boldsymbol{x}}_2 - \dot{\boldsymbol{x}}_2^a \tag{6.68}$$

Using the errors in (6.66)–(6.67), a new error transformation variable $\boldsymbol{s}$ is proposed as

$$\boldsymbol{s} = \boldsymbol{K}_1\, \boldsymbol{e} + \boldsymbol{K}_2\, \dot{\boldsymbol{e}} \tag{6.69}$$

where $\boldsymbol{K}_1$ and $\boldsymbol{K}_2$ are positive definite diagonal matrices. Since the matrices are diagonal in nature, the ith component of the error transformation can be written as

$$s_i = k_{1_i} e_i + k_{2_i} \dot{e}_i \tag{6.70}$$

It can be noted that the error transformation is linear in nature, and when $s_i \to 0$, errors $e_i \to 0$ and $\dot{e}_i \to 0$, $\forall i$. The time derivative of the transformed error $\boldsymbol{s}$ can be obtained as

$$\dot{\boldsymbol{s}} = \boldsymbol{K}_1\, \dot{\boldsymbol{e}} + \boldsymbol{K}_2\, \ddot{\boldsymbol{e}} \tag{6.71}$$

Using the system errors (6.66)–(6.68) and proposed error transformation in (6.69), the approximated system dynamics is constructed as

$$\begin{aligned} \dot{\boldsymbol{x}}_1^a &= \boldsymbol{x}_2^a \\ \dot{\boldsymbol{x}}_2^a &= \boldsymbol{f}(\boldsymbol{x}) + \boldsymbol{G}(\boldsymbol{x})\boldsymbol{u} + \hat{\boldsymbol{d}}(\boldsymbol{x}) + \boldsymbol{K}_2^{-1}\boldsymbol{K}_3\boldsymbol{s} + \boldsymbol{K}_1\boldsymbol{K}_2^{-1}\dot{\boldsymbol{e}} + \boldsymbol{K}_2\boldsymbol{s} \end{aligned} \tag{6.72}$$

with the same initial condition as the actual system, where $\boldsymbol{K}_3$ is a diagonal matrix with positive elements and term $\hat{\boldsymbol{d}}(\boldsymbol{x})$ is the approximation of modeling uncertainties $\boldsymbol{d}(\boldsymbol{x})$.

By substituting $\ddot{\boldsymbol{e}}$ from (6.68), $\dot{\boldsymbol{s}}$ in (6.71) is modified as

$$\dot{\boldsymbol{s}} = \boldsymbol{K}_1\dot{\boldsymbol{e}} + \boldsymbol{K}_2\big(\dot{\boldsymbol{x}}_2 - \dot{\boldsymbol{x}}_2^a\big) \tag{6.73}$$

Using (6.65) and (6.72), $\dot{\boldsymbol{s}}$ can be written as

$$\dot{\boldsymbol{s}} = \boldsymbol{K}_1\dot{\boldsymbol{e}} + \boldsymbol{K}_2\Big(\boldsymbol{d}(\boldsymbol{x}) - \hat{\boldsymbol{d}}(\boldsymbol{x}) - \boldsymbol{K}_2^{-1}\boldsymbol{K}_3\boldsymbol{s} - \boldsymbol{K}_1\boldsymbol{K}_2^{-1}\dot{\boldsymbol{e}} - \boldsymbol{K}_2\boldsymbol{s}\Big) \tag{6.74}$$

By using the definitions of $\boldsymbol{d}(\boldsymbol{x})$ and $\hat{\boldsymbol{d}}(\boldsymbol{x})$ from (6.30) and (6.31), the following is obtained:

$$\dot{\boldsymbol{s}} = \boldsymbol{K}_1\dot{\boldsymbol{e}} + \boldsymbol{K}_2\Big(\tilde{\boldsymbol{W}}^T\boldsymbol{\phi}(\boldsymbol{x}) + \boldsymbol{\epsilon} - \boldsymbol{K}_2^{-1}\boldsymbol{K}_3\boldsymbol{s} - \boldsymbol{K}_1\boldsymbol{K}_2^{-1}\dot{\boldsymbol{e}} - \boldsymbol{K}_2\boldsymbol{s}\Big) \tag{6.75}$$

where $\tilde{\boldsymbol{W}}$ is defined as $\tilde{\boldsymbol{W}} \triangleq \boldsymbol{W} - \hat{\boldsymbol{W}}$ and $\boldsymbol{\phi}(\boldsymbol{x})$ represents basis function vector. Since the matrices are diagonal in nature and an NN approximation can be used for individual components of the dynamics, the expression of $\dot{\boldsymbol{s}}$ in (6.75) can be expressed in i^{th} components as

$$\dot{s}_i = k_{1_i}\dot{e}_i + k_{2_i}\left(\tilde{\boldsymbol{w}}_i^T\boldsymbol{\phi}(\boldsymbol{x}) + \epsilon_i - \frac{k_{3_i}s_i}{k_{2_i}} - \frac{k_{1_i}\dot{e}_i}{k_{2_i}} - k_{2_i}s_i\right), \quad \forall i \tag{6.76}$$

where $\tilde{\boldsymbol{w}}_i^T$ is the ith row of $\tilde{\boldsymbol{W}}^T$ and k_{1_i}, k_{2_i}, and k_{3_i} are the ith diagonal component of matrices $\boldsymbol{K}_1$, $\boldsymbol{K}_2$, and $\boldsymbol{K}_3$, respectively.

To analyze the stability of (6.76), the following Lyapunov function is chosen:

$$v_i = \frac{1}{2}s_i^2 + \frac{1}{2\gamma_i}\tilde{\boldsymbol{w}}_i^T\tilde{\boldsymbol{w}}_i \tag{6.77}$$

where $\gamma_i > 0$ is learning rate constants. By taking the time derivative of the Lyapunov function, the following is yielded:

$$\dot{v}_i = s_i\dot{s}_i + \frac{\tilde{\boldsymbol{w}}_i^T\dot{\tilde{\boldsymbol{w}}}_i}{\gamma_i} \tag{6.78}$$

By substituting $\dot{s}_i$ from (6.76) and $\dot{\tilde{\boldsymbol{w}}}_i = -\dot{\hat{\boldsymbol{w}}}_i$, the time derivative of the Lyapunov function $\dot{v}_i$ can be expressed as

$$\begin{aligned}
\dot{v}_i &= s_i\left[k_{1_i}\dot{e}_i + k_{2_i}\left(\tilde{\boldsymbol{w}}_i^T\boldsymbol{\phi}(\boldsymbol{x}) + \epsilon_i - \frac{k_{3_i}s_i}{k_{2_i}} - \frac{k_{1_i}\dot{e}_i}{k_{2_i}} - k_{2_i}s_i\right)\right] - \frac{\tilde{\boldsymbol{w}}_i^T\dot{\hat{\boldsymbol{w}}}_i}{\gamma_i} \\
&= s_ik_{2_i}\left(\tilde{\boldsymbol{w}}_i^T\boldsymbol{\phi}(\boldsymbol{x}) + \epsilon_i - \frac{k_{3_i}s_i}{k_{2_i}} - k_{2_i}s_i\right) - \frac{\tilde{\boldsymbol{w}}_i^T\dot{\hat{\boldsymbol{w}}}_i}{\gamma_i} \\
&= s_ik_{2_i}\tilde{\boldsymbol{w}}_i^T\boldsymbol{\phi}(\boldsymbol{x}) + s_ik_{2_i}\epsilon_i - k_{3_i}s_i^2 - k_{2_i}^2s_i^2 - \frac{\tilde{\boldsymbol{w}}_i^T\dot{\hat{\boldsymbol{w}}}_i}{\gamma_i}
\end{aligned} \tag{6.79}$$

Using Young's inequality, the following is obtained:

$$\begin{aligned}\dot{v}_i &\leq s_i k_{2_i} \tilde{\boldsymbol{w}}_i^T \boldsymbol{\phi}(\boldsymbol{x}) + s_i^2 k_{2_i}^2 + \frac{\epsilon_i^2}{4} - k_{3_i} s_i^2 - k_{2_i}^2 s_i^2 - \frac{\tilde{\boldsymbol{w}}_i^T \dot{\hat{\boldsymbol{w}}}_i}{\gamma_i} \\ &\leq -k_{3_i} s_i^2 + \frac{\epsilon_i^2}{4} + s_i k_{2_i} \tilde{\boldsymbol{w}}_i^T \boldsymbol{\phi}(\boldsymbol{x}) - \frac{\tilde{\boldsymbol{w}}_i^T \dot{\hat{\boldsymbol{w}}}_i}{\gamma_i} \end{aligned} \tag{6.80}$$

By choosing the weight update rule as

$$\dot{\hat{\boldsymbol{w}}}_i = \gamma_i k_{2_i} s_i \boldsymbol{\phi}(\boldsymbol{x}) - \alpha_i \hat{\boldsymbol{w}}_i \tag{6.81}$$

where α_i is a positive constant, which is required to ensure the closed-loop stability of the weight dynamics. By substituting the weight update rule, the expression of $\dot{v}_i$ is modified as

$$\begin{aligned}\dot{v}_i &\leq -k_{3_i} s_i^2 + \frac{\epsilon_i^2}{4} + s_i k_{2_i} \tilde{\boldsymbol{w}}_i^T \boldsymbol{\phi}(\boldsymbol{x}) - s_i k_{2_i} \tilde{\boldsymbol{w}}_i^T \boldsymbol{\phi}(\boldsymbol{x}) + \frac{\alpha_i}{\gamma_i} \tilde{\boldsymbol{w}}_i^T \hat{\boldsymbol{w}}_i \\ &\leq -k_{3_i} s_i^2 + \frac{\epsilon_i^2}{4} + \frac{\alpha_i}{\gamma_i} \tilde{\boldsymbol{w}}_i^T \hat{\boldsymbol{w}}_i \end{aligned} \tag{6.82}$$

Again, using Young's inequality and following the definition of $\tilde{\boldsymbol{w}}_i^T$, the term $\frac{\alpha_i \tilde{\boldsymbol{w}}_i^T \hat{\boldsymbol{w}}_i}{\gamma_i}$ can be expressed as $\frac{\alpha_i}{\gamma_i} \tilde{\boldsymbol{w}}_i^T \hat{\boldsymbol{w}}_i \leq -\frac{\alpha_i}{2\gamma_i} \tilde{\boldsymbol{w}}_i^T \tilde{\boldsymbol{w}}_i + \frac{\alpha_i}{2\gamma_i} \boldsymbol{w}_i^T \boldsymbol{w}_i$. Furthermore, assuming that ϵ_i is upper bounded as $\epsilon_i^2 \leq \epsilon_{m_i}$, the Lyapunov derivative $\dot{v}_i$ is modified as

$$\begin{aligned}\dot{v}_i &\leq -k_{3_i} s_i^2 + \frac{\epsilon_i^2}{4} - \frac{\alpha_i}{2\gamma_i} \tilde{\boldsymbol{w}}_i^T \tilde{\boldsymbol{w}}_i + \frac{\alpha_i}{2\gamma_i} \boldsymbol{w}_i^T \boldsymbol{w}_i \\ &\leq -k_{4_i} \left(\frac{s_i^2}{2} + \frac{1}{2\gamma_i} \tilde{\boldsymbol{w}}_i^T \tilde{\boldsymbol{w}}_i \right) + \frac{\epsilon_{m_i}}{4} + \frac{\alpha_i}{2\gamma_i} \boldsymbol{w}_i^T \boldsymbol{w}_i \\ &\leq -k_{4_i} v_i + k_{5_i} \end{aligned} \tag{6.83}$$

where the constants k_{4_i} and k_{5_i} are defined as $k_{4_i} = \min\left(2k_{3_i}, \alpha_i\right)$ and $k_{5_i} = \frac{\epsilon_{m_i}}{4} + \frac{\alpha_i \boldsymbol{w}_i^T \boldsymbol{w}_i}{2\gamma_i}$, respectively. By integrating (6.83) in time, the following is obtained:

$$v_i(t) \leq v_i(0) \exp(-k_{4_i} t) + \frac{k_{5_i}}{k_{4_i}} \left[1 - \exp(-k_{4_i} t)\right] \tag{6.84}$$

where $v_i(0)$ is the initial value of the Lyapunov function (6.77) and 'exp' represents exponential function. By analyzing (6.84), the upper bound on $v_i(t)$ is obtained as

$$v_i(t) \leq v_i(0) + \frac{k_{5_i}}{k_{4_i}} \tag{6.85}$$

Since $v_i(t)$ is bounded, the stability of the transformed error dynamics in (6.76) can be concluded. Further, using this relation, the upper bound error of the transformation variable s can be obtained as $|s_i| \leq \sqrt{2v_i(0) + \frac{2k_{5_i}}{k_{4_i}}}$. Since s_i is bounded, from the characteristics of transform, e_i and $\dot{e}_i$ will remain bounded.

After approximating modeling uncertainties, the next step is to design a tracking controller for the approximated system in (6.72) to track the reference signal. An asymptotically stable tracking controller is designed in the following subsection.

6.5.2 *Tracking Control Design*

The error between approximated output $\boldsymbol{x}_1^a$ and reference signal $\boldsymbol{x}^*$ is defined as

$$\boldsymbol{e}^a = \boldsymbol{x}_1^a - \boldsymbol{x}^* \tag{6.86}$$

The time derivatives of the error (6.86) are calculated as

$$\begin{aligned} \dot{\boldsymbol{e}}^a &= \boldsymbol{x}_2^a - \dot{\boldsymbol{x}}^* \\ \ddot{\boldsymbol{e}}^a &= \dot{\boldsymbol{x}}_2^a - \ddot{\boldsymbol{x}}^* \end{aligned} \tag{6.87}$$

By substituting $\dot{\boldsymbol{x}}_2^a$ from (6.72), the error dynamics is modified as

$$\ddot{\boldsymbol{e}}^a = \boldsymbol{f}(\boldsymbol{x}) + \boldsymbol{G}(\boldsymbol{x})\boldsymbol{u} + \hat{\boldsymbol{d}}(\boldsymbol{x}) + \boldsymbol{K}_2^{-1}\boldsymbol{K}_3\boldsymbol{s} + \boldsymbol{K}_1\boldsymbol{K}_2^{-1}\dot{\boldsymbol{e}} + \boldsymbol{K}_2\boldsymbol{s} - \ddot{\boldsymbol{x}}^* \tag{6.88}$$

To stabilize the error dynamics (6.88), the state feedback closed-loop tracking controller can be designed as

$$\boldsymbol{u} = \boldsymbol{G}(\boldsymbol{x})^{-1}\begin{pmatrix} -\boldsymbol{f}(\boldsymbol{x}) - \hat{\boldsymbol{d}}(\boldsymbol{x}) - \boldsymbol{K}_2^{-1}\boldsymbol{K}_3\boldsymbol{s} - \boldsymbol{K}_1\boldsymbol{K}_2^{-1}\dot{\boldsymbol{e}} \\ -\boldsymbol{K}_2\boldsymbol{s} - \boldsymbol{K}_1^a\boldsymbol{e}^a - \boldsymbol{K}_2^a\dot{\boldsymbol{e}}^a + \ddot{\boldsymbol{x}}^* \end{pmatrix} \tag{6.89}$$

where matrices $\boldsymbol{K}_1^a$ and $\boldsymbol{K}_2^a$ are positive definite diagonal matrices. Further, by substituting the control expression (6.89) in (6.88), the following closed-loop error dynamics is obtained

$$\ddot{\boldsymbol{e}}^a = -\boldsymbol{K}_1^a\boldsymbol{e}^a - \boldsymbol{K}_2^a\dot{\boldsymbol{e}}^a \tag{6.90}$$

which is an exponentially stable linear closed-loop dynamics. Hence, error $\boldsymbol{e}^a \to \boldsymbol{0}$ and $\dot{\boldsymbol{e}}^a \to \boldsymbol{0}$ exponentially. Thus, both steps of control design will ensure that the actual system output tracks the reference output with small approximation errors.

Example 6.4 The same Van der Pol system is used to demonstrate the efficacy of the control design. The dynamic equation of the Van der Pol system is given by

$$\begin{aligned}\dot{x}_1 &= x_2\\ \dot{x}_2 &= (1-x_1^2)x_2 - x_1 + d(x_1,x_2) + (1+x_1^2+x_2^2)u\end{aligned}\tag{6.91}$$

where x_1 and x_2 are states, u is the control input for the system, and $d(x_1, x_2)$ is the unknown function present in the system dynamics. The objective of the controller is to stabilize the plant.

The desired known system dynamics extracted from the actual system is defined by

$$\begin{aligned}\dot{x}_1^d &= x_2^d\\ \dot{x}_2^d &= (1-(x_1^d)^2)x_2^d - x_1^d + (1+(x_1^d)^2+(x_2^d)^2)u^d\end{aligned}\tag{6.92}$$

The stabilizing desired controller for the desired system can be designed using dynamic inversion control design. The desired control expression is given as

$$u^d = -\frac{(1-(x_1^d)^2)x_2^d - x_1^d + 2\zeta\omega_n x_2^d + \omega_n^2 x_1^d}{(1+(x_1^d)^2+(x_2^d)^2)}$$

where $\omega_n = \frac{20}{2\pi}$ and $\zeta = 0.7$. Following the adaptive control design steps, the following adaptive control expression is obtained:

$$u = \frac{1}{(1+x_1^2+x_2^2)}\begin{pmatrix}-(1-x_1^2)x_2 + x_1 - \hat{d}(x_1,x_2) - \frac{k_3}{k_2}s - k_2 s\\ -\frac{k_1}{k_2}(x_2 - x_2^a) - k_1^a(x_1^a - x_1^d) - k_2^a(x_2^a - x_2^d) + \dot{x}_2^d\end{pmatrix}$$

where $s = k_1(x_1 - x_1^a) + k_2(x_2 - x_2^a)$. The term $\hat{d}(x_1, x_2)$ is calculated as $\hat{d}(x_1, x_2) = \hat{\boldsymbol{w}}^T\boldsymbol{\phi}(x_1, x_2)$, where the weights are updated from the learning rule $\dot{\hat{w}} = \gamma k_2 s\boldsymbol{\phi}(x_1, x_2) - \alpha\hat{w}$.

To simulate the controller response, the unknown function $d(x_1, x_2)$ is assumed as $d(x) = 2\cos(x_2)$. The controller gains $k_1, k_2, k_1^a, k_2^a, k_3, \alpha$, and γ are selected as $k_1 = 1$, $k_2 = 5$, $k_1^a = 1$, $k_2^a = 5$, $k_3 = 1$, $\alpha = 10^{-3}$, and $\gamma = 10$, respectively. The basis functions to approximate the unknown function are chosen as $\boldsymbol{\phi}(x) = kron(\boldsymbol{c}_1, \boldsymbol{c}_2)$, where $\boldsymbol{c}_1 = [1, \sin(x_1), \cos(x_1)]^T$ and $\boldsymbol{c}_2 = [1, \sin(x_2), \cos(x_2)]^T$.

The position, velocity, and control histories of the system are illustrated in Figs. 6.9a and b, and 6.10a. These figures show that while the desired

(continued)

Example 6.4 (continued)
controller performs effectively for the designed system, it does not achieve the same level of performance for the actual system. In contrast, the adaptive controller ensures the desired tracking with almost zero error, demonstrating its effectiveness.

Additionally, the approximation of the unknown function is shown in Fig. 6.10b, where the neural network approximates the unknown function well and compensates for the uncertainties in the system dynamics. It is also evident that, with the same gain settings, this controller is more robust and better than the controller presented in the previous section, particularly in handling unstructured uncertainties.

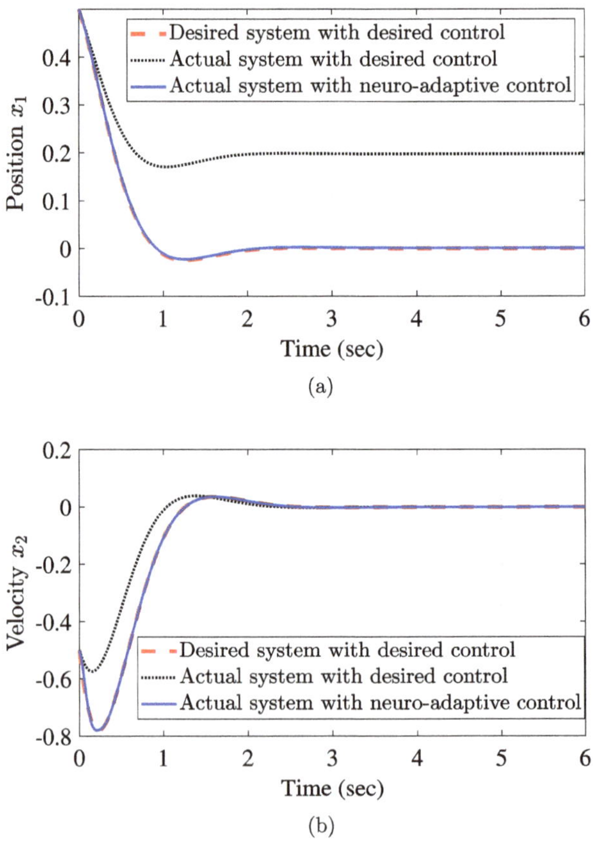

Fig. 6.9 (**a**) Position histories. (**b**) Velocity histories

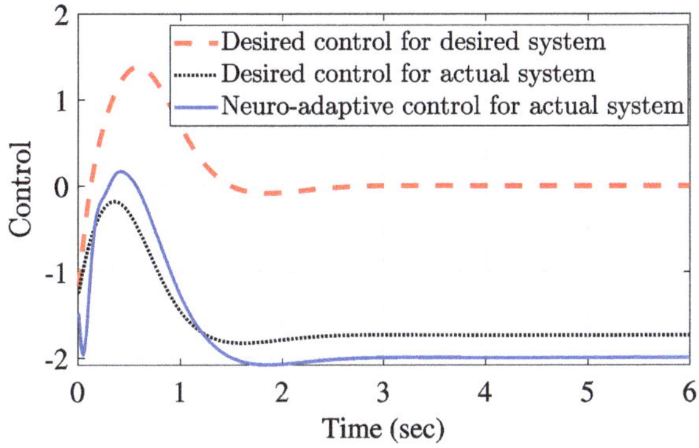

(a)

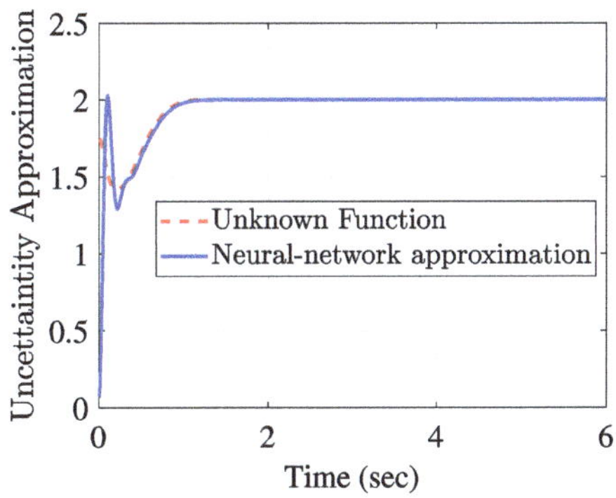

(b)

Fig. 6.10 (**a**) Control histories. (**b**) Unknown function and its approximation

It should be emphasized that the aforementioned neuro-adaptive control laws have certain design limitations. One key limitation is that the control solution in the tracking control design step ($\boldsymbol{x}^a \rightarrow \boldsymbol{x}^*$) is highly dependent on the specific type of nonlinear system. As demonstrated in previous sections, if the nonlinear system is in a control-affine form, the solution can be obtained relatively easily due to the straightforward relationship between the state and control input. However, for nonlinear systems that are not in control-affine form, deriving an appropriate control solution becomes significantly more challenging. This issue is addressed in the following section, where robust control design strategies are presented for more generic nonlinear systems.

6.6 Generic Controller Design Procedure

In this section, the controller is designed for four different important classes of nonlinear systems. These classes include (i) square and affine systems, (ii) non-square and affine systems, (iii) square and non-affine systems, and (iv) the most generic class: non-square and non-affine systems.

To begin with, consider the desired nonlinear system represented by

$$\dot{\boldsymbol{x}}^d = \boldsymbol{f}^*(\boldsymbol{x}^d, \boldsymbol{u}^d) \tag{6.93}$$

where $\boldsymbol{x}^d \in \Re^n$ is the state vector and $\boldsymbol{u}^d \in \Re^m (m \leq n)$ is the control input to the desired system. It is assumed that the desired system is completely known, and the controller is synthesized using the available nonlinear control techniques to meet the control design and performance requirements. However, the desired nonlinear system does not represent the actual system due to modeling uncertainties and unmodeled dynamics. Hence, the actual plant is assumed to have the following dynamics:

$$\dot{\boldsymbol{x}} = \boldsymbol{f}(\boldsymbol{x}, \boldsymbol{u}) + \boldsymbol{d}(\boldsymbol{x}) \tag{6.94}$$

where $\boldsymbol{x} \in \Re^n$ and $\boldsymbol{u} \in \Re^m (m \leq n)$ are the state vector and control input to the actual system, respectively. The term $\boldsymbol{d}(\boldsymbol{x}) \in \Re^n$ represents the unknown part of the system dynamics.

The objective is to design controller $\boldsymbol{u}$ in such a way that the states of the actual system track the respective states of the desired system, i.e., $\boldsymbol{x} \to \boldsymbol{x}^d$ as $t \to \infty$. This ensures that the actual system performs like the desired system.

To achieve this objective, an approximated system is constructed as an intermediate step to capture the unknown function $\boldsymbol{d}(\boldsymbol{x})$ using a neural network approximation $\hat{\boldsymbol{d}}(\boldsymbol{x})$. The dynamics of the approximated system is represented as

$$\dot{\boldsymbol{x}}^a = \boldsymbol{f}(\boldsymbol{x}, \boldsymbol{u}) + \hat{\boldsymbol{d}}(\boldsymbol{x}) + \boldsymbol{K}(\boldsymbol{x} - \boldsymbol{x}^a) \tag{6.95}$$

with the same initial condition as the actual system, i.e., $\boldsymbol{x}^a(0) = \boldsymbol{x}(0)$. Here, $\boldsymbol{K}$ is a positive definite diagonal matrix. Following the neuro-adaptive control design philosophy, the control objective will be achieved in two steps: (i) approximation of model uncertainties $\boldsymbol{d}(\boldsymbol{x})$, which leads to $\boldsymbol{x} \to \boldsymbol{x}^a$ as $t \to \infty$, and (ii) tracking controller design, which ensures $\boldsymbol{x}^a \to \boldsymbol{x}^d$ as $t \to \infty$. Both control design steps are explained in detail as follows.

It is worth mentioning that, unlike the approach taken in previous sections, where the weight update rule was derived first and the tracking control was designed subsequently, this section adopts the opposite methodology. Here, the tracking controller is designed first under the assumption that the unknown function is available, and then, the weight update rule is constructed. This approach is chosen because, depending on the system model, the model uncertainties may depend on different variables. However, it is important to note that the sequence of these steps can be chosen based on the preference of the designer.

6.6.1 Tracking Control Design

Assuming that the unknown function $\hat{\boldsymbol{d}}(\boldsymbol{x})$ is available, the objective of this step is to drive $\boldsymbol{x}^a \rightarrow \boldsymbol{x}^d$ as $t \rightarrow \infty$. In order to achieve this, consider the following stable linear error dynamics:

$$\dot{\boldsymbol{e}}^a + \boldsymbol{K}^a \boldsymbol{e}^a = \boldsymbol{0} \tag{6.96}$$

where $\boldsymbol{e}^a$ is the error, defined as $\boldsymbol{e}^a = \boldsymbol{x}^a - \boldsymbol{x}^d$, and $\boldsymbol{K}^a$ is a diagonal positive definite gain matrix.

Further, by substituting $\boldsymbol{e}^a = \boldsymbol{x}^a - \boldsymbol{x}^d$, (6.96) is

$$\left(\dot{\boldsymbol{x}}^a - \dot{\boldsymbol{x}}^d\right) + \boldsymbol{K}^a(\boldsymbol{x}^a - \boldsymbol{x}^d) = \boldsymbol{0} \tag{6.97}$$

By substituting $\dot{\boldsymbol{x}}^d$ and $\dot{\boldsymbol{x}}^a$, (6.97) can be expressed as follows:

$$\boldsymbol{f}(\boldsymbol{x}, \boldsymbol{u}) + \hat{\boldsymbol{d}}(\boldsymbol{x}) + \boldsymbol{K}(\boldsymbol{x} - \boldsymbol{x}^a) - \boldsymbol{f}^*(\boldsymbol{x}^d, \boldsymbol{u}^d) + \boldsymbol{K}^a(\boldsymbol{x}^a - \boldsymbol{x}^d) = \boldsymbol{0} \tag{6.98}$$

which can be solved for $\boldsymbol{f}(\boldsymbol{x}, \boldsymbol{u})$ as

$$\boldsymbol{f}(\boldsymbol{x}, \boldsymbol{u}) = \boldsymbol{\lambda}(\boldsymbol{x}, \boldsymbol{x}^a, \boldsymbol{x}^d, \boldsymbol{u}^d) \tag{6.99}$$

where

$$\boldsymbol{\lambda}(\boldsymbol{x}, \boldsymbol{x}^a, \boldsymbol{x}^d, \boldsymbol{u}^d) \triangleq \boldsymbol{f}^*(\boldsymbol{x}^d, \boldsymbol{u}^d) - \boldsymbol{K}^a(\boldsymbol{x}^a - \boldsymbol{x}^d) - \boldsymbol{K}(\boldsymbol{x} - \boldsymbol{x}^a) - \hat{\boldsymbol{d}}(\boldsymbol{x})$$

To derive the control solution, Eq. (6.99) must be solved for $\boldsymbol{u}$. As previously mentioned, the nature of the solution depends on the specific type of system under consideration. In order to obtain a unique and meaningful control solution from Eq. (6.99), it is necessary to utilize different solution strategies appropriate for each type of nonlinear system. The following subsections present these solution strategies for various categories of systems.

6.6.1.1 Square and Affine System

If the system is square and affine, then the following conditions are satisfied: (i) The state and control input dimensions are equal, i.e., $m = n$, and (ii) the system $\boldsymbol{f}(\boldsymbol{x}, \boldsymbol{u})$ can be expressed in the form $\boldsymbol{f}(\boldsymbol{x}, \boldsymbol{u}) = \boldsymbol{f}_1(\boldsymbol{x}) + \boldsymbol{G}_1(\boldsymbol{x})\boldsymbol{u}$, where $\boldsymbol{G}_1(\boldsymbol{x})$ is a non-singular matrix for all time.

If the system follows these properties, the control solution can be obtained as

$$\boldsymbol{u} = [\boldsymbol{G}_1(\boldsymbol{x})]^{-1}\left(\boldsymbol{\lambda}(\boldsymbol{x}, \boldsymbol{x}^a, \boldsymbol{x}^d, \boldsymbol{u}^d) - \boldsymbol{f}_1(\boldsymbol{x})\right) \tag{6.100}$$

6.6.1.2 Non-Square-Affine System

If the system is non-square but affine in control, two cases may arise: either $(m > n)$ or $(m < n)$.

Case 1 $(m > n)$ This scenario indicates that there are more control variables than states. In such cases, any optimization framework, such as linear programming, can be utilized to determine the control solution. Linear programming involves optimizing a linear objective function subject to a finite number of linear equality and inequality constraints. By formulating an appropriate objective function, one can find an optimal control solution that satisfies the control objectives while utilizing the available control variables effectively.

Case 2 $(m < n)$ In this scenario, the number of system states exceeds the number of control variables, which is a more realistic and commonly encountered situation in real applications. To derive the control solution in this case, a slack variable is introduced in the control term to make the system square. The detailed strategy for implementing this method is explored in the following paragraphs.

Since $(m < n)$, (6.98) cannot be solved for $\boldsymbol{u}$. In order to find a solution, (i) a slack-variable vector $\boldsymbol{u}_s$ is introduced, (ii) an $n \times (n-m)$ matrix $\boldsymbol{\Psi}(\boldsymbol{x})$ is designed, and (iii) $\boldsymbol{\Psi}(\boldsymbol{x})\boldsymbol{u}_s$ is added to the approximated system (6.95). Hence, the modified approximated system is defined as

$$\dot{\boldsymbol{x}}^a = \left[\boldsymbol{f}(\boldsymbol{x}, \boldsymbol{u}) + \boldsymbol{\Psi}(\boldsymbol{x})\boldsymbol{u}_s\right] + \hat{\boldsymbol{d}}(\boldsymbol{x}) - \boldsymbol{\Psi}(\boldsymbol{x})\boldsymbol{u}_s + \boldsymbol{K}(\boldsymbol{x} - \boldsymbol{x}^a) \tag{6.101}$$

with $\boldsymbol{x}^a(0) = \boldsymbol{x}(0)$. Here, $\boldsymbol{K}$ is a positive definite diagonal matrix. Furthermore, the following new variables are defined:

$$\overline{\boldsymbol{u}} \triangleq \left[\boldsymbol{u}^T, \boldsymbol{u}_s^T\right]^T \tag{6.102}$$

$$\boldsymbol{f}_a(\boldsymbol{x}, \overline{\boldsymbol{u}}) \triangleq [\boldsymbol{f}(\boldsymbol{x}, \boldsymbol{u}) + \boldsymbol{\Psi}(\boldsymbol{x})\boldsymbol{u}_s]^T \tag{6.103}$$

$$\hat{\boldsymbol{d}}_a(\boldsymbol{x}, \boldsymbol{u}_s) \triangleq \hat{\boldsymbol{d}}(\boldsymbol{x}) - \boldsymbol{\Psi}(\boldsymbol{x})\boldsymbol{u}_s \tag{6.104}$$

Using these variable definitions, (6.101) can be modified as

$$\dot{\boldsymbol{x}}^a = \boldsymbol{f}_a(\boldsymbol{x}, \overline{\boldsymbol{u}}) + \hat{\boldsymbol{d}}_a(\boldsymbol{x}, \boldsymbol{u}_s) + \boldsymbol{K}(\boldsymbol{x} - \boldsymbol{x}^a), \quad \boldsymbol{x}^a(0) = \boldsymbol{x}(0) \tag{6.105}$$

It can be noticed that (6.105) is a square system in $\boldsymbol{x}$ and $\overline{\boldsymbol{u}}$, and hence, a feasible solution for $\overline{\boldsymbol{u}}$ can be obtained. In the solution, first m elements of $\overline{\boldsymbol{u}}$ represent $\boldsymbol{u}$.

The modified uncertainty vector $\hat{\boldsymbol{d}}_a(\boldsymbol{x}, \boldsymbol{u}_s)$ is obtained as an output of the neural network, represented as $\hat{\boldsymbol{d}}_{a_i}(\boldsymbol{x}, \boldsymbol{u}_s) = \hat{\boldsymbol{w}}_i^T \boldsymbol{\phi}_i(\boldsymbol{x}, \boldsymbol{u}_s), \quad \forall i$. Here, $\hat{\boldsymbol{w}}_i$ and $\boldsymbol{\phi}_i$ are the weight vector and basis function vector of the neural network, and $[\boldsymbol{x}^T, \boldsymbol{u}_s^T]$ is the input of the neural network. The subscript i stands for each state of the system; i.e., each state equation has a separate neural network associated with it.

Similar to (6.97), the error dynamic equation for the non-square control-affine system $\boldsymbol{f}(\boldsymbol{x}, \boldsymbol{u}) = \boldsymbol{f}_1(\boldsymbol{x}) + \boldsymbol{G}_1(\boldsymbol{x})\boldsymbol{u}$ can be written as

$$\boldsymbol{f}_1(\boldsymbol{x}) + \left[\boldsymbol{G}_1(\boldsymbol{x})\ \boldsymbol{\Psi}(\boldsymbol{x})\right]\begin{bmatrix}\boldsymbol{u}\\ \boldsymbol{u}_s\end{bmatrix} + \hat{\boldsymbol{d}}_a(\boldsymbol{x}, \boldsymbol{u}_s) + \boldsymbol{K}(\boldsymbol{x} - \boldsymbol{x}^a) - \boldsymbol{f}^*(\boldsymbol{x}^d, \boldsymbol{u}^d) + \boldsymbol{K}^a(\boldsymbol{x}^a - \boldsymbol{x}^d) = \boldsymbol{0} \tag{6.106}$$

$$\boldsymbol{f}_1(\boldsymbol{x}) + \left[\boldsymbol{G}_1(\boldsymbol{x})\ \boldsymbol{\Psi}(\boldsymbol{x})\right]\overline{\boldsymbol{u}} + \hat{\boldsymbol{d}}_a(\boldsymbol{x}, \boldsymbol{u}_s) + \boldsymbol{K}(\boldsymbol{x} - \boldsymbol{x}^a) - \boldsymbol{f}^*(\boldsymbol{x}^d, \boldsymbol{u}^d) + \boldsymbol{K}^a(\boldsymbol{x}^a - \boldsymbol{x}^d) = \boldsymbol{0} \tag{6.107}$$

This leads to

$$\overline{\boldsymbol{u}} = \left[\boldsymbol{G}(\boldsymbol{x})\right]^{-1}\boldsymbol{\lambda}_s\left(\boldsymbol{x}, \boldsymbol{x}^a, \boldsymbol{x}^d, \boldsymbol{u}^d, \boldsymbol{u}_s\right) \tag{6.108}$$

where

$$\boldsymbol{G}(\boldsymbol{x}) \triangleq \left[\boldsymbol{G}_1(\boldsymbol{x})\ \boldsymbol{\Psi}(\boldsymbol{x})\right] \tag{6.109}$$

$$\boldsymbol{\lambda}_s(\boldsymbol{x}, \boldsymbol{x}^a, \boldsymbol{x}^d, \boldsymbol{u}^d, \boldsymbol{u}_s) \triangleq \boldsymbol{f}^*(\boldsymbol{x}^d, \boldsymbol{u}^d) - \boldsymbol{K}^a(\boldsymbol{x}^a - \boldsymbol{x}^d) - \hat{\boldsymbol{d}}_a(\boldsymbol{x}, \boldsymbol{u}_s) - \boldsymbol{K}(\boldsymbol{x} - \boldsymbol{x}^a) - \boldsymbol{f}_1(\boldsymbol{x}) \tag{6.110}$$

$$\hat{d}_{a_i}(\boldsymbol{x}, \boldsymbol{u}_s) \triangleq \hat{\boldsymbol{w}}_i^T\boldsymbol{\phi}_i(\boldsymbol{x}, \boldsymbol{u}_s), \quad i = 1, \ldots, n \tag{6.111}$$

It is important to choose the function $\boldsymbol{\Psi}(\boldsymbol{x})$ carefully to ensure that the square matrix $\boldsymbol{G}(\boldsymbol{x})$ does not become singular. However, selecting such a function $\boldsymbol{\Psi}(\boldsymbol{x})$ is problem-dependent, so careful consideration is required during the selection process.

It should be noted that this formulation results in a fixed-point problem in the control solution because the control vector $\overline{\boldsymbol{u}}$ contains the vector $\boldsymbol{u}_s$, and the control solution Eq. (6.108) also includes $\boldsymbol{u}_s$ on the right-hand side. In Eq. (6.108), $\boldsymbol{u}_s$ serves as an input to the neural network that approximates the uncertain function.

The solution for $\overline{\boldsymbol{u}}$ is obtained numerically using the iterative formula $\overline{\boldsymbol{u}}_{\tau+1} = \boldsymbol{G}^{-1}\left(\boldsymbol{h} - \hat{\boldsymbol{d}}_a(\boldsymbol{x}, \overline{\boldsymbol{u}}_\tau)\right)$, where $\tau = 0, 1, 2, \ldots$ and τ represent the iteration number. Here, $\boldsymbol{h}$ is defined as $\boldsymbol{h} \triangleq \boldsymbol{f}^*(\boldsymbol{x}^d, \boldsymbol{u}^d) - \boldsymbol{K}^a(\boldsymbol{x}^a - \boldsymbol{x}^d) - \boldsymbol{K}(\boldsymbol{x} - \boldsymbol{x}^a) - \boldsymbol{f}_1(\boldsymbol{x})$. The validity of this solution has been proven using the contraction mapping theorem (Padhi et al. 2007).

6.6.1.3 Square Non-Affine System

When the system is square but non-affine, the following three options are available for control design: (*i*) The form of the equation may still allow for a closed-form solution for the control variable, depending on the specific characteristics of the system. (ii) Another option is to use numerical techniques, such as the standard Newton-Raphson method, provided a good initial guess is available. (iii) For designing a controller for a non-affine system of the form $\dot{\boldsymbol{x}} = \boldsymbol{f}(\boldsymbol{x}, \boldsymbol{u})$, where $\boldsymbol{f}$ is a smooth mapping and satisfies $\boldsymbol{f}(\boldsymbol{0}, \boldsymbol{0}) = \boldsymbol{0}$, a novel method introduced in the literature can be employed. This method is elaborated as follows.

If the unforced dynamic equation $\dot{\boldsymbol{x}} = \boldsymbol{f}(\boldsymbol{x}, \boldsymbol{0}) \triangleq \boldsymbol{f}_0(\boldsymbol{x})$ is Lyapunov stable, then the system equation can be represented as

$$\dot{\boldsymbol{x}} = \boldsymbol{f}_0(\boldsymbol{x}) + \boldsymbol{G}_0(\boldsymbol{x})\boldsymbol{u} + \sum_{i=1}^{m} u_i\big(\boldsymbol{R}_i(\boldsymbol{x}, \boldsymbol{u})\boldsymbol{u}\big) \tag{6.112}$$

where $\boldsymbol{f}_0(\boldsymbol{x}) \triangleq \boldsymbol{f}(\boldsymbol{x}, \boldsymbol{0})$, $\boldsymbol{G}_0(\boldsymbol{x}) \triangleq \frac{\partial \boldsymbol{f}}{\partial \boldsymbol{u}}(\boldsymbol{x}, \boldsymbol{0}) = \big[\boldsymbol{g}_1^0(\boldsymbol{x}), \ldots, \boldsymbol{g}_m^0(\boldsymbol{x})\big] \in \Re^{n\times m}$, and $\boldsymbol{R}_i(\boldsymbol{x}, \boldsymbol{u}) \in \Re^{n\times n}$, $1 \le i \le m$.

Using the aforementioned method, the actual non-affine plant equation $\dot{\boldsymbol{x}} = \boldsymbol{f}(\boldsymbol{x}, \boldsymbol{u}) + \boldsymbol{d}(\boldsymbol{x})$ can be expressed as

$$\dot{\boldsymbol{x}} = \boldsymbol{f}_0(\boldsymbol{x}) + \boldsymbol{G}_0(\boldsymbol{x})\boldsymbol{u} + \sum_{i=1}^{m} u_i\big(\boldsymbol{R}_i(\boldsymbol{x}, \boldsymbol{u})\boldsymbol{u}\big) + \boldsymbol{d}(\boldsymbol{x}) \tag{6.113}$$

The approximate system can now be constructed as

$$\dot{\boldsymbol{x}}^a = \boldsymbol{f}_0(\boldsymbol{x}) + \boldsymbol{G}_0(\boldsymbol{x})\boldsymbol{u} + \sum_{i=1}^{m} u_i\big(\boldsymbol{R}_i(\boldsymbol{x}, \boldsymbol{u})\boldsymbol{u}\big) + \hat{\boldsymbol{d}}(\boldsymbol{x}, \boldsymbol{u}) + \boldsymbol{K}(\boldsymbol{x} - \boldsymbol{x}^a) \tag{6.114}$$

where $\boldsymbol{K}$ is a positive definite diagonal matrix. In this case, the online neural network output $\hat{\boldsymbol{d}}(\boldsymbol{x}, \boldsymbol{u})$ captures the uncertainty $\sum_{i=1}^{m} u_i\big(\boldsymbol{R}_i(\boldsymbol{x}, \boldsymbol{u})\boldsymbol{u}\big) + \boldsymbol{d}(\boldsymbol{x})$. The control solution can then be derived from the error dynamic Eq. (6.97) between the approximate state and the desired state as follows:

$$\boldsymbol{u} = [\boldsymbol{G}_0(\boldsymbol{x})]^{-1}\Big(\boldsymbol{\lambda}(\boldsymbol{x}, \boldsymbol{x}^a, \boldsymbol{x}^d, \boldsymbol{u}^d, \boldsymbol{u}) - \boldsymbol{f}_0(\boldsymbol{x})\Big) \tag{6.115}$$

where $[\boldsymbol{G}_0(\boldsymbol{x})]$ is assumed to be non-singular for all time, and

$$\begin{aligned}\boldsymbol{\lambda}(\boldsymbol{x}, \boldsymbol{x}^a, \boldsymbol{x}^d, \boldsymbol{u}^d, \boldsymbol{u}) \triangleq\ & \boldsymbol{f}^*(\boldsymbol{x}^d, \boldsymbol{u}^d) - \boldsymbol{K}^a(\boldsymbol{x}^a - \boldsymbol{x}^d) \\ & - \boldsymbol{K}(\boldsymbol{x} - \boldsymbol{x}^a) - \hat{\boldsymbol{d}}(\boldsymbol{x}, \boldsymbol{u})\end{aligned} \tag{6.116}$$

It can be observed that (6.115) presents a fixed-point problem. The control solution is determined numerically using the iterative formula:

$$\boldsymbol{u}_{\tau+1} = [\boldsymbol{G}_0(\boldsymbol{x})]^{-1}\Big(\boldsymbol{\lambda}(\boldsymbol{x}, \boldsymbol{x}^a, \boldsymbol{x}^d, \boldsymbol{u}^d, \boldsymbol{u}_\tau) - \boldsymbol{f}_0(\boldsymbol{x})\Big) \tag{6.117}$$

where $\tau = 0, 1, 2, \ldots$ and τ denote the iteration number. The validity of this solution is confirmed by proving that the mapping in (6.115) is a contraction mapping (Padhi et al. 2007).

6.6.1.4 Non-Square Non-Affine System

If the plant dynamics is both non-square and non-affine in nature, then by combining the previously explained strategies, the approximation system for the plant is constructed as follows:

$$\dot{\boldsymbol{x}}^a = \boldsymbol{f}_0(\boldsymbol{x}) + \boldsymbol{G}_0(\boldsymbol{x})\boldsymbol{u} + \hat{\boldsymbol{d}}_a(\boldsymbol{x}, \boldsymbol{u}, \boldsymbol{u}_s) + \boldsymbol{K}(\boldsymbol{x} - \boldsymbol{x}^a) + \boldsymbol{\Psi}(\boldsymbol{x})\boldsymbol{u}_s \tag{6.118}$$

which can be rewritten as

$$\dot{\boldsymbol{x}}^a = \boldsymbol{f}_0(\boldsymbol{x}) + [\boldsymbol{G}_0(\boldsymbol{x}) \quad \boldsymbol{\Psi}(\boldsymbol{x})] \begin{bmatrix} \boldsymbol{u} \\ \boldsymbol{u}_s \end{bmatrix} + \hat{\boldsymbol{d}}_a(\boldsymbol{x}, \boldsymbol{u}, \boldsymbol{u}_s) + \boldsymbol{K}(\boldsymbol{x} - \boldsymbol{x}^a) \tag{6.119}$$

where $\boldsymbol{K}$ is a positive definite diagonal matrix and

$$\hat{\boldsymbol{d}}_a(\boldsymbol{x}, \boldsymbol{u}, \boldsymbol{u}_s) \triangleq \boldsymbol{d}(\boldsymbol{x}) - \boldsymbol{\Psi}(\boldsymbol{x})\boldsymbol{u}_s + \sum_{i=1}^{m} u_i\big(\boldsymbol{R}_i(\boldsymbol{x}, \boldsymbol{u})\boldsymbol{u}\big) \tag{6.120}$$

By defining $\overline{\boldsymbol{u}} \triangleq [\boldsymbol{u}^T \quad \boldsymbol{u}_s^T]^T$ and $\boldsymbol{G}(\boldsymbol{x}) \triangleq [\boldsymbol{G}_0(\boldsymbol{x}) \quad \boldsymbol{\Psi}(\boldsymbol{x})]$, the error dynamic equation can be expressed as

$$\begin{aligned} &\boldsymbol{f}_0(\boldsymbol{x}) + \boldsymbol{G}(\boldsymbol{x})\overline{\boldsymbol{u}} + \hat{\boldsymbol{d}}_a(\boldsymbol{x}, \boldsymbol{u}, \boldsymbol{u}_s) + \boldsymbol{K}(\boldsymbol{x} - \boldsymbol{x}^a) \\ &\qquad - \boldsymbol{f}^*(\boldsymbol{x}^d, \boldsymbol{u}^d) + \boldsymbol{K}^a(\boldsymbol{x}^a - \boldsymbol{x}^d) = \boldsymbol{0} \end{aligned} \tag{6.121}$$

The control expression can be extracted as

$$\overline{\boldsymbol{u}} = [\boldsymbol{G}(\boldsymbol{x})]^{-1}\, \boldsymbol{\lambda}_s(\boldsymbol{x}, \boldsymbol{x}^a, \boldsymbol{x}^d, \boldsymbol{u}^d, \boldsymbol{u}, \boldsymbol{u}_s) \tag{6.122}$$

where

$$\begin{aligned} \boldsymbol{\lambda}_s(\boldsymbol{x}, \boldsymbol{x}^a, \boldsymbol{x}^d, \boldsymbol{u}^d, \boldsymbol{u}, \boldsymbol{u}_s) &\triangleq \boldsymbol{f}^*(\boldsymbol{x}^d, \boldsymbol{u}^d) - \boldsymbol{f}_0(\boldsymbol{x}) - \hat{\boldsymbol{d}}_a(\boldsymbol{x}, \boldsymbol{u}, \boldsymbol{u}_s) \\ &\quad - \boldsymbol{K}(\boldsymbol{x} - \boldsymbol{x}^a) - \boldsymbol{K}^a(\boldsymbol{x}^a - \boldsymbol{x}^d) \end{aligned} \tag{6.123}$$

and $\boldsymbol{G}(\boldsymbol{x})$ is assumed to be non-singular at all times. Only the first m elements of $\overline{\boldsymbol{u}}$ are required for the implementation of the control on the actual system.

The solution for $\overline{\boldsymbol{u}}$ is obtained numerically using the iterative formula $\overline{\boldsymbol{u}}_{\tau+1} = \boldsymbol{G}^{-1}\Big(\boldsymbol{h} - \hat{\boldsymbol{d}}_a(\boldsymbol{x}, \overline{\boldsymbol{u}}_\tau)\Big)$, where $\tau = 0, 1, 2, \ldots$ and τ represent the iteration number.

Here, $\boldsymbol{h}$ is defined as $\boldsymbol{h} \triangleq \left\{ \boldsymbol{f}^*(\boldsymbol{x}^d, \boldsymbol{u}^d) - \boldsymbol{K}^a(\boldsymbol{x}^a - \boldsymbol{x}^d) - \boldsymbol{K}(\boldsymbol{x} - \boldsymbol{x}^a) - \boldsymbol{f}_0(\boldsymbol{x}) \right\}$. The validity of this solution has been established using the contraction mapping theorem. The relevant proofs and conditions supporting the validity of this solution are detailed in the referenced literature Padhi et al. (2007).

6.6.2 Approximation of Model Uncertainties

This step involves approximating the model uncertainties of the system using a neural network. To develop a weight update rule for adjusting the neural network weights, the error between the actual and approximated systems is defined as follows:

$$\boldsymbol{e} = \boldsymbol{x} - \boldsymbol{x}^a \tag{6.124}$$

The time differentiation of (6.124) gives

$$\dot{\boldsymbol{e}} = \dot{\boldsymbol{x}} - \dot{\boldsymbol{x}}^a \tag{6.125}$$

Substituting (6.94) and (6.95) into (6.125) leads to

$$\dot{\boldsymbol{e}} = \boldsymbol{d}(\boldsymbol{x}) - \hat{\boldsymbol{d}}_a(\boldsymbol{x}, \boldsymbol{u}, \boldsymbol{u}_s) - \boldsymbol{K}\boldsymbol{e} \tag{6.126}$$

where note that the $\hat{\boldsymbol{d}}_a(\boldsymbol{x}, \boldsymbol{u}, \boldsymbol{u}_s)$ represents the most general form of the neural network output. Depending on the system type, the model uncertainties and the corresponding neural network outputs vary. These differences are summarized in Table 6.1.

Since the matrix $\boldsymbol{K}$ is a positive definite diagonal matrix, (6.126) can be analyzed in terms of its ith component. Here, i^{th} components of $\boldsymbol{e}$, $\boldsymbol{d}(\boldsymbol{x})$, and $\hat{\boldsymbol{d}}(\boldsymbol{x}, \boldsymbol{u}, \boldsymbol{u}_s)$ are represented by e_i, $d_i(\boldsymbol{x})$, and $\hat{d}_{a_i}(\boldsymbol{x}, \boldsymbol{u}, \boldsymbol{u}_s)$, respectively. Therefore, the ith component of the error dynamics in Eq. (6.126) can be expressed as follows:

$$\dot{e}_i = d_i(\boldsymbol{x}) - \hat{d}_{a_i}(\boldsymbol{x}, \boldsymbol{u}, \boldsymbol{u}_s) - k_i e_i \tag{6.127}$$

Table 6.1 Uncertainties and neural network (NN) outputs for different types of systems

System type	Uncertainty	NN output
Square, affine	$\boldsymbol{d}(\boldsymbol{x})$	$\hat{\boldsymbol{d}}(\boldsymbol{x})$
Non-square, affine	$\boldsymbol{d}(\boldsymbol{x}) - \boldsymbol{\Psi}(\boldsymbol{x})\,\boldsymbol{u}_s$	$\hat{\boldsymbol{d}}_a(\boldsymbol{x}, \boldsymbol{u}_s)$
Square, non-affine	$\boldsymbol{d}(\boldsymbol{x}) + \sum_{i=1}^{m} u_i(\boldsymbol{R}_i(\boldsymbol{x}, \boldsymbol{u})\boldsymbol{x})$	$\hat{\boldsymbol{d}}(\boldsymbol{x}, \boldsymbol{u})$
Non-square, non-affine	$\boldsymbol{d}(\boldsymbol{x}) - \boldsymbol{\Psi}(\boldsymbol{x})\,\boldsymbol{u}_s$ $+ \sum_{i=1}^{m} u_i(\boldsymbol{R}_i(\boldsymbol{x}, \boldsymbol{u})\boldsymbol{u})$	$\hat{\boldsymbol{d}}_a(\boldsymbol{x}, \boldsymbol{u}, \boldsymbol{u}_s)$

Utilizing linear-in-weight architecture of neural networks, the uncertainties $d_i(\boldsymbol{x})$ and the approximation of uncertainties $\hat{d}_{a_i}(\boldsymbol{x}, \boldsymbol{u}, \boldsymbol{u}_s)$ are defined as

$$d_i(\boldsymbol{x}) = \boldsymbol{w}_i^T \boldsymbol{\phi}(\boldsymbol{x}, \boldsymbol{u}, \boldsymbol{u}_s) + \epsilon_i \tag{6.128}$$

$$\hat{d}_{a_i}(\boldsymbol{x}, \boldsymbol{u}, \boldsymbol{u}_s) = \hat{\boldsymbol{w}}_i^T \boldsymbol{\phi}(\boldsymbol{x}, \boldsymbol{u}, \boldsymbol{u}_s) \tag{6.129}$$

where $\boldsymbol{w}_i$ and $\hat{\boldsymbol{w}}_i$ are the optimal and approximated weights and $\boldsymbol{\phi}(\boldsymbol{x})$ is the known basis function vector. The term ϵ_i is the finite approximation error. Note that $\boldsymbol{w}_i$ is a vector with constant elements.

Using these definitions, the modified error dynamic equation is obtained as

$$\dot{e}_i = \tilde{\boldsymbol{w}}_i^T \boldsymbol{\phi}(\boldsymbol{x}, \boldsymbol{u}, \boldsymbol{u}_s) + \epsilon_i - k_i e_i \tag{6.130}$$

where weight error $\tilde{\boldsymbol{w}}_i^T$ is defined as $\tilde{\boldsymbol{w}}_i^T \triangleq \boldsymbol{w}_i^T - \hat{\boldsymbol{w}}_i^T$.

Similar to the previous sections, the weight update rule is constructed using the Lyapunov stability theory. Consider the following quadratic Lyapunov function:

$$V_i = \frac{\beta_i e_i^2}{2} + \frac{\tilde{\boldsymbol{w}}_i^T \tilde{\boldsymbol{w}}_i}{2\gamma_i}$$

where β_i is a positive constant and $\gamma_i > 0$ is known as the learning rate constant.

The time derivative of the Lyapunov function in (6.131) is

$$\dot{V}_i = \beta_i e_i \dot{e}_i + \frac{\tilde{\boldsymbol{w}}_i^T \dot{\tilde{\boldsymbol{w}}}_i}{\gamma_i} \tag{6.131}$$

Further, using error dynamics (6.130), the time derivative of the Lyapunov function is modified as

$$\begin{aligned}\dot{V}_i &= \beta_i e_i \left(\tilde{\boldsymbol{w}}_i^T \boldsymbol{\phi}(\boldsymbol{x}, \boldsymbol{u}, \boldsymbol{u}_s) + \epsilon_i - k_i e_i\right) - \frac{\tilde{\boldsymbol{w}}_i^T \dot{\hat{\boldsymbol{w}}}_i}{\gamma_i} \\ &= \tilde{\boldsymbol{w}}_i^T \left[\beta_i e_i \boldsymbol{\phi}(\boldsymbol{x}, \boldsymbol{u}, \boldsymbol{u}_s) - \frac{\dot{\hat{\boldsymbol{w}}}_i}{\gamma_i}\right] + \beta_i e_i \epsilon_i - k_i \beta_i e_i^2\end{aligned}$$

By selecting the weight update rule as

$$\dot{\hat{\boldsymbol{w}}}_i = \gamma_i \beta_i e_i \boldsymbol{\phi}(\boldsymbol{x}, \boldsymbol{u}, \boldsymbol{u}_s) - \sigma_i \hat{\boldsymbol{w}}_i \tag{6.132}$$

where $\sigma_i > 0$ is a constant. With the selection of this learning rule, the weights will be bounded even if the error is nonzero. The bound of nonzero error due to ϵ_i can be calculated from the resultant $\dot{V}_i$ expression as follows:

$$\dot{V}_i = -k_i\beta_i e_i^2 + \beta_i e_i \epsilon_i + \frac{\sigma_i \tilde{\boldsymbol{w}}_i^T \hat{\boldsymbol{w}}_i}{\gamma_i} \tag{6.133}$$

By analyzing the term $\tilde{\boldsymbol{w}}_i^T \hat{\boldsymbol{w}}_i$, the following is obtained:

$$\tilde{\boldsymbol{w}}_i^T \hat{\boldsymbol{w}}_i = \tilde{\boldsymbol{w}}_i^T \boldsymbol{w}_i - \tilde{\boldsymbol{w}}_i^T \tilde{\boldsymbol{w}}_i \tag{6.134}$$

where

$$\tilde{\boldsymbol{w}}_i^T \boldsymbol{w}_i = \frac{1}{2}\left(-\hat{\boldsymbol{w}}_i^T \hat{\boldsymbol{w}}_i + \tilde{\boldsymbol{w}}_i^T \tilde{\boldsymbol{w}}_i + \boldsymbol{w}_i^T \boldsymbol{w}_i\right) \tag{6.135}$$

Hence, $\tilde{\boldsymbol{w}}_i^T \hat{\boldsymbol{w}}_i$ can be rewritten as

$$\tilde{\boldsymbol{w}}_i^T \hat{\boldsymbol{w}}_i = \frac{1}{2}\left(-\hat{\boldsymbol{w}}_i^T \hat{\boldsymbol{w}}_i - \tilde{\boldsymbol{w}}_i^T \tilde{\boldsymbol{w}}_i + \boldsymbol{w}_i^T \boldsymbol{w}_i\right) \tag{6.136}$$

$$\leq \frac{1}{2}\left(-\|\tilde{\boldsymbol{w}}_i\|^2 - \|\hat{\boldsymbol{w}}_i\|^2 + \|\boldsymbol{w}_i\|^2\right) \tag{6.137}$$

Therefore, the term $\frac{\sigma_i \tilde{\boldsymbol{w}}_i^T \hat{\boldsymbol{w}}_i}{\gamma_i}$ in (6.133) can be modified as

$$\frac{\sigma_i \tilde{\boldsymbol{w}}_i^T \hat{\boldsymbol{w}}_i}{\gamma_i} \leq \frac{\sigma_i}{2\gamma_i}\left(-\|\tilde{\boldsymbol{w}}_i\|^2 - \|\hat{\boldsymbol{w}}_i\|^2 + \|\boldsymbol{w}_i\|^2\right) \tag{6.138}$$

Following this, the Lyapunov derivative $\dot{V}_i$ can be expressed as

$$\dot{V}_i \leq -k_i\beta_i e_i^2 + \beta_i e_i \epsilon_i - \frac{\sigma_i}{2\gamma_i}\|\tilde{\boldsymbol{w}}_i\|^2 - \frac{\sigma_i}{2\gamma_i}\|\hat{\boldsymbol{w}}_i\|^2 + \frac{\sigma_i}{2\gamma_i}\|\boldsymbol{w}_i\|^2 \tag{6.139}$$

$$\leq \frac{\beta_i^2 e_i^2}{2} + \frac{\epsilon_i^2}{2} - k_i\beta_i e_i^2 - \frac{\sigma_i}{2\gamma_i}\|\tilde{\boldsymbol{w}}_i\|^2 - \frac{\sigma_i}{2\gamma_i}\|\hat{\boldsymbol{w}}_i\|^2 + \frac{\sigma_i}{2\gamma_i}\|\boldsymbol{w}_i\|^2 \tag{6.140}$$

$$\leq -c_1 e_i^2 + c_2 - \frac{\sigma_i}{2\gamma_i}\|\tilde{\boldsymbol{w}}_i\|^2 - \frac{\sigma_i}{2\gamma_i}\|\hat{\boldsymbol{w}}_i\|^2 \tag{6.141}$$

where constants c_1 and c_2 are defined as $c_1 \triangleq \left(k_i\beta_i - \frac{\beta_i^2}{2}\right)$ and $c_2 \triangleq \left(\frac{\epsilon_i^2}{2} + \frac{\sigma_i}{2\gamma_i}\|\boldsymbol{w}_i\|^2\right)$, respectively. The Lyapunov derivative $\dot{V}_i < 0$ if $(-c_1 e_i^2 + c_2) < 0$. This leads to $|e_i| > \sqrt{\frac{c_2}{c_1}}$.

Thus, it can be observed that selecting a sufficiently small σ_i and choosing an appropriate set of basis functions will minimize the approximation error ϵ_i, which will help in maintaining a small error bound. Furthermore, carefully selecting gains k_i and β_i will further reduce the error bound. Ultimately, the error bound for the weight update scheme is given by $\sqrt{\frac{c_2}{c_1}}$.

Example 6.5 Similar to the previous sections, the same Van der Pol system is used to demonstrate the efficacy of the control design. The dynamic equation of the Van der Pol system is given by

$$\begin{aligned}\dot{x}_1 &= x_2 \\ \dot{x}_2 &= (1 - x_1^2)x_2 - x_1 + d(x_1, x_2) + (1 + x_1^2 + x_2^2)u\end{aligned} \tag{6.142}$$

where x_1 and x_2 are states, u is the control input for the system, and $d(x_1, x_2)$ is the unknown function present in the system dynamics. The objective of the controller is to stabilize the plant.

The desired known system dynamic equation of the plant is defined by

$$\begin{aligned}\dot{x}_1^d &= x_2^d \\ \dot{x}_2^d &= (1 - (x_1^d)^2)x_2^d - x_1^d + (1 + (x_1^d)^2 + (x_2^d)^2)u^d\end{aligned} \tag{6.143}$$

The stabilizing desired controller for the desired system can be designed using dynamic inversion control design. The desired control expression is given as

$$u^d = -\frac{(1 - (x_1^d)^2)x_2^d - x_1^d + 2\zeta\omega_n x_2^d + \omega_n^2 x_1^d}{(1 + (x_1^d)^2 + (x_2^d)^2)}$$

where $\omega_n = \frac{20}{2\pi}$ and $\zeta = 0.7$.

Following the adaptive control design steps, the approximated system is constructed as

$$\begin{aligned}\dot{x}_1^a &= x_2 + k_1(x_1 - x_1^a) \\ \dot{x}_2^a &= (1 - x_1^2)x_2 - x_1 + \hat{d}(x_1, x_2) + (1 + x_1^2 + x_2^2)u + k_2(x_2 - x_2^a)\end{aligned} \tag{6.144}$$

(continued)

Example 6.5 (continued)
Since the problem is in non-square form, $\boldsymbol{\Psi}$ is selected as $\boldsymbol{\Psi} = [-10 \quad 10]^T$. Using $\boldsymbol{\Psi}$, the approximated system is modified as

$$\begin{aligned}\dot{x}_1^a &= x_2 + k_1(x_1 - x_1^a) - 10u_s + \hat{d}_1^a(x_1, x_2, u_s)\\ \dot{x}_2^a &= (1 - x_1^2)x_2 - x_1 + 10u_s + \hat{d}_2^a(x_1, x_2, u_s)\\ &\quad +(1 + x_1^2 + x_2^2)u + k_2(x_2 - x_2^a)\end{aligned} \tag{6.145}$$

where $10u_s$ represents the uncertainty introduced to make the system square, and it will be approximated by $\hat{d}_1^a(x_1, x_2, u_s)$. The sum of the model uncertainties and $10u_s$ present in the second state will be approximated by $\hat{d}_2^a(x_1, x_2, u_s)$. The augmented controller can be expressed as follows:

$$\overline{\boldsymbol{u}} = \begin{bmatrix} 0 & -10 \\ 1 + x_1^2 + x_2^2 & 10 \end{bmatrix}^{-1} \left(\begin{bmatrix} -k_1^a \left(x_1^a - x_1^d\right) \\ -k_2^a \left(x_2^a - x_2^d\right) \end{bmatrix} - \begin{bmatrix} \Delta f_1 \\ \Delta f_2 \end{bmatrix} - \begin{bmatrix} k_1 \left(x_1 - x_1^a\right) \\ k_2 \left(x_2 - x_2^a\right) \end{bmatrix} \right) \tag{6.146}$$

where

$$\begin{bmatrix} \Delta f_1 \\ \Delta f_2 \end{bmatrix} = \begin{bmatrix} (x_2 - x_2^d) + \hat{d}_1^a(x_1, x_2, u_s) \\ (1 - x_1^2)x_2 - x_1 + \hat{d}_2^a(x_1, x_2, u_s) - (1 - x_1^2)x_2^d + x_1^d \\ -(1 + (x_1^d)^2 + (x_2^d)^2)u^d \end{bmatrix} \tag{6.147}$$

where k_1, k_2, k_1^a, and k_2^a are positive gain constants. As previously mentioned, the control input $\boldsymbol{u}$ can be derived from $\overline{\boldsymbol{u}}$. Furthermore, the solution is obtained iteratively, with the initial step assuming $u_s = 0$.

The terms $\hat{d}_1^a(x_1, x_2, u_s)$ and $\hat{d}_2^a(x_1, x_2, u_s)$ are calculated as $\hat{d}_1^a(x_1, x_2, u_s) = \hat{\boldsymbol{w}}_1^T \boldsymbol{\phi}(x_1, x_2, u_s)$ and $\hat{d}_2^a(x_1, x_2, u_s) = \hat{\boldsymbol{w}}_2^T \boldsymbol{\phi}(x_1, x_2, u_s)$, respectively. The weights are updated from the learning rules $\dot{\hat{\boldsymbol{w}}}_1 = \gamma_1 \beta_1 (x_1 - x_1^a) \boldsymbol{\phi}(x_1, x_2, u_s) - \sigma_1 \hat{\boldsymbol{w}}_1$ and $\dot{\hat{\boldsymbol{w}}}_2 = \gamma_2 \beta_2 (x_2 - x_2^a) \boldsymbol{\phi}(x_1, x_2, u_s) - \sigma_2 \hat{\boldsymbol{w}}_2$, respectively.

To simulate the controller response, the unknown function $d(x_1, x_2)$ is assumed as $d(x) = 2\cos(x_2)$. The controller gains $k_1, k_2, k_1^a, k_2^a, k_3, \sigma_1, \sigma_2, \gamma_1$, and γ_2 are selected as $k_1 = 1, k_2 = 5, k_1^a = 1, k_2^a = 5, k_3 = 1, \sigma_1 = 10^{-6}$, $\sigma_2 = 10^{-6}, \gamma_1 = 2$, and $\gamma_2 = 1$, respectively. The basis functions to approximate the unknown function are chosen as $\boldsymbol{\phi}(x) = \text{kron}(\boldsymbol{c}_1, \text{kron}(\boldsymbol{c}_2, \boldsymbol{c}_3))$, where "kron" is the Kronecker product, $\boldsymbol{c}_1 = [1, \sin(x_1), \cos(x_1)]^T$, $\boldsymbol{c}_2 = [1, \sin(x_2), \cos(x_2)]^T$, and $\boldsymbol{c}_3 = [1, \sin(u_s), \cos(u_s)]^T$.

(continued)

Example 6.5 (continued)
The position, velocity, and control histories of the system are illustrated in Figs. 6.11a and b, and 6.12a. These figures show that while the desired controller performs effectively for the designed system, it does not achieve the same level of performance for the actual system. In contrast, the adaptive controller ensures the desired tracking with almost zero error, demonstrating its effectiveness. Additionally, the approximation of the unknown function is shown in Fig. 6.12b, where the neural network approximates the unknown function with a finite approximation error and compensates for the uncertainties in the system dynamics.

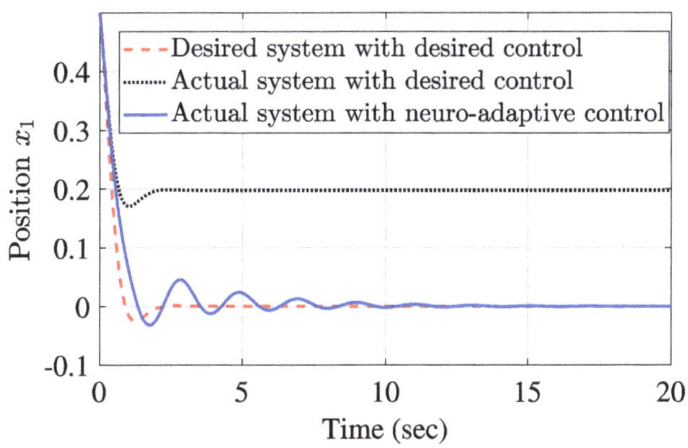

(a)

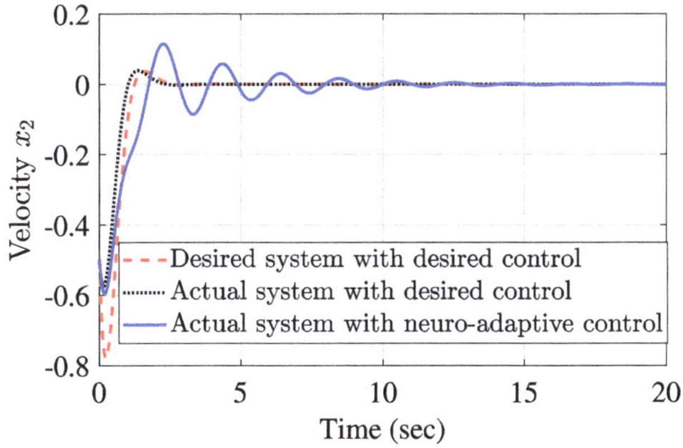

(b)

Fig. 6.11 (**a**) Position histories. (**b**) Velocity histories

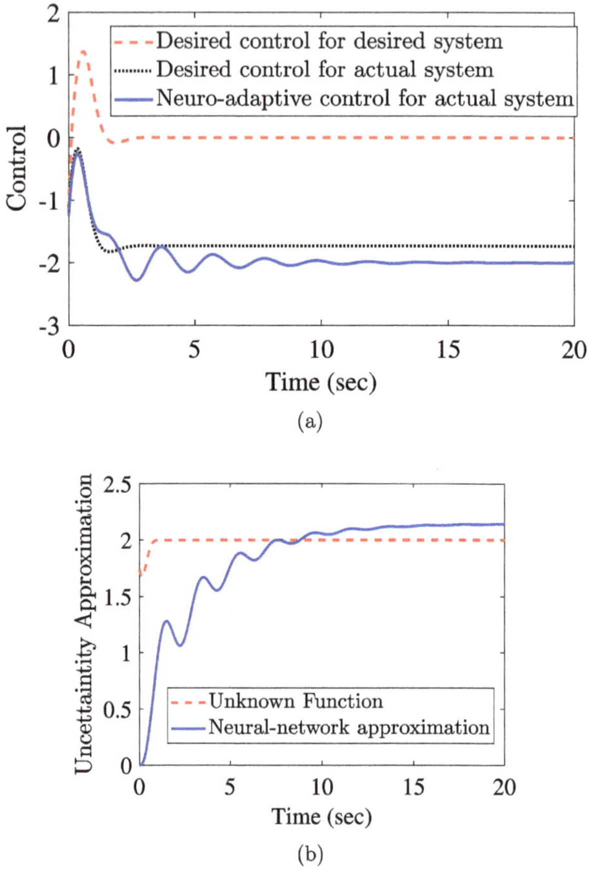

Fig. 6.12 (**a**) Control histories. (**b**) Unknown function and its approximation

Example 6.6 In this example, the control design for a double inverted pendulum system with uncertainties is chosen to demonstrate the effectiveness of the neuro-adaptive controller. The non-square and non-affine dynamic equation of the system is given by

$$\dot{x}_1 = x_2$$

$$\dot{x}_2 = (\alpha_1 + \Delta\alpha_1)\sin(x_1) + \beta_1 + \xi_1 \tanh(u_1) + \sigma_1 \sin(x_4) + \tilde{f}_1(\boldsymbol{x})$$

$$\dot{x}_3 = x_4$$

$$\dot{x}_4 = (\alpha_2 + \Delta\alpha_2)\sin(x_3) + \beta_2 + \xi_2 \tanh(u_2) + \sigma_2 \sin(x_2) + \tilde{f}_2(\boldsymbol{x})$$

(continued)

Example 6.6 (continued)
where x_1 and x_2 denote the position and velocity of the first mass, respectively. Similarly, x_3 and x_4 denote the position and velocity of the second mass, respectively. The state vector $\boldsymbol{x}$ is defined as $\boldsymbol{x} = [x_1, x_2, x_3, x_4]^T$, and the control torque input applied by the servo motors is represented by u_1 and u_2. Moreover, for $i = 1, 2$, the parameters are defined as

$$\alpha_i \triangleq \left(\frac{m_i g r}{J_i} - \frac{kr^2}{4J_i}\right)$$
$$\beta_i \triangleq \frac{kr}{2J_i}(l - b)$$
$$\xi_i \triangleq \frac{u_{i_{\max}}}{J_i}$$
$$\sigma_i \triangleq \frac{kr^2}{4J_i}$$

The system parameters and their values are listed in Table 6.2.

The desired system for the actual system is given by

$$\dot{x}_1^d = x_2^d$$
$$\dot{x}_2^d = \alpha_1 \sin(x_1^d) + \beta_1 + \xi_1 \tanh(u_1^d) + \sigma_1 \sin(x_4^d)$$
$$\dot{x}_3^d = x_4^d$$
$$\dot{x}_4^d = \alpha_2 \sin(x_3^d) + \beta_2 + \xi_2 \tanh(u_2^d) + \sigma_2 \sin(x_2^d)$$

The objective of the desired controller is to ensure that $x_1^d \to x_1^*$ and $x_3^d \to x_3^*$, where reference signals x_1^* and x_3^* are defined as $x_1^* = x_3^* = \sin(2\pi t/10)$. A dynamic inversion controller, defined by the closed-loop error dynamics $(\ddot{\boldsymbol{x}}^d - \ddot{\boldsymbol{x}}^*) + \boldsymbol{K}_d(\dot{\boldsymbol{x}}^d - \dot{\boldsymbol{x}}^*) + \boldsymbol{K}_p(\boldsymbol{x}^d - \boldsymbol{x}^*) = \boldsymbol{0}$ is used to achieve the desired controller design objectives, where $\boldsymbol{x}^d = [x_1^d \quad x_3^d]^T$, $\boldsymbol{x}^* = [x_1^* \quad x_3^*]^T$, and gain matrices $\boldsymbol{K}_d$ and $\boldsymbol{K}_p$ are selected as identity matrix.

The controller effectiveness is demonstrated by selecting $\Delta\alpha_1$ and $\Delta\alpha_2$ as 20% of α_1 and α_2, respectively. Uncertainties $\tilde{f}_1(\boldsymbol{x})$ and $\tilde{f}_2(\boldsymbol{x})$ are assumed as exponential functions and selected as $\tilde{f}_1(\boldsymbol{x}) = 0.1e^{0.01x_1}$ and $\tilde{f}_2(\boldsymbol{x}) = 0.1e^{0.01x_1}$, respectively. The uncertainties of the system can be represented as $\boldsymbol{d}(\boldsymbol{x}) = [d_1(\boldsymbol{x}), d_2(\boldsymbol{x}), d_3(\boldsymbol{x}), d_4(\boldsymbol{x})]^T$, where $d_1(\boldsymbol{x}) = 0$, $d_2(\boldsymbol{x}) = \Delta\alpha_1 \sin(x_1) + 0.1e^{0.01x_1}$, $d_3(\boldsymbol{x}) = 0$, and $d_4(\boldsymbol{x}) = \Delta\alpha_2 \sin(x_3) + 0.1e^{0.01x_3}$, respectively.

(continued)

Example 6.6 (continued)

Since the system is in non-square and non-affine form with fewer control variables than the states, the variables $\boldsymbol{f}_0(\boldsymbol{x})$ and $\boldsymbol{G}_0(\boldsymbol{x})$ can be defined as

$$\boldsymbol{f}_0(\boldsymbol{x}) \triangleq \begin{bmatrix} x_2 \\ \alpha_1 \sin(x_1) + \beta_1 + \sigma_1 \sin(x_4) \\ x_4 \\ \alpha_2 \sin(x_3) + \beta_2 + \sigma_2 \sin(x_2) \end{bmatrix} \tag{6.148}$$

$$\boldsymbol{G}_0(\boldsymbol{x}) \triangleq \begin{bmatrix} 0 & 0 \\ \xi_1 & 0 \\ 0 & 0 \\ 0 & \xi_2 \end{bmatrix} \tag{6.149}$$

The actual plant can be expressed as

$$\dot{\boldsymbol{x}} = \boldsymbol{f}_0(\boldsymbol{x}) + \boldsymbol{G}_0(\boldsymbol{x})\boldsymbol{u} + \sum_{i=1}^{m} u_i(\boldsymbol{R}_i(\boldsymbol{x}, \boldsymbol{u})\boldsymbol{u}) + \boldsymbol{d}(\boldsymbol{x}) \tag{6.150}$$

Since $\boldsymbol{G}_0(\boldsymbol{x})$ is not square, a slack variable is added and $\boldsymbol{\Psi}(\boldsymbol{x})$ is selected as $\boldsymbol{\Psi}(\boldsymbol{x}) = \begin{bmatrix} -10 & 10 & 0 & 0 \\ 10 & -10 & 10 & -10 \end{bmatrix}$.

In simulations, the gain matrix for stable linear error dynamic equation is chosen as $\boldsymbol{K} = \operatorname{diag}\left(\frac{1}{\tau_1}, \frac{1}{\tau_2}, \frac{1}{\tau_3}, \frac{1}{\tau_4}\right)$ with $\tau_1 = \tau_2 = \tau_3 = \tau_4 = 0.2$. The control solution vector was obtained as

$$\boldsymbol{v} = \left[\boldsymbol{G}_0(\boldsymbol{x})\boldsymbol{\Psi}^{-1}\right]\left(-\left(\boldsymbol{f}_0(\boldsymbol{x}) + \hat{\boldsymbol{d}}_a(\boldsymbol{x}, \boldsymbol{u}, \boldsymbol{u}_s)\right) + (\boldsymbol{u} - \boldsymbol{u}^a) - \dot{\boldsymbol{x}}^d + \boldsymbol{K}(\boldsymbol{x}^a - \boldsymbol{x}^d)\right)$$

The numerical solution was obtained using

$$\boldsymbol{v}_{k+1} = \left[\boldsymbol{G}_0(\boldsymbol{x})\boldsymbol{\Psi}^{-1}\right]\left(-\left(\boldsymbol{f}_0(\boldsymbol{x}) + \hat{\boldsymbol{d}}^a(\boldsymbol{x}, \boldsymbol{v}_k)\right) + (\boldsymbol{v} - \boldsymbol{x}^a) - \dot{\boldsymbol{x}}^d + \boldsymbol{K}(\boldsymbol{x}^a - \boldsymbol{x}^d)\right)$$

After solving for $\boldsymbol{v}$, control $\boldsymbol{u}$ can be extracted from the first two elements. The basis function vectors $\boldsymbol{\phi}_1(\boldsymbol{x})$, $\boldsymbol{\phi}_2(\boldsymbol{x})$, $\boldsymbol{\phi}_3(\boldsymbol{x})$, and $\boldsymbol{\phi}_4(\boldsymbol{x})$ are selected as

$$\boldsymbol{\phi}_1(\boldsymbol{x}) = \boldsymbol{\phi}_2(\boldsymbol{x}) = \operatorname{kron}(\boldsymbol{c}_{11}, \operatorname{kron}(\boldsymbol{c}_{12}, \operatorname{kron}(\boldsymbol{c}_3, \operatorname{kron}(\boldsymbol{c}_4, \operatorname{kron}(\boldsymbol{c}_5, \boldsymbol{c}_6)))))$$

$$\boldsymbol{\phi}_3(\boldsymbol{x}) = \boldsymbol{\phi}_4(\boldsymbol{x}) = \operatorname{kron}(\boldsymbol{c}_{21}, \operatorname{kron}(\boldsymbol{c}_{22}, \operatorname{kron}(\boldsymbol{c}_3, \operatorname{kron}(\boldsymbol{c}_4, \operatorname{kron}(\boldsymbol{c}_5, \boldsymbol{c}_6)))))$$

(continued)

Example 6.6 (continued)
where

$$c_{11} = \left[1 \ \sin(x_1) \ \cos(x_1)\right]^T \tag{6.151}$$

$$c_{12} = \left[1 \ \sin(x_2) \ \cos(x_2)\right]^T \tag{6.152}$$

$$c_{21} = \left[1 \ \sin(x_3) \ \cos(x_3)\right]^T \tag{6.153}$$

$$c_{22} = \left[1 \ \sin(x_4) \ \cos(x_4)\right]^T \tag{6.154}$$

$$c_3 = \left[1 \ \sin(u_1)\right]^T \tag{6.155}$$

$$c_4 = \left[1 \ \sin(u_2)\right]^T \tag{6.156}$$

$$c_5 = \left[1 \ \sin(u_{s_1})\right]^T \tag{6.157}$$

$$c_6 = \left[1 \ \sin(u_{s_2})\right]^T \tag{6.158}$$

The weight update rule parameters selected were $\sigma_1 = \sigma_2 = \sigma_3 = \sigma_4 = 1 \times 10^{-6}$ and $\gamma_1 = \gamma_2 = \gamma_3 = \gamma_4 = \gamma_s = 20$. For the first iteration on the control solution, the scheme $\boldsymbol{v} = [u_{1d} \quad u_{2d} \quad 0 \quad 0]^T$ is used.

The position, velocity, and control histories of the system are illustrated in Figs. 6.13a–b, 6.14a–b, and 6.15a–b. These figures show that while the desired controller performs effectively for the designed system, it does not achieve the same level of performance for the actual system. In contrast, the adaptive controller ensures the desired tracking with almost zero error, demonstrating its effectiveness. Additionally, the approximation of the unknown function is shown in Fig. 6.16a and b, where the neural network approximates the unknown function well and compensates for the uncertainties in the system dynamics.

6.7 Summary

This chapter provides an overview of neuro-adaptive control design for different classes of nonlinear systems. To illustrate the fundamental concept, neuro-adaptive control design is first presented for a relative degree one nonlinear system with structured uncertainties. This design is further improved by incorporating gradient information into the learning rule, leading to better approximation. Subsequently, the neuro-adaptive control design is extended to a practically relevant Euler-Lagrange (EL) system with structured and unstructured uncertainties. Here, the con-

Table 6.2 System parameters and their values

System parameter	Value	Units
End mass of pendulum 1 (m_1)	2	kg
End mass of pendulum 1 (m_2)	2.5	kg
Moment of inertia (J_1)	0.5	kg m^2
Moment of inertia (J_2)	0.625	kg m^2
Spring constant of connecting spring (k)	100	N/m
Pendulum height (r)	0.5	m
Natural length of spring (l)	0.5	m
Gravitational acceleration (g)	9.81	m/s^2
Distance between pendulum hinges (b)	0.4	m
Maximum torque input ($u_{1_{\max}}$)	20	Nm
Maximum torque input ($u_{2_{\max}}$)	20	Nm

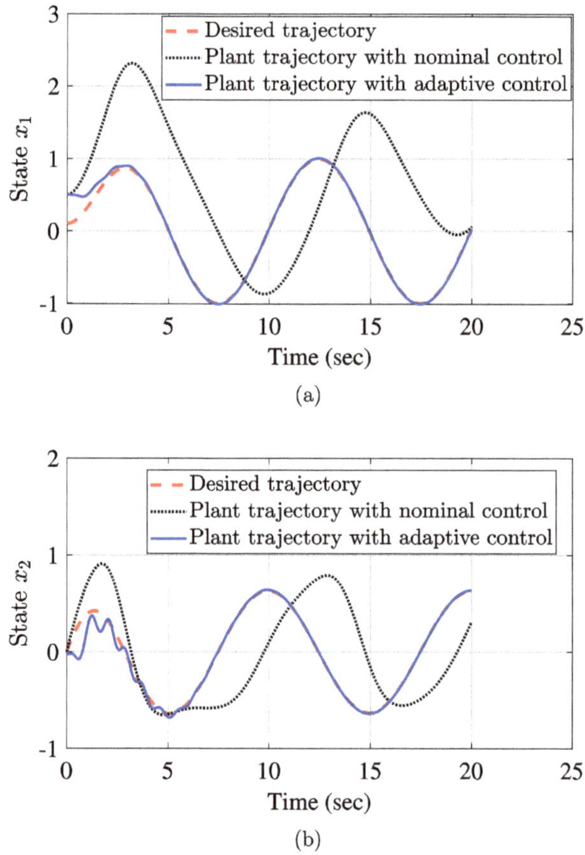

Fig. 6.13 (**a**) Position x_1 histories. (**b**) Velocity x_1 histories

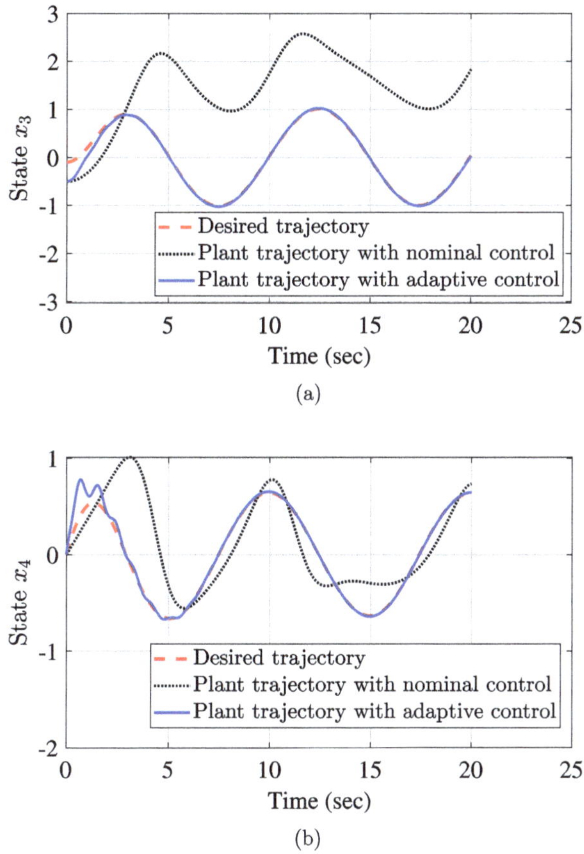

Fig. 6.14 (**a**) Position x_3 histories. (**b**) Velocity x_4 histories

cept of system transformation is also explored to enhance controller performance. Finally, a neuro-adaptive control design is introduced for a generic, non-affine, non-square nonlinear system. Each design concept is explained through examples for better understanding. Readers interested in a more in-depth understanding of neuro-adaptive control design may refer to additional sources Padhi et al. (2007), Sachan and Padhi (2020a), Padhi and Kothari (2007), Ambati and Padhi (2017), Rajasekaran et al. (2009), Padhi and Kothari (2007), Sachan and Padhi (2020b), Maity et al. (2018), and Mathavaraj and Padhi (2021).

6.8 Exercise Problems

1. Design the stabilizing controller for the following nonlinear systems $(a - h)$ using the neuro-adaptive control design philosophy. The desired controller for these systems can be designed using the methods given in Chap. 3.

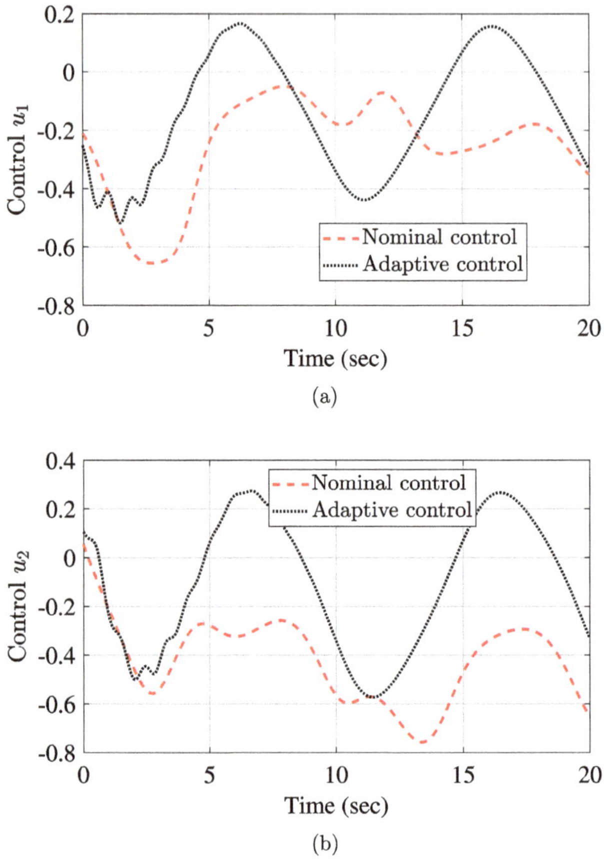

Fig. 6.15 (**a**) Torque control u_1. (**b**) Torque control u_2

(a) $\dot{x} = -x^2 + (x-3)\,u + d(x)$
where $d(x)$ is structured uncertainty, defined as $d(x) = 4x$.

(b) $\ddot{x} = -\sin x + u + d(x)$
where $d(x)$ is structured uncertainty, defined as $d(x) = 0.4 \sin x$.

(c) $\ddot{x} = \left(1 - x^2 - \dot{x}^2\right)\dot{x} + u + d(x)$
where $d(x)$ is an unstructured uncertainty, defined as $d(x) = 0.3\dot{x} + 0.1x^2$.

(d) $\ddot{x} - 0.7(1-x^2)\dot{x} - x^3 + (1+\dot{x}^2)u + d(x) = 0$
where $d(x)$ is an unstructured uncertainty, defined as $d(x) = 0.1x\dot{x} + 0.1x^3$.

(e) $\ddot{x} + u + \frac{x}{1+\sin^2 x} + d(x) = 0$
where $d(x)$ is an unstructured uncertainty, defined as $d(x) = 0.1\dot{x}\cos x$.

(f) $\begin{bmatrix} \dot{x}_1 \\ \dot{x}_2 \end{bmatrix} = \begin{bmatrix} x_2 \\ -2\sin x_1 \cos x_2 + x_1 + \tanh(u) + d(x) \end{bmatrix}$
where, $d(x)$ is an unstructured uncertainty, defined as $d(x) = 0.1 \sin x_1 \cos x_2 + 0.2x_1$.

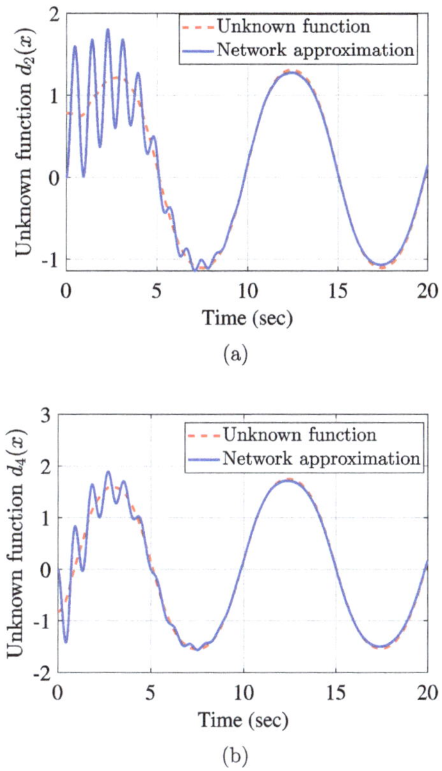

Fig. 6.16 (**a**) Unknown function $d_2(\boldsymbol{x})$ and its approximation. (**b**) Unknown function $d_4(\boldsymbol{x})$ and its approximation

(g) $\begin{bmatrix} \dot{x}_1 \\ \dot{x}_2 \end{bmatrix} = \begin{bmatrix} x_2 + u \\ -x_2 - x_1 - (x_1 + x_2)^2 x_2 + u + d(x) \end{bmatrix}$
where, $d(x)$ is an unstructured uncertainty, defined as $d(x) = x_2^2 - 0.3x_1$.

(h) $\begin{bmatrix} \dot{x}_1 \\ \dot{x}_2 \end{bmatrix} = \begin{bmatrix} -x_1^3 - x_1 x_2^2 \\ -x_2^3 - x_1^2 x_2 + x_2 + u + d(x) \end{bmatrix}$ where, $d(x)$ is an unstructured uncertainty, defined as $d(x) = 0.3x_1^2 x_2^2 - 0.1x_2$.

2. Following the definition of weights in (6.30) and (6.31), prove the following inequality:

$$\frac{\alpha_i \tilde{\boldsymbol{w}}_i^T \hat{\boldsymbol{w}}_i}{\gamma_i} \leq -\frac{\alpha_i \tilde{\boldsymbol{w}}_i^T \tilde{\boldsymbol{w}}_i}{2\gamma_i} + \frac{\alpha_i \boldsymbol{w}_i^T \boldsymbol{w}_i}{2\gamma_i}$$

where $\boldsymbol{w}_i$ and $\hat{\boldsymbol{w}}_i$ are the optimal and approximated weights, respectively, and $\tilde{\boldsymbol{w}}_i$ is defined as $\tilde{\boldsymbol{w}}_i \triangleq \boldsymbol{w}_i - \hat{\boldsymbol{w}}_i$.

3. Consider the error transformation variable $\boldsymbol{s}$ is defined as

$$\boldsymbol{s} = \boldsymbol{K}_1 \, \boldsymbol{e} + \boldsymbol{K}_2 \, \dot{e} \tag{6.159}$$

where $\boldsymbol{K}_1$ and $\boldsymbol{K}_2$ are positive definite diagonal matrices, and errors are defined in (6.66)–(6.67). Show that the error transformation has the following properties:

(a) When $\boldsymbol{s} \to \boldsymbol{0}$, errors $\boldsymbol{e} \to \boldsymbol{0}$ and $\dot{\boldsymbol{e}} \to \boldsymbol{0}$ simultaneously.
(b) If $\boldsymbol{s}$ is bounded, errors $\boldsymbol{e} \to \boldsymbol{0}$ and $\dot{\boldsymbol{e}} \to \boldsymbol{0}$ will remain bounded.

4. Consider the weight update rule with σ modification defined as

$$\dot{\hat{\boldsymbol{w}}}_i = \gamma_i \beta_i e_i \boldsymbol{\phi}(\boldsymbol{x}, \boldsymbol{u}, \boldsymbol{u}_s) - \sigma_i \hat{\boldsymbol{w}}_i \tag{6.160}$$

where all variables can be referred to from (6.132). Show that for nonzero error e_i, the weight update rule is stable, and time histories of weights will remain bounded.

References

P. R. Ambati, R. Padhi, Robust auto-landing of fixed-wing $UAVs$ using neuro-adaptive design. Control Eng. Pract. **60**, 218–232 (2017)

S. S. Haykin, *Neural Networks and Learning Machines*, 3rd edn. (Prentice Hall, New York, 2009)

A. Maity, L. Höcht, C. Heise, F. Holzapfel, Adaptive optimal control using frequency selective information of the system uncertainty with application to unmanned aircraft. IEEE Trans. Cybern. **48**(1), 165–177 (2018)

S. Mathavaraj, R. Padhi, *Adaptive Dynamic Inversion for Satellite Formation Flying* (Springer Singapore, Singapore, 2021), pp. 83–98

R. Padhi, M. Kothari, An optimal dynamic inversion-based neuro-adaptive approach for treatment of chronic myelogenous leukemia. Computer Methods Program. Biomed. **87**(3), 208–224 (2007)

R. Padhi, N. Unnikrishnan, S. N. Balakrishnan, Model-following neuro-adaptive control design for non-square, non-affine nonlinear systems. IET Control Theory Appl. **1**(6), 1650–1661 (2007)

J. Rajasekaran, A. Chunodkar, R. Padhi, Structured model-following neuro-adaptive design for attitude maneuver of rigid bodies. Control Eng. Pract. **17**(6), 676–689 (2009)

K. Sachan, R. Padhi, Nonlinear robust neuro-adaptive flight control for hypersonic vehicles with state constraints. Control Eng. Pract. **102**, 104526 (2020a)

K. Sachan, R. Padhi, Safety-constrained robust adaptive control for a class of MIMO non-linear systems. IET Control Theory Appl. **14**(7), 982–990 (2020b)

Chapter 7
BLF-Based State-Constrained Neuro-Adaptive Control

The preceding chapters provided an overview of existing nonlinear adaptive control methodologies, with a particular emphasis on neuro-adaptive control for various classes of nonlinear systems. Nominal state-constrained control strategies for a perfectly known system model were presented in Chap. 4. However, the presence of modeling uncertainties and external disturbances significantly complicates the control design process for constrained systems. To address this challenge, in this chapter, the neuro-adaptive control design framework is extended to distinct classes of nonlinear systems, assuming the presence of structured modeling uncertainties. In the next chapter, the neuro-adaptive control framework is developed to address unstructured modeling uncertainties and bounded external disturbances. Thus, the current chapter serves as a foundational introduction to state-constrained neuro-adaptive control design.

7.1 Adaptive Control of Nonlinear System with Relative Degree One

A nonlinear system with relative degree one is given by

$$\dot{\boldsymbol{x}} = \boldsymbol{f}(\boldsymbol{x}, \boldsymbol{u}) + \boldsymbol{d}(\boldsymbol{x}) \tag{7.1}$$

where $\boldsymbol{x} \in \Re^n$ and $\boldsymbol{u} \in \Re^n$ are the states and control input, respectively. The vector $\boldsymbol{f}(\boldsymbol{x}, \boldsymbol{u}) \in \Re^n$ represents the known part of the dynamics, and $\boldsymbol{d}(\boldsymbol{x})$ denotes the uncertain part. The model uncertainties of the system are assumed to be bounded and structured in nature.

The objective of the control design is to design an adaptive state feedback controller for the system in (7.1), which ensures the desired tracking of a given reference signal $\boldsymbol{x}^*(t)$, without violation of user-imposed state constraints. The

K. Sachan et al., *Intelligent Adaptive Control with System Constraints*,
https://doi.org/10.1007/978-981-95-8220-4_7

imposed constraints (asymmetric) on the ith component of state is defined as $\overline{\zeta}_{x_i}$ and $\underline{\zeta}_{x_i}$, i.e., $\underline{\zeta}_{x_i} < x_i < \overline{\zeta}_{x_i}, \forall i$, where $\underline{\zeta}_{x_i}$ and $\overline{\zeta}_{x_i}$ are lower and upper bounds on state x_i, respectively.

To mitigate the conflict between reference signal tracking and constraint satisfaction, it is assumed that there exist positive constants $\underline{\zeta}_{x_i^*}, \overline{\zeta}_{x_i^*}$ satisfying $-\underline{\zeta}_{x_i} < -\underline{\zeta}_{x_i^*} \leq x_i^* \leq \overline{\zeta}_{x_i^*} < \overline{\zeta}_{x_i}, \forall i$, where x_i^* denotes the ith component of the reference signal $\boldsymbol{x}^*$. Moreover, the reference signal and its time derivatives are assumed to be smooth and bounded.

7.1.1 Philosophy of Gradient Learning-Based Adaptive Control

The control design philosophy used here aligns with the neuro-adaptive control approach outlined in the previous chapter. However, due to the presence of state constraints and utilization of gradient learning in control, the design methodology and stability analysis differ significantly.

In the control synthesis process, the states of the actual system, the states of the approximated system, the states of the known part of the system dynamics (desired system), and the reference tracking signals are denoted by $\boldsymbol{x}, \boldsymbol{x}^a, \boldsymbol{x}^d$, and $\boldsymbol{x}^*$, respectively. The control design framework is illustrated in Fig. 7.1. The design process involves the following steps: (i) an asymptotically stable desired controller is developed for the known part of the system dynamics to ensure $\boldsymbol{x}^d \rightarrow \boldsymbol{x}^*$; ($ii$) a neural network-based system is constructed to approximate system uncertainties, enabling $\boldsymbol{x} \rightarrow \boldsymbol{x}^a$; and ($iii$) an asymptotically stable tracking controller is designed to minimize the error between the approximated system states and the desired system states, ensuring $\boldsymbol{x}^a \rightarrow \boldsymbol{x}^d$. In all three steps, the system errors are constrained by the respective bounds, calculated using the user-imposed constraints (as discussed in (7.38) and (7.39)). These steps ensure that the actual system states track the reference signal with the desired accuracy ($\boldsymbol{x} \rightarrow \boldsymbol{x}^*$) without violating the imposed state constraints. The steps involved in the control design are explained in the subsequent subsections.

The proposed adaptive control design can also be implemented without using step (i), i.e., $\boldsymbol{x}^d \rightarrow \boldsymbol{x}^*$. In that scenario, tracking control is designed directly to

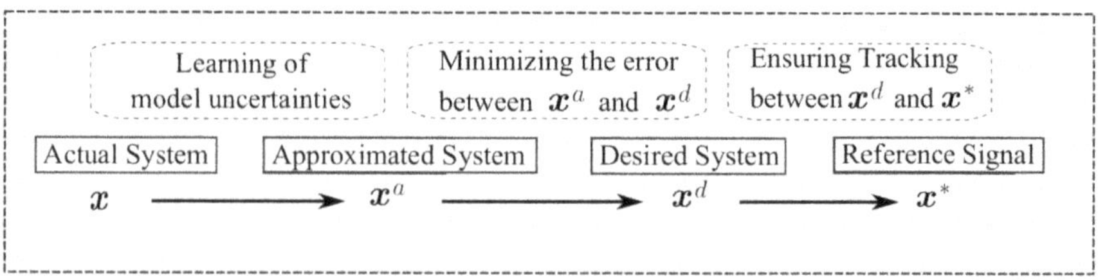

Fig. 7.1 Philosophy of state-constrained neuro-adaptive control design

minimize the error between approximated states and the reference signal, i.e., $\boldsymbol{x}^a \rightarrow \boldsymbol{x}^*$. However, step (i) leads to a relatively better transients-free and smooth response.

7.1.2 Desired State-Constrained Control Design

The controller developed in this step ensures the asymptotic tracking between $\boldsymbol{x}^d$ and $\boldsymbol{x}^*$, i.e., $\boldsymbol{x}^d \rightarrow \boldsymbol{x}^*$ for the known part of system dynamics (desired system). The dynamic of the desired system is given by

$$\dot{\boldsymbol{x}}^d = \boldsymbol{f}\left(\boldsymbol{x}^d, \boldsymbol{u}^d\right) \tag{7.2}$$

where, $\boldsymbol{x}^d, \boldsymbol{u}^d \in \Re^n$ are states and inputs of the system. The objective of desired control design is to ensure asymptotic tracking of the reference signal $\boldsymbol{x}^*$ without violation of asymmetric constraints on desired states of the system in (7.2), where the upper and lower bounds on the ith component of state are assumed as $\underline{\zeta}_{x_i^d}$ and $\overline{\zeta}_{x_i^d}$, respectively, i.e., $\underline{\zeta}_{x_i^d} < x_i^d < \overline{\zeta}_{x_i^d}, \forall i$. The selection strategy for these constraints is obtained from (7.38) and (7.39).

The tracking error between the states of the desired system and the reference signal is defined as

$$\boldsymbol{e}^d = \boldsymbol{x}^d - \boldsymbol{x}^* \tag{7.3}$$

By taking the time derivative of the error in (7.3) and substituting $\dot{\boldsymbol{x}}^d$ from (7.2), the following error dynamics is obtained:

$$\dot{\boldsymbol{e}}^d = \boldsymbol{f}\left(\boldsymbol{x}^d, \boldsymbol{u}^d\right) - \dot{\boldsymbol{x}}^* \tag{7.4}$$

To construct the controller, the following asymptotically stable closed-loop error dynamics is selected (the proof of asymptotic stability is given in Theorem 7.1):

$$\dot{\boldsymbol{e}}^d = \left[\boldsymbol{K}^d\left(\boldsymbol{e}^d\right)\right]\boldsymbol{e}^d \tag{7.5}$$

where $\boldsymbol{K}^d\left(\boldsymbol{e}^d\right)$ is diagonal symmetric matrix, defined as

$$\boldsymbol{K}^d\left(\boldsymbol{e}^d\right) = \text{diag}\left[\gamma_i^*\left(\frac{c_i^d}{\overline{\zeta}_{e_i^d}^2 - e_i^{d2}}\right) + (1-\gamma_i^*)\left(\frac{c_i^d}{\underline{\zeta}_{e_i^d}^2 - e_i^{d2}}\right)\right], \ \forall i \tag{7.6}$$

where constant c_i^d is positive controller gain and e_i^d is the ith component of error term $\boldsymbol{e}^d$. The terms $\overline{\zeta}_{e_i^d}$ and $\underline{\zeta}_{e_i^d}$ are defined as $\overline{\zeta}_{e_i^d} = \overline{\zeta}_{x_i^d} - \overline{\zeta}_{x_i^*}$ and $-\underline{\zeta}_{e_i^d} = \underline{\zeta}_{x_i^d} - \underline{\zeta}_{x_i^*}$, respectively, and γ_i^* is defined as $\gamma_i^*(e_i) = \begin{cases} 1, & \text{if } 0 < e_i^d < \overline{\zeta}_{e_i^d} \\ 0, & \text{if } -\underline{\zeta}_{e_i^d} < e_i^d \leq 0 \end{cases}, \forall i.$

By comparing (7.4) and (7.5), the following expression is obtained:

$$\boldsymbol{f}\left(\boldsymbol{x}^d, \boldsymbol{u}^d\right) + \left[\boldsymbol{K}^d\left(\boldsymbol{e}^d\right)\right]\boldsymbol{e}^d - \dot{\boldsymbol{x}}^* = 0 \tag{7.7}$$

Based on the structure of the desired system in (7.2), the desired control expression $\boldsymbol{u}^d$ can be derived from (7.7). If the desired system is in a control-affine form, the desired control expression can be explicitly obtained in a closed-form solution. Conversely, if the system is represented in a more generic form, fast numerical algorithms, such as the Newton-Raphson method, can be employed to compute the control solution efficiently.

Since closed-loop error dynamics in (7.5) is nonlinear in nature, the asymptotic stability of the closed-loop desired system is required to be shown. This requirement is fulfilled in the following theorem.

Theorem 7.1 *Consider the closed-loop error dynamics (7.5) of the desired system (7.2), with initial errors satisfying* $-\underline{\zeta}_{e_i^d} < e_i^d(0) < \overline{\zeta}_{e_i^d}, \forall i$. *Then, the closed-loop error dynamics are asymptotically stable and remain bounded, such that* $-\underline{\zeta}_{e_i^d} < e_i^d(t) < \overline{\zeta}_{e_i^d}, \forall i$. *Furthermore, the states of the closed-loop desired system are constrained within the imposed state bounds, i.e.,* $\underline{\zeta}_{x_i^d} < x_i^d < \overline{\zeta}_{x_i^d}, \forall i$.

Proof Since the gain matrix (7.6) is diagonal in nature, the closed-loop error dynamics (7.5) can be expressed in terms of ith components, i.e.,

$$\dot{e}_i^d + \left[\gamma_i^* \frac{c_i^d}{\left(\overline{\zeta}_{e_i^d}^2 - e_i^{d2}\right)} + (1-\gamma_i^*)\frac{c_i^d}{\left(\underline{\zeta}_{e_i^d}^2 - e_i^{d2}\right)}\right] e_i^d = 0, \forall i. \tag{7.8}$$

To show the asymptotic stability of (7.5), the barrier Lyapunov function is chosen as

$$V^d = \frac{1}{2}\sum_{i=1}^{n} c_i^d \left[\gamma_i^* \log\left(\frac{\overline{\zeta}_{e_i^d}^2}{\overline{\zeta}_{e_i^d}^2 - e_i^{d2}}\right) + (1-\gamma_i^*)\log\left(\frac{\underline{\zeta}_{e_i^d}^2}{\underline{\zeta}_{e_i^d}^2 - e_i^{d2}}\right)\right] \tag{7.9}$$

where constants c_i^d, $\overline{\zeta}_{e_i^d}$, and $\underline{\zeta}_{e_i^d}$ are already defined in (7.6) and "log" is natural logarithm. Note that the Lyapunov function is valid when $e_i^d(0)$ satisfies $-\underline{\zeta}_{e_i^d} < e_i^d(0) < \overline{\zeta}_{e_i^d}, \forall i$. The time derivative of Lyapunov function is

$$\dot{V}^d = \sum_{i=1}^{n} c_i^d \left[\gamma_i^* \frac{e_i^d \dot{e}_i^d}{\left(\overline{\zeta}_{e_i^d}^2 - e_i^{d2}\right)} + (1-\gamma_i^*) \frac{e_i^d \dot{e}_i^d}{\left(\underline{\zeta}_{e_i^d}^2 - e_i^{d2}\right)} \right] \tag{7.10}$$

By substituting $\dot{e}_i^d$ from (7.8), $\dot{V}^d$ becomes

$$\dot{V}^d = -\sum_{i=1}^{n} \left[\gamma_i^* \frac{c_i^d}{\left(\overline{\zeta}_{e_i^d}^2 - e_i^{d2}\right)} + (1-\gamma_i^*) \frac{c_i^d}{\left(\underline{\zeta}_{e_i^d}^2 - e_i^{d2}\right)} \right]^2 e_i^{d2} \tag{7.11}$$

Since $\dot{V}^d$ is negative definite, the asymptotic stability of the closed-loop error dynamics in (7.5) can be concluded, i.e., $e_i^d(t) \to 0$ as $t \to \infty, \ \forall i$.

Since $\dot{V}^d < 0$, $V^d(t) \leq V_0^d, \forall t \geq 0$, where V_0^d is the value of the selected barrier Lyapunov function in (7.9) at $t = 0$. Since each individual element of V^d in (7.9) is positive definite, the following expression is true:

$$\frac{1}{2} c_i^d \left[\gamma_i^* \log \frac{\overline{\zeta}_{e_i^d}^2}{\overline{\zeta}_{e_i^d}^2 - e_i^{d2}} + (1-\gamma_i^*) \log \frac{\underline{\zeta}_{e_i^d}^2}{\underline{\zeta}_{e_i^d}^2 - e_i^{d2}} \right] \leq V_0^d \tag{7.12}$$

By solving (7.12) for e_i^d, and performing mathematical simplifications, the following inequality is obtained:

$$-\underline{\zeta}_{e_i^d} \sqrt{\left(1 - \exp\left(\frac{-2V_0^d}{c_i^d}\right)\right)} \leq e_i^d \leq \overline{\zeta}_{e_i^d} \sqrt{\left(1 - \exp\left(\frac{-2V_0^d}{c_i^d}\right)\right)}, \quad \forall i \tag{7.13}$$

where "exp" represents the exponential function. Since the term $\frac{2V_0^d}{c_i^d}$ is a positive quantity for nonzero $e_i^d(0)$, the term $\sqrt{\left(1 - \exp\left(\frac{-2V_0^d}{c_i^d}\right)\right)}$ is bounded as

$$0 < \sqrt{\left(1 - \exp\left(\frac{-2V_0^d}{c_i^d}\right)\right)} < 1$$

Thus, the closed-loop errors e_i^d will remain bounded as $-\underline{\zeta}_{e_i^d} < e_i^d(t) < \overline{\zeta}_{e_i^d}, \forall i$.

The error e_i^d is defined as $e_i^d = x_i^d - x_i^*$ and hence, $x_i^d = e_i^d + x_i^*$. Since the desired output x_i^* and e_i^d are bounded as $\underline{\zeta}_{x_i^*} \leq x_i^* \leq \overline{\zeta}_{x_i^*}$ and $-\underline{\zeta}_{e_i^d} < e_i^d < \overline{\zeta}_{e_i^d}, \forall i$ respectively, the bounds on the state x_i^d are obtained as $\left(-\underline{\zeta}_{e_i^d} + \underline{\zeta}_{x_i^*}\right) < x_i^d < \left(\overline{\zeta}_{e_i^d} + \overline{\zeta}_{x_i^*}\right)$. By the definition of $\underline{\zeta}_{e_i^d}$ and $\overline{\zeta}_{e_i^d}$ in (7.6), $\underline{\zeta}_{x_i^d}$ and $\overline{\zeta}_{x_i^d}$ are defined as $\underline{\zeta}_{x_i^d} = -\underline{\zeta}_{e_i^d} + \underline{\zeta}_{x_i^*}$ and $\overline{\zeta}_{x_i^d} = \overline{\zeta}_{e_i^d} + \overline{\zeta}_{x_i^*}$, respectively. Hence, the following can be concluded:

$$\underline{\zeta}_{x_i^d} < x_i^d < \overline{\zeta}_{x_i^d}, \quad \forall i. \tag{7.14}$$

Thus, the closed-loop states of the desired system will remain bounded by the imposed state constraints. □

7.1.3 *Approximation of Uncertainties with Gradient Learning*

In this step, the model uncertainties of the actual system are approximated by constructing an approximated system. Using the BLF-based stability analysis, a neural network-based weight update rule is developed to ensure that $\boldsymbol{x} \to \boldsymbol{x}^a$.

The dynamics of actual system (7.1) is given by

$$\dot{\boldsymbol{x}} = \boldsymbol{f}(\boldsymbol{x}, \boldsymbol{u}) + \boldsymbol{d}(\boldsymbol{x}) \tag{7.15}$$

The corresponding approximated system is constructed as

$$\dot{\boldsymbol{x}}^a = \boldsymbol{f}(\boldsymbol{x}, \boldsymbol{u}) + \hat{\boldsymbol{d}}(\boldsymbol{x}) + \boldsymbol{K}(\boldsymbol{e})\,\boldsymbol{e} \tag{7.16}$$

where $\boldsymbol{x}^a$ is the state of the approximated system with initial condition identical to the actual system, $\boldsymbol{e} = (\boldsymbol{x} - \boldsymbol{x}^a)$, $\hat{\boldsymbol{d}}(\boldsymbol{x})$ is the approximation of $\boldsymbol{d}(\boldsymbol{x})$, and $\boldsymbol{K}(\boldsymbol{e})$ is a diagonal matrix. The ith component of $\boldsymbol{K}(\boldsymbol{e})$ is defined as $k_i(e_i) = \frac{c_i}{\zeta_{e_i}^2 - e_i^2}$, where c_i is a positive constant, e_i is the ith component of error $\boldsymbol{e}$, and ζ_{e_i} is a bound on e_i (obtained from (7.38) and (7.39)). The use of $\boldsymbol{K}(\boldsymbol{e})$ is inspired by observer design theory, and the bound ζ_{e_i} is used for improving the transient performance.

To design a weight update rule to update the weights of the neural network, the error between the actual and approximated system is defined as

$$\boldsymbol{e} = \boldsymbol{x} - \boldsymbol{x}^a \tag{7.17}$$

The time differentiation of (7.17) gives

$$\dot{\boldsymbol{e}} = \dot{\boldsymbol{x}} - \dot{\boldsymbol{x}}^a \tag{7.18}$$

Substituting (7.15) and (7.16) into (7.18) leads to

$$\dot{\boldsymbol{e}} = \boldsymbol{d}(\boldsymbol{x}) - \hat{\boldsymbol{d}}(\boldsymbol{x}) - \boldsymbol{K}(\boldsymbol{e})\,\boldsymbol{e} \tag{7.19}$$

Since the matrix $\boldsymbol{K}(\boldsymbol{e})$ is a diagonal matrix, (7.19) can be analyzed in terms of its ith component, where ith components of $\boldsymbol{e}$, $\boldsymbol{d}(\boldsymbol{x})$, and $\hat{\boldsymbol{d}}(\boldsymbol{x})$ are represented by e_i, $d_i(\boldsymbol{x})$, and $\hat{d}_i(\boldsymbol{x})$, respectively. Thus, the ith component of the error dynamics in (7.19) can be written as

$$\dot{e}_i = d_i(\boldsymbol{x}) - \hat{d}_i(\boldsymbol{x}) - \frac{c_i e_i}{\zeta_{e_i}^2 - e_i^2} \tag{7.20}$$

Utilizing linear-in-weight architecture of neural networks, the structured uncertainties $d_i(\boldsymbol{x})$ (approximation error $\epsilon_i = 0$) and the approximation of uncertainties $\hat{d}_i(\boldsymbol{x})$ are defined as

$$d_i(\boldsymbol{x}) = \boldsymbol{w}_i^T \boldsymbol{\phi}(\boldsymbol{x}) \tag{7.21}$$

$$\hat{d}_i(\boldsymbol{x}) = \hat{\boldsymbol{w}}_i^T \boldsymbol{\phi}(\boldsymbol{x}) \tag{7.22}$$

where $\boldsymbol{w}_i$ and $\hat{\boldsymbol{w}}_i$ are the optimal and approximated weights and $\boldsymbol{\phi}(\boldsymbol{x})$ is known basis function vector. Note that $\boldsymbol{w}_i$ is a vector with constant elements.

By substituting (7.21) and (7.22) into (7.20), the following can be obtained:

$$\dot{e}_i = \tilde{\boldsymbol{w}}_i^T \boldsymbol{\phi}(\boldsymbol{x}) - \frac{c_i e_i}{\zeta_{e_i}^2 - e_i^2} \tag{7.23}$$

where $\tilde{\boldsymbol{w}}_i$ is defined as $\tilde{\boldsymbol{w}}_i = \boldsymbol{w}_i - \hat{\boldsymbol{w}}_i$.

In order to derive weight update rule while ensuring the stability of closed-loop error dynamics in (7.23), the following combination of barrier and quadratic Lyapunov function is selected (Sachan and Padhi, 2020d)

$$\begin{aligned} V_i = {} & \frac{\beta_i}{2}\log\left(\frac{\zeta_{e_i}^2}{\zeta_{e_i}^2 - e_i^2}\right) + \frac{\tilde{\boldsymbol{w}}_i^T \tilde{\boldsymbol{w}}_i}{2\gamma_i} \\ & + \left[\frac{\partial d_i(\boldsymbol{x})}{\partial \boldsymbol{x}} - \frac{\partial \hat{d}_i(\boldsymbol{x})}{\partial \boldsymbol{x}}\right]^T \frac{\Theta_i}{2}\left[\frac{\partial d_i(\boldsymbol{x})}{\partial \boldsymbol{x}} - \frac{\partial \hat{d}_i(\boldsymbol{x})}{\partial \boldsymbol{x}}\right] \end{aligned} \tag{7.24}$$

where Θ_i and β_i are positive constants and $\gamma_i > 0$ is known as learning rate constant. The terms $\frac{\partial d_i(\boldsymbol{x})}{\partial \boldsymbol{x}}$ and $\frac{\partial \hat{d}_i(\boldsymbol{x})}{\partial \boldsymbol{x}}$ are gradients of model uncertainties and approximated model uncertainties, respectively. The advantage of using gradient information in the Lyapunov function can be seen in the construction of the weight update rule in (7.28).

The time derivative of the Lyapunov function in (7.24) is

$$\dot{V}_i = \frac{\beta_i e_i \dot{e}_i}{\zeta_{e_i}^2 - e_i^2} + \frac{\tilde{\boldsymbol{w}}_i^T \dot{\tilde{\boldsymbol{w}}}_i}{\gamma_i} + \frac{d}{dt}\left[\tilde{\boldsymbol{w}}_i^T \left[\frac{\partial \boldsymbol{\phi}}{\partial \boldsymbol{x}}\right] \frac{\Theta_i}{2} \left[\frac{\partial \boldsymbol{\phi}}{\partial \boldsymbol{x}}\right]^T \tilde{\boldsymbol{w}}_i\right] \tag{7.25}$$

$$= \frac{\beta_i e_i \dot{e}_i}{\zeta_{e_i}^2 - e_i^2} + \frac{\tilde{\boldsymbol{w}}_i^T \dot{\tilde{\boldsymbol{w}}}_i}{\gamma_i} + \tilde{\boldsymbol{w}}_i^T \left[\frac{\partial \boldsymbol{\phi}}{\partial \boldsymbol{x}}\right] \Theta_i \left[\left[\frac{\partial \boldsymbol{\phi}}{\partial \boldsymbol{x}}\right]^T \dot{\tilde{\boldsymbol{w}}}_i + \frac{d}{dt}\left[\frac{\partial \boldsymbol{\phi}}{\partial \boldsymbol{x}}\right]^T \tilde{\boldsymbol{w}}_i\right] \tag{7.26}$$

where $\frac{\partial \boldsymbol{\phi}}{\partial \boldsymbol{x}}$ is a Jacobian matrix. Further, using error dynamics (7.23), the time derivative of the Lyapunov function is modified as

$$\begin{aligned}
\dot{V}_i &= \frac{\beta_i e_i}{\zeta_{e_i}^2 - e_i^2}\left(\tilde{\boldsymbol{w}}_i^T \boldsymbol{\phi}(\boldsymbol{x}) - \frac{c_i e_i}{\zeta_{e_i}^2 - e_i^2}\right) - \frac{\tilde{\boldsymbol{w}}_i^T \dot{\hat{\boldsymbol{w}}}_i}{\gamma_i} \\
&\quad - \tilde{\boldsymbol{w}}_i^T \left[\frac{\partial \boldsymbol{\phi}}{\partial \boldsymbol{x}}\right] \Theta_i \left[\frac{\partial \boldsymbol{\phi}}{\partial \boldsymbol{x}}\right]^T \dot{\hat{\boldsymbol{w}}}_i + \tilde{\boldsymbol{w}}_i^T \left[\frac{\partial \boldsymbol{\phi}}{\partial \boldsymbol{x}}\right] \Theta_i \frac{d}{dt}\left[\frac{\partial \boldsymbol{\phi}}{\partial \boldsymbol{x}}\right]^T \tilde{\boldsymbol{w}}_i \\
&= \tilde{\boldsymbol{w}}_i^T \left[\frac{\beta_i e_i \boldsymbol{\phi}(\boldsymbol{x})}{\zeta_{e_i}^2 - e_i^2} - \frac{\dot{\hat{\boldsymbol{w}}}_i}{\gamma_i} - \left[\frac{\partial \boldsymbol{\phi}}{\partial \boldsymbol{x}}\right] \Theta_i \left[\frac{\partial \boldsymbol{\phi}}{\partial \boldsymbol{x}}\right]^T \dot{\hat{\boldsymbol{w}}}_i\right] \\
&\quad - \frac{c_i \beta_i e_i^2}{\left(\zeta_{e_i}^2 - e_i^2\right)^2} + \tilde{\boldsymbol{w}}_i^T \left[\frac{\partial \boldsymbol{\phi}}{\partial \boldsymbol{x}}\right] \Theta_i \frac{d}{dt}\left[\frac{\partial \boldsymbol{\phi}}{\partial \boldsymbol{x}}\right]^T \tilde{\boldsymbol{w}}_i \\
&= \tilde{\boldsymbol{w}}_i^T \left[\frac{\beta_i e_i \boldsymbol{\phi}(\boldsymbol{x})}{\zeta_{e_i}^2 - e_i^2} - \left(\frac{I_P}{\gamma_i} + \left[\frac{\partial \boldsymbol{\phi}}{\partial \boldsymbol{x}}\right] \Theta_i \left[\frac{\partial \boldsymbol{\phi}}{\partial \boldsymbol{x}}\right]^T\right) \dot{\hat{\boldsymbol{w}}}_i\right] \\
&\quad - \frac{c_i \beta_i e_i^2}{\left(\zeta_{e_i}^2 - e_i^2\right)^2} + \tilde{\boldsymbol{w}}_i^T \left[\frac{\partial \boldsymbol{\phi}}{\partial \boldsymbol{x}}\right] \Theta_i \frac{d}{dt}\left[\frac{\partial \boldsymbol{\phi}}{\partial \boldsymbol{x}}\right]^T \tilde{\boldsymbol{w}}_i
\end{aligned} \tag{7.27}$$

where I_P is an identity matrix.

By selecting the weight update rule as

$$\dot{\hat{\boldsymbol{w}}}_i = \left[\frac{I_P}{\gamma_i} + \left[\frac{\partial \boldsymbol{\phi}}{\partial \boldsymbol{x}}\right] \Theta_i \left[\frac{\partial \boldsymbol{\phi}}{\partial \boldsymbol{x}}\right]^T\right]^{-1} \frac{\beta_i e_i \boldsymbol{\phi}(\boldsymbol{x})}{\zeta_{e_i}^2 - e_i^2} \tag{7.28}$$

The resultant $\dot{V}_i$ expression in (7.27) can be written as

$$\dot{V}_i = -\frac{c_i \beta_i e_i^2}{\left(\zeta_{e_i}^2 - e_i^2\right)^2} + \tilde{\boldsymbol{w}}_i^T \left[\frac{\partial \boldsymbol{\phi}}{\partial \boldsymbol{x}}\right] \Theta_i \frac{d}{dt}\left[\frac{\partial \boldsymbol{\phi}}{\partial \boldsymbol{x}}\right]^T \tilde{\boldsymbol{w}}_i \tag{7.29}$$

Since the weight update rule in (7.28) uses the information of unknown function and its gradients, it leads to fast and smooth approximation. Interested readers can find more information about gradient learning-based function approximation in Czarnecki et al. (2017), Ambati and Padhi (2017) and references therein.

To analyze further, the expression of $\dot{V}_i$ can be expressed as

$$\dot{V}_i = -\frac{\beta_i}{\left(\zeta_{e_i}^2 - e_i^2\right)^2}\left[c_i e_i^2 - \left(\zeta_{e_i}^2 - e_i^2\right)^2 \alpha_{\boldsymbol{w}_i}\right] \tag{7.30}$$

where $\alpha_{\boldsymbol{w}_i} = \frac{\tilde{\boldsymbol{w}}_i^T}{\beta_i}\left[\frac{\partial \boldsymbol{\phi}}{\partial \boldsymbol{x}}\right]\Theta_i \frac{d}{dt}\left[\frac{\partial \boldsymbol{\phi}}{\partial \boldsymbol{x}}\right]^T \tilde{\boldsymbol{w}}_i$.

The condition for which $\dot{V}_i$ becomes negative definite is

$$\left[c_i e_i^2 - \left(\zeta_{e_i}^2 - e_i^2\right)^2 \alpha_{\boldsymbol{w}_i}\right] > 0 \tag{7.31}$$

Since $e_i(0)$ is always zero at $t = 0$ (the initial conditions for both actual and approximated system are equal), the error e_i may grow toward either sides of zero based on the sign of $\dot{V}_i$. It can be noticed that as $e_i \to \zeta_{e_i}$ or $e_i \to -\zeta_{e_i}$, the term $\left(\zeta_{e_i}^2 - e_i^2\right)^2 \alpha_{\boldsymbol{w}_i} \to 0$, and $c_i e_i^2 \to R^+$ (positive). Hence, $\dot{V}_i$ eventually becomes negative definite. Moreover, since $\dot{V}_i < 0$ as $e_i \to \zeta_{e_i}$ or $e_i \to -\zeta_{e_i}$, the boundedness of the error e_i in the domain $\left(-\zeta_{e_i}, \zeta_{e_i}\right)$ can also be concluded.

Furthermore, the term $\frac{d}{dt}\left[\frac{\partial \boldsymbol{\phi}}{\partial \boldsymbol{x}}\right] = \left[\frac{\partial^2 \boldsymbol{\phi}}{\partial \boldsymbol{x}^2}\right]\dot{\boldsymbol{x}}$ contains a double partial derivative of basis function $\boldsymbol{\phi}$. Since typical basis functions are commonly selected as smooth functions, the Taylor series expansion of $\boldsymbol{\phi}(\boldsymbol{x})$ about zero results in the cancellation of dominant (linear and constant) terms in the entries of the matrix $\left[\frac{\partial^2 \boldsymbol{\phi}}{\partial \boldsymbol{x}^2}\right]$. This leads to smaller $\alpha_{\boldsymbol{w}_i}$. Additionally, another objective in the Lyapunov function selection is to drive $\tilde{\boldsymbol{w}}_i$ toward zero, and hence, when $\tilde{\boldsymbol{w}}_i$ becomes small, the term $\alpha_{\boldsymbol{w}_i}$ reduces further and eventually diminishes to zero (Kim and Calise, 1997). This ensures the stability of weight update rule in (7.28). Finally, using the weight update rule (7.28), model uncertainties of the system can be approximated as $\hat{d}_i(\boldsymbol{x}) = \hat{\boldsymbol{w}}_i^T \boldsymbol{\phi}(\boldsymbol{x})$. This completes the first step of the neuro-adaptive control design.

7.1.4 Tracking Control Design with Improved Model

After capturing model uncertainties, the next step is to drive the error between approximated and desired systems to zero. Note that, using the nominal controller (Theorem 7.1), states of the desired system track the reference signal asymptotically, i.e., $\boldsymbol{x}^d \to \boldsymbol{x}^*$. Moreover, using the learning rule in (7.28), $\boldsymbol{x} \to \boldsymbol{x}^a$ with desired accuracy. Thus, the minimization of error between $\boldsymbol{x}^a$ and $\boldsymbol{x}^d$ eventually leads to $\boldsymbol{x} \to \boldsymbol{x}^a \to \boldsymbol{x}^d \to \boldsymbol{x}^*$.

To achieve $\boldsymbol{x}^a \rightarrow \boldsymbol{x}^d$, consider the following closed-loop nonlinear stable error dynamics (similar to (7.5)):

$$\dot{\boldsymbol{e}}^a + \left[\boldsymbol{K}^a\left(\boldsymbol{e}^a\right)\right]\boldsymbol{e}^a = 0 \tag{7.32}$$

where $\boldsymbol{e}^a$ is the error, defined as $\boldsymbol{e}^a = \boldsymbol{x}^a - \boldsymbol{x}^d$, and $\left[\boldsymbol{K}^a\left(\boldsymbol{e}^a\right)\right]$ is a diagonal gain matrix, defined as

$$\left[\boldsymbol{K}^a\left(\boldsymbol{e}^a\right)\right] = \text{diag}\left[\gamma_i^{**}\left(\frac{c_i^a}{\overline{\zeta}_{e_i^a}^2 - e_i^{a2}}\right) + (1-\gamma_i^{**})\left(\frac{c_i^a}{\underline{\zeta}_{e_i^a}^2 - e_i^{a2}}\right)\right]$$

The term γ_i^{**} is defined as $\gamma_i^{**} = \begin{cases} 1, & \text{if } 0 < e_i^a < \overline{\zeta}_{e_i^a}, \forall i \\ 0, & \text{if } -\underline{\zeta}_{e_i^a} < e_i^a \leq 0, \forall i \end{cases}$

where $e_i^a, \forall i$ are the ith component of error $\boldsymbol{e}^a$. The terms $c_i^a, \forall i$ are a constant positive controller gain, and constants $\underline{\zeta}_{e_i^a}, \overline{\zeta}_{e_i^a}$ are the limits on the errors $e_i^a, \forall i$. Note that the initial values of errors $e_i^a(0)$ are assumed to be chosen from the interval $\left(-\underline{\zeta}_{e_i^a}, \overline{\zeta}_{e_i^a}\right)$, and hence, similar to Theorem 7.1, $e_i^a(t)$ will remain bounded as $-\underline{\zeta}_{e_i^a} < e_i^a(t) < \overline{\zeta}_{e_i^a}$ can be concluded. Further, by substituting $\boldsymbol{e}^a = \boldsymbol{x}^a - \boldsymbol{x}^d$, (7.32) is

$$\left(\dot{\boldsymbol{x}}^a - \dot{\boldsymbol{x}}^d\right) + \left[\boldsymbol{K}^a\left(\boldsymbol{e}^a\right)\right]\left(\boldsymbol{x}^a - \boldsymbol{x}^d\right) = \boldsymbol{0} \tag{7.33}$$

By substituting $\dot{\boldsymbol{x}}^a$ and $\dot{\boldsymbol{x}}^d$ from (7.16) and (7.2), (7.33) can be expressed as follows:

$$\boldsymbol{f}\left(\boldsymbol{x}, \boldsymbol{x}^a, \boldsymbol{x}^d, \boldsymbol{u}, \boldsymbol{u}^d, \hat{\boldsymbol{d}}(\boldsymbol{x})\right) = \boldsymbol{0} \tag{7.34}$$

where $\boldsymbol{u}^d$ is obtained in (7.7) and $\hat{\boldsymbol{d}}(\boldsymbol{x})$ is already obtained as $\hat{d}_i(\boldsymbol{x}) = \hat{\boldsymbol{w}}_i^T \boldsymbol{\phi}(\boldsymbol{x})$ in Sect. 7.1.3. Finally, the adaptive control $\boldsymbol{u}$ can be extracted using (7.34). Note that if the expression is not in control-affine form, any fast numerical solving algorithm, e.g., the Newton-Raphson method, can be used to obtain the required control solution, as indicated in Padhi et al. (2007).

The overall objective of the control design is to enforce state constraints for all time while ensuring practical stability (asymptotic stability in presence of structured uncertainties) of the actual closed-loop system. Since the application of adaptive control leads to $\boldsymbol{x} \rightarrow \boldsymbol{x}^a$ and $\boldsymbol{x}^a \rightarrow \boldsymbol{x}^d$, and using desired control $\boldsymbol{x}^d \rightarrow \boldsymbol{x}^*$ asymptotically, the states of the actual system $\boldsymbol{x} \rightarrow \boldsymbol{x}^*$ with desired accuracy. Thus, the closed-loop stability of the actual system is ensured by the proposed control architecture.

Recall that the constraints on ith component of actual state x_i are enforced as $-\underline{\zeta}_{x_i} < x_i < \overline{\zeta}_{x_i}$, where $\underline{\zeta}_{x_i}$ and $\overline{\zeta}_{x_i}$ are user defined lower and upper bounds. Thus, by subtracting x_i^*, the following can be achieved:

$$-\underline{\zeta}_{x_i} - x_i^* < x_i - x_i^* < \overline{\zeta}_{x_i} - x_i^* \tag{7.35}$$

Since x_i^* is assumed to be bounded as $\underline{\zeta}_{x_i^*} \le x_i^* \le \overline{\zeta}_{x_i^*}, \forall i$, the following is obtained:

$$-\underline{\zeta}_{x_i} - \underline{\zeta}_{x_i^*} < \left(x_i - x_i^*\right) < \overline{\zeta}_{x_i} + \overline{\zeta}_{x_i^*} \tag{7.36}$$

which can be manipulated as

$$-\underline{\zeta}_{x_i} - \underline{\zeta}_{x_i^*} < \left(x_i - x_i^a\right) + \left(x_i^a - x_i^d\right) + \left(x_i^d - x_i^*\right) < \overline{\zeta}_{x_i} + \overline{\zeta}_{x_i^*} \tag{7.37}$$

Using error definitions, (7.37) can be expressed as

$$-\underline{\zeta}_{x_i} - \underline{\zeta}_{x_i^*} < e_i + e_i^a + e_i^d < \overline{\zeta}_{x_i} + \overline{\zeta}_{x_i^*} \tag{7.38}$$

Since errors are already shown to be bounded as $|e_i| < \zeta_{e_i}$, $-\underline{\zeta}_{e_i^a} < e_i^a < \overline{\zeta}_{e_i^a}$, and $-\underline{\zeta}_{e_i^d} < e_i^d < \overline{\zeta}_{e_i^d}, \forall i$, the following inequality holds:

$$-\zeta_{e_i} - \underline{\zeta}_{e_i^a} - \underline{\zeta}_{e_i^d} < e_i + e_i^a + e_i^d < \zeta_{e_i} + \overline{\zeta}_{e_i^a} + \overline{\zeta}_{e_i^d} \tag{7.39}$$

Hence, by comparing (7.38) and (7.39), the constraints $\underline{\zeta}_{x_i}$ and $\overline{\zeta}_{x_i}$ can be determined as $-\underline{\zeta}_{x_i} \le -\zeta_{e_i} - \underline{\zeta}_{e_i^a} - \underline{\zeta}_{e_i^d} + \underline{\zeta}_{x_i^*}$ and $\overline{\zeta}_{x_i} \ge \zeta_{e_i} + \overline{\zeta}_{e_i^a} + \overline{\zeta}_{e_i^d} - \overline{\zeta}_{x_i^*}$, respectively. Since $\overline{\zeta}_{x_i^*}$ and $\underline{\zeta}_{x_i^*}$ are fixed bounds on reference signal, and $\underline{\zeta}_{x_i}$ and $\overline{\zeta}_{x_i}$ are given by the user, the other constraints can be obtained by any feasible combination that satisfy above inequalities. From application point of view, these constants are usually chosen as $\underline{\zeta}_{e_i^d} > \underline{\zeta}_{e_i^a} > \zeta_{e_i}$ and $\overline{\zeta}_{e_i^d} > \overline{\zeta}_{e_i^a} > \zeta_{e_i}$, respectively.

Example 7.1 Consider the nonlinear system given by

$$\dot{x_1} = x_1 + x_2^2 + d_{x_1} + \left(1 + x_2^2\right) u_1 + u_2 \tag{7.40a}$$

$$\dot{x_2} = x_2 + x_3^2 + d_{x_2} + \left(1 + x_3^2\right) u_2 + u_3 \tag{7.40b}$$

$$\dot{x_3} = x_3 + x_1^2 + d_{x_3} + \left(1 + x_1^2\right) u_3 + u_1 \tag{7.40c}$$

(continued)

Example 7.1 (continued)
where x_1, x_2, and x_3 are states of the system; u_1, u_2, and u_3 are control inputs; and d_{x_1}, d_{x_2}, and d_{x_3} represent unknown additive modeling uncertainties of the system, respectively. The control design objectives are the following:

(i) Design a stabilizing nominal constrained controller for the system to ensure that $x_1 \to 0$, $x_2 \to 0$, and $x_3 \to 0$ as $t \to \infty$, without violation of imposed state constraints, defined as $|\zeta_{x_i}| < 0.25$ for $i = 1, 2, 3$ assuming $d_{x_1} = 0$, $d_{x_2} = 0$ and $d_{x_3} = 0$. Here, controller gains are chosen as $c_i^d = 0.25\ \forall i$. Simulation results are carried out for randomly selected initial conditions from the open set $(-0.25, 0.25)$. The states and corresponding desired control profiles are shown in Figs. 7.2, 7.3, 7.4, 7.5, 7.6, 7.7. The desired objectives have been achieved in all cases, without violation of any state constraint. All states and control histories are smooth and continuous, which demonstrates the effectiveness of the desired controller.

(ii) Since the formulation of the desired and DI controllers are similar, design an DI controller for the system and compare it with the desired controller in the presence of modeling uncertainties for the same gains as in (i) and random initial conditions from the same set. In simulations, the modeling uncertainties are chosen as $d_{x_1} = -1.5 \cos x_2$, $d_{x_2} = -2 \cos x_3$, and $d_{x_3} = 1.8 \sin x_1$.

The state and control trajectories are presented in Figs. 7.8, 7.9, 7.10, 7.11, 7.12, 7.13. It can be seen that the state histories obtained using the desired controller are stable and within the imposed state constraints. However, the performance of the DI controller is not adequate as closed-loop trajectories violate imposed constraints. Thus, despite similar formulations, the performance of the state-constrained desired controller is better in the presence of modeling uncertainties.

(iii) It can be seen from Figs. 7.8, 7.9, 7.10, 7.11, 7.12, 7.13 that the state trajectories obtained using the desired controller in the presence of modeling uncertainties are only stable. However, the desired tracking response is not guaranteed using the desired control. Therefore, the next objective is to design a neuro-adaptive adaptive controller to achieve the tracking response. In simulations, random initial conditions are chosen from the same set, and initial conditions for weight update dynamics are chosen as zero; gains are selected as $c_i = 0.1, c_i^a = 0.25\ \forall i$, state constraints are imposed as $|\zeta_{x_i}| < 0.25$ for $i = 1, 2, 3$, and ζ_{e_i} is assumed as $|\zeta_{e_i}| < 0.05$; modeling uncertainties are chosen as $d_{x_1} = -1.5 \cos x_2$, $d_{x_2} = -2 \cos x_3$, and $d_{x_3} = 1.8 \sin x_1$; and learning gains are selected as $\gamma_1 = 5, \gamma_2 = 5, \gamma_3 = 10, \beta_1 = 1, \beta_2 = 1, \beta_3 = 5, \Theta_i = 0.1\ \forall i$, respectively.

(continued)

Example 7.1 (continued)
The state and control trajectories obtained using the proposed adaptive control approach are illustrated in Figs. 7.14, 7.15, 7.16, 7.17, 7.18, 7.19. It is evident from these results that the desired tracking performance is achieved in all scenarios. Furthermore, the closed-loop system states remain within the imposed constraints. The modeling uncertainties, along with their approximations using gradient learning-based neural networks rule, are depicted in Figs. 7.20, 7.21, 7.22. In each case, the modeling uncertainties are successfully approximated within a finite time. Additionally, the control histories are smooth and continuous, further demonstrating the practical viability of the proposed adaptive control strategy.

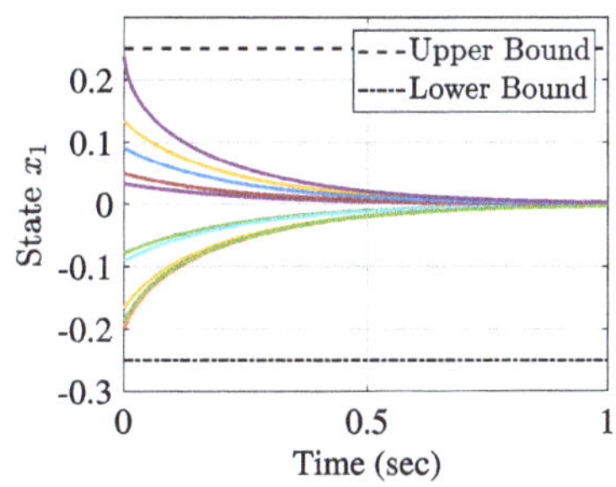

Fig. 7.2 State histories obtained using desired control

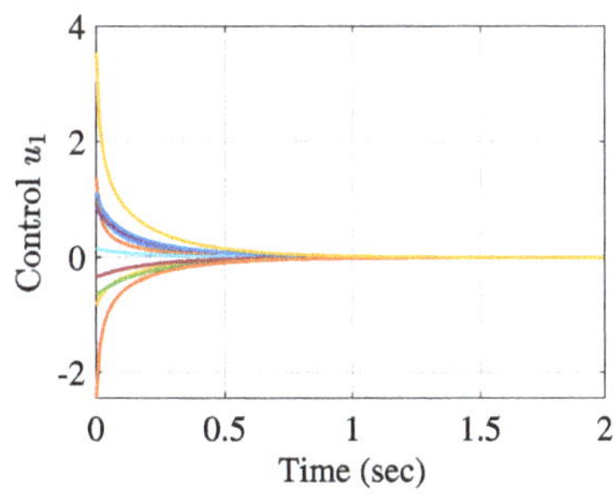

Fig. 7.3 Desired control histories for various initial conditions

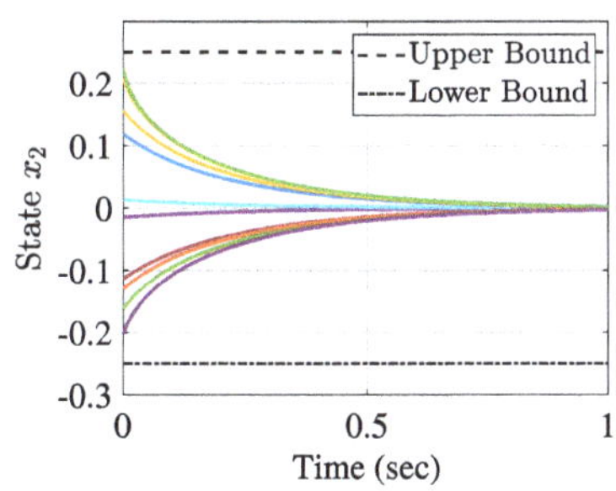

Fig. 7.4 State histories obtained using desired control

Fig. 7.5 Desired control histories for various initial conditions

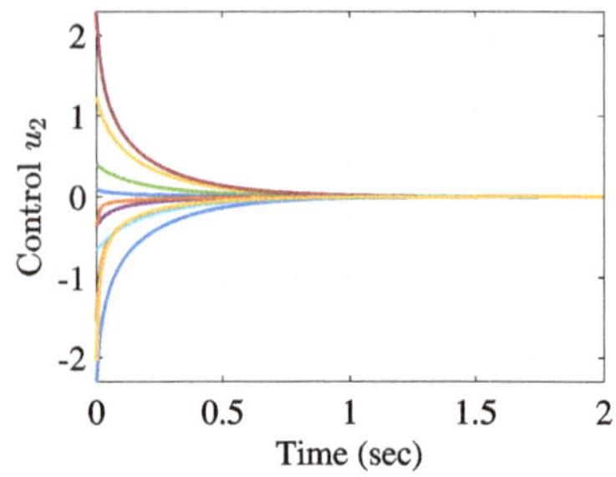

Fig. 7.6 State histories obtained using desired control

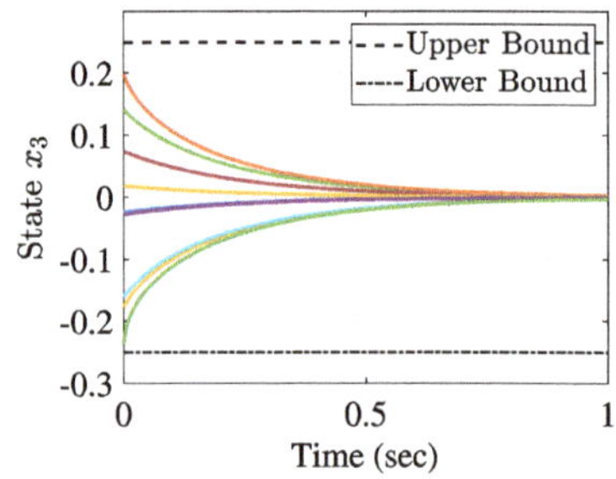

Fig. 7.7 Desired control histories for various initial conditions

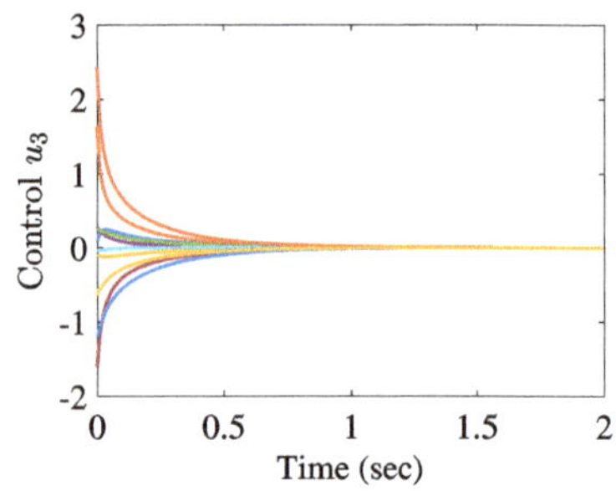

Fig. 7.8 State histories obtained using desired and DI controllers

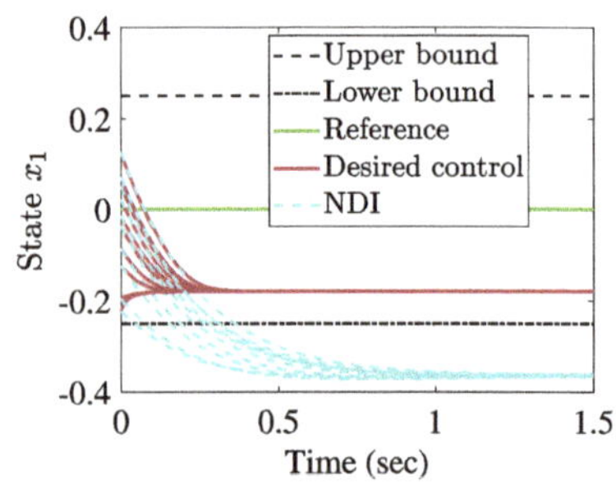

In this section, a nonlinear indirect neuro-adaptive control design is presented for nonlinear systems with relative degree one. The proposed controller ensures closed-loop stability, achieves desired tracking accuracy, and maintains imposed state constraints. It is applicable to practical systems with inherent timescale separation, as demonstrated through a hypersonic vehicle cruise control as a case study in

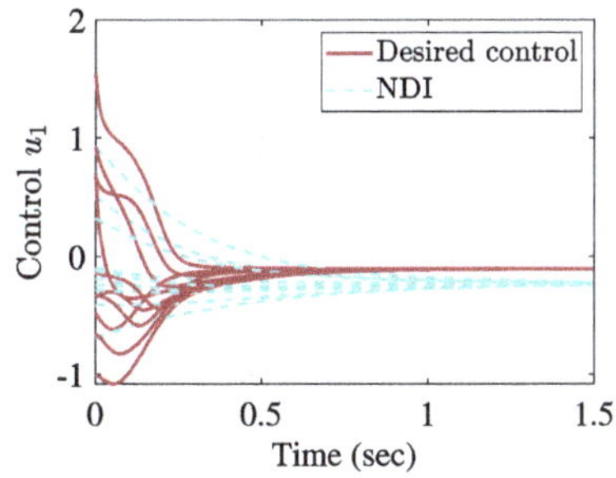

Fig. 7.9 Desired and DI control histories for various initial conditions

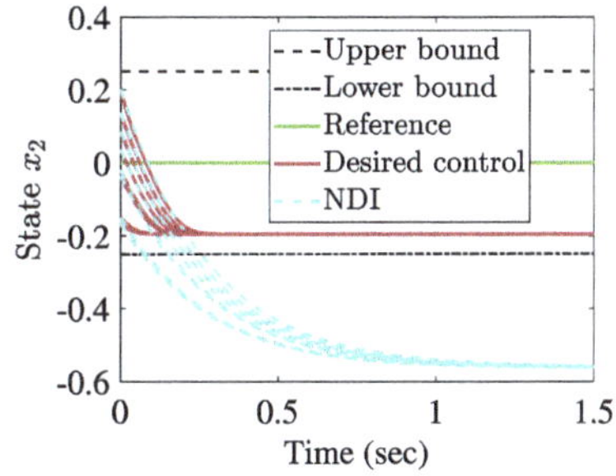

Fig. 7.10 State histories obtained using desired and DI controllers

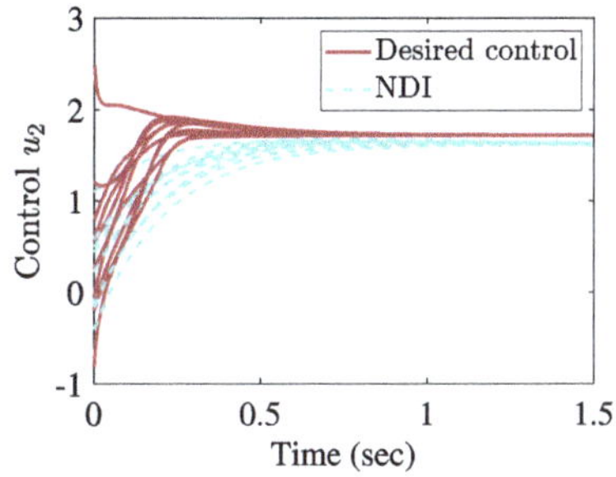

Fig. 7.11 Desired and DI control histories for various initial conditions

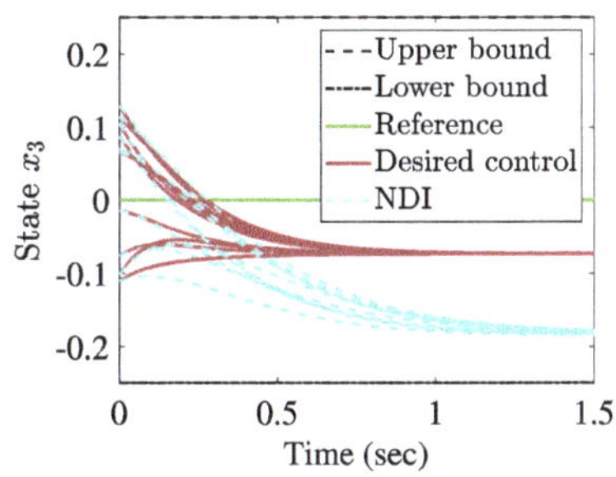

Fig. 7.12 State histories obtained using desired and DI controllers

Sect. 12.2 of Chap. 12 at the end of this book. However, its applicability is limited to specific system structures. To address this, the next section extends the adaptive control design to Euler-Lagrange systems, broadening its applicability to a wider range of practical systems.

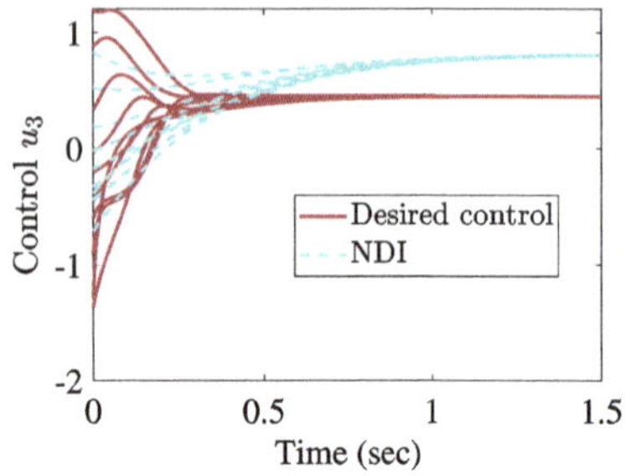

Fig. 7.13 Desired and DI control histories for various initial conditions

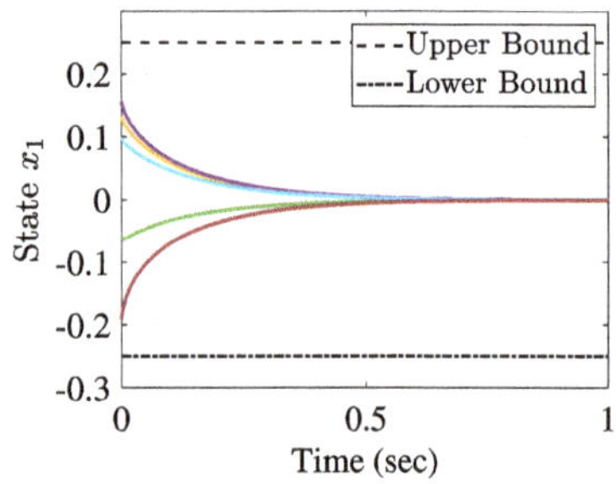

Fig. 7.14 State histories obtained using adaptive control

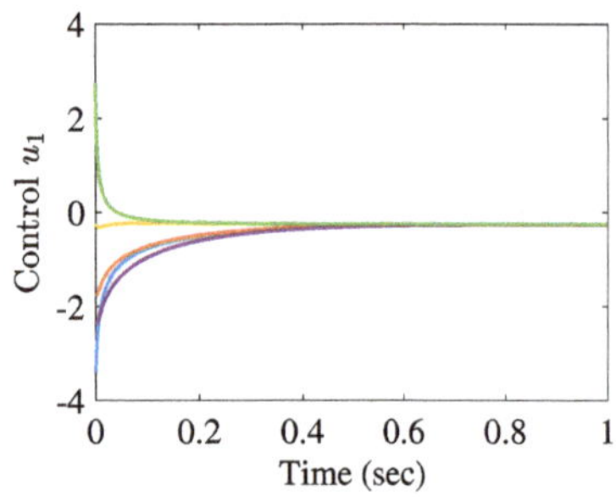

Fig. 7.15 Adaptive control histories for various initial conditions

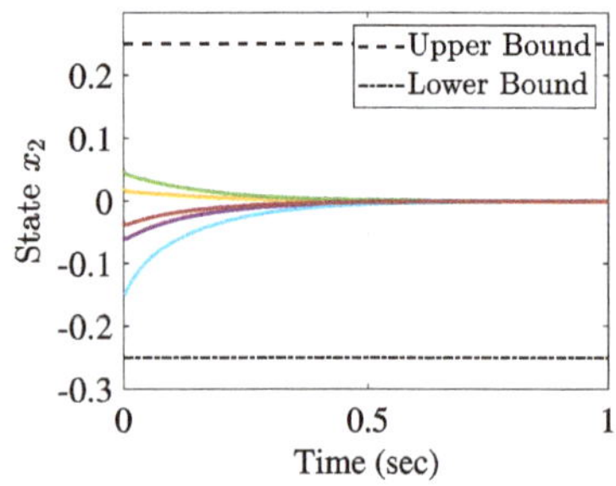

Fig. 7.16 State histories obtained using adaptive control

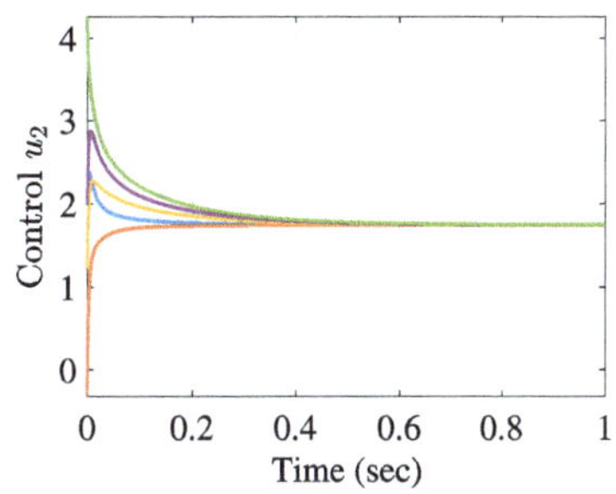

Fig. 7.17 Adaptive control histories for various initial conditions

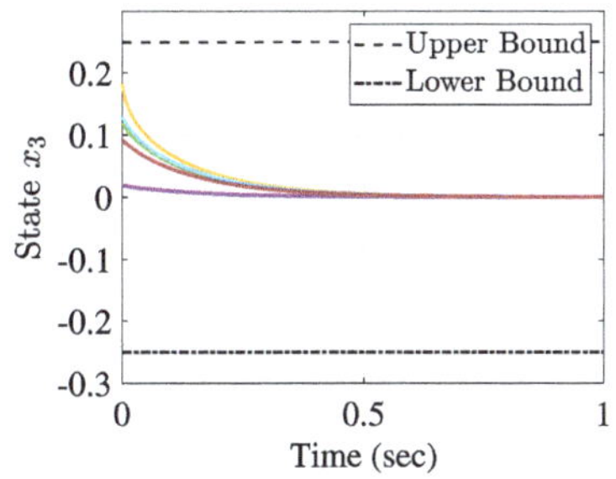

Fig. 7.18 State histories obtained using adaptive control

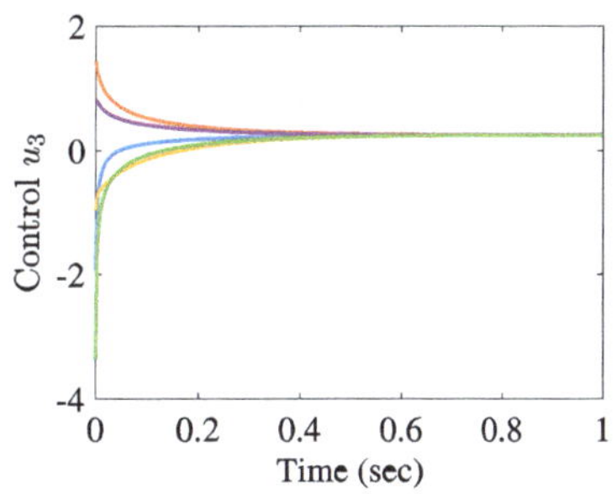

Fig. 7.19 Adaptive control histories for various initial conditions

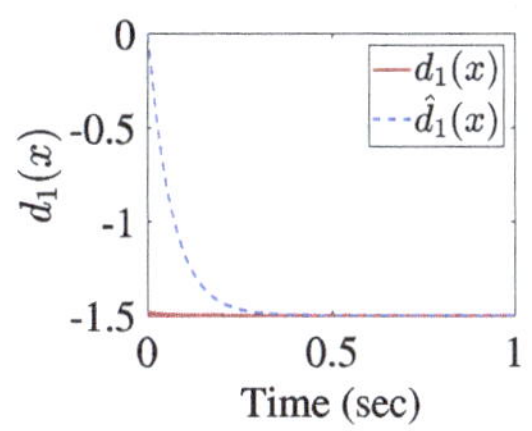

Fig. 7.20 Learning of modeling uncertainties

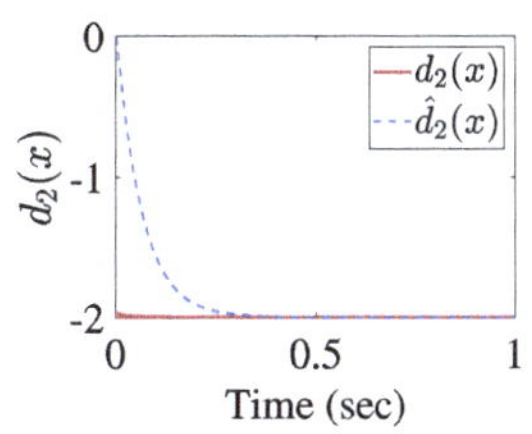

Fig. 7.21 Learning of modeling uncertainties

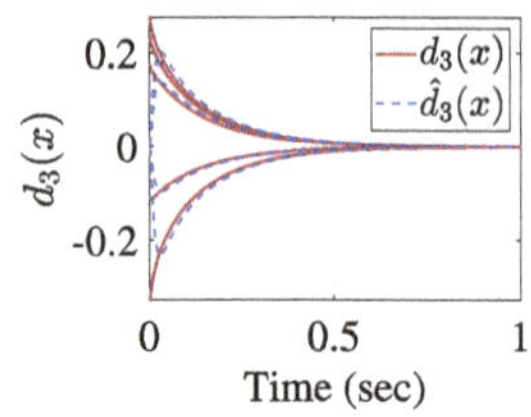

Fig. 7.22 Learning of modeling uncertainties

7.2 Adaptive Control of Euler-Lagrange Systems

This section introduces a barrier Lyapunov function-based neuro-adaptive control design for uncertain Euler-Lagrange (EL) systems, a critical class of MIMO nonlinear systems representing the dynamics of various practical systems, including mechanical and electrical systems. The state-constrained adaptive controller is developed in two steps following a neuro-adaptive control framework. It is demonstrated that the closed-loop system states remain within the user-imposed state constraints, and the steady-state error converges to a user-defined domain. The detailed control formulation is presented in the following discussion.

The EL systems, defined in (1.7), can be represented in state-space form as

$$\begin{aligned} \dot{\boldsymbol{x}}_1 &= \boldsymbol{x}_2 \\ \dot{\boldsymbol{x}}_2 &= \boldsymbol{f}(\boldsymbol{x}) + \boldsymbol{G}(\boldsymbol{x})\boldsymbol{u} + \boldsymbol{d}(\boldsymbol{x}) \end{aligned} \tag{7.41}$$

where state $\boldsymbol{x}$ is defined as $\boldsymbol{x} \triangleq \left[\boldsymbol{x}_1^T, \boldsymbol{x}_2^T\right]^T$ and $\boldsymbol{x}_1 \in \Re^n$ and $\boldsymbol{x}_2 \in \Re^n$ are defined as kinematic and dynamic states of the system. The ith element of the states $\boldsymbol{x}_1$ and $\boldsymbol{x}_2$ are represented as x_{1i} and x_{2i}, respectively, $\forall\ i = 1, \ldots, n$. System output is selected as $\boldsymbol{x}_1 \in \Re^n$ and the control input to the system is denoted by $\boldsymbol{u} \in \Re^n$. The system is assumed to be associated with structured model uncertainties $\boldsymbol{d}(\boldsymbol{x})$.

The objective of the control design is to develop a stable controller that satisfies the following criteria: (i) The output error, $\left(x_{1i} - x_i^*\right), \forall i$, asymptotically converges to a user-defined open set, $\left(-\zeta_{e_i}, \zeta_{e_i}\right), \forall i$. (ii) The closed-loop signals adhere to the user-imposed state constraints, ensuring no violations. The position and velocity constraints of the system are defined as $\underline{\zeta}_{x_{1i}} < x_{1i} < \overline{\zeta}_{x_{1i}}, \forall i$ and $\underline{\zeta}_{x_{2i}} < x_{2i} < \overline{\zeta}_{x_{2i}}, \forall i$, respectively.

All four types of constraints are demonstrated in Fig. 7.23 for a kinematic state $x_{1i}(t)$, where the imposed state constraints on $x_{1i}(t)$ are represented by $\underline{\zeta}_{x_{1i}}$ and $\overline{\zeta}_{x_{1i}}$, and the steady-state tracking error constraints are shown by $\pm\zeta_{e_i}$.

The reference signal $\boldsymbol{x}^*$ and its time derivative $\dot{\boldsymbol{x}}^*$ are assumed to be bounded as $-\underline{\zeta}_{x_i^*} \leq x_i^* \leq \overline{\zeta}_{x_i^*}, \forall i$ and $-\underline{\zeta}_{\dot{x}_i^*} \leq \dot{x}_i^* \leq \overline{\zeta}_{\dot{x}_i^*}, \forall i$, where $\underline{\zeta}_{x_{1i}} > 0, \overline{\zeta}_{x_{1i}} > 0, \underline{\zeta}_{x_{2i}} > 0$ and $\overline{\zeta}_{x_{2i}} > 0$ are known constants. The term x_i^* and $\dot{x}_i^*$ are ith components of $\boldsymbol{x}^*$ and $\dot{\boldsymbol{x}}^*$, respectively. To ensure consistency between the imposed state constraints and the desired tracking objectives, it is further assumed that $-\underline{\zeta}_{x_{1i}} < -\underline{\zeta}_{x_i^*} \leq x_i^* \leq$

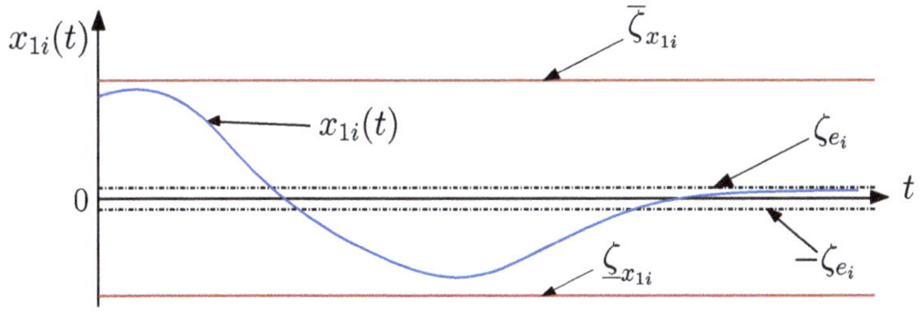

Fig. 7.23 The demonstration of imposed user-defined state constraints and steady-state tracking error bounds for the kinematic state $x_{1i}(t)$ with reference trajectory as zero

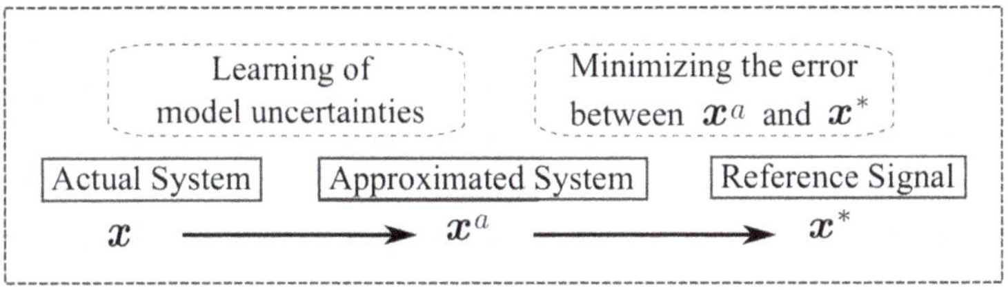

Fig. 7.24 State-constrained adaptive control design philosophy

$\overline{\zeta}_{x_i^*} < \overline{\zeta}_{x_{1i}}, \forall i$ and $-\underline{\zeta}_{x_{2i}} < -\underline{\zeta}_{\dot{x}_i^*} \le \dot{x}_i^* \le \overline{\zeta}_{\dot{x}_i^*} < \overline{\zeta}_{x_{2i}}, \forall i$, respectively. These conditions ensure that the desired reference signals and their derivatives remain within the bounds defined by the imposed state constraints.

7.2.1 Neuro-adaptive Control Design Philosophy

The control design in this section follows a similar philosophy to the previous one but is adapted for Euler-Lagrange (EL) systems, necessitating a new approximated model for system uncertainties. The adaptive control structure involves two steps: (i) constructing a neural network-based approximated system to handle model uncertainties, with weight updates derived using Lyapunov stability theory, and (ii) designing an asymptotically stable tracking controller. As illustrated in Fig. 7.24, the proposed controller ensures that the overall system error asymptotically converges to a user-defined open set $\left(-\zeta_{e_i}, \zeta_{e_i}\right), \forall i$. The synthesis of the control design is detailed below.

7.2.2 Approximation of Structured Model Uncertainties

The first step in the neuro-adaptive control design is to construct an approximated system that captures the uncertainties of the actual system. Accordingly, a novel approximated system dynamics is developed as follows:

$$
\begin{aligned}
\dot{\boldsymbol{x}}_1^a &= \boldsymbol{x}_2^a \\
\dot{\boldsymbol{x}}_2^a &= \boldsymbol{f}(\boldsymbol{x}) + \boldsymbol{G}(\boldsymbol{x})\boldsymbol{u} + \hat{\boldsymbol{d}}(\boldsymbol{x}) + \boldsymbol{K}_1\left(\boldsymbol{e}, \dot{\boldsymbol{e}}\right)\boldsymbol{e} + \boldsymbol{K}_2\left(\dot{\boldsymbol{e}}\right)\dot{\boldsymbol{e}}
\end{aligned}
\tag{7.42}
$$

where $\boldsymbol{x}_1^a$ and $\boldsymbol{x}_2^a$ are state variables of the approximated system with initial conditions identical to those of the actual system. Error variables $\boldsymbol{e}$ and $\dot{\boldsymbol{e}}$ are defined as $\boldsymbol{e} = \boldsymbol{x}_1 - \boldsymbol{x}_1^a$ and $\dot{\boldsymbol{e}} = \boldsymbol{x}_2 - \boldsymbol{x}_2^a$, respectively. Error-dependent gain matrices $\boldsymbol{K}_2\left(\dot{\boldsymbol{e}}\right)$ and $\boldsymbol{K}_1\left(\boldsymbol{e}, \dot{\boldsymbol{e}}\right)$ are symmetric and diagonal, defined as

$$
\boldsymbol{K}_2\left(\dot{\boldsymbol{e}}\right) = \mathrm{diag}\left[\frac{c_{2_1}}{\left(\zeta_{\dot{e}_1}^2 - \dot{e}_1^2\right)}, \ldots, \frac{c_{2_n}}{\left(\zeta_{\dot{e}_n}^2 - \dot{e}_n^2\right)}\right] \tag{7.43}
$$

$$
\boldsymbol{K}_1\left(\boldsymbol{e}, \dot{\boldsymbol{e}}\right) = \mathrm{diag}\left[\frac{c_{1_1}\left(\zeta_{\dot{e}_1}^2 - \dot{e}_1^2\right)}{\left(\zeta_{e_1}^2 - e_1^2\right)}, \ldots, \frac{c_{1_n}\left(\zeta_{\dot{e}_n}^2 - \dot{e}_n^2\right)}{\left(\zeta_{e_n}^2 - e_n^2\right)}\right] \tag{7.44}
$$

where $c_{1_i}, \forall i$ and $c_{2_i}, \forall i$ are positive constants. The variables $\zeta_{\dot{e}_i}, \forall i$ and $\zeta_{e_i}, \forall i$ are user-defined positive constants. The term $\dot{\hat{d}}(\boldsymbol{x})$ is the approximation of model uncertainties, obtained as $\hat{d}_i(\boldsymbol{x}) = \hat{\boldsymbol{w}}_i^T \boldsymbol{\phi}(\boldsymbol{x}), \forall i$.

Using systems (7.41) and (7.42), the error is defined as

$$
\boldsymbol{e} = \boldsymbol{x}_1 - \boldsymbol{x}_1^a \tag{7.45}
$$

The double time derivative of error in (7.45) is

$$
\ddot{\boldsymbol{e}} = \dot{\boldsymbol{x}}_2 - \dot{\boldsymbol{x}}_2^a \tag{7.46}
$$

By substituting $\dot{\boldsymbol{x}}_2$ and $\dot{\boldsymbol{x}}_2^a$ from (7.41) and (7.42), the following is obtained:

$$
\ddot{\boldsymbol{e}} = \boldsymbol{d}(\boldsymbol{x}) - \hat{\boldsymbol{d}}(\boldsymbol{x}) - \boldsymbol{K}_1(\boldsymbol{e}, \dot{\boldsymbol{e}})\boldsymbol{e} - \boldsymbol{K}_2(\dot{\boldsymbol{e}})\dot{\boldsymbol{e}} \tag{7.47}
$$

Since $\boldsymbol{K}_1$ in (7.44) and $\boldsymbol{K}_2$ in (7.43) are diagonal matrices, the ith component of closed-loop error dynamics (7.47) can be expressed as

$$
\ddot{e}_i = d_i(\boldsymbol{x}) - \hat{d}_i(\boldsymbol{x}) - \frac{c_{1_i}\left(\zeta_{\dot{e}_i}^2 - \dot{e}_i^2\right)e_i}{\left(\zeta_{e_i}^2 - e_i^2\right)} - \frac{c_{2_i}\dot{e}_i}{\left(\zeta_{\dot{e}_i}^2 - \dot{e}_i^2\right)} \tag{7.48}
$$

Using the structures of $d_i(\boldsymbol{x}) = \boldsymbol{w}_i^T \boldsymbol{\phi}(\boldsymbol{x})$ and $\hat{d}_i(\boldsymbol{x}) = \hat{\boldsymbol{w}}_i^T \boldsymbol{\phi}(\boldsymbol{x})$, the following is obtained:

$$\ddot{e}_i = \tilde{\boldsymbol{w}}_i^T \boldsymbol{\phi}(\boldsymbol{x}) - \frac{c_{1_i}\left(\zeta_{\dot{e}_i}^2 - \dot{e}_i^2\right) e_i}{\left(\zeta_{e_i}^2 - e_i^2\right)} - \frac{c_{2_i}\dot{e}_i}{\left(\zeta_{\dot{e}_i}^2 - \dot{e}_i^2\right)} \tag{7.49}$$

where $\tilde{\boldsymbol{w}}_i^T$ is defined as $\tilde{\boldsymbol{w}}_i^T = \boldsymbol{w}_i^T - \hat{\boldsymbol{w}}_i^T$. To analyze the stability of error dynamics (7.49), the following logarithm barrier Lyapunov function (Tee et al., 2009) is selected:

$$V_i = \frac{c_{1_i}c_{2_i}}{2}\log\left(\frac{\zeta_{e_i}^2}{\left(\zeta_{e_i}^2 - e_i^2\right)}\right) + \frac{c_{2_i}}{2}\log\left(\frac{\zeta_{\dot{e}_i}^2}{\left(\zeta_{\dot{e}_i}^2 - \dot{e}_i^2\right)}\right) + \frac{\tilde{\boldsymbol{w}}_i^T \tilde{\boldsymbol{w}}_i}{2\gamma_i} \tag{7.50}$$

where $\gamma_i > 0$ is learning rate constant. All other terms are already defined in (7.43) and (7.44). The time derivative of (7.50) is obtained as

$$\dot{V}_i = \frac{c_{1_i}c_{2_i}e_i\dot{e}_i}{\left(\zeta_{e_i}^2 - e_i^2\right)} + \frac{c_{2_i}\dot{e}_i\ddot{e}_i}{\left(\zeta_{\dot{e}_i}^2 - \dot{e}_i^2\right)} - \frac{\tilde{\boldsymbol{w}}_i^T \dot{\hat{\boldsymbol{w}}}_i}{\gamma_i} \tag{7.51}$$

Using (7.49), the derivative of the Lyapunov function in (7.51) can be expressed as

$$\begin{aligned}\dot{V}_i = {} & \frac{c_{1_i}c_{2_i}e_i\dot{e}_i}{\left(\zeta_{e_i}^2 - e_i^2\right)} - \frac{c_{1_i}c_{2_i}e_i\dot{e}_i}{\left(\zeta_{e_i}^2 - e_i^2\right)} - \left(\frac{c_{2_i}\dot{e}_i}{\left(\zeta_{\dot{e}_i}^2 - \dot{e}_i^2\right)}\right)^2 + \frac{\tilde{\boldsymbol{w}}_i^T \boldsymbol{\phi}(\boldsymbol{x})}{c}{}_{2_i}\dot{e}_i\left(\zeta_{\dot{e}_i}^2 - \dot{e}_i^2\right) \\ & - \frac{\tilde{\boldsymbol{w}}_i^T \dot{\hat{\boldsymbol{w}}}_i}{\gamma_i}\end{aligned} \tag{7.52}$$

which can be simplified as

$$\dot{V}_i = -\left(\frac{c_{2_i}\dot{e}_i}{\left(\zeta_{\dot{e}_i}^2 - \dot{e}_i^2\right)}\right)^2 + \tilde{\boldsymbol{w}}_i^T\left(\frac{\boldsymbol{\phi}(\boldsymbol{x})c_{2_i}\dot{e}_i}{\left(\zeta_{\dot{e}_i}^2 - \dot{e}_i^2\right)} - \frac{\dot{\hat{\boldsymbol{w}}}_i}{\gamma_i}\right) \tag{7.53}$$

By selecting the weight learning rule as

$$\dot{\hat{\boldsymbol{w}}}_i = \frac{\gamma_i c_{2_i}\dot{e}_i\boldsymbol{\phi}(\boldsymbol{x})}{\left(\zeta_{\dot{e}_i}^2 - \dot{e}_i^2\right)}, \tag{7.54}$$

The time derivative of Lyapunov function in (7.53) is modified as

$$\dot{V}_i = -\left(\frac{c_{2_i}\dot{e}_i}{\left(\zeta_{\dot{e}_i}^2 - \dot{e}_i^2\right)}\right)^2 \tag{7.55}$$

Since $\dot{V}_i \leq 0$, e_i, $\dot{e}_i$, and $\tilde{\boldsymbol{w}}_i$ are bounded. Furthermore, as basis functions $\boldsymbol{\phi}(\boldsymbol{x})$ are chosen to be bounded in nature, by analyzing (7.49), $\ddot{e}_i$ is also bounded. Therefore, all closed-loop signals are bounded and stable in the sense of Lyapunov.

The nature of $\dot{e}_i$ can be further analyzed by calculating the time derivative of $\dot{V}_i$ in (7.55) as

$$\ddot{V}_i = \frac{-2c_{2_i}\dot{e}_i}{\left(\zeta_{\dot{e}_i}^2 - \dot{e}_i^2\right)^3}\left[c_{2_i}\ddot{e}_i\left(\zeta_{\dot{e}_i}^2 - \dot{e}_i^2\right) + 2c_{2_i}\dot{e}_i^2\ddot{e}_i\right] \tag{7.56}$$

Since each term of $\ddot{V}_i$ is bounded, using Barbalat's Lemma, $\dot{V}_i \rightarrow 0$ as $t \rightarrow \infty$. Thus, $\dot{e}_i \rightarrow 0$ as $t \rightarrow \infty$.

The bounds on e_i can also be determined using Lyapunov analysis. Since $\dot{V}_i \leq 0$ in (7.55), $V_i(t) \leq V_0, \forall t$, where V_0 is the value of V_i at $t = 0$. As each term of the selected Lyapunov function is positive definite, the following expression is true for an individual term:

$$\frac{c_{1_i}c_{2_i}}{2}\log\left(\frac{\zeta_{e_i}^2}{\left(\zeta_{e_i}^2 - e_i^2\right)}\right) \leq V_0 \tag{7.57}$$

By taking the exponential on both sides and following the necessary algebra, it can be shown that

$$\zeta_{e_i}^2\exp\left(\frac{-2V_0}{c_{1_i}c_{2_i}}\right) \leq \left(\zeta_{e_i}^2 - e_i^2\right) \tag{7.58}$$

which can be simplified as

$$|e_i| \leq \zeta_{e_i}\sqrt{1 - \exp\left(-2V_0/\left(c_{1_i}c_{2_i}\right)\right)} \tag{7.59}$$

where "exp" is the exponential function. Since V_0 is always positive except at the equilibrium point zero, the exponential term $\exp\left(-2V_0/\left(c_{1_i}c_{2_i}\right)\right)$ is bounded in the domain (0, 1). Therefore, the errors e_i will remain bounded as $|e_i| < \zeta_{e_i}, \forall i$. Similar analysis can be performed for $\dot{e}_i$ and bounds on $\dot{e}_i$ can be shown as $|\dot{e}_i| < \zeta_{\dot{e}_i}, \forall i$.

After approximating model uncertainties, the next step is to design an asymptotically stable controller for the approximated system to track the reference signal, which is presented as follows.

7.2.3 Asymptotically Stable Tracking Control Design

In order to design the tracking controller, the error between the approximated system in (7.42) and reference output $\boldsymbol{x}^*$ is defined as

$$\boldsymbol{e}^a = \boldsymbol{x}_1^a - \boldsymbol{x}^* \tag{7.60}$$

The time derivatives of (7.60) can be calculated as

$$\dot{\boldsymbol{e}}^a = \boldsymbol{x}_2^a - \dot{\boldsymbol{x}}^* \tag{7.61}$$

$$\ddot{\boldsymbol{e}}^a = \dot{\boldsymbol{x}}_2^a - \ddot{\boldsymbol{x}}^* \tag{7.62}$$

By substituting $\dot{\boldsymbol{x}}_2^a$ from (7.42), the error dynamics can be written as

$$\ddot{\boldsymbol{e}}^a = \boldsymbol{f}(\boldsymbol{x}) + \boldsymbol{G}(\boldsymbol{x})\boldsymbol{u} + \hat{\boldsymbol{d}}(\boldsymbol{x}) + \boldsymbol{K}_1(\boldsymbol{e}, \dot{\boldsymbol{e}})\boldsymbol{e} + \boldsymbol{K}_2(\dot{\boldsymbol{e}})\dot{\boldsymbol{e}} - \ddot{\boldsymbol{x}}^* \tag{7.63}$$

To stabilize the error dynamics (7.63), a closed-form adaptive control law is proposed as

$$\boldsymbol{u} = [\boldsymbol{G}(\boldsymbol{x})]^{-1}\begin{pmatrix} -\boldsymbol{f}(\boldsymbol{x}) - \boldsymbol{K}_1^a(\boldsymbol{e}^a, \dot{\boldsymbol{e}}^a)\boldsymbol{e}^a - \boldsymbol{K}_2^a(\dot{\boldsymbol{e}}^a)\dot{\boldsymbol{e}}^a \\ -\boldsymbol{K}_1(\boldsymbol{e}, \dot{\boldsymbol{e}})\boldsymbol{e} - \boldsymbol{K}_2(\dot{\boldsymbol{e}})\dot{\boldsymbol{e}} - \hat{\boldsymbol{d}}(\boldsymbol{x}) + \ddot{\boldsymbol{x}}_d^* \end{pmatrix} \tag{7.64}$$

where $[\boldsymbol{G}(\boldsymbol{x})]^{-1}$ will always exist due to the properties of EL systems. The gain matrices $\boldsymbol{K}_2^a(\dot{\boldsymbol{e}}^a)$ and $\boldsymbol{K}_1^a(\boldsymbol{e}^a, \dot{\boldsymbol{e}}^a)$ are symmetric and diagonal matrices. Matrix $\boldsymbol{K}_2^a(\dot{\boldsymbol{e}}^a)$ is defined as

$$\boldsymbol{K}_2^a(\dot{\boldsymbol{e}}^a) = \operatorname{diag}\left[k_{21}^a\left(\dot{e}_1^a\right),\ k_{22}^a\left(\dot{e}_2^a\right), \ldots,\ k_{2n}^a\left(\dot{e}_n^a\right)\right] \tag{7.65}$$

where the ith component of matrix $\boldsymbol{K}_2^a(\dot{\boldsymbol{e}}^a)$ is

$$k_{2i}^a\left(\dot{e}_i^a\right) = \left[\frac{\gamma_i^* c_{2_i}^a}{\left(\overline{\zeta}_{\dot{e}_i^a}^2 - \dot{e}_i^{a^2}\right)} + \left(\underline{\zeta}_{\dot{e}_i^a} / \overline{\zeta}_{\dot{e}_i^a}\right)^2 \frac{\left(1 - \gamma_i^*\right) c_{2_i}^a}{\left(\underline{\zeta}_{\dot{e}_i^a}^2 - \dot{e}_i^{a^2}\right)}\right] \tag{7.66}$$

with $\quad \gamma_i^*(\dot{e}_i^a) = \begin{cases} 1, & \text{if } 0 < \dot{e}_i^a < \overline{\zeta}_{\dot{e}_i^a} \\ 0, & \text{if } -\underline{\zeta}_{\dot{e}_i^a} < \dot{e}_i^a \leq 0 \end{cases}$

where $c_{2_i}^a > 0, \forall i$ is a constant. Constants $\overline{\zeta}_{\dot{e}_i^a}$ and $\underline{\zeta}_{\dot{e}_i^a}$ represent upper and lower bounds on error $\dot{e}_i^a$, obtained as

$$\overline{\zeta}_{\dot{e}_i^a} = \overline{\zeta}_{x_{2i}} - \overline{\zeta}_{\dot{x}_i^*} - \zeta_{\dot{e}_i} \tag{7.67}$$

$$\underline{\zeta}_{\dot{e}_i^a} = \underline{\zeta}_{\dot{x}_i^*} - \underline{\zeta}_{x_{2i}} - \zeta_{\dot{e}_i} \tag{7.68}$$

where, $\overline{\zeta}_{x_{2i}}$ and $\underline{\zeta}_{x_{2i}}$ are imposed state constraints on the dynamic states of actual system, known constant $\zeta_{\dot{e}_i}$ is defined in (7.43), and $\overline{\zeta}_{\dot{x}_i^*}$ and $\underline{\zeta}_{\dot{x}_i^*}$ are known bounds on the derivative of reference signal $\dot{x}_i^*$. Matrix $\boldsymbol{K}_1^a(\boldsymbol{e}^a, \dot{\boldsymbol{e}}^a)$ is defined as

$$\boldsymbol{K}_1^a(\boldsymbol{e}^a, \dot{\boldsymbol{e}}^a) = \text{diag}\left[k_{11}^a\left(e_1^a, \dot{e}_1^a\right),\ k_{12}^a\left(e_2^a, \dot{e}_2^a\right), \ldots,\ k_{1n}^a\left(e_n^a, \dot{e}_n^a\right)\right] \tag{7.69}$$

where the ith component of matrix $\boldsymbol{K}_1^a(\boldsymbol{e}^a, \dot{\boldsymbol{e}}^a)$ is given as

$$k_{1i}^a\left(e_i^a, \dot{e}_i^a\right) = \left[\frac{\gamma_i^{**} c_{1_i}^a}{\left(\overline{\zeta}_{e_i^a}^2 - e_i^{a2}\right)} + \left[\frac{\underline{\zeta}_{e_i^a}^2}{\overline{\zeta}_{e_i^a}^2}\right] \frac{\left(1-\gamma_i^{**}\right) c_{1_i}^a}{\left(\underline{\zeta}_{e_i^a}^2 - e_i^{a2}\right)}\right]\left(\frac{1}{k_{2i}^a\left(\dot{e}_i^a\right)}\right) \tag{7.70}$$

with $\quad \gamma_i^{**}(e_i^a) = \begin{cases} 1, & \text{if } 0 < e_i^a < \overline{\zeta}_{e_i^a} \\ 0, & \text{if } -\underline{\zeta}_{e_i^a} < e_i^a \leq 0 \end{cases}$

where $c_{1_i}^a > 0, \forall i$ is constant. Note that the term $k_{2i}^a\left(\dot{e}_i^a\right)$, defined in (7.66), is always positive. Constants $\overline{\zeta}_{e_i^a}$ and $\underline{\zeta}_{e_i^a}$ are upper and lower bounds on e_i^a, calculated as

$$\overline{\zeta}_{e_i^a} = \overline{\zeta}_{x_{1i}} - \overline{\zeta}_{x_i^*} - \zeta_{e_i} \tag{7.71}$$

$$\underline{\zeta}_{e_i^a} = \underline{\zeta}_{x_i^*} - \underline{\zeta}_{x_{1i}} - \zeta_{e_i} \tag{7.72}$$

where $\overline{\zeta}_{x_{1i}}$ and $\underline{\zeta}_{x_{1i}}$ are imposed state constraints on kinematic states of the actual system dynamics, known constant ζ_{e_i} is defined in (7.44), and $\overline{\zeta}_{x_i^*}$, $\underline{\zeta}_{x_i^*}$ are the bounds on reference signal x_i^*.

By substituting the control expression (7.64) in (7.63), the following closed-loop error dynamics is obtained as

$$\ddot{\boldsymbol{e}}^a = -\boldsymbol{K}_1^a(\boldsymbol{e}^a, \dot{\boldsymbol{e}}^a)\boldsymbol{e}^a - \boldsymbol{K}_2^a(\dot{\boldsymbol{e}}^a)\dot{\boldsymbol{e}}^a \tag{7.73}$$

Since matrices $\boldsymbol{K}_1^a(\boldsymbol{e}^a, \dot{\boldsymbol{e}}^a)$ and $\boldsymbol{K}_2^a(\dot{\boldsymbol{e}}^a)$ are diagonal in nature, the ith component of closed-loop error dynamics can be written as

$$\ddot{e}_i^a = -k_{1i}^a\left(e_i^a, \dot{e}_i^a\right) e_i^a - k_{2i}^a\left(\dot{e}_i^a\right) \dot{e}_i^a \tag{7.74}$$

To analyze the stability of (7.74), the following asymmetric logarithm BLF is chosen

$$V_i^a = \frac{c_{1_i}^a}{2}\left[\gamma_i^{**}\log\left(\frac{\overline{\zeta}_{e_i^a}^2}{\overline{\zeta}_{e_i^a}^2 - e_i^{a2}}\right) + \frac{\left(1-\gamma_i^{**}\right)\underline{\zeta}_{e_i^a}^2}{\overline{\zeta}_{e_i^a}^2}\log\left(\frac{\underline{\zeta}_{e_i^a}^2}{\underline{\zeta}_{e_i^a}^2 - e_i^{a2}}\right)\right]$$

$$+\frac{c^a_{2_i}}{2}\left[\gamma_i^*\log\left(\frac{\overline{\zeta}^2_{\dot{e}^a_i}}{\overline{\zeta}^2_{\dot{e}^a_i}-\dot{e}^{a^2}_i}\right)+\frac{\left(1-\gamma_i^*\right)\underline{\zeta}^2_{\dot{e}^a_i}}{\overline{\zeta}^2_{\dot{e}^a_i}}\log\left(\frac{\underline{\zeta}^2_{\dot{e}^a_i}}{\underline{\zeta}^2_{\dot{e}^a_i}-\dot{e}^{a^2}_i}\right)\right] \tag{7.75}$$

where all variables used in (7.75) are already defined in (7.66) and (7.70). Note that the presence of switching terms γ_i^{**} and γ_i^* does not affect the differentiability and continuity of the asymmetric BLF, which is also shown by Tee et al. (2009). The time derivative of (7.75) can be obtained as

$$\begin{aligned}\dot{V}^a_i &= c^a_{1_i}\left[\frac{\gamma_i^{**}e^a_i\dot{e}^a_i}{\left(\overline{\zeta}^2_{e^a_i}-e^{a^2}_i\right)}+\frac{\left(1-\gamma_i^{**}\right)\underline{\zeta}^2_{e^a_i}e^a_i\dot{e}^a_i}{\overline{\zeta}^2_{e^a_i}\left(\underline{\zeta}^2_{e^a_i}-e^{a^2}_i\right)}\right]\\ &+c^a_{2_i}\left[\frac{\gamma_i^*\dot{e}^a_i\ddot{e}^a_i}{\left(\overline{\zeta}^2_{\dot{e}^a_i}-\dot{e}^{a^2}_i\right)}+\frac{\left(1-\gamma_i^*\right)\underline{\zeta}^2_{\dot{e}^a_i}\dot{e}^a_i\ddot{e}^a_i}{\overline{\zeta}^2_{\dot{e}^a_i}\left(\underline{\zeta}^2_{\dot{e}^a_i}-\dot{e}^{a^2}_i\right)}\right]\end{aligned} \tag{7.76}$$

which can rewritten as

$$\begin{aligned}\dot{V}^a_i &= e^a_i\dot{e}^a_i\left[\frac{\gamma_i^{**}c^a_{1_i}}{\left(\overline{\zeta}^2_{e^a_i}-e^{a^2}_i\right)}+\frac{\left(1-\gamma_i^{**}\right)c^a_{1_i}\underline{\zeta}^2_{e^a_i}}{\overline{\zeta}^2_{e^a_i}\left(\underline{\zeta}^2_{e^a_i}-e^{a^2}_i\right)}\right]\\ &+\dot{e}^a_i\ddot{e}^a_i\left[\frac{\gamma_i^*c^a_{2_i}}{\left(\overline{\zeta}^2_{\dot{e}^a_i}-\dot{e}^{a^2}_i\right)}+\frac{\left(1-\gamma_i^*\right)c^a_{2_i}\underline{\zeta}^2_{\dot{e}^a_i}}{\overline{\zeta}^2_{\dot{e}^a_i}\left(\underline{\zeta}^2_{\dot{e}^a_i}-\dot{e}^{a^2}_i\right)}\right]\end{aligned} \tag{7.77}$$

Using the definitions of $k^a_{1i}\left(e^a_i,\dot{e}^a_i\right)$ and $k^a_{2i}\left(\dot{e}^a_i\right)$ from (7.70) and (7.66), $\dot{V}^a_i$ is expressed as

$$\dot{V}^a_i = k^a_{1i}\left(e^a_i,\dot{e}^a_i\right)\,k^a_{2i}\left(\dot{e}^a_i\right)\,e^a_i\,\dot{e}^a_i + k^a_{2i}\left(\dot{e}^a_i\right)\,\dot{e}^a_i\,\ddot{e}^a_i \tag{7.78}$$

By substituting the value of $\ddot{e}^a_i$ from (7.74), the following simplified expression of $\dot{V}^a_i$ is obtained:

$$\dot{V}^a_i = -k^2_{a2i}\dot{e}^{a^2}_i \tag{7.79}$$

Since $\dot{V}^a_i \leq 0$, the asymptotic stability of closed-loop error dynamics in (7.74) cannot be concluded. However, $\dot{V}^a_i = 0$ only when $\dot{e}^a_i = 0, \forall i$ for all time, which implies that $\ddot{e}^a_i = 0, \forall i$. Using (7.74), when $\dot{e}^a_i = \ddot{e}^a_i = 0$, one can conclude that $e^a_i = 0, \forall i$ (provided $e^a_i \in \left(-\underline{\zeta}_{e^a_i}, \overline{\zeta}_{e^a_i}\right)$ and $\dot{e}^a_i \in \left(-\underline{\zeta}_{\dot{e}^a_i}, \overline{\zeta}_{\dot{e}^a_i}\right)$ satisfy, which is proven next). Thus, the equilibrium point is the only invariant set, and hence, using LaSalle's principle, the asymptotic stability of the closed-loop error dynamics can be concluded. Furthermore, as $\dot{V}_i^a \leq 0$, it can be concluded that all closed-loop signals will be bounded.

The upper bound on selected Lyapunov is obtained as $V_i^a \leq V_0^a$, $\forall t$, i.e.,

$$\frac{c_{1_i}^a}{2}\left[\gamma_i^{**}\log\left(\frac{\overline{\zeta}_{e_i^a}^2}{\overline{\zeta}_{e_i^a}^2 - e_i^{a2}}\right) + \frac{\left(1-\gamma_i^{**}\right)\underline{\zeta}_{e_i^a}^2}{\overline{\zeta}_{e_i^a}^2}\log\left(\frac{\underline{\zeta}_{e_i^a}^2}{\underline{\zeta}_{e_i^a}^2 - e_i^{a2}}\right)\right]$$
$$+\frac{c_{2_i}^a}{2}\left[\gamma_i^{*}\log\left(\frac{\overline{\zeta}_{\dot{e}_i^a}^2}{\overline{\zeta}_{\dot{e}_i^a}^2 - \dot{e}_i^{a2}}\right) + \frac{\left(1-\gamma_i^{*}\right)\underline{\zeta}_{\dot{e}_i^a}^2}{\overline{\zeta}_{\dot{e}_i^a}^2}\log\left(\frac{\underline{\zeta}_{\dot{e}_i^a}^2}{\underline{\zeta}_{\dot{e}_i^a}^2 - \dot{e}_i^{a2}}\right)\right] \leq V_0^a \tag{7.80}$$

where V_0^a is the value of the Lyapunov function at $t = 0$. Since the left-hand side terms are positive definite, the individual term is upper-bounded as

$$\frac{c_{1_i}^a}{2}\left[\gamma_i^{**}\log\left(\frac{\overline{\zeta}_{e_i^a}^2}{\overline{\zeta}_{e_i^a}^2 - e_i^{a2}}\right) + \frac{\left(1-\gamma_i^{**}\right)\underline{\zeta}_{e_i^a}^2}{\overline{\zeta}_{e_i^a}^2}\log\left(\frac{\underline{\zeta}_{e_i^a}^2}{\underline{\zeta}_{e_i^a}^2 - e_i^{a2}}\right)\right] \leq V_0^a \tag{7.81}$$

which can be modified as

$$\frac{2V_0^a}{c_{1_i}^a} \geq \begin{cases} \log\left(\frac{\overline{\zeta}_{e_i^a}^2}{\left(\overline{\zeta}_{e_i^a}^2 - e_i^{a2}\right)}\right), & 0 < e_i^a < \overline{\zeta}_{e_i^a} \\ \frac{\underline{\zeta}_{e_i^a}^2}{\overline{\zeta}_{e_i^a}^2}\log\left(\frac{\underline{\zeta}_{e_i^a}^2}{\left(\underline{\zeta}_{e_i^a}^2 - e_i^{a2}\right)}\right), & -\underline{\zeta}_{e_i^a} < e_i^a \leq 0 \end{cases} \tag{7.82}$$

By taking the exponent on both sides and considering the fact that $V_0^a > 0$ for nonzero e_i^a and $\dot{e}_{a_i}$, the following inequality can be obtained after simplification:

$$-\underline{\zeta}_{e_i^a} < e_i^a < \overline{\zeta}_{e_i^a} \tag{7.83}$$

Similar analysis can be carried out for $\dot{e}_i^a$, and the following bounds can be obtained

$$-\underline{\zeta}_{\dot{e}_i^a} < \dot{e}_i^a < \overline{\zeta}_{\dot{e}_i^a} \tag{7.84}$$

Thus, the error e_i^a and its derivative $\dot{e}_i^a$ will remain bounded in open sets, defined as $\left(-\underline{\zeta}_{e_i^a}, \overline{\zeta}_{e_i^a}\right)$ and $\left(-\underline{\zeta}_{\dot{e}_i^a}, \overline{\zeta}_{\dot{e}_i^a}\right)$, $\forall i$, respectively.

The proposed control formulation has the following properties: (i) by selecting $\underline{\zeta}_{x1i} = \overline{\zeta}_{x1_i}$, $\forall i$ and $\underline{\zeta}_{x2i} = \overline{\zeta}_{x2_i}$, $\forall i$, the controller can function as a symmetric state-constrained controller; (ii) it can be converted into an output-constrained control by choosing the elements of the matrix $\boldsymbol{K}_2^a(\dot{\boldsymbol{e}}^a)$ as positive constants; and (iii) it reduces to a nonlinear dynamic inversion control when the elements of the matrices $\boldsymbol{K}_1^a(\boldsymbol{e}^a, \dot{\boldsymbol{e}}^a)$ and $\boldsymbol{K}_2^a(\dot{\boldsymbol{e}}^a)$ are selected as positive constants.

The above analysis demonstrates that all closed-loop signals remain bounded in both steps of the control design. However, the behavior of the closed-loop state of the actual system has not yet been analyzed. Therefore, the closed-loop response of the system is examined in the following theorem.

Theorem 7.2 *Consider the actual systems (7.41) under the controller (7.64) with the weight update learning rule in (7.54), and if initial conditions are chosen as* $e_i^a(0) \in \left(-\underline{\zeta}_{e_i^a}, \overline{\zeta}_{e_i^a}\right)$, $\dot{e}_i^a(0) \in \left(-\underline{\zeta}_{\dot{e}_i^a}, \overline{\zeta}_{\dot{e}_i^a}\right) \forall i$, *then following properties hold:*

(i) The closed-loop tracking error $\left(x_{1i} - x_i^*\right)$, $\forall i$ *will asymptotically converge to the open set* $\left(-\zeta_{e_i}, \zeta_{e_i}\right)$, $\forall i$, *where* ζ_{e_i} *is a user-defined positive constant, defined in (7.44).*
(ii) Closed-loop kinematic and dynamic states of the system will remain bounded as $\underline{\zeta}_{x_{1i}} < x_{1i} < \overline{\zeta}_{x_{1i}}$ *and* $\underline{\zeta}_{x_{2i}} < x_{2i} < \overline{\zeta}_{x_{2i}}$, $\forall t \geq 0, \forall i$.

Proof

(i) The closed-loop tracking error of the actual system is defined as $\left(x_{1i} - x_i^*\right)$. By adding and subtracting x_{1i}^a, the error can be modified as $\left[\left(x_{1i} - x_{1i}^a\right) + \left(x_{1i}^a - x_i^*\right)\right]$, which is $\left(e_i + e_i^a\right)$. Since $e_i \in \left(-\zeta_{e_i}, \zeta_{e_i}\right)$ and e_i^a is asymptotically stable, i.e., $e_i^a \to 0$ as $t \to \infty$, the tracking error $\left(x_{1i} - x_i^*\right)$ will asymptotically converge to the open set $\left(-\zeta_{e_i}, \zeta_{e_i}\right)$, $\forall i$.
(ii) Since the errors e_i^a and e_i are bounded as $-\underline{\zeta}_{e_i^a} < e_i^a < \overline{\zeta}_{e_i^a}$ and $|e_i| < \zeta_{e_i}, \forall i$, respectively, the following is true:

$$-\underline{\zeta}_{e_i^a} - \zeta_{e_i} < e_i^a + e_i < \overline{\zeta}_{e_i^a} + \zeta_{e_i} \tag{7.85}$$

Substituting $e_i^a = x_{1i}^a - x_i^*$ and $e_i = x_{1i} - x_{1i}^a$, (7.85) is modified as

$$-\underline{\zeta}_{e_i^a} - \zeta_{e_i} < x_{1i} - x_i^* < \overline{\zeta}_{e_i^a} + \zeta_{e_i} \tag{7.86}$$

By substituting bounds of x_i^*, the following is obtained:

$$-\underline{\zeta}_{e_i^a} - \zeta_{e_i} + \underline{\zeta}_{x_i^*} < x_{1_i} < \overline{\zeta}_{e_i^a} + \zeta_{e_i} + \overline{\zeta}_{x_i^*} \tag{7.87}$$

Using the definitions of $\overline{\zeta}_{x_{1i}}$ and $\underline{\zeta}_{x_{1i}}$ from (7.71) and (7.72), inequality (7.87) is modified as

$$\underline{\zeta}_{x_{1i}} < x_{1i} < \overline{\zeta}_{x_{1i}} \tag{7.88}$$

Similar analysis can be performed using the bounds of $\dot{e}_i^a$ and $\dot{e}_i$. Thus, by using (7.67) and (7.68), the following bounds can be obtained:

$$\underline{\zeta}_{x_{2i}} < x_{2i} < \overline{\zeta}_{x_{2i}} \tag{7.89}$$

Therefore, closed-loop states of the actual system will remain bounded by the imposed state constraints. This completes the proof.

□

7.2.4 Application to a Two-Link Robotic Manipulator

In this section, the applicability of the proposed control design is validated through simulation results. Simulation results are presented for a two-link robot manipulator model. The dynamics of the two-link manipulator system is given by Slotine and Li (1991)

$$\begin{bmatrix} \ddot{q}_1 \\ \ddot{q}_2 \end{bmatrix} = \left[H\right]^{-1} \begin{bmatrix} h\dot{q}_2 & h\left(\dot{q}_1 + \dot{q}_2\right) \\ -h\dot{q}_1 & 0 \end{bmatrix} \begin{bmatrix} \dot{q}_1 \\ \dot{q}_2 \end{bmatrix} + \left[H\right]^{-1} \begin{bmatrix} \tau_1 \\ \tau_2 \end{bmatrix} + \begin{bmatrix} d_1 \\ d_2 \end{bmatrix}$$

where

$$\begin{aligned} H &= \begin{bmatrix} H_{11} & H_{12} \\ H_{21} & H_{22} \end{bmatrix} \\ H_{11} &= a_1 + 2a_3 \cos q_2 + 2a_4 \sin q_2 \\ H_{12} &= H_{21} = a_2 + a_3 \cos q_2 + a_4 \sin q_2 \\ H_{22} &= a_2, \quad h = a_3 \sin q_2 - a_4 \cos q_2 \end{aligned}$$

with

$$\begin{aligned} a_1 &= I_1 + m_1 {l_{c1}}^2 + I_e + m_e {l_{ce}}^2 + m_e {l_1}^2 \\ a_2 &= I_e + m_e {l_{ce}}^2 \\ a_3 &= m_e l_1 l_{ce} \cos \delta_e \\ a_4 &= m_e l_1 l_{ce} \sin \delta_e \end{aligned}$$

where $q_1 \in \Re$ and $q_2 \in \Re$ represent joint positions (in deg) and $\dot{q}_1 \in \Re$ and $\dot{q}_2 \in \Re$ represent joint velocities (in deg/sec) of the manipulator. Modeling uncertainties are represented by d_1 and d_2, respectively. The torque τ_j $(j = 1, 2)$ are control inputs to the system. All parameters used in the dynamics have their standard dimensions and usual meanings and can be found in Slotine and Li (1991). In simulations, the parameter values of the model are taken as $m_1 = 1$, $l_1 = 1$, $m_e = 2$, $\delta_e = 30°$, $I_1 = 0.12$, $I_{c1} = 0.5$, $I_e = 0.25$, and $I_{ce} = 0.6$, respectively.

The states of the system can be defined as $x_{11} \triangleq q_1$, $x_{12} \triangleq q_2$, $x_{21} \triangleq \dot{q}_1$, and $x_{22} \triangleq \dot{q}_2$. Thus, the control objective is to generate the required control $u_i \triangleq \tau_i$ $(i = 1, 2)$ to track the desired outputs (joint positions) with desired accuracy while maintaining the constraints on kinematic and dynamic states, i.e., constraints on joint positions and velocities. The initial conditions for the simulations are selected as $x_{11}(0) = 4.7°$, $x_{21}(0) = 9$ deg/sec, $x_{12}(0) = -4°$, and $x_{22}(0) = -8$

deg/sec, respectively. Initial conditions for weight update dynamics are chosen as zero. The controller gains are chosen as $c^a_{1_i} = 1$, $c^a_{2_i} = 0.1$, $c_{1_i} = 0.001$, and $c_{2_i} = 0.15$, respectively, and learning rate γ_i is chosen as $\gamma_i = 1$, where $i = 1, 2$. Model uncertainties are assumed as $d_1 = 0.5\sin(x_{12}) - 0.3\cos(x_{11})$ and $d_2 = -0.4\cos(x_{12}) = 0.7\sin(x_{11})$, and structured basis functions are chosen as $\boldsymbol{\phi}_1 = [\sin(x_{12}), \ \cos(x_{11})]^T$ and $\boldsymbol{\phi}_2 = [\cos(x_{12}), \ \sin(x_{11})]^T$, respectively.

State constraints on the system are selected as $\underline{\zeta}_{x_{11}} = \overline{\zeta}_{x_{11}} = 8°$, $\underline{\zeta}_{x_{12}} = \overline{\zeta}_{x_{12}} = 8°$, $\underline{\zeta}_{x_{21}} = \overline{\zeta}_{x_{21}} = 15°$, and $\underline{\zeta}_{x_{22}} = \overline{\zeta}_{x_{22}} = 15°$, respectively. The constraints on learning error dynamics are chosen as $\zeta_{e_i} = 1°$ and $\zeta_{\dot{e}_i} = 0.01°$. The performance of the proposed controller is demonstrated through sinusoidal output tracking. The desired output for x_{11} is selected as $x_1^* = 2° \sin(t)$ and for x_{12} is selected as $x_2^* = 1.5° \cos(2t)$, respectively.

In simulations, the kinematic and dynamic state trajectories are shown in Figs. 7.25 and 7.26, respectively. It can be seen that all profiles are bounded by the imposed constraints and effectively track the reference signal. Transients are almost negligible in the state trajectories, which is happening due to imposed constraints on learning error. The control histories are presented in Fig. 7.27, where the profiles are smooth and continuous. Position error histories are presented in Fig. 7.28, where the final values of the errors are within the predefined domains.

The histories of uncertainty learning errors are shown in Figs. 7.29 and 7.30, respectively. As shown in theory, the error is bounded, and the derivative of the error is bounded as well as asymptotically stable. The weight update profiles are shown in Fig. 7.31, where all trajectories are bounded and stable. The capability of the neural network in approximating the model uncertainties is shown in Fig. 7.32, where

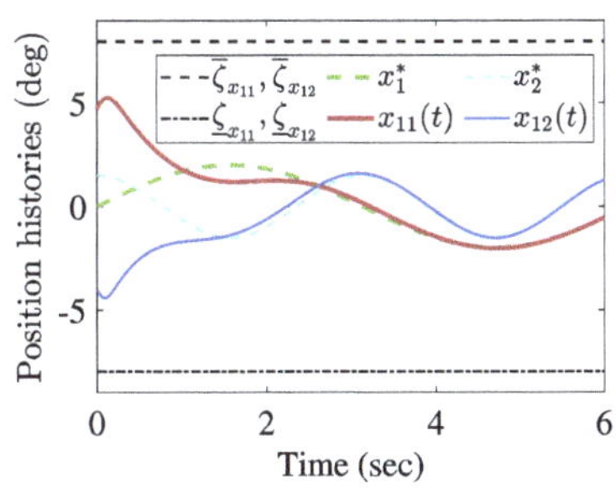

Fig. 7.25 Kinematic state trajectories

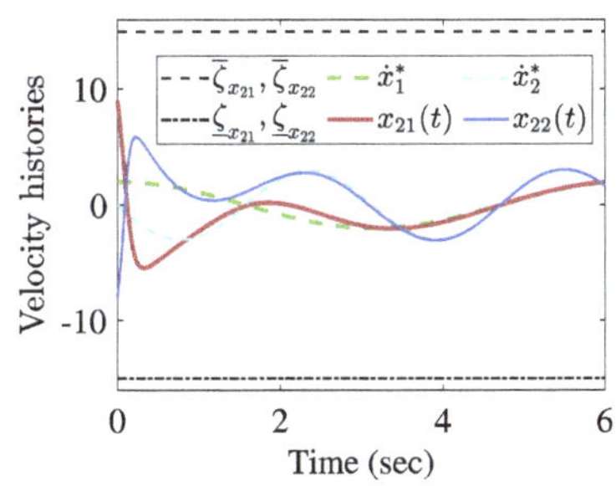

Fig. 7.26 Dynamic state trajectories

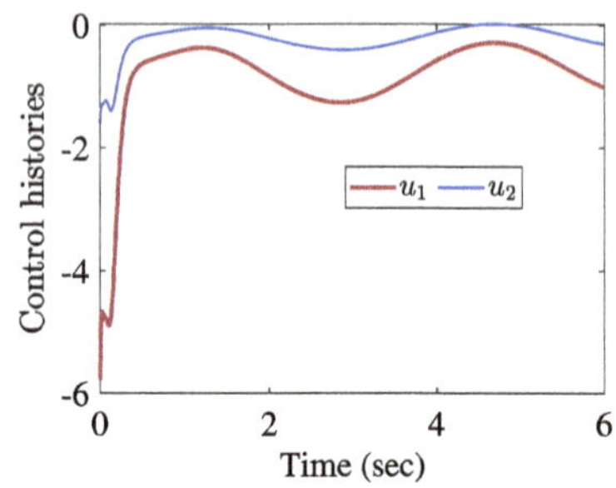

Fig. 7.27 Control histories

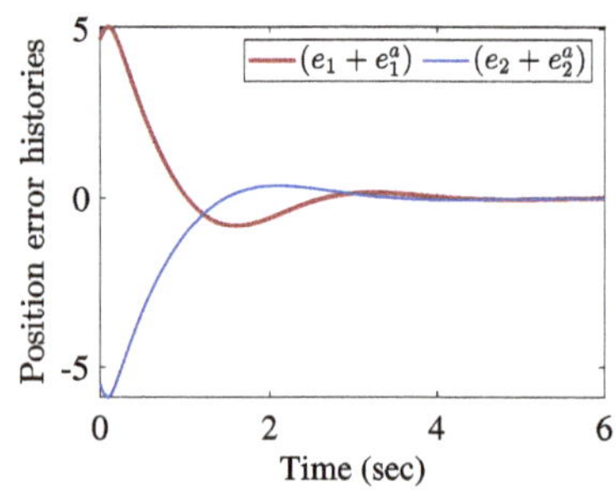

Fig. 7.28 Position error histories

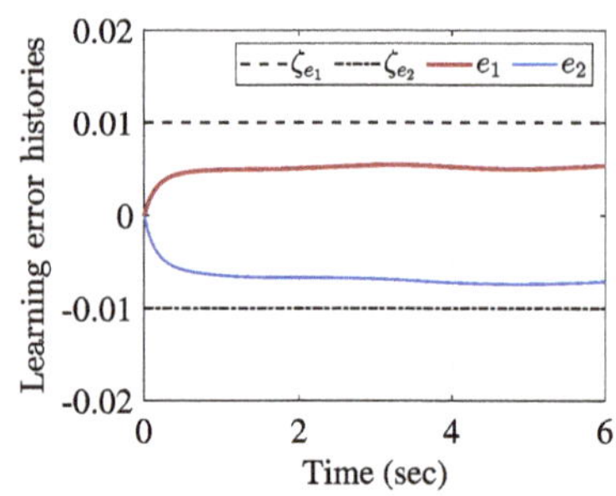

Fig. 7.29 Constrained learning error trajectories

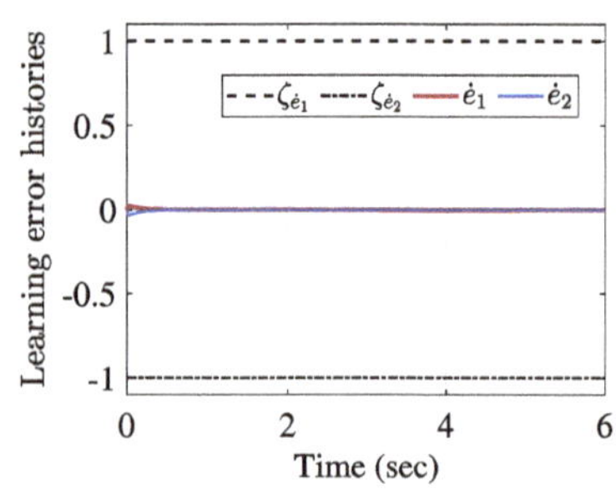

Fig. 7.30 Constrained histories of time derivative of learning errors

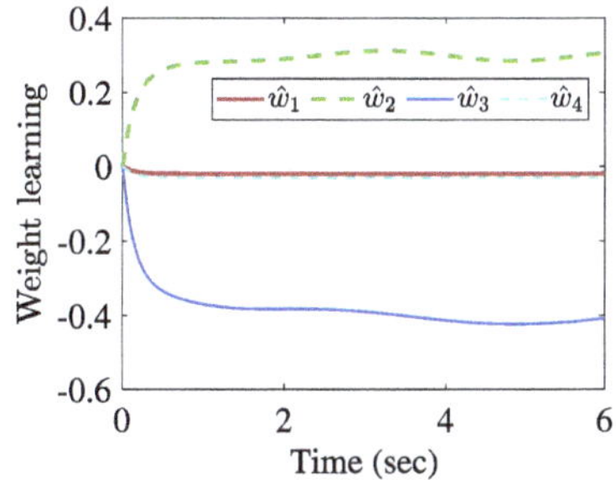

Fig. 7.31 Weight update histories during the approximation of modeling uncertainties

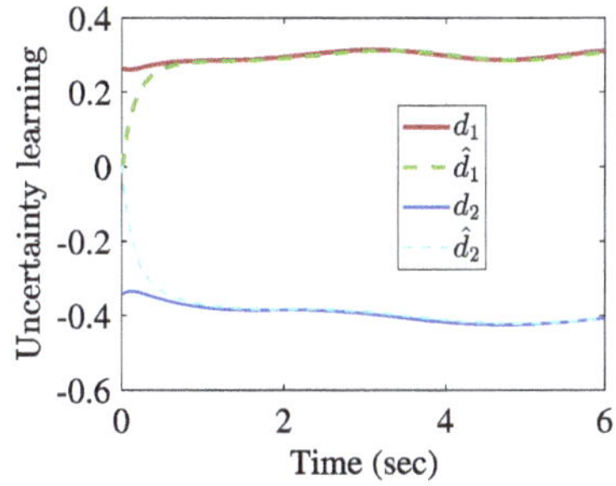

Fig. 7.32 Trajectories of modeling uncertainties and their approximation using neural network

the approximation of modeling uncertainties follows the actual model uncertainties. These results also demonstrate the effectiveness of the proposed control algorithm for complex systems.

7.3 Summary

In this chapter, barrier Lyapunov function (BLF)-based state-constrained adaptive control strategies are presented for relative degree one nonlinear systems and Euler-Lagrange systems with structured modeling uncertainties. The stability of the closed-loop system is established using Lyapunov stability theory. It is demonstrated that the closed-loop states remain bounded within the imposed state constraints and ultimately converge to a user-defined domain. To validate the effectiveness of the proposed controller, the dynamics of a challenging two-link robot manipulator are utilized, and extensive simulation results are presented to support the theoretical claims. Readers aspiring to a more in-depth understanding of BLF-based neuro-adaptive control design may refer to additional sources (Sachan and Padhi, 2020d,a, 2018a,b,c, 2020b,c).

7.4 Exercise Problems

1. Design the state-constrained neuro-adaptive controllers using the BLF-based approach for the following nonlinear systems to stabilize the system response considering structured model uncertainties. The desired controller for these systems can be designed using the methods given in Chap. 4.
   (a) $\dot{x} = -x + x^3 + u + d(x)$ where $d(x)$ is defined as $d(x) = 0.3x^2$. The user-imposed state constraints are imposed as ζ is $\zeta = 0.5$.
   (b) $\dot{x} = -x - x^3 + u + d(x)$ where $d(x)$ is defined as $d(x) = x + 0.2\,x^3$. The user-imposed state constraint is imposed as ζ is $\zeta = 0.5$.
   (c) $\dot{x} = -x^2 + \left(x^2 - 3\right)u + d(x)$ where $d(x)$ is defined as $d(x) = 0.2\,x^2$. The user-imposed state constraint is imposed as ζ is $\zeta = 0.5$.
   (d) $\begin{bmatrix} \dot{x}_1 \\ \dot{x}_2 \end{bmatrix} = \begin{bmatrix} x_2 \\ -x_2 - x_1 - (x_1 + x_2)^2 x_2 + u \end{bmatrix}$ where $d(x)$ is defined as $d(x) = 0.2x_1^2$. The user-imposed state constraints are imposed as ζ_1 and ζ_2 are $\zeta_1 = 0.3$ and $\zeta_2 = 0.8$, respectively.
   (e) $\ddot{x} = -\sin x + u + d(x)$ where $d(x)$ is defined as $d(x) = 0.1\cos x$. The user-imposed state constraints are imposed as ζ_1 and ζ_2 are $\zeta_1 = 0.3$ and $\zeta_2 = 0.8$, respectively.
   (f) $\ddot{x} - 0.7(1 - x^2)\dot{x} - x^3 + (1 + \dot{x}^2)u + d(x) = 0$ where $d(x)$ is defined as $d(x) = 0.2x^2 - 0.15x^2$. The user-imposed state constraints are imposed as ζ_1 and ζ_2 are $\zeta_1 = 0.3$ and $\zeta_2 = 0.8$, respectively.

2. Design a state-constrained controller using transformation-based approach for the following nonlinear systems assuming imposed state constraints ζ_1, ζ_2, and ζ_3 are $\zeta_1 = 0.3$, $\zeta_2 = 0.6$, and $\zeta_3 = 0.8$, respectively. (Assume the problem is a stabilization problem.)
   (a) $\begin{bmatrix} \dot{x}_1 \\ \dot{x}_2 \\ \dot{x}_3 \end{bmatrix} = \begin{bmatrix} x_1 + x_2 \\ x_3 \\ u + d(x) \end{bmatrix}$ where $d(x)$ is defined as $d(x) = 0.2x_1 - 0.15x_2 + 0.1x_1x_2$. The user-imposed state constraints ζ_1, ζ_2, and ζ_3 are $\zeta_1 = 0.3$, $\zeta_2 = 0.6$, and $\zeta_3 = 0.8$, respectively.
   (b) $\begin{bmatrix} \dot{x}_1 \\ \dot{x}_2 \\ \dot{x}_3 \end{bmatrix} = \begin{bmatrix} -x_1 + x_2 \\ x_3 \\ u \end{bmatrix}$ where $d(x)$ is defined as $d(x) = 0.2x_1 - 0.15x_2 + 0.1x_1x_2$. The user-imposed state constraints ζ_1, ζ_2, and ζ_3 are $\zeta_1 = 0.3$, $\zeta_2 = 0.6$, and $\zeta_3 = 0.8$, respectively.

3. Refer to the Eqs. (7.12) and (7.13), and show that the relation

$$-\underline{\zeta}_{e_i^d}\sqrt{\left(1 - \exp\left(\frac{-2V_0^d}{c_i^d}\right)\right)} \leq e_i^d \leq \overline{\zeta}_{e_i^d}\sqrt{\left(1 - \exp\left(\frac{-2V_0^d}{c_i^d}\right)\right)}, \quad \forall i$$

can be obtained from

$$\frac{1}{2}c_i^d\left[\gamma_i^*\log\frac{\overline{\zeta}_{e_i^d}^2}{\overline{\zeta}_{e_i^d}^2-e_i^{d2}}+(1-\gamma_i^*)\log\frac{\underline{\zeta}_{e_i^d}^2}{\underline{\zeta}_{e_i^d}^2-e_i^{d2}}\right]\le V_0^d$$

where "exp" represents the exponential function and other variables are defined in Eqs. (7.12) and (7.13).

4. Shown that for any positive constant $\zeta_{e_i^d}$, the following inequalities holds for e_i^d in an interval $|e_i^d| < \zeta_{e_i^d}$

$$\log\left(\frac{\zeta_{e_i^d}^2}{\zeta_{e_i^d}^2-e_i^{d2}}\right)\le\frac{e_i^{d2}}{\zeta_{e_i^d}^2-e_i^{d2}}$$

$$\log\left(\frac{\zeta_{e_i^d}^2}{\zeta_{e_i^d}^2-e_i^{d2}}\right)\le\left(\frac{\zeta_{e_i^d}\,e_i^d}{\zeta_{e_i^d}^2-e_i^{d2}}\right)^2$$

where all variables are defined in Eqs. (7.12) and (7.13).

5. Consider the desired system with bounded system uncertainties $\boldsymbol{d}(\boldsymbol{x})$ as

$$\dot{\boldsymbol{x}}^d=\boldsymbol{f}\left(\boldsymbol{x}^d,\boldsymbol{u}^d\right)+\boldsymbol{d}(\boldsymbol{x})$$

Show that with the application of the desired controller constructed in (7.7), the state of the desired system with uncertainties will remain bounded as $-\underline{\zeta}_{x_i} < x_i < \overline{\zeta}_{x_i}$, where $\underline{\zeta}_{x_i}$ and $\overline{\zeta}_{x_i}$ are user-defined lower and upper bounds. To understand the definition of variables, one can refer to Sect. 7.1.2.
Hint: Solve the closed-loop error dynamics with uncertainties, which can be obtained as $\dot{\boldsymbol{e}}^d=-\left[\boldsymbol{K}^d\left(\boldsymbol{e}^d\right)\right]\boldsymbol{e}^d+\boldsymbol{d}(\boldsymbol{x})$, and show that it is bounded.

References

P.R. Ambati, R. Padhi, Robust auto-landing of fixed-wing $UAVs$ using neuro-adaptive design. Control Eng. Pract. **60**, 218–232 (2017)

W.M. Czarnecki, S. Osindero, M. Jaderberg, G. Swirszcz, R. Pascanu, Sobolev training for neural networks, in *Advances in Neural Information Processing Systems 30* (Curran Associates, Inc., 2017), pp. 4278–4287

B.S. Kim, A.J. Calise, Nonlinear flight control using neural networks. J. Guid. Control Dynam. **20**(1), 26–33 (1997)

R. Padhi, N. Unnikrishnan, S.N. Balakrishnan, Model-following neuro-adaptive control design for non-square, non-affine nonlinear systems. IET Control Theory Appl. **1**(6), 1650–1661 (2007)

K. Sachan, R. Padhi, Barrier Lyapunov function based output-constrained control of nonlinear Euler-Lagrange systems, in *2018 15th International Conference on Control, Automation, Robotics and Vision, ICARCV 2018* (2018a)

K. Sachan, R. Padhi, Barrier Lyapunov function based state-constrained control for a class of nonlinear systems, in *IFAC-PapersOnLine*, vol. 51(1) (2018b)

K. Sachan, R. Padhi, State-constrained robust adaptive cruise control design for air-breathing hypersonic vehicles, in *2018 AIAA Guidance, Navigation, and Control Conference* (Jan. 2018c)

K. Sachan, R. Padhi, Safety-constrained robust adaptive control for a class of MIMO non-linear systems. IET Control Theory Appl. **14**(7), 982–990 (2020a)

K. Sachan, R. Padhi, Robust control design for euler-lagrange systems with output constraints, in *IFAC-PapersOnLine*, vol. 53(1), pp. 313–318 (2020b), 6th Conference on Advances in Control and Optimization of Dynamical Systems ACODS 2020

K. Sachan, R. Padhi, Synthesis of an adaptive state-constrained control for mimo euler-lagrange systems, in *IFAC-PapersOnLine*, vol. 53(2), pp. 5518–5523 (2020c)

K. Sachan, R. Padhi, Nonlinear robust neuro-adaptive flight control for hypersonic vehicles with state constraints. Control Eng. Pract. **102**, 104526 (2020d)

J.-J.E. Slotine, W. Li et al., *Applied Nonlinear Control*, vol. 199(1) (Prentice hall Englewood Cliffs, NJ, 1991)

K.P. Tee, S.S. Ge, E.H. Tay, Barrier Lyapunov Functions for the control of output-constrained nonlinear systems. Automatica **45**(4), 918–927 (2009)

Chapter 8
Transformation-Based State-Constrained Adaptive Control

In the previous chapter, the neuro-adaptive controller was designed under the assumption of structured modeling uncertainties, which requires precise knowledge of the basis functions. However, this assumption is restrictive and may not be feasible for many practical applications. To address unstructured modeling uncertainties and bounded external disturbances, a new error transformation-based adaptive controller is developed in this chapter. The adaptive controller is designed following both the neuro-adaptive control philosophy (indirect) and direct adaptive control design philosophy, applicable to Euler-Lagrange (EL) systems and more general nonlinear systems.[1]

8.1 Neuro-adaptive Control of Euler-Lagrange Systems

In this section, a state-constrained neuro-adaptive controller is designed for EL systems subject to unstructured model uncertainties and bounded external disturbances. The neuro-adaptive control design philosophy follows a similar approach, as illustrated in Fig. 7.24 in the previous chapter. The transformation-based controller is developed in two steps. First, a neural network-based approximation of the system dynamics is constructed using a novel error transformation that converts the constrained system into an equivalent unconstrained system. Lyapunov stability analysis is then used to design a stable weight update rule for the neural network. After approximating the modeling uncertainties, a tracking controller is designed for the approximated system to track the reference signal. The use of a new error

[1] Contents of this chapter are partially used from Publication "Safety-constrained robust adaptive control for a class of MIMO non-linear systems", IET Control Theory and Applications, vol. 14, pp. 982–990, Apr 2020, with permission from John Wiley and Sons (License Number: 6156890630508).

K. Sachan et al., *Intelligent Adaptive Control with System Constraints*,
https://doi.org/10.1007/978-981-95-8220-4_8

transformation in this control formulation leads to a different construction of the approximated system compared to the previous chapter. Additionally, instead of a barrier Lyapunov function (BLF), a standard quadratic Lyapunov function (QLF) is used to assess system stability. The stability and boundedness of the system are proven in the presence of unstructured modeling uncertainties and external disturbances. The control design ensures that the closed-loop states remain within the imposed state constraints and asymptotically converge to a predefined open set. The control architecture is detailed in the following sections.

The state-space model of EL systems is given by

$$\begin{aligned} \dot{\boldsymbol{x}}_1 &= \boldsymbol{x}_2 \\ \dot{\boldsymbol{x}}_2 &= \boldsymbol{f}(\boldsymbol{x}) + \boldsymbol{G}(\boldsymbol{x})\boldsymbol{u} + \boldsymbol{d}(\boldsymbol{x}, \boldsymbol{u}) + \boldsymbol{\eta}(t) \end{aligned} \tag{8.1}$$

where states of the system are defined as $\boldsymbol{x} \triangleq \left[\boldsymbol{x}_1^T, \boldsymbol{x}_2^T\right]^T$ and kinematic and dynamic states of the system are defined as $\boldsymbol{x}_1 \triangleq [x_{11}, x_{12}, \ldots, x_{1n}]^T \in \Re^n$ and $\boldsymbol{x}_2 \triangleq [x_{21}, x_{22}, \ldots, x_{2n}]^T \in \Re^n$, respectively. The system output is selected as $\boldsymbol{x}_1 \in \Re^n$, and the control input to the system is denoted by $\boldsymbol{u} \in \Re^n$. The control effectiveness matrix is denoted by $\boldsymbol{G}(\boldsymbol{x})$, assumed to be known, invertible, and positive definite in nature. Terms $\boldsymbol{d}(\boldsymbol{x}, \boldsymbol{u})$ and $\boldsymbol{\eta}(t)$ represent lumped unstructured modeling uncertainties and possible external disturbances, respectively. The components of external disturbance $\eta_i(t),\ \forall i$ are assumed to be smooth and bonded, i.e., $|\eta_i(t)| \leq \overline{\eta}_i,\ \forall i$, where $\overline{\eta}_i$ is an unknown positive constant.

The system in (8.1) is assumed to be operated in a constrained environment. Thus, components of states x_{1i} and x_{2i} are constrained, i.e., $\underline{\zeta}_{x_{1i}} < x_{1i} < \overline{\zeta}_{x_{1i}}$ and $\underline{\zeta}_{x_{2i}} < x_{2i} < \overline{\zeta}_{x_{2i}},\ \forall i$, respectively, where $\underline{\zeta}_{x_{1i}}$ and $\overline{\zeta}_{x_{1i}}$ are the user-defined asymmetric lower and upper constant bounds on kinematic state x_{1i} and $\underline{\zeta}_{x_{2i}}$ and $\overline{\zeta}_{x_{2i}}$ are the constant asymmetric lower and upper bounds on dynamic state x_{2i}. The user-defined tracking error bounds on the kinematic and dynamic states are defined as $\left(-\zeta_{e_i}, \zeta_{e_i}\right)$ and $\left(-\zeta_{\dot{e}_i}, \zeta_{\dot{e}_i}\right)$, respectively, where ζ_{e_i} and $\zeta_{\dot{e}_i}$ are positive constants. The qualitative representation of the constraints is presented in Fig. 8.1.

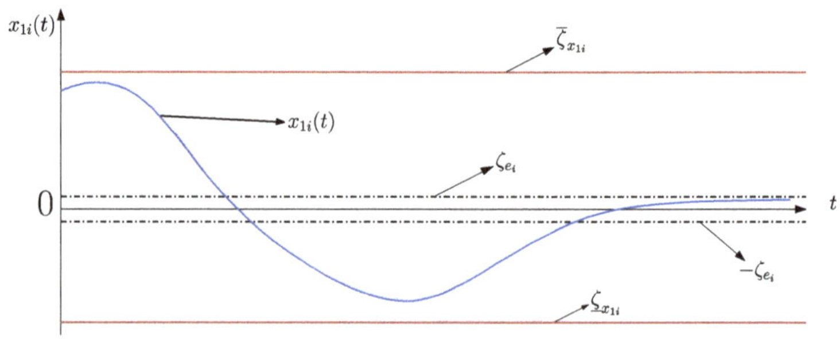

Fig. 8.1 The demonstration of imposed user-defined state constraints and steady-state tracking error bounds for the kinematic state $x_{1i}(t)$ with reference trajectory as zero

The objectives of the proposed control design are twofold. (i) The controller should enforce the system output x_{1i} to track the reference signal x_i^* asymptotically with a desired accuracy $\pm\zeta_{e_i}$, i.e., $|\left(x_{1i} - x_i^*\right)| < \zeta_{e_i}, \forall i$ as $t \to \infty$, where x_i^* is the ith component of reference trajectory $\boldsymbol{x}^*$. (ii) Closed-loop states of the system (8.1) should remain bounded by the imposed constraints in presence of model uncertainties and bounded external disturbances, i.e., $x_{1i}(t) \in \left(\underline{\zeta}_{x_{1i}}, \overline{\zeta}_{x_{1i}}\right)$ and $x_{2i}(t) \in \left(\underline{\zeta}_{x_{2i}}, \overline{\zeta}_{x_{2i}}\right), \forall t$.

Similar to the previous chapter, the reference signal $\boldsymbol{x}^*$ and its time derivative $\dot{\boldsymbol{x}}^*$ are assumed to be bounded as $-\underline{\zeta}_{x_i^*} \leq x_i^* \leq \overline{\zeta}_{x_i^*}, \forall i$ and $-\underline{\zeta}_{\dot{x}_i^*} \leq \dot{x}_i^* \leq \overline{\zeta}_{\dot{x}_i^*}, \forall i$, where $\underline{\zeta}_{x_{1i}} > 0$, $\overline{\zeta}_{x_{1i}} > 0$, $\underline{\zeta}_{x_{2i}} > 0$, and $\overline{\zeta}_{x_{2i}} > 0$ are known constants. The term x_i^* and $\dot{x}_i^*$ are ith components of $\boldsymbol{x}^*$ and $\dot{\boldsymbol{x}}^*$, respectively. To ensure consistency between the imposed state constraints and the desired tracking objectives, it is further assumed that $-\underline{\zeta}_{x_{1i}} < -\underline{\zeta}_{x_i^*} \leq x_i^* \leq \overline{\zeta}_{x_i^*} < \overline{\zeta}_{x_{1i}}, \forall i$ and $-\underline{\zeta}_{x_{2i}} < -\underline{\zeta}_{\dot{x}_i^*} \leq \dot{x}_i^* \leq \overline{\zeta}_{\dot{x}_i^*} < \overline{\zeta}_{x_{2i}}, \forall i$, respectively. These conditions ensure that the desired reference signals and their derivatives remain within the bounds defined by the imposed state constraints.

8.1.1 *Proposed Error Transformation*

In this subsection, an error transformation is introduced to convert the error-constrained system into an equivalent unconstrained system (Sachan and Padhi, 2020). To define the error transformation, consider the kinematic and dynamic states of the approximated system as $\boldsymbol{x}_1^a$ and $\boldsymbol{x}_2^a$, respectively (the complete dynamics of the approximated system are defined in (8.12)). The error between $\boldsymbol{x}_1$ and $\boldsymbol{x}_1^a$ is then given by

$$\boldsymbol{e} \triangleq \boldsymbol{x}_1 - \boldsymbol{x}_1^a \tag{8.2}$$

The time derivatives of the approximation error (8.2) is obtained as

$$\dot{\boldsymbol{e}} = \boldsymbol{x}_2 - \boldsymbol{x}_2^a \tag{8.3}$$

$$\ddot{\boldsymbol{e}} = \dot{\boldsymbol{x}}_2 - \dot{\boldsymbol{x}}_2^a \tag{8.4}$$

Using the errors in (8.3)-(8.4), a new error transformation $\boldsymbol{s}$ is proposed as

$$\boldsymbol{s} = \boldsymbol{P}_1(\boldsymbol{e})\boldsymbol{e} + \boldsymbol{P}_2(\dot{\boldsymbol{e}})\dot{\boldsymbol{e}} \tag{8.5}$$

where $\boldsymbol{P}_1(\boldsymbol{e})$ and $\boldsymbol{P}_2(\dot{\boldsymbol{e}})$ are diagonal matrices, defined as

$$\boldsymbol{P}_1(\boldsymbol{e}) = \text{diag}\left[\frac{c_{1_1}}{\left(\zeta_{e_1}^2 - e_1^2\right)}, \ldots, \frac{c_{1_n}}{\left(\zeta_{e_n}^2 - e_n^2\right)}\right] \tag{8.6}$$

$$P_2(\dot{e}) = \operatorname{diag}\left[\frac{c_{2_1}}{\left(\zeta_{\dot{e}_1}^2 - \dot{e}_1^2\right)}, \ldots, \frac{c_{2_n}}{\left(\zeta_{\dot{e}_n}^2 - \dot{e}_n^2\right)}\right] \tag{8.7}$$

where e_i and $\dot{e}_i$ are the ith components of $\boldsymbol{e}$ and $\dot{\boldsymbol{e}}$, respectively. The constants ζ_{e_i} and $\zeta_{\dot{e}_i}$ are the imposed error constraints on e_i and $\dot{e}_i$, respectively. The gains c_{1_i} and c_{2_i}, $\forall i$ are positive constants. Since the matrices are diagonal in nature, the ith component of the error transformation can be written as

$$s_i = \frac{c_{1_i} e_i}{\left(\zeta_{e_i}^2 - e_i^2\right)} + \frac{c_{2_i} \dot{e}_i}{\left(\zeta_{\dot{e}_i}^2 - \dot{e}_i^2\right)} \tag{8.8}$$

The proposed error transformation has the following properties.

Lemma 8.1 *If s_i is bounded, then $e_i(t) \in \left(-\zeta_{e_i}, \zeta_{e_i}\right)$ and $\dot{e}_i(t) \in \left(-\zeta_{\dot{e}_i}, \zeta_{\dot{e}_i}\right)$, $\forall t \geq 0, \forall i$.*

Proof Since s_i is bounded, the expression $\left(\frac{c_{1_i} e_i}{\left(\zeta_{e_i}^2 - e_i^2\right)} + \frac{c_{2_i} \dot{e}_i}{\left(\zeta_{\dot{e}_i}^2 - \dot{e}_i^2\right)}\right)$ is bounded. The initial conditions for e_i and $\dot{e}_i$ are always zero because the actual and approximated system are initialized with the same initial conditions. Now, the boundedness of e_i and $\dot{e}_i$ can be proven using contradiction. If s_i is bounded then e_i and $\dot{e}_i$ can go unbounded only if $e_i \rightarrow \zeta_{e_i}$ and $\dot{e}_i \rightarrow -\zeta_{\dot{e}_i}$ or $e_i \rightarrow -\zeta_{e_i}$ and $\dot{e}_i \rightarrow \zeta_{\dot{e}_i}$. However, both conditions are not possible because if the derivative of e_i, which is $\dot{e}_i$, is positive, then e_i cannot decrease and vice versa. Therefore, e_i and $\dot{e}_i$ will remain bounded by the user-defined error-constrained open set, i.e., $e_i(t) \in \left(-\zeta_{e_i}, \zeta_{e_i}\right)$ and $\dot{e}_i(t) \in \left(-\zeta_{\dot{e}_i}, \zeta_{\dot{e}_i}\right)$, $\forall t \geq 0, \forall i$. □

Lemma 8.2 *If $s_i \rightarrow 0$ uniformly, then $e_i(t) \rightarrow 0$ and $\dot{e}_i(t) \rightarrow 0$, $\forall i$.*

Proof From Lemma 8.1, since $e_i(t) \in \left(-\zeta_{e_i}, \zeta_{e_i}\right)$ and $\dot{e}_i(t) \in \left(-\zeta_{\dot{e}_i}, \zeta_{\dot{e}_i}\right)$, $\forall t \geq 0, \forall i$, it is obvious that the expression $\left(\frac{c_{1_i} e_i}{\left(\zeta_{e_i}^2 - e_i^2\right)} + \frac{c_{2_i} \dot{e}_i}{\left(\zeta_{\dot{e}_i}^2 - \dot{e}_i^2\right)}\right) \rightarrow 0$ only when both $e_i(t) \rightarrow 0$ and $\dot{e}_i(t) \rightarrow 0$ uniformly. Thus, as $s_i \rightarrow 0$ uniformly, $e_i(t) \rightarrow 0$ and $\dot{e}_i(t) \rightarrow 0, \forall i$. □

Note that the proposed error transformation $\boldsymbol{s}$ can be reduced to design an output-constrained control by selecting $\boldsymbol{P}_2(\dot{\boldsymbol{e}})$ as an identity matrix. Moreover, an unconstrained controller can be designed by selecting both $\boldsymbol{P}_1(\boldsymbol{e})$ and $\boldsymbol{P}_2(\dot{\boldsymbol{e}})$ matrices as identity matrices.

The time derivative of the transformed error $\boldsymbol{s}$ can be obtained as

$$\dot{s} = \boldsymbol{P}_3(\boldsymbol{e})\dot{\boldsymbol{e}} + \boldsymbol{P}_4(\dot{\boldsymbol{e}})\ddot{\boldsymbol{e}} \tag{8.9}$$

where $\boldsymbol{P}_3(\boldsymbol{e})$ and $\boldsymbol{P}_4(\dot{\boldsymbol{e}})$ are diagonal matrices, defined as

$$\boldsymbol{P}_3(\boldsymbol{e}) = \text{diag}\left[\frac{c_{1_1}\left(\zeta_{e_1}^2 + e_1^2\right)}{\left(\zeta_{e_1}^2 - e_1^2\right)^2}, \ldots, \frac{c_{1_n}\left(\zeta_{e_n}^2 + e_n^2\right)}{\left(\zeta_{e_n}^2 - e_n^2\right)^2}\right] \tag{8.10}$$

$$\boldsymbol{P}_4(\dot{\boldsymbol{e}}) = \text{diag}\left[\frac{c_{2_1}\left(\zeta_{\dot{e}_1}^2 + \dot{e}_1^2\right)}{\left(\zeta_{\dot{e}_1}^2 - \dot{e}_1^2\right)^2}, \ldots, \frac{c_{2_n}\left(\zeta_{\dot{e}_n}^2 + \dot{e}_n^2\right)}{\left(\zeta_{\dot{e}_n}^2 - \dot{e}_n^2\right)^2}\right] \tag{8.11}$$

Without loss of generality, in the further analysis, the matrices $\boldsymbol{P}_1(\boldsymbol{e})$, $\boldsymbol{P}_2(\dot{\boldsymbol{e}})$, $\boldsymbol{P}_3(\boldsymbol{e})$, and $\boldsymbol{P}_4(\dot{\boldsymbol{e}})$ are represented as $\boldsymbol{P}_1$, $\boldsymbol{P}_2$, $\boldsymbol{P}_3$, and $\boldsymbol{P}_4$, respectively.

8.1.2 Approximation of System Uncertainties

Using the system errors (8.2)-(8.4) and proposed error transformation in (8.5), the approximated system dynamics is constructed as

$$\begin{aligned} \dot{\boldsymbol{x}}_1^a &= \boldsymbol{x}_2^a \\ \dot{\boldsymbol{x}}_2^a &= \boldsymbol{f}(\boldsymbol{x}) + \boldsymbol{G}(\boldsymbol{x})\boldsymbol{u} + \hat{\boldsymbol{d}}(\boldsymbol{x}, \boldsymbol{u}) + \boldsymbol{P}_4^{-1}\boldsymbol{P}_5\boldsymbol{s} + \boldsymbol{P}_3\boldsymbol{P}_4^{-1}\dot{\boldsymbol{e}} + \boldsymbol{P}_4\boldsymbol{s} \end{aligned} \tag{8.12}$$

where $\boldsymbol{P}_5$ is a diagonal matrix with positive elements and term $\hat{\boldsymbol{d}}(\boldsymbol{x}, \boldsymbol{u})$ is the approximation of modeling uncertainties $\boldsymbol{d}(\boldsymbol{x}, \boldsymbol{u})$. By substituting $\ddot{\boldsymbol{e}}$ from (8.4), $\dot{\boldsymbol{s}}$ in (8.9) is modified as

$$\dot{\boldsymbol{s}} = \boldsymbol{P}_3\dot{\boldsymbol{e}} + \boldsymbol{P}_4\left(\dot{\boldsymbol{x}}_2 - \dot{\boldsymbol{x}}_2^a\right) \tag{8.13}$$

Using (8.1) and (8.12), $\dot{\boldsymbol{s}}$ can be written as

$$\begin{aligned} \dot{\boldsymbol{s}} &= \boldsymbol{P}_3\,\dot{\boldsymbol{e}} \\ &+ \boldsymbol{P}_4\Big(\boldsymbol{d}(\boldsymbol{x}, \boldsymbol{u}) - \hat{\boldsymbol{d}}(\boldsymbol{x}, \boldsymbol{u}) + \eta(t) - \boldsymbol{P}_4^{-1}\boldsymbol{P}_5\boldsymbol{s} - \boldsymbol{P}_3\boldsymbol{P}_4^{-1}\dot{\boldsymbol{e}} - \boldsymbol{P}_4\boldsymbol{s}\Big) \end{aligned} \tag{8.14}$$

where $\hat{\boldsymbol{d}}(\boldsymbol{x}, \boldsymbol{u})$ and $\boldsymbol{d}(\boldsymbol{x}, \boldsymbol{u})$ are defined as $\hat{\boldsymbol{d}}(\boldsymbol{x}, \boldsymbol{u}) = \hat{\boldsymbol{W}}^T\boldsymbol{\phi}(z)$ and $\boldsymbol{d}(\boldsymbol{x}, \boldsymbol{u}) = \boldsymbol{W}^T\boldsymbol{\phi}(z) + \boldsymbol{\epsilon}$, respectively. The term $\boldsymbol{\phi}(z)$ represents the basis function vector, which is a combination of states. The approximation error is denoted by $\boldsymbol{\epsilon}$, which is assumed to be bounded as $\epsilon_i^2 \leq \epsilon_{m_i}$, $\forall i$, where ϵ_{m_i} is an unknown positive constant. By using the definitions of $\hat{\boldsymbol{d}}(\boldsymbol{x}, \boldsymbol{u})$ and $\boldsymbol{d}(\boldsymbol{x}, \boldsymbol{u})$, the following is obtained:

$$\begin{aligned} \dot{\boldsymbol{s}} &= \boldsymbol{P}_3\dot{\boldsymbol{e}} \\ &+ \boldsymbol{P}_4\left(\tilde{\boldsymbol{W}}^T\boldsymbol{\phi}(z(t)) + \boldsymbol{\epsilon} + \eta(t) - \boldsymbol{P}_4^{-1}\boldsymbol{P}_5\boldsymbol{s} - \boldsymbol{P}_3\boldsymbol{P}_4^{-1}\dot{\boldsymbol{e}} - \boldsymbol{P}_4\boldsymbol{s}\right) \end{aligned} \tag{8.15}$$

where $\tilde{W}$ is defined as $\tilde{W} \triangleq W - \hat{W}$. Since the matrices are diagonal in nature and NN approximation can be used for individual components of the dynamics, the expression of $\dot{s}$ in (8.15) can be expressed in ith components as

$$\dot{s}_i = p_{3_i}\dot{e}_i + p_{4_i}\left(\tilde{w}_i^T \phi(z) + \epsilon_i + \eta_i(t) - \frac{p_{5_i}s_i}{p_{4_i}} - \frac{p_{3_i}\dot{e}_i}{p_{4_i}} - p_{4_i}s_i\right), \ \forall i \quad (8.16)$$

where $\tilde{w}_i^T$ is the ith row of $\tilde{W}^T$ and p_{3_i}, p_{4_i} and p_{5_i} are the ith diagonal component of matrices P_3, P_4, and P_5, respectively. The external disturbance in the ith dynamic state is represented by $\eta_i(t)$.

To analyze the stability of (8.16), the following Lyapunov function is chosen:

$$v_i = \frac{1}{2}s_i^2 + \frac{1}{2\gamma_i}\tilde{w}_i^T \tilde{w}_i \quad (8.17)$$

where $\gamma_i > 0$ is learning rate constant. By taking the time derivative of the Lyapunov function, the following is obtained:

$$\dot{v}_i = s_i\dot{s}_i + \frac{\tilde{w}_i^T \dot{\tilde{w}}_i}{\gamma_i} \quad (8.18)$$

By substituting $\dot{s_i}$ from (8.16) and $\dot{\tilde{w}}_i = -\dot{\hat{w}}_i$, the time derivative of Lyapunov function $\dot{v}_i$ can be expressed as

$$\begin{aligned}
\dot{v}_i &= s_i\left[p_{3_i}\dot{e}_i + p_{4_i}\left(\tilde{w}_i^T \phi(z) + \epsilon_i + \eta_i(t) - \frac{p_{5_i}s_i}{p_{4_i}} - \frac{p_{3_i}\dot{e}_i}{p_{4_i}} - p_{4_i}s_i\right)\right] - \frac{\tilde{w}_i^T \dot{\hat{w}}_i}{\gamma_i} \\
&= s_i p_{4_i}\left(\tilde{w}_i^T \phi(z) + \epsilon_i + \eta_i(t) - \frac{p_{5_i}s_i}{p_{4_i}} - p_{4_i}s_i\right) - \frac{\tilde{w}_i^T \dot{\hat{w}}_i}{\gamma_i} \\
&= s_i p_{4_i}\tilde{w}_i^T \phi(z) + s_i p_{4_i}\epsilon_i + s_i p_{4_i}\eta_i(t) - p_{5_i}s_i^2 - p_{4_i}^2 s_i^2 - \frac{\tilde{w}_i^T \dot{\hat{w}}_i}{\gamma_i} \quad (8.19)
\end{aligned}$$

Using Young's inequality (Zhao et al., 2018) in (8.19), the following is obtained:

$$\begin{aligned}
\dot{v}_i &\le s_i\, p_{4_i}\, \tilde{w}_i^T\, \phi(z) + \frac{s_i^2\, p_{4_i}^2}{2} + \frac{\epsilon_i^2}{2} + \frac{s_i^2\, p_{4_i}^2}{2} + \frac{\eta_i^2(t)}{2} - p_{5_i}s_i^2 - p_{4_i}^2 s_i^2 - \frac{\tilde{w}_i^T\, \dot{\hat{w}}_i}{\gamma_i} \\
&\le -\, p_{5_i}s_i^2 + \frac{\epsilon_i^2}{2} + \frac{\eta_i^2(t)}{2} + s_i\, p_{4_i}\, \tilde{w}_i^T\, \phi(z) - \frac{\tilde{w}_i^T\, \dot{\hat{w}}_i}{\gamma_i} \quad (8.20)
\end{aligned}$$

The weight update rule can be selected as

$$\dot{\hat{\boldsymbol{w}}}_i = \gamma_i p_{4_i} s_i \boldsymbol{\phi}(z) - \alpha_i \hat{\boldsymbol{w}}_i \tag{8.21}$$

where α_i is a positive constant, which is required to ensure the closed-loop stability of the weight update dynamics. By substituting the weight update rule, the expression of $\dot{v}_i$ is modified as

$$\begin{aligned}\dot{v}_i &\leq -p_{5_i} s_i^2 + \frac{\epsilon_i^2}{2} + \frac{\eta_i^2(t)}{2} + s_i p_{4_i} \tilde{\boldsymbol{w}}_i^T \boldsymbol{\phi}(z) - s_i p_{4_i} \tilde{\boldsymbol{w}}_i^T \boldsymbol{\phi}(z) + \frac{\alpha_i \tilde{\boldsymbol{w}}_i^T \hat{\boldsymbol{w}}_i}{\gamma_i} \\ &\leq -p_{5_i} s_i^2 + \frac{\epsilon_i^2}{2} + \frac{\eta_i^2(t)}{2} + \frac{\alpha_i \tilde{\boldsymbol{w}}_i^T \hat{\boldsymbol{w}}_i}{\gamma_i}\end{aligned} \tag{8.22}$$

Again, using Young's inequality and following the definition of $\tilde{\boldsymbol{w}}_i^T$, the term $\frac{\alpha_i \tilde{\boldsymbol{w}}_i^T \hat{\boldsymbol{w}}_i}{\gamma_i}$ can be expressed as

$$\frac{\alpha_i \tilde{\boldsymbol{w}}_i^T \hat{\boldsymbol{w}}_i}{\gamma_i} \leq -\frac{\alpha_i \tilde{\boldsymbol{w}}_i^T \tilde{\boldsymbol{w}}_i}{2\gamma_i} + \frac{\alpha_i \boldsymbol{w}_i^T \boldsymbol{w}_i}{2\gamma_i}$$

Substituting the upper bound of ϵ_i^2 and $\eta_i(t)$ as $\epsilon_i^2 \leq \epsilon_{m_i}$ and $\eta_i(t) \leq \overline{\eta}_i$, respectively, the expression of $\dot{v}_i$ is modified as

$$\begin{aligned}\dot{v}_i &\leq -p_{5_i} s_i^2 + \frac{\epsilon_i^2}{2} + \frac{\eta_i^2(t)}{2} - \frac{\alpha_i \tilde{\boldsymbol{w}}_i^T \tilde{\boldsymbol{w}}_i}{2\gamma_i} + \frac{\alpha_i \boldsymbol{w}_i^T \boldsymbol{w}_i}{2\gamma_i} \\ &\leq -p_{6_i} \left(\frac{s_i^2}{2} + \frac{\tilde{\boldsymbol{w}}_i^T \tilde{\boldsymbol{w}}_i}{2\gamma_i} \right) + \frac{\epsilon_{m_i}}{2} + \frac{\overline{\eta}_i^2}{2} + \frac{\alpha_i \boldsymbol{w}_i^T \boldsymbol{w}_i}{2\gamma_i} \\ &\leq -p_{6_i} v_i + p_{7_i}\end{aligned} \tag{8.23}$$

where the constants p_{6_i} and p_{7_i} are defined as $p_{6_i} = \min\left(2p_{5_i}, \alpha_i\right)$ and $p_{7_i} = \frac{\epsilon_{m_i}}{2} + \frac{\overline{\eta}_i^2}{2} + \frac{\alpha_i \boldsymbol{w}_i^T \boldsymbol{w}_i}{2\gamma_i}$, respectively. By integrating (8.23) in time, the following is obtained:

$$v_i(t) \leq v_i(0)\exp(-p_{6_i} t) + \frac{p_{7_i}}{p_{6_i}} \left[1 - \exp(-p_{6_i} t)\right] \tag{8.24}$$

where "exp" is the exponential function and $v_i(0)$ is the initial value of Lyapunov function (8.17). By analyzing (8.24), the upper bound on $v_i(t)$ can be obtained as

$$v_i(t) \leq v_i(0) + \frac{p_{7_i}}{p_{6_i}} \tag{8.25}$$

Since $v_i(t)$ is bounded, the stability of the transformed error dynamics in (8.16) can be concluded. Further, from (8.25), by substituting the terms of $v_i(t)$, the following relation is obtained:

$$\frac{1}{2}s_i^2 + \frac{1}{2\gamma_i}\tilde{\boldsymbol{w}}_i^T\tilde{\boldsymbol{w}}_i \leq v_i(0) + \frac{p_{7_i}}{p_{6_i}} \tag{8.26}$$

Since $\frac{1}{2\gamma_i}\tilde{\boldsymbol{w}}_i^T\tilde{\boldsymbol{w}}_i \geq 0$, the error transformation variable s will remain bounded as $|s_i| \leq \sqrt{2v_i(0) + \frac{2p_{7_i}}{p_{6_i}}}$. Since s_i is bounded, form Lemma 8.1, e_i and $\dot{e}_i$ will remain bounded by the user-defined error-constrained set, i.e., $e_i(t) \in \left(-\zeta_{e_i}, \zeta_{e_i}\right)$ and $\dot{e}_i(t) \in \left(-\zeta_{\dot{e}_i}, \zeta_{\dot{e}_i}\right)$, $\forall t \geq 0$, $\forall i$, respectively.

After approximating the modeling uncertainties, the next step is to design a tracking controller for the approximated system in (8.12) to track the reference signal. Therefore, in the following subsection, an asymptotically stable tracking controller is designed.

8.1.3 Asymptotically Stable Tracking Control Design

The error between approximated output $\boldsymbol{x}_1^a$ and reference output $\boldsymbol{x}^*$ is defined as

$$\boldsymbol{e}^a = \boldsymbol{x}_1^a - \boldsymbol{x}^* \tag{8.27}$$

The time derivatives of the error (8.27) are calculated as

$$\begin{aligned} \dot{\boldsymbol{e}}^a &= \boldsymbol{x}_2^a - \dot{\boldsymbol{x}}^* \\ \ddot{\boldsymbol{e}}^a &= \dot{\boldsymbol{x}}_2^a - \ddot{\boldsymbol{x}}^* \end{aligned} \tag{8.28}$$

By substituting $\dot{\boldsymbol{x}}_2^a$ from (8.12), the error dynamics can be modified as

$$\ddot{\boldsymbol{e}}^a = \boldsymbol{f}(\boldsymbol{x}) + \boldsymbol{G}(\boldsymbol{x})\boldsymbol{u} + \hat{\boldsymbol{d}}(\boldsymbol{x}, \boldsymbol{u}) + \boldsymbol{P}_4^{-1}\boldsymbol{P}_5\boldsymbol{s} + \boldsymbol{P}_3\boldsymbol{P}_4^{-1}\dot{\boldsymbol{e}} + \boldsymbol{P}_4\boldsymbol{s} - \ddot{\boldsymbol{x}}^* \tag{8.29}$$

To stabilize the error dynamics (8.29), the state feedback closed-loop tracking controller is designed as

$$\boldsymbol{u} = \boldsymbol{G}(\boldsymbol{x})^{-1}\begin{pmatrix} -\boldsymbol{f}(\boldsymbol{x}) - \hat{\boldsymbol{d}}(\boldsymbol{x}, \boldsymbol{u}) - \boldsymbol{P}_4^{-1}\boldsymbol{P}_5\boldsymbol{s} - \boldsymbol{P}_3\boldsymbol{P}_4^{-1}\dot{\boldsymbol{e}} \\ -\boldsymbol{P}_4\boldsymbol{s} - \boldsymbol{K}_1^a(\boldsymbol{e}^a, \dot{\boldsymbol{e}}^a)\boldsymbol{e}^a - \boldsymbol{K}_2^a(\dot{\boldsymbol{e}}^a)\dot{\boldsymbol{e}}^a + \ddot{\boldsymbol{x}}^* \end{pmatrix} \tag{8.30}$$

where, due to the positive definite nature of $\boldsymbol{G}(\boldsymbol{x})$, $[\boldsymbol{G}(\boldsymbol{x})]^{-1}$ will always exist. The matrices $\boldsymbol{K}_2^a(\dot{\boldsymbol{e}}^a)$ and $\boldsymbol{K}_1^a(\boldsymbol{e}^a, \dot{\boldsymbol{e}}^a)$ are symmetric and diagonal matrices. The matrix $\boldsymbol{K}_2^a(\dot{\boldsymbol{e}}^a)$ is defined as

$$\boldsymbol{K}_2^a(\dot{\boldsymbol{e}}^a) = \mathrm{diag}\left[k_{21}^a\left(\dot{e}_1^a\right),\ k_{22}^a\left(\dot{e}_2^a\right),\ldots,\ k_{2n}^a\left(\dot{e}_n^a\right)\right] \tag{8.31}$$

where the ith component of matrix $\boldsymbol{K}_2^a(\dot{\boldsymbol{e}}^a)$ is

$$k_{2i}^a\left(\dot{e}_i^a\right) = \left[\frac{\gamma_i^* c_{2_i}^a}{\left(\overline{\zeta}_{\dot{e}_i^a}^2 - \dot{e}_i^{a^2}\right)} + \left(\underline{\zeta}_{\dot{e}_i^a} / \overline{\zeta}_{\dot{e}_i^a}\right)^2 \frac{\left(1-\gamma_i^*\right) c_{2_i}^a}{\left(\underline{\zeta}_{\dot{e}_i^a}^2 - \dot{e}_i^{a^2}\right)}\right] \tag{8.32}$$

with $\gamma_i^*(\dot{e}_i^a) = \begin{cases} 1, & \text{if } 0 < \dot{e}_i^a < \overline{\zeta}_{\dot{e}_i^a} \\ 0, & \text{if } -\underline{\zeta}_{\dot{e}_i^a} < \dot{e}_i^a \le 0 \end{cases}$

where $c_{2_i}^a > 0, \forall i$ is a constant, and constants $\overline{\zeta}_{\dot{e}_i^a}$ and $\underline{\zeta}_{\dot{e}_i^a}$ represent upper and lower bounds on error $\dot{e}_i^a$, obtained as

$$\overline{\zeta}_{\dot{e}_i^a} = \overline{\zeta}_{x2i} - \overline{\zeta}_{\dot{x}_i^*} - \zeta_{\dot{e}_i} \tag{8.33}$$

$$\underline{\zeta}_{\dot{e}_i^a} = \underline{\zeta}_{\dot{x}_i^*} - \underline{\zeta}_{x2i} - \zeta_{\dot{e}_i} \tag{8.34}$$

where $\overline{\zeta}_{x2_i}$ and $\underline{\zeta}_{x2_i}$ are imposed state constraints on the dynamic states of the actual system, known constant $\zeta_{\dot{e}_i}$ is defined in (8.7), and $\overline{\zeta}_{\dot{x}_i^*}$ and $\underline{\zeta}_{\dot{x}_i^*}$ are the known bounds on the derivative of reference signal $\dot{x}_i^*$. The matrix $\boldsymbol{K}_1^a(\boldsymbol{e}^a, \dot{\boldsymbol{e}}^a)$ is defined as

$$\boldsymbol{K}_1^a(\boldsymbol{e}^a, \dot{\boldsymbol{e}}^a) = \mathrm{diag}\left[k_{11}^a\left(e_1^a, \dot{e}_1^a\right),\ k_{12}^a\left(e_2^a, \dot{e}_2^a\right),\ldots,\ k_{1n}^a\left(e_n^a, \dot{e}_n^a\right)\right] \tag{8.35}$$

where the ith component of matrix $\boldsymbol{K}_1^a(\boldsymbol{e}^a, \dot{\boldsymbol{e}}^a)$ is given as

$$k_{1i}^a\left(e_i^a, \dot{e}_i^a\right) = \left[\frac{\gamma_i^{**} c_{1_i}^a}{\left(\overline{\zeta}_{e_i^a}^2 - e_i^{a^2}\right)} + \left[\frac{\underline{\zeta}_{e_i^a}^2}{\overline{\zeta}_{e_i^a}^2}\right] \frac{\left(1-\gamma_i^{**}\right) c_{1_i}^a}{\left(\underline{\zeta}_{e_i^a}^2 - e_i^{a^2}\right)}\right]\left(\frac{1}{k_{2i}^a\left(\dot{e}_i^a\right)}\right) \tag{8.36}$$

with $\gamma_i^{**}(e_i^a) = \begin{cases} 1, & \text{if } 0 < e_i^a < \overline{\zeta}_{e_i^a} \\ 0, & \text{if } -\underline{\zeta}_{e_i^a} < e_i^a \le 0 \end{cases}$

where $c_{1_i}^a > 0, \forall i$ is a constant. Note that the term $k_{2i}^a\left(\dot{e}_i^a\right)$, defined in (8.32), is always positive. The constants $\overline{\zeta}_{e_i^a}$ and $\underline{\zeta}_{e_i^a}$ are upper and lower bounds on e_i^a, calculated as

$$\overline{\zeta}_{e_i^a} = \overline{\zeta}_{x1i} - \overline{\zeta}_{x_i^*} - \zeta_{e_i} \tag{8.37}$$

$$\underline{\zeta}_{e_i^a} = \underline{\zeta}_{x_i^*} - \underline{\zeta}_{x1i} - \zeta_{e_i} \tag{8.38}$$

where $\overline{\zeta}_{x_{1i}}$ and $\underline{\zeta}_{x_{1i}}$ are imposed state constraints on kinematic states of the actual system dynamics, known constant ζ_{e_i} is defined in (8.6), and $\overline{\zeta}_{x_i^*}$ and $\underline{\zeta}_{x_i^*}$ are the bounds on reference signal x_i^*.

Further, by substituting the control expression (8.30) in (8.29), the following closed-loop error dynamics is obtained:

$$\ddot{\boldsymbol{e}}^a = -\boldsymbol{K}_1^a(\boldsymbol{e}^a, \dot{\boldsymbol{e}}^a)\, \boldsymbol{e}^a - \boldsymbol{K}_2^a(\dot{\boldsymbol{e}}^a)\, \dot{\boldsymbol{e}}^a \tag{8.39}$$

The asymptotic stability of the error dynamics in (8.39) has already been proven in the Sect. 7.2.3 of Chap. 7. The proof can be followed from (7.73)–(7.79).

The above analysis shows that all closed-loop signals will remain bounded in both steps of the control design. However, the behavior of the actual closed-loop system is not analyzed yet. Thus, the closed-loop response of the actual system is analyzed in the following theorem.

Theorem 8.1 *Consider the actual systems (8.1) under the controller (8.30) with the weight update learning rule in (8.21), and if the initial conditions are chosen as* $e_i^a(0) \in \left(-\underline{\zeta}_{e_i^a}, \overline{\zeta}_{e_i^a}\right)$, $\dot{e}_i^a(0) \in \left(-\underline{\zeta}_{\dot{e}_i^a}, \overline{\zeta}_{\dot{e}_i^a}\right)$ $\forall i$, *then following properties hold:*

(i) The tracking error $\left(x_{1i} - x_i^*\right)$ *will asymptotically converge to the open set* $\left(-\zeta_{e_i}, \zeta_{e_i}\right)$, $\forall i$, *where* ζ_{e_i} *is a user-defined positive constant, defined in (8.6).*

(ii) The closed-loop kinematic and dynamic states of the system will remain bounded as $\underline{\zeta}_{x_{1i}} < x_{1i} < \overline{\zeta}_{x_{1i}}$ *and* $\underline{\zeta}_{x_{2i}} < x_{2i} < \overline{\zeta}_{x_{2i}}, \forall t \geq 0, \forall i$.

Proof The proof of Theorem 8.1 is exactly similar to the proof presented in Theorem 7.2 of Chap. 7. Therefore, the reader can refer to Chap. 7 for complete proof. □

8.1.4 Application to a Two-Link Robotic Manipulator

In this section, the effectiveness of the proposed state-constrained adaptive controller is demonstrated using a two-link robot manipulator model (Slotine and Li, 1991). The dynamics of a two-link manipulator is given by

$$\begin{bmatrix} \ddot{q}_1 \\ \ddot{q}_2 \end{bmatrix} = \left[\, H \,\right]^{-1} \begin{bmatrix} h\dot{q}_2 & h\,(\dot{q}_1 + \dot{q}_2) \\ -h\dot{q}_1 & 0 \end{bmatrix} \begin{bmatrix} \dot{q}_1 \\ \dot{q}_2 \end{bmatrix} + \left[\, H \,\right]^{-1} \begin{bmatrix} \tau_1 \\ \tau_2 \end{bmatrix} + \begin{bmatrix} d_1 + \eta_1(t) \\ d_2 + \eta_2(t) \end{bmatrix}$$

where $\boldsymbol{f}(\boldsymbol{x}) := \boldsymbol{f}\,(q_1, q_2, \dot{q}_1, \dot{q}_2) = \left[\, H \,\right]^{-1} \begin{bmatrix} h\dot{q}_2 & h\,(\dot{q}_1 + \dot{q}_2) \\ -h\dot{q}_1 & 0 \end{bmatrix} \begin{bmatrix} \dot{q}_1 \\ \dot{q}_2 \end{bmatrix}$

The complete description of the two-link manipulator model can be found in Sect. 7.2.4 of Chap. 7.

States of the system can be defined as $x_{11} \triangleq q_1$, $x_{12} \triangleq q_2$, $x_{21} \triangleq \dot{q}_1$, and $x_{22} \triangleq \dot{q}_2$. The controller objective is to generate the required control $u_i \triangleq \tau_i$ $(i =$

1, 2) to track reference outputs while maintaining the imposed constraints on the closed-loop states. In simulations, the initial conditions are chosen as $x_{11}(0) = 4.7°$, $x_{21}(0) = 9$ deg/sec, $x_{12}(0) = -4°$, and $x_{22}(0) = -8$ deg/sec, respectively. Initial conditions for weight update dynamics are chosen as zero. The reference profiles for $x_{11}(t)$ and $x_{12}(t)$ are selected as $x_1^* = 2° \sin(t)$ and $x_2^* = 1.5° \cos(2t)$, respectively. To ensure the safety of the system, state constraints on the system are selected as $\underline{\zeta}_{x11} = \overline{\zeta}_{x_{11}} = 8°$, $\underline{\zeta}_{x12} = \overline{\zeta}_{x_{12}} = 8°$, $\underline{\zeta}_{x21} = \overline{\zeta}_{x_{21}} = 15°$, and $\underline{\zeta}_{x22} = \overline{\zeta}_{x_{22}} = 15°$, respectively. The steady-state error bounds for joint positions and velocities are chosen as $|\zeta_{e_i}| < 0.5°$ and $|\zeta_{\dot{e}_i}| < 0.5$, respectively, where $i = 1, 2$.

The controller gains are selected as $c^a_{1_1} = 10$, $c^a_{2_1} = 3$, $c^a_{1_2} = 30$, $c^a_{2_2} = 5$, $c_{1_i} = 0.001$, $c_{2_i} = 0.0005$, $p_{5_i} = 1$, and $\alpha_i = 0.01$, respectively, and learning rate γ_i is chosen as $\gamma_i = 3$, where $i = 1, 2$. The model uncertainties are assumed as 50% of $\boldsymbol{f}(\boldsymbol{x})$, i.e., $[d_1, d_2]^T = 0.5\boldsymbol{f}(\boldsymbol{x})$, where $\boldsymbol{f}(\boldsymbol{x})$ is defined in model description. External disturbances $\eta_1(t)$ and $\eta_2(t)$ are assumed as $\eta_1(t) = 0.8 \sin(\pi(t - 1))$ for time interval $2 \leq t \leq 6$ and $\eta_2(t) = 1.2 \sin(2\pi/3(t - 4))$ for time interval $4 \leq t \leq 7$, respectively. To approximate model uncertainties, the basis functions are selected as $\boldsymbol{\phi}_i = [x_{11}, x_{12}, x_{21}, x_{22}, \sin(x_{12}), \cos(x_{12}), \sin(x_{11}), \cos(x_{11})]^T, \forall i$.

The position and velocity histories of the joints of the two-link robot manipulator are shown in Figs. 8.2 and 8.3, respectively. All trajectories accurately follow the reference trajectories while remaining constrained within the safety domain. The control histories required to stabilize the closed-loop system are illustrated in Fig. 8.4, where the profiles are smooth and continuous, making them suitable for practical applications.

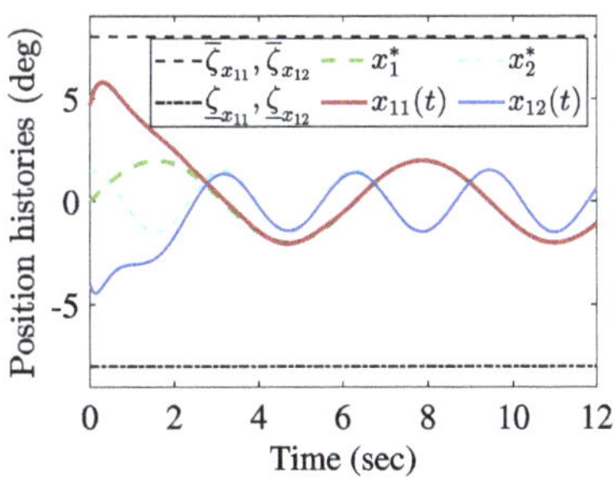

Fig. 8.2 Kinematic state trajectories

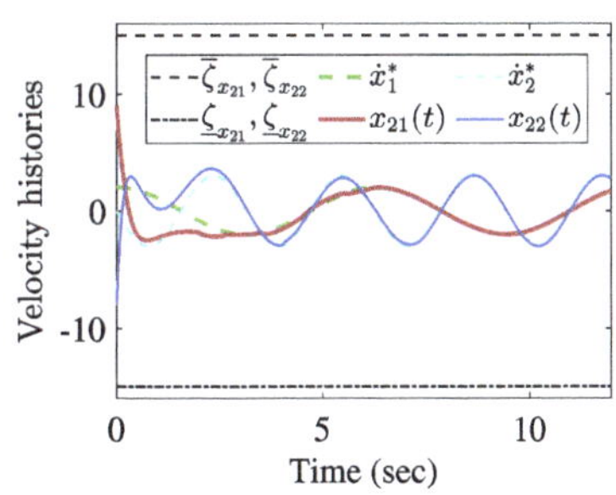

Fig. 8.3 Dynamic state trajectories

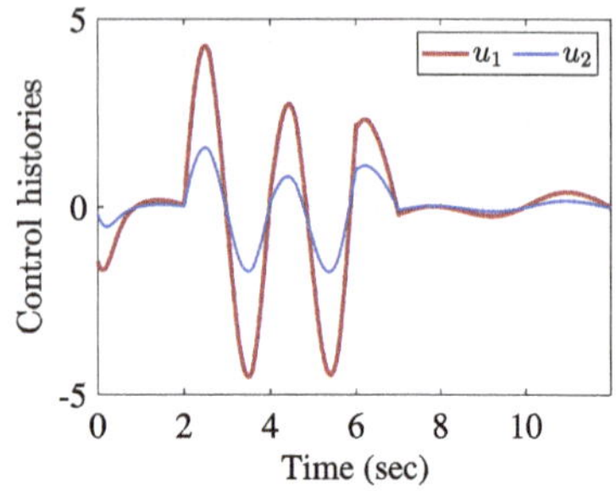

Fig. 8.4 Control histories

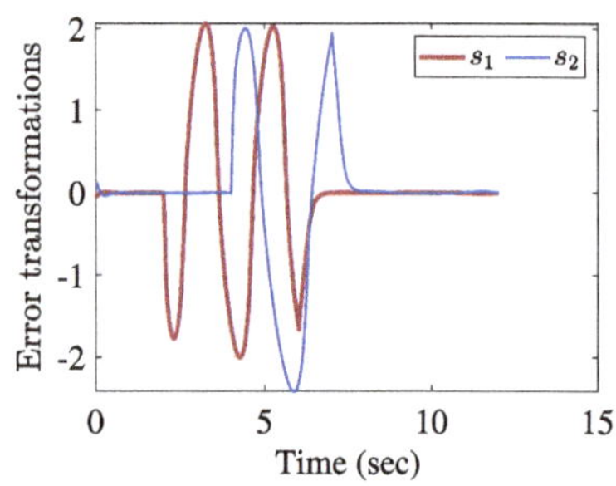

Fig. 8.5 Trajectories of proposed error transformation

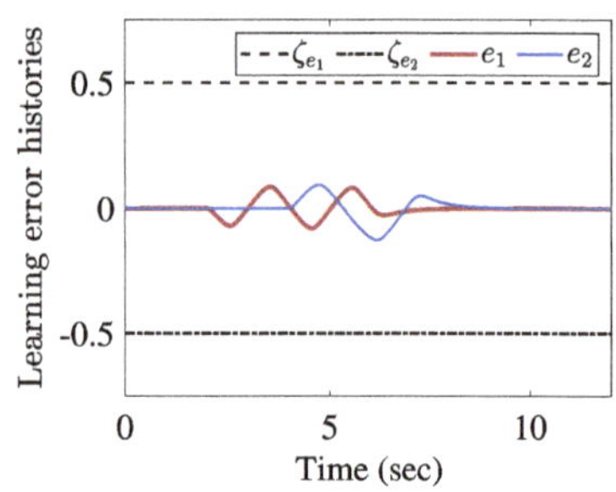

Fig. 8.6 Constrained learning error trajectories

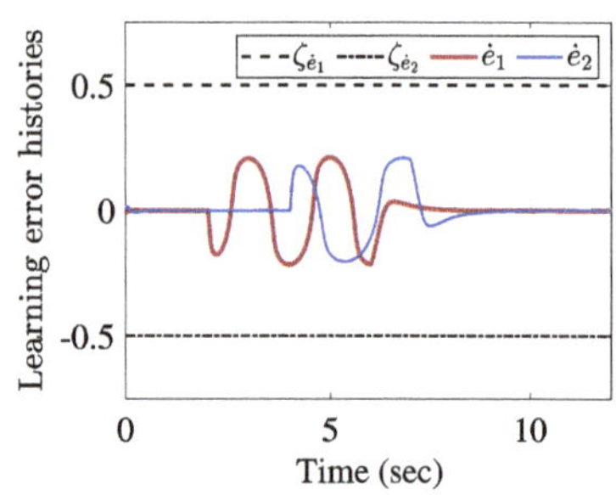

Fig. 8.7 Constrained histories of time derivative of learning errors

The profiles of the transformed error variables s_1 and s_2 are presented in Fig. 8.5, showing bounded and stable behavior as proven in theory. The state histories exhibit minimal transients, attributed to the imposed constraints on the learning errors, as depicted in Figs. 8.6 and 8.7. As shown in theory, the learning error and its derivative are bounded by the specified error constraints. The weight

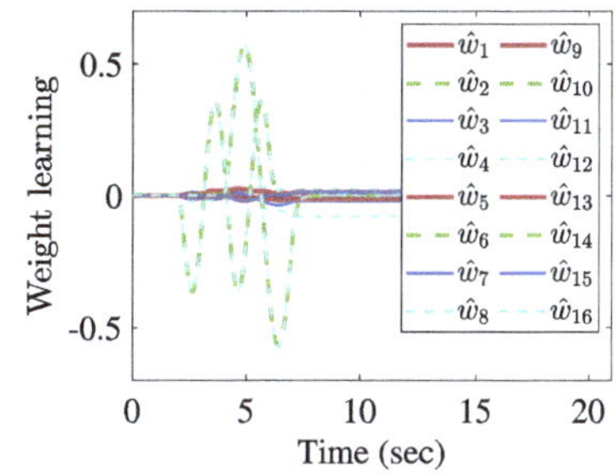

Fig. 8.8 Weight update histories during the approximation of modeling uncertainties

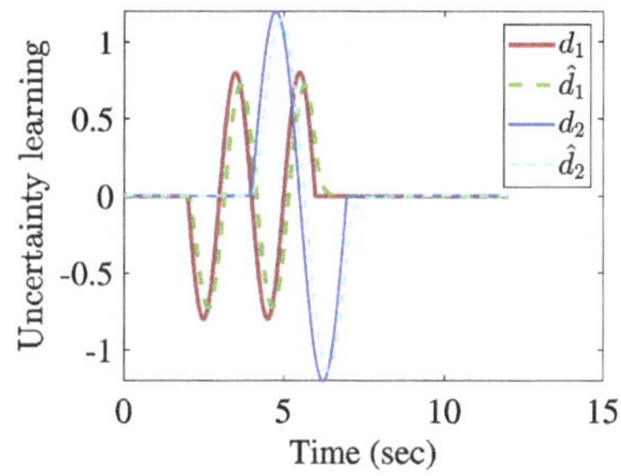

Fig. 8.9 Trajectories of modeling uncertainties and their approximation using neural network

update histories, depicted in Fig. 8.8, remain bounded despite the presence of external disturbances. The effectiveness of neural networks in approximating model uncertainties is demonstrated in Fig. 8.9, where the approximated profiles closely follow the modeling uncertainties.

8.2 Adaptive Control of Nonlinear System with Relative Degree Two

In the previous section, an error transformation-based state-constrained adaptive control design was presented for uncertain EL systems. However, the design assumed that the control effectiveness matrix of the system $\boldsymbol{G}(\boldsymbol{x})$ is known, which may not be practical in certain challenging applications. Additionally, the sign of $\boldsymbol{G}(\boldsymbol{x})$ (positive or negative definite) may be unknown, presenting the well-known issue of unknown control directions (UCD). Furthermore, asymptotic tracking was not achieved with the earlier approach. To address these limitations, this section presents a new adaptive control design for generic second-order nonlinear systems subjected to state constraints, unstructured uncertainties, external disturbances, and UCD.

The design employs the previously proposed error transformation to enforce state constraints and utilizes the Nussbaum gain to handle UCD. A single-parameter adaptive approach is introduced to approximate modeling uncertainties and external disturbances, significantly reducing computational cost. Lyapunov stability analysis demonstrates the asymptotic stability of the closed-loop system, ensuring that all signals remain bounded and comply with imposed state and input constraints. The problem formulation and control architecture are detailed in the following sections.

The dynamics of a second-order nonlinear MIMO system is given by

$$\begin{aligned}\dot{\boldsymbol{x}}_1 &= \boldsymbol{x}_2 \\ \dot{\boldsymbol{x}}_2 &= \boldsymbol{f}(\boldsymbol{x}) + \boldsymbol{G}(\boldsymbol{x})\boldsymbol{u} + \boldsymbol{d}_1(\boldsymbol{x}) + \boldsymbol{\eta}(t)\end{aligned} \tag{8.40}$$

where states $\boldsymbol{x}_1$ and $\boldsymbol{x}_1$ are defined as $\boldsymbol{x}_1 \triangleq [x_{11}, x_{12}, \ldots, x_{1n}]^T \in \Re^n$ and $\boldsymbol{x}_2 \triangleq [x_{21}, x_{22}, \ldots, x_{2n}]^T \in \Re^n$, respectively, and total states are defined as $\boldsymbol{x} \triangleq \left[\boldsymbol{x}_1^T, \boldsymbol{x}_2^T\right]^T \in \Re^{2n}$. Components of $\boldsymbol{f}(\boldsymbol{x}) \in \Re^n$ and $\boldsymbol{G}(\boldsymbol{x}) \in \Re^{n\times n}$ are assumed to be smooth and continuous. In the plant, $\boldsymbol{f}(\boldsymbol{x})$ is assumed to be a known part of dynamics, and unknown unstructured modeling uncertainties are represented by $\boldsymbol{d}_1(\boldsymbol{x})$. External disturbances to the system are denoted by $\boldsymbol{\eta}(t)$. The control input to the system is represented by $\boldsymbol{u} \in \Re^n$. By substituting $\boldsymbol{d}(\boldsymbol{x}, t) = \boldsymbol{d}_1(\boldsymbol{x}) + \boldsymbol{\eta}(t)$, the actual system in (8.40) can be modified as

$$\begin{aligned}\dot{\boldsymbol{x}}_1 &= \boldsymbol{x}_2 \\ \dot{\boldsymbol{x}}_2 &= \boldsymbol{f}(\boldsymbol{x}) + \boldsymbol{G}(\boldsymbol{x})\boldsymbol{u} + \boldsymbol{d}(\boldsymbol{x}, t)\end{aligned} \tag{8.41}$$

The term $\boldsymbol{d}(\boldsymbol{x}, t)$ can be approximated as $\boldsymbol{d}(\boldsymbol{x}, t) = \boldsymbol{W}^T\boldsymbol{\phi}(\boldsymbol{x}) + \boldsymbol{\epsilon}(t)$, where $\boldsymbol{\phi}(\boldsymbol{x})$ is the known basis function vector, $\boldsymbol{W}$ are the optimum weights, and $\boldsymbol{\epsilon}(t)$ is the approximation error. The norm of optimal NN weights $\|\boldsymbol{W}\|$ and approximation error $\boldsymbol{\epsilon}(t)$ are assumed to be bounded as $\|\boldsymbol{W}\|^2 \le w_m$ and $\|\boldsymbol{\epsilon}\|^2 \le \epsilon_m$, respectively, where w_m and ϵ_m are unknown positive constants. Moreover, for the control effectiveness matrix $\boldsymbol{G}(\boldsymbol{x})$, it is assumed that the matrix $(1/2)\left[\boldsymbol{G}(\boldsymbol{x}) + \boldsymbol{G}(\boldsymbol{x})^T\right]$ is either uniformly positive definite or uniformly negative definite for all $\boldsymbol{x} \in \Re^{2n}$, i.e., $\underline{\sigma}\left(\frac{\boldsymbol{G}(\boldsymbol{x})+\boldsymbol{G}(\boldsymbol{x})^T}{2}\right) \ge b^* > 0$ where $\underline{\sigma}(.)$ represents the smallest singular value of the matrix inside the bracket and b^* is its lower bound. With this property of $\boldsymbol{G}(\boldsymbol{x})$, the following Lemma will be useful in the control design.

Lemma 8.3 (Xu and Ioannou, 2003) *Let us define $\alpha(\boldsymbol{x}) \triangleq \boldsymbol{s}_e^T\boldsymbol{G}(\boldsymbol{x})\boldsymbol{s}_e$, where $\boldsymbol{s}_e \triangleq \boldsymbol{s}/\|\boldsymbol{s}\|$ is the unit vector. Let the matrix $(1/2)\left[\boldsymbol{G}(\boldsymbol{x}) + \boldsymbol{G}(\boldsymbol{x})^T\right]$ is known to be either uniformly positive definite or uniformly negative definite for all $\boldsymbol{x} \in \Omega_x$ where $\Omega_x \subset \Re^{2n}$ is a compact set, i.e.,*

$$\underline{\sigma}\left(\frac{\boldsymbol{G}(\boldsymbol{x}) + \boldsymbol{G}(\boldsymbol{x})^T}{2}\right) \ge b^* > 0$$

where $\underline{\sigma}(.)$ represents the smallest singular value of the matrix inside the bracket and b^* is its lower bound. Then we have $|\alpha(\boldsymbol{x})| > b^*$ and we have $\boldsymbol{s}^T(\boldsymbol{x})\boldsymbol{G}(\boldsymbol{x})\boldsymbol{s}(\boldsymbol{x}) = \alpha(\boldsymbol{x})\|\boldsymbol{s}(\boldsymbol{x})\|^2$.

The control design objective is to design a state-feedback closed-form adaptive control law for uncertain system (8.40) such that the states of the system track the reference signal asymptotically, i.e., $x_{1i} \rightarrow x_i^*$ and $x_{2i} \rightarrow \dot{x}_i^*$, without violation of imposed state constraints while ensuring the overall stability of the closed-loop system, where x_i^* is the ith component of reference trajectory $\boldsymbol{x}^*$. The imposed state constraints on the system are defined as $|x_{1i}(t)| < \zeta_{x_{1i}}$ and $|x_{2i}(t)| < \zeta_{x_{2i}}$, where $\zeta_{x_{1i}}$ and $\zeta_{x_{2i}}$ are positive user-defined constants.

The reference signal and its time derivatives are assumed to be smooth and bounded. Moreover, for any $\zeta_{x_{1i}} > 0$ and $\zeta_{x_{2i}} > 0$, there exist positive constants $\zeta_{x_i^*}, \zeta_{\dot{x}_i^*}$ satisfying $-\zeta_{x_{1i}} < -\zeta_{x_i^*} \leq x_i^* \leq \zeta_{x_i^*} < \zeta_{x_{1i}}, \forall i$ and $-\zeta_{x_{2i}} < -\zeta_{\dot{x}_i^*} \leq \dot{x}_i^* \leq \zeta_{\dot{x}_i^*} < \zeta_{x_{2i}}, \forall i$, where x_i^* and $\dot{x}_i^*$ are the ith components of the reference signal $\boldsymbol{x}^*$ and its time derivative $\dot{\boldsymbol{x}}^*$, respectively.

8.2.1 Adaptive Control Design

To fulfill control objectives, the system error is defined as

$$\boldsymbol{e} = \boldsymbol{x}_1 - \boldsymbol{x}^* \tag{8.42}$$

By using system dynamics (8.40), the derivatives of the error are obtained as

$$\dot{\boldsymbol{e}} = \boldsymbol{x}_2 - \dot{\boldsymbol{x}}^* \tag{8.43}$$

$$\ddot{\boldsymbol{e}} = \dot{\boldsymbol{x}}_2 - \ddot{\boldsymbol{x}}^* \tag{8.44}$$

The error transformation variable $\boldsymbol{s}$, defined in Sect. 8.1.1, can be written as

$$\boldsymbol{s} \triangleq \boldsymbol{P}_1(\boldsymbol{e})\,\boldsymbol{e} + \boldsymbol{P}_2(\dot{\boldsymbol{e}})\,\dot{\boldsymbol{e}} \tag{8.45}$$

where $\boldsymbol{P}_1(\boldsymbol{e})$ and $\boldsymbol{P}_2(\dot{\boldsymbol{e}})$ are symmetric diagonal matrices, defined as

$$\boldsymbol{P}_1(\boldsymbol{e}) = \operatorname{diag}\left[\frac{c_{1_i}}{\left(\zeta_{e_i}^2 - e_i^2\right)}\right], \ \forall\, i = 1, 2, \ldots, n \tag{8.46}$$

$$\boldsymbol{P}_2(\dot{\boldsymbol{e}}) = \operatorname{diag}\left[\frac{c_{2_i}}{\left(\zeta_{\dot{e}_i}^2 - \dot{e}_i^2\right)}\right], \ \forall\, i = 1, 2, \ldots, n \tag{8.47}$$

where c_{1_i} and $c_{2_i}, \forall i = 1, \ldots, n$ are positive constants. The constants, $\zeta_{\dot{e}_i}$ and $\zeta_{e_i}, \forall i$ are maximum bound on the variation of errors e_i and $\dot{e}_i, \forall i$, respectively. The error bound $\zeta_{\dot{e}_i}$ and ζ_{e_i} are obtained using imposed state constraints, calculated as

$$\zeta_{e_i} = \zeta_{x_{1i}} - \zeta_{x_i^*} \tag{8.48}$$

$$\zeta_{\dot{e}_i} = \zeta_{x_{2i}} - \zeta_{\dot{x}_i^*} \tag{8.49}$$

where $\zeta_{x_{1i}}$ and $\zeta_{x_{2i}}$ are imposed state constraints and $\zeta_{x_i^*}$ and $\zeta_{\dot{x}_i^*}$ are bounds on reference signal.

The time derivative of $\boldsymbol{s}$ in (8.45) is obtained as

$$\dot{\boldsymbol{s}} = \boldsymbol{P}_3(\boldsymbol{e})\,\dot{\boldsymbol{e}} + \boldsymbol{P}_4(\dot{\boldsymbol{e}})\,\ddot{\boldsymbol{e}} \tag{8.50}$$

where $\boldsymbol{P}_3(\boldsymbol{e})$ and $\boldsymbol{P}_4(\dot{\boldsymbol{e}})$ are symmetric diagonal matrices, defined as

$$\boldsymbol{P}_3(\boldsymbol{e}) = \text{diag}\left[\frac{c_{1_i}\left(\zeta_{e_i}^2 + e_i^2\right)}{\left(\zeta_{e_i}^2 - e_i^2\right)^2}\right], \ \forall\, i = 1, 2, \ldots, n \tag{8.51}$$

$$\boldsymbol{P}_4(\dot{\boldsymbol{e}}) = \text{diag}\left[\frac{c_{2_i}\left(\zeta_{\dot{e}_i}^2 + \dot{e}_i^2\right)}{\left(\zeta_{\dot{e}_i}^2 - \dot{e}_i^2\right)^2}\right], \ \forall\, i = 1, 2, \ldots, n \tag{8.52}$$

To design the state-feedback asymptotically stable controller for the new transformed system, the following quadratic Lyapunov function is chosen:

$$V = \frac{1}{2}\boldsymbol{s}^T\boldsymbol{s} + \frac{1}{2\gamma}\tilde{w}_m^2 \tag{8.53}$$

where $\tilde{w}_m$ is defined as $\tilde{w}_m = w_m - \hat{w}_m$, w_m is defined (8.41), and $\hat{w}_m$ is the estimation of w_m. The constant $\gamma > 0$ is learning rate constant.

The time derivative of Lyapunov function in (8.53) is given as

$$\dot{V} = \boldsymbol{s}^T\dot{\boldsymbol{s}} - \frac{1}{\gamma}\tilde{w}_m\dot{\hat{w}}_m \tag{8.54}$$

By substituting $\dot{\boldsymbol{s}}$ from (8.50), $\dot{V}$ in (8.54) can be modified as

$$\dot{V} = \boldsymbol{s}^T\left[\boldsymbol{P}_3(\boldsymbol{e})\,\dot{\boldsymbol{e}} + \boldsymbol{P}_4(\dot{\boldsymbol{e}})\,\ddot{\boldsymbol{e}}\right] - \frac{1}{\gamma}\tilde{w}_m\dot{\hat{w}}_m \tag{8.55}$$

Without loss of generality, in the further analysis, $\boldsymbol{f}(\boldsymbol{x})$, $\boldsymbol{G}(\boldsymbol{x})$, $\boldsymbol{d}(\boldsymbol{x}, t)$, $\boldsymbol{P}_1(\boldsymbol{e})$, $\boldsymbol{P}_2(\dot{\boldsymbol{e}})$, $\boldsymbol{P}_3(\boldsymbol{e})$, and $\boldsymbol{P}_4(\dot{\boldsymbol{e}})$ are denoted as $\boldsymbol{f}, \boldsymbol{G}, \boldsymbol{d}, \boldsymbol{P}_1, \boldsymbol{P}_2, \boldsymbol{P}_3$, and $\boldsymbol{P}_4$, respectively. By substituting $\boldsymbol{e}$ and $\dot{\boldsymbol{e}}$ from (8.42) and (8.43), and $\dot{\boldsymbol{x}}_2$ from (8.41) into (8.55), the following analysis can be carried out:

$$\begin{aligned}\dot{V} &= \boldsymbol{s}^T\left[\boldsymbol{P}_3\left[\boldsymbol{x}_2 - \dot{\boldsymbol{x}}^*\right] + \boldsymbol{P}_4\left[\dot{\boldsymbol{x}}_2 - \ddot{\boldsymbol{x}}^*\right]\right] - \frac{1}{\gamma}\tilde{w}_m\dot{\hat{w}}_m \\ &= \boldsymbol{s}^T\left[\boldsymbol{P}_3\boldsymbol{x}_2 - \boldsymbol{P}_3\dot{\boldsymbol{x}}^* + \boldsymbol{P}_4\boldsymbol{f} + \boldsymbol{P}_4\boldsymbol{G}\boldsymbol{u} + \boldsymbol{P}_4\boldsymbol{d} - \boldsymbol{P}_4\ddot{\boldsymbol{x}}_d\right] - \frac{1}{\gamma}\tilde{w}_m\dot{\hat{w}}_m\end{aligned}$$

$$= \boldsymbol{s}^T \boldsymbol{P}_3 \boldsymbol{x}_2 - \boldsymbol{s}^T \boldsymbol{P}_3 \dot{\boldsymbol{x}}^* + \boldsymbol{s}^T \boldsymbol{P}_4 \boldsymbol{f} + \boldsymbol{s}^T \boldsymbol{P}_4 \boldsymbol{G} \boldsymbol{u} + \boldsymbol{s}^T \boldsymbol{P}_4 \boldsymbol{d} - \boldsymbol{s}^T \boldsymbol{P}_4 \ddot{\boldsymbol{x}}_d$$
$$-\frac{1}{\gamma} \tilde{w}_m \dot{\hat{w}}_m \tag{8.56}$$

By substituting $\boldsymbol{d}(\boldsymbol{x}, t) = \boldsymbol{W}^T \boldsymbol{\phi}(\boldsymbol{x}) + \boldsymbol{\epsilon}(t)$, the following is obtained:

$$\dot{V} = \boldsymbol{s}^T \boldsymbol{P}_3 \boldsymbol{x}_2 - \boldsymbol{s}^T \boldsymbol{P}_3 \dot{\boldsymbol{x}}^* + \boldsymbol{s}^T \boldsymbol{P}_4 \boldsymbol{f} + \boldsymbol{s}^T \boldsymbol{P}_4 \boldsymbol{G} \boldsymbol{u} + \boldsymbol{s}^T \boldsymbol{P}_4 \boldsymbol{W}^T \boldsymbol{\phi}(\boldsymbol{x})$$
$$+\boldsymbol{s}^T \boldsymbol{P}_4 \boldsymbol{\epsilon}(t) - \boldsymbol{s}^T \boldsymbol{P}_4 \dot{\boldsymbol{x}}^* - \frac{1}{\gamma} \tilde{w}_m \dot{\hat{w}}_m \tag{8.57}$$

which can be written as

$$\dot{V} = \boldsymbol{s}^T \left(\boldsymbol{P}_4 [\boldsymbol{P}_4]^{-1} \right) \boldsymbol{P}_3 \boldsymbol{x}_2 - \boldsymbol{s}^T \left(\boldsymbol{P}_4 [\boldsymbol{P}_4]^{-1} \right) \boldsymbol{P}_3 \dot{\boldsymbol{x}}^* + \boldsymbol{s}^T \boldsymbol{P}_4 \boldsymbol{f}$$
$$+\boldsymbol{s}^T \boldsymbol{P}_4 \boldsymbol{G} \boldsymbol{u} + \boldsymbol{s}^T \boldsymbol{P}_4 \boldsymbol{W}^T \boldsymbol{\phi}$$
$$+\boldsymbol{s}^T \boldsymbol{P}_4 \boldsymbol{\epsilon}(t) - \boldsymbol{s}^T \boldsymbol{P}_4 \dot{\boldsymbol{x}}^* - \frac{1}{\gamma} \tilde{w}_m \dot{\hat{w}}_m \tag{8.58}$$

With the application of Young's inequality (Zhao et al., 2018), terms in (8.58) can be expressed as

$$\dot{V} \leq \|\boldsymbol{P}_4 \boldsymbol{s}\|^2 \left\| [\boldsymbol{P}_4]^{-1} \boldsymbol{P}_3 \boldsymbol{x}_2 \right\|^2 + \frac{1}{4} + \|\boldsymbol{P}_4 \boldsymbol{s}\|^2 \left\| [\boldsymbol{P}_4]^{-1} \boldsymbol{P}_3 \dot{\boldsymbol{x}}^* \right\|^2 + \frac{1}{4}$$
$$+ \|\boldsymbol{P}_4 \boldsymbol{s}\|^2 \|\boldsymbol{f}\|^2 + \frac{1}{4} + \|\boldsymbol{P}_4 \boldsymbol{s}\|^2 \|\boldsymbol{W}\|^2 \|\boldsymbol{\phi}\|^2 + \frac{1}{4} + \frac{\|\boldsymbol{P}_4 \boldsymbol{s}\|^2}{2} + \frac{\|\boldsymbol{\epsilon}(t)\|^2}{2}$$
$$+ \|\boldsymbol{P}_4 \boldsymbol{s}\|^2 \left\| \dot{\boldsymbol{x}}^* \right\|^2 + \frac{1}{4} + \boldsymbol{s}^T \boldsymbol{P}_4 \boldsymbol{G} \boldsymbol{u} - \frac{1}{\gamma} \tilde{w}_m \dot{\hat{w}}_m \tag{8.59}$$
$$\leq \|\boldsymbol{P}_4 \boldsymbol{s}\|^2 \left[\left\| [\boldsymbol{P}_4]^{-1} \boldsymbol{P}_3 \boldsymbol{x}_2 \right\|^2 + \left\| [\boldsymbol{P}_4]^{-1} \boldsymbol{P}_3 \dot{\boldsymbol{x}}^* \right\|^2 + \|\boldsymbol{f}\|^2 + \left\| \dot{\boldsymbol{x}}^* \right\|^2 + \frac{1}{2} \right]$$
$$+ \|\boldsymbol{P}_4 \boldsymbol{s}\|^2 w_m \|\boldsymbol{\phi}\|^2 + \frac{5}{4} + \frac{\epsilon_m}{2} + \boldsymbol{s}^T \boldsymbol{P}_4 \boldsymbol{G} \boldsymbol{u} - \frac{1}{\gamma} \tilde{w}_m \dot{\hat{w}}_m \tag{8.60}$$

where the relations $\|\boldsymbol{W}\|^2 \leq w_m$ and $\|\boldsymbol{\epsilon}\|^2 \leq \epsilon_m$ are used in the analysis. The control variable $\boldsymbol{u}$ and the learning rule for approximation variable $\hat{w}_m$ are chosen as

$$\boldsymbol{u} = \boldsymbol{P}_4 \boldsymbol{s} \mathcal{N}(\omega) \left[u' + \hat{w}_m \|\boldsymbol{\phi}\|^2 + c_2 \right] \tag{8.61}$$

$$\dot{\omega} = \|\boldsymbol{P}_4 \boldsymbol{s}\|^2 \left[u' + \hat{w}_m \|\boldsymbol{\phi}\|^2 + c_2 \right] \tag{8.62}$$

$$u' = \left\| [\boldsymbol{P}_4]^{-1} \boldsymbol{P}_3 \boldsymbol{x}_2 \right\|^2 + \left\| [\boldsymbol{P}_4]^{-1} \boldsymbol{P}_3 \dot{\boldsymbol{x}}^* \right\|^2 + \|\boldsymbol{f}\|^2 + \left\| \dot{\boldsymbol{x}}^* \right\|^2$$
$$\dot{\hat{w}}_m = \gamma \|\boldsymbol{P}_4 \boldsymbol{s}\|^2 \|\boldsymbol{\phi}\|^2 - \sigma \hat{w}_m \tag{8.63}$$

where c_2 and σ are positive constants and $\hat{w}_m(0) \geq 0$, $\omega(0) \geq 0$, and $\mathcal{N}(\bullet)$ are Nussbaum gain function. The Nussbaum gain function is defined as follows.

Definition 8.1 (Ge et al., 2004; Nussbaum, 1983) Any continuous function $\mathcal{N}(s) : \Re \to \Re$ is a function of Nussbaum type if it has the following properties:

$$\lim_{s\to+\infty} \sup \frac{1}{s}\int_0^s \mathcal{N}(\omega)d\omega = +\infty \tag{8.64}$$

$$\lim_{s\to+\infty} \inf \frac{1}{s}\int_0^s \mathcal{N}(\omega)d\omega = -\infty \tag{8.65}$$

For example, the continuous functions $\omega^2\cos(\omega)$, $\omega^2\sin(\omega)$, and $e^{\omega^2}\cos\left(\frac{\omega\pi}{2}\right)$ belong to the Nussbaum function category.

Note that from learning rule (8.63), for $\hat{w}_m(0) \geq 0$, $\hat{w}_m(t) \geq 0$, $\forall t$. By substituting $\boldsymbol{u}$, u' and $\dot{\hat{w}}_m$ in (8.60), and by adding and subtracting $\dot{\omega}$, the following is obtained:

$$\begin{aligned}
\dot{V} \leq\ & \|\boldsymbol{P}_4\boldsymbol{s}\|^2\left[u' + \frac{1}{2}\right] + \|\boldsymbol{P}_4\boldsymbol{s}\|^2\, w_m\, \|\boldsymbol{\phi}\|^2 - \dot{\omega} + \frac{5}{4} + \frac{\epsilon_m}{2} \\
& + \boldsymbol{s}^T\boldsymbol{P}_4\boldsymbol{G}\boldsymbol{P}_4\boldsymbol{s}\mathcal{N}(\omega)\left[u' + \hat{w}_m\, \|\boldsymbol{\phi}\|^2 + c_2\right] - \frac{1}{\gamma}\tilde{w}_m\left(\gamma\, \|\boldsymbol{P}_4\boldsymbol{s}\|^2\, \|\boldsymbol{\phi}\|^2 - \sigma\,\hat{w}_m\right) + \dot{\omega} \\
\leq\ & \|\boldsymbol{P}_4\boldsymbol{s}\|^2\left[u' + \frac{1}{2}\right] + \|\boldsymbol{P}_4\boldsymbol{s}\|^2\, w_m\, \|\boldsymbol{\phi}\|^2 - \|\boldsymbol{P}_4\boldsymbol{s}\|^2\left[u' + \hat{w}_m\, \|\boldsymbol{\phi}\|^2 + c_2\right] \\
& - \|\boldsymbol{P}_4\boldsymbol{s}\|^2\,\tilde{w}_m\, \|\boldsymbol{\phi}\|^2 + \frac{\sigma\,\tilde{w}_m\hat{w}_m}{\gamma} + \boldsymbol{s}^T\boldsymbol{P}_4\boldsymbol{G}\boldsymbol{P}_4\boldsymbol{s}\mathcal{N}(\omega)\left[u' + \hat{w}_m\, \|\boldsymbol{\phi}\|^2 + c_2\right] \\
& + \|\boldsymbol{P}_4\boldsymbol{s}\|^2\left[u' + \hat{w}_m\, \|\boldsymbol{\phi}\|^2 + c_2\right] + \frac{5}{4} + \frac{\epsilon_m}{2}
\end{aligned} \tag{8.66}$$

Applying Lemma 8.3, which is $[\boldsymbol{P}_4\boldsymbol{s}]^T\,\boldsymbol{G}\,[\boldsymbol{P}_4\boldsymbol{s}] = \alpha(\boldsymbol{x}(t))\,\|\boldsymbol{P}_4\boldsymbol{s}\|^2$, the following simplified expression is obtained:

$$\begin{aligned}
\dot{V} \leq\ & \left[\frac{1}{2} - c_2\right][\boldsymbol{P}_4\boldsymbol{s}]^T[\boldsymbol{P}_4\boldsymbol{s}] + \|\boldsymbol{P}_4\boldsymbol{s}\|^2\, w_m\, \|\boldsymbol{\phi}\|^2 - \|\boldsymbol{P}_4\boldsymbol{s}\|^2\,\hat{w}_m\, \|\boldsymbol{\phi}\|^2 \\
& - \|\boldsymbol{P}_4\boldsymbol{s}\|^2\,\tilde{w}_m\, \|\boldsymbol{\phi}\|^2 + \frac{\sigma\,\tilde{w}_m\hat{w}_m}{\gamma} + \alpha(\boldsymbol{x}(t))\,\|\boldsymbol{P}_4\boldsymbol{s}\|^2\mathcal{N}(\omega)\left[u' + \hat{w}_m\, \|\boldsymbol{\phi}\|^2 + c_2\right] \\
& + \|\boldsymbol{P}_4\boldsymbol{s}\|^2\left[u' + \hat{w}_m\, \|\boldsymbol{\phi}\|^2 + c_2\right] + \frac{5}{4} + \frac{\epsilon_m}{2}
\end{aligned}$$

$$\leq \left[\frac{1}{2} - c_2\right][\boldsymbol{P}_4\boldsymbol{s}]^T[\boldsymbol{P}_4\boldsymbol{s}] + \frac{\sigma\,\tilde{w}_m\hat{w}_m}{\gamma} + \alpha(\boldsymbol{x}(t))\mathcal{N}(\omega)\dot{\omega} + \dot{\omega} + \frac{5}{4} + \frac{\epsilon_m}{2} \tag{8.67}$$

$$\leq -\left[c_2 - \frac{1}{2}\right]\boldsymbol{s}^T\boldsymbol{P}_4\boldsymbol{P}_4\boldsymbol{s} + \frac{5}{4} + \frac{\epsilon_m}{2} + \frac{\sigma\,\tilde{w}_m\hat{w}_m}{\gamma} + [\alpha(t)\mathcal{N}(\omega) + 1]\,\dot{\omega} \tag{8.68}$$

since $\boldsymbol{P}_4$ is a diagonal symmetric matrix with positive elements. From the definition of $\boldsymbol{P}_4$ in (8.52), it can be observed that the minimum value of the elements of $\boldsymbol{P}_4$ is obtained at $\dot{e}_i = 0$, which is $c_{2_i}/\zeta_{\dot{e}_i}^2, \forall i$. By utilizing the matrix property $-\boldsymbol{s}^T \boldsymbol{P}_4 \boldsymbol{P}_4 \boldsymbol{s} \leq -\lambda_{\min}(\boldsymbol{P}_4 \boldsymbol{P}_4) \boldsymbol{s}^T \boldsymbol{s}$, where $\lambda_{\min}(\boldsymbol{P}_4 \boldsymbol{P}_4) = c_{2_i}^2/\zeta_{\dot{e}_i}^4, \forall i$. By selecting $c_2 > \frac{1}{2}$, and defining $c_3 = \left(c_2 - \frac{1}{2}\right) \lambda_{\min}(\boldsymbol{P}_4 \boldsymbol{P}_4)$, the following expression is obtained:

$$\dot{V} \leq -c_3 \boldsymbol{s}^T \boldsymbol{s} + \frac{5}{4} + \frac{\epsilon_m}{2} + \frac{\sigma \tilde{w}_m \hat{w}_m}{\gamma} + [\alpha(t)\mathcal{N}(\omega) + 1]\dot{\omega} \tag{8.69}$$

By applying Young's inequality, $\frac{\sigma \tilde{w}_m \hat{w}_m}{\gamma} \leq -\frac{\sigma \tilde{w}_m^2}{2\gamma} + \frac{\sigma w_m^2}{2\gamma}$, (8.69) can be expressed as

$$\dot{V} \leq -c_3 \boldsymbol{s}^T \boldsymbol{s} + \frac{5}{4} + \frac{\epsilon_m}{2} - \frac{\sigma \tilde{w}_m^2}{2\gamma} + \frac{\sigma w_m^2}{2\gamma} + [\alpha(t)\mathcal{N}(\omega) + 1]\dot{\omega} \tag{8.70}$$

$$\begin{aligned} &\leq -c_3 \boldsymbol{s}^T \boldsymbol{s} - \frac{\sigma \tilde{w}_m^2}{2\gamma} + c_4 + [\alpha(t)\mathcal{N}(\omega) + 1]\dot{\omega} \\ &\leq -c_5 V + [\alpha(t)\mathcal{N}(\omega) + 1]\dot{\omega} + c_4 \end{aligned} \tag{8.71}$$

where $c_4 = \left(\frac{5}{4} + \frac{\epsilon_m}{2} + \frac{\sigma w_m^2}{2\gamma}\right)$, and $c_5 = \min\left(2c_3, \frac{\sigma}{\gamma}\right)$.

Integrating (8.71) over [0,t], we have

$$V(t) \leq c_6 + \exp(-c_5 t) \int_0^t [\alpha(\tau)\mathcal{N}(\omega) + 1]\dot{\omega} \exp(c_5 \tau)\, d\tau \tag{8.72}$$

where $c_6 = V(0) + \frac{c_4}{c_5}$ and "exp" is the exponential function. In the context of the Nussbaum function, the following theorem is useful.

Theorem 8.2 (Ge et al., 2004; Wang et al., 2016) *Let $V(\bullet)$ and $\omega(\bullet)$ be smooth functions defined on $\left[0, t_f\right)$ with $V(t) \geq 0, \ \forall t \in \left[0, t_f\right)$, and $\mathcal{N}(\bullet)$ be an even smooth Nussbaum-type function. If the following inequality holds:*

$$V(t) \leq c_0 + e^{-c_1 t} \int_0^t [\alpha(\tau)\mathcal{N}(\omega) + 1]\dot{\omega} e^{c_1 \tau} d\tau, \tag{8.73}$$

where c_0 represents some suitable constant, c_1 is a positive constant, and $\alpha(\tau)$ is a time varying parameter which takes values in the unknown closed interval $I \triangleq \left[l^-, l^+\right]$ with $0 \notin I$, then $V(t)$, $\omega(t)$ and $\int_0^t [\alpha(\tau)\mathcal{N}(\zeta) + 1]\dot{\omega} e^{c_1 \tau} d\tau$ must be bounded on $\left[0, t_f\right)$.

To generalize the stable closed-loop solution for all time, the following Lemma is useful.

Lemma 8.4 (Ryan, 1991) *Consider the system $\dot{x}(t) = f(x(t))$ under the initial condition $x(0) = x_0$ with solution $x(t)$, $\forall t \in [0, \tau)$. If the solution $x(t)$ is a bounded solution then $\tau = \infty$.*

Applying Theorem 8.2, the boundedness of $V(t)$ and $\omega(t)$ can be concluded from (8.72). Since $V(t)$ is bounded, $\boldsymbol{s}$ and $\hat{w}_m$ are bounded; the closed-loop system is bounded. Therefore. according to Lemma 8.4, if the solution of the closed-loop system is bounded, then $t_f \to +\infty$.

8.2.2 Closed-Loop System Analysis

In the previous subsection, the boundedness of the closed-loop system was demonstrated using Lyapunov stability theory. However, the asymptotic stability and satisfaction of user-imposed state constraints have not yet been addressed. Therefore, this subsection focuses on analyzing the closed-loop system to establish its stability and verify compliance with the imposed state constraints.

Using Lemma 8.1, since $\boldsymbol{s}$ is bounded, errors will remain bounded, i.e., $|e_i| < \zeta_{e_i}, \forall t \geq 0$ and $|\dot{e}_i| < \zeta_{\dot{e}_i}, \forall t \geq 0$. Further, by using the definition of $e_i = x_{1i} - x_i^*$, the bounds on x_{1i} are expressed as $-\zeta_{e_i} + x_i^* < x_{1i} < x_i^* + \zeta_{e_i}$. Since $-\zeta_{x_i^*} \leq x_i^* \leq \zeta_{x_i^*}$, the bounds on x_{1i} are modified as

$$-\zeta_{e_i} - \zeta_{x_i^*} < x_{1i} < \zeta_{x_i^*} + \zeta_{e_i} \tag{8.74}$$

From (8.48), $\zeta_{x_{1i}} = \zeta_{x_i^*} + \zeta_{e_i}$. Therefore $-\zeta_{x_{1i}} < x_{1i} < \zeta_{x_{1i}}$. Similar analysis can be carried out to obtain the bounds on x_{2i}. Thus, the closed-loop states of the system will remain bounded by the imposed constraints, i.e., $|x_{1i}| < \zeta_{x_{1i}}$ and $|x_{2i}| < \zeta_{x_{2i}}, \forall i$.

To show the asymptotic stability of the system $\dot{\omega}$ is used. From (8.62), $\dot{\omega}$ can be written as

$$\dot{\omega} = \|\boldsymbol{P}_4\boldsymbol{s}\|^2 u' + \|\boldsymbol{P}_4\boldsymbol{s}\|^2 \hat{w}_m \|\boldsymbol{\phi}\|^2 + \|\boldsymbol{P}_4\boldsymbol{s}\|^2 c_2 \tag{8.75}$$

Since right-hand side terms are positive, $\dot{\omega} \geq [\boldsymbol{P}_4\boldsymbol{s}]^T [\boldsymbol{P}_4\boldsymbol{s}] c_2$. By integrating, we have $c_2 \int_0^t [\boldsymbol{P}_4\boldsymbol{s}]^T [\boldsymbol{P}_4\boldsymbol{s}] d\tau \leq \omega(t) - \omega(0)$. Since $\omega(t)$ is bounded, $(\boldsymbol{P}_4\boldsymbol{s}) \in \mathscr{L}_2$, which means $\boldsymbol{s} \in \mathscr{L}_2$. Furthermore, from the stability analysis in (8.72), $\boldsymbol{s}$ is bounded and hence $\boldsymbol{s} \in \mathscr{L}_2 \cap \mathscr{L}_\infty$. Then, by using Barbalat Lemma, $\boldsymbol{s} \to 0$ can be ensured (Ge et al., 2004; Wang et al., 2016). Now, from Lemma 8.2, since $\boldsymbol{s} \to 0$ uniformly, $e_i \to 0$ and $\dot{e}_i \to 0$. Thus, asymptotic tracking of the closed-loop system can be concluded, i.e., $\boldsymbol{x}_1 \to \boldsymbol{x}^*$ and $\boldsymbol{x}_2 \to \dot{\boldsymbol{x}}^*$ as $t \to \infty$.

8.2.3 Application to a Two-Link Robotic Manipulator

In this section, the effectiveness of the developed adaptive controller is demonstrated using a two-link robot manipulator model (Slotine and Li, 1991). The dynamics of a two-link manipulator are given as

$$\begin{bmatrix} \ddot{q}_1 \\ \ddot{q}_2 \end{bmatrix} = [H]^{-1} \begin{bmatrix} h\dot{q}_2 & h(\dot{q}_1 + \dot{q}_2) \\ -h\dot{q}_1 & 0 \end{bmatrix} \begin{bmatrix} \dot{q}_1 \\ \dot{q}_2 \end{bmatrix} + [H]^{-1} \begin{bmatrix} \tau_1 \\ \tau_2 \end{bmatrix} + \begin{bmatrix} d_1 + \eta_1(t) \\ d_2 + \eta_2(t) \end{bmatrix}$$

where $\boldsymbol{f}(\boldsymbol{x}) \triangleq \boldsymbol{f}(q_1, q_2, \dot{q}_1, \dot{q}_2) = [H]^{-1} \begin{bmatrix} h\dot{q}_2 & h(\dot{q}_1 + \dot{q}_2) \\ -h\dot{q}_1 & 0 \end{bmatrix} \begin{bmatrix} \dot{q}_1 \\ \dot{q}_2 \end{bmatrix}$

The complete description of the two-link manipulator model can be found in the Sect. 7.2.4 of Chap. 7.

States of the system can be defined as $x_{11} \triangleq q_1$, $x_{12} \triangleq q_2$, $x_{21} \triangleq \dot{q}_1$, and $x_{22} \triangleq \dot{q}_2$. The controller objective is to generate the required control $u_i \triangleq \tau_i$ $(i = 1, 2)$ to track reference outputs while maintaining the imposed constraints on the closed-loop states. The initial conditions for the simulations are randomly selected from the open set $x_{11}(0) \in (-6°, 6°)$, $x_{21}(0) \in (-12, 12)$, $x_{12}(0) \in (-6°, 6°)$, and $x_{22}(0) \in (-12, 12)$. Initial conditions for the weight update rule and Nussbaum dynamic are chosen as $\hat{w}_m(0) = 0$ and $\omega(0) = 2$, respectively. Without loss of generality, the problem is considered as a regulation problem, i.e., $x_{11} \to 0$ and $x_{12} \to 0$ as $t \to \infty$. The controller gains are selected as $c_{1_i} = 0.003$, $c_{2_i} = 0.03$, $c_2 = 15$, $\sigma_i = 1$ and learning rate $\gamma_i = 1$, where $i = 1, 2$. To ensure the safety of the system, the constraints are imposed as $-6° < x_{11} < 6°$, $-6° < x_{12} < 6°$, $-12 < x_{21} < 12$, and $-12 < x_{22} < 12$, respectively.

The model uncertainties are assumed as 30% of $\boldsymbol{f}(\boldsymbol{x})$, i.e., $[d_1, d_2]^T = 0.3\boldsymbol{f}(\boldsymbol{x})$, where $\boldsymbol{f}(\boldsymbol{x})$ is defined in model description. External disturbances $\eta_1(t)$ and $\eta_2(t)$ are assumed as $\eta_1(t) = 0.1\sin(\pi(t-1))$ for time interval $2 \le t \le 6$ and $\eta_2(t) = 0.1\sin(2\pi/3(t-4))$ for time interval $4 \le t \le 7$, respectively. To approximate model uncertainties, generic Gaussian RBFNN with width two are used as basis functions, defined as $\boldsymbol{\phi}(z)$, where

$$z = [1, x_{11}, x_{21}, x_{11}^2, x_{21}^2, x_{11}x_{21}, x_{11}^3, x_{21}x_{11}^2, x_{11}x_{21}^2, x_{21}^3, x_{11}^2 x_{21}^2]^T.$$

The output histories of the system under the proposed controller are illustrated in Fig. 8.10, demonstrating that the controller asymptotically drives the system output x_{1i} to zero while maintaining the imposed constraints. The dynamic state trajectories, shown in Fig. 8.11, are smooth and remain bounded within the specified constraints. The corresponding control input histories are provided in Fig. 8.12, highlighting smooth and feasible profiles. The trajectory of the approximation variable $\hat{w}_m$ is depicted in Fig. 8.13. As expected, for positive initial conditions, $\hat{w}_m(t)$ remains positive and bounded throughout the simulation. The transformed error trajectories s_i, shown in Fig. 8.14, are bounded and converge to zero as theoretically proven. Finally, the trajectories of the Nussbaum gain $\mathcal{N}(\omega)$ and its argument ω are presented in Fig. 8.15, showing continuous profiles that rapidly stabilize to constant values.

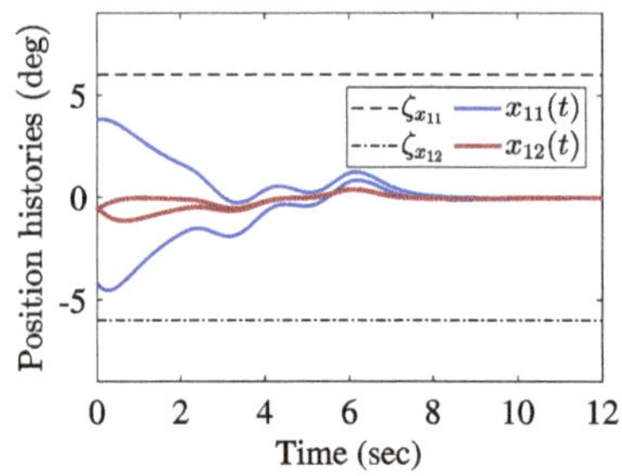

Fig. 8.10 Output histories under output constraints for two random initial conditions

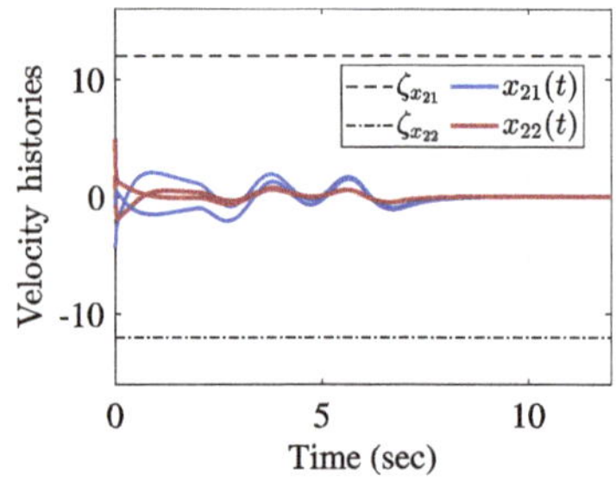

Fig. 8.11 Trajectories of dynamic states x_{21}, x_{22} for two random initial conditions

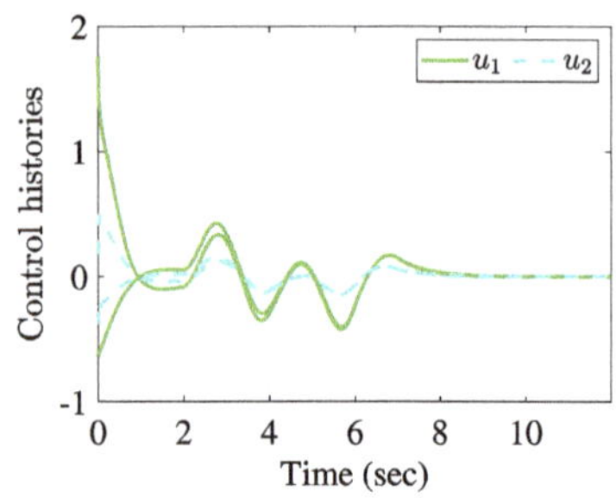

Fig. 8.12 Control histories of the system for two random initial conditions

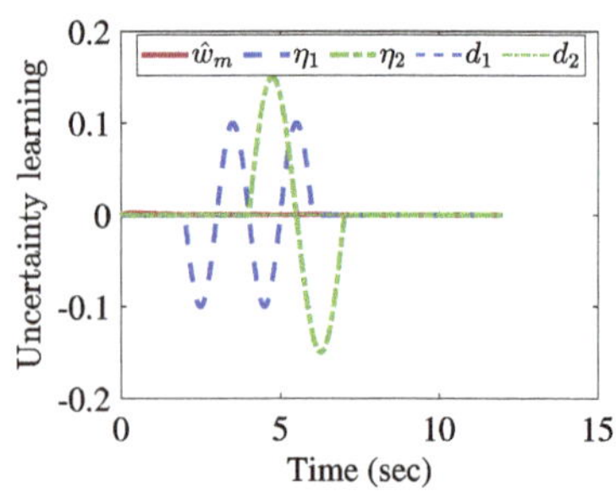

Fig. 8.13 Approximation of plant uncertainties for two random initial conditions

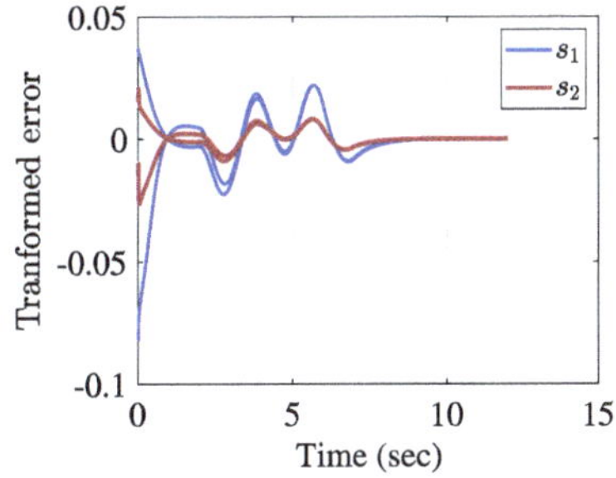

Fig. 8.14 Transformed error for two random initial conditions

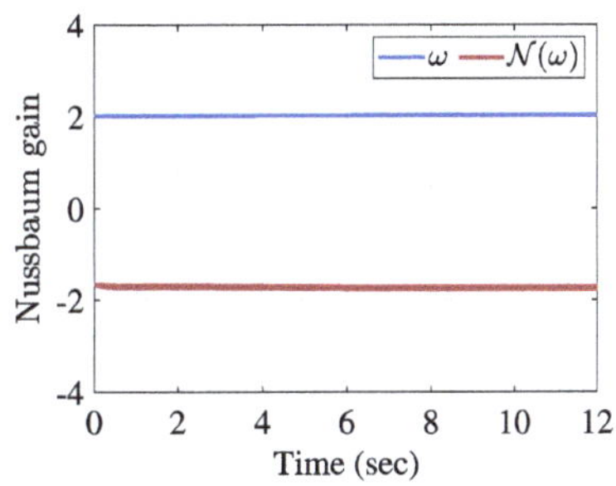

Fig. 8.15 Nussbaum gain for two random initial conditions

8.3 Summary

This chapter presents a transformation-based state-constrained adaptive controller for nonlinear systems with unstructured modeling uncertainties and bounded external disturbances. The adaptive controllers are presented following both direct and indirect adaptive control design philosophies, and the stability of the closed-loop system is analyzed using the Lyapunov stability theory, demonstrating that the system is asymptotically stable and its states remain bounded within the user-imposed constraints. A two-link robot manipulator is considered a case study to validate the effectiveness of the proposed control approach.

8.4 Exercise Problems

1. Design the transformation-based adaptive controllers using all the discussed approaches for the following nonlinear systems to stabilize the system response considering unstructured model uncertainties. The desired controller for these systems can be designed using the methods given in Chap. 4.

   (a) $\begin{bmatrix} \dot{x}_1 \\ \dot{x}_2 \end{bmatrix} = \begin{bmatrix} x_2 \\ -x_2 - x_1 - (x_1 + x_2)^2 x_2 + u \end{bmatrix}$ where $d(x)$ is defined as $d(x) = 0.2x_1^2 x_2 + x_2^3$. The user-imposed state constraints are imposed as ζ_1 and ζ_2 are $\zeta_1 = 0.3$ and $\zeta_2 = 0.8$, respectively.

(b) $\ddot{x} = -\sin x + u + d(x)$ where $d(x)$ is defined as $d(x) = x^2\dot{x} + 0.1\sin\dot{x}\cos x$. The user-imposed state constraints are imposed as ζ_1 and ζ_2 are $\zeta_1 = 0.3$ and $\zeta_2 = 0.8$, respectively.

(c) $\ddot{x} - 0.7(1 - x^2)\dot{x} - x^3 + (1 + \dot{x}^2)u + d(x) = 0$ where $d(x)$ is defined as $d(x) = x^2 - 0.15\dot{x}^2 + 0.2x\dot{x}$. The user-imposed state constraints are imposed as ζ_1 and ζ_2 are $\zeta_1 = 0.3$ and $\zeta_2 = 0.8$, respectively.

(d) $\begin{bmatrix} \dot{x}_1 \\ \dot{x}_2 \\ \dot{x}_3 \end{bmatrix} = \begin{bmatrix} x_1 + x_2 \\ x_3 \\ u + d(x) \end{bmatrix}$ where $d(x)$ is defined as $d(x) = -0.2x_1x_2^2 + 0.4x_1^2x_2^2$. The user-imposed state constraints ζ_1, ζ_2, and ζ_3 are $\zeta_1 = 0.3$, $\zeta_2 = 0.6$, and $\zeta_3 = 0.8$, respectively.

(e) $\begin{bmatrix} \dot{x}_1 \\ \dot{x}_2 \\ \dot{x}_3 \end{bmatrix} = \begin{bmatrix} -x_1 + x_2 \\ x_3 \\ u \end{bmatrix}$ where $d(x)$ is defined as $d(x) = -0.2x_1x_2^2 + 0.4x_1^2x_2^2$. The user-imposed state constraints ζ_1, ζ_2, and ζ_3 are $\zeta_1 = 0.3$, $\zeta_2 = 0.6$, and $\zeta_3 = 0.8$, respectively.

2. Refer to the Eqs. (8.23) and (8.24), and show that the mathematical algebra of deriving the expression

$$v_i(t) \le v_i(0)\exp(-p_{6_i}t) + \frac{p_{7_i}}{p_{6_i}}\left[1 - \exp(-p_{6_i}t)\right]$$

from the relation

$$\dot{v}_i \le -p_{6_i}v_i + p_{7_i}$$

where “exp” is the exponential function and $v_i(0)$, p_{6_i}, and p_{7_i} are constants. These variables are defined in Eqs. (8.23) and (8.24).

3. Refer to the Eq. (8.63)

$$\dot{\hat{w}}_m = \gamma \left\| \boldsymbol{P}_4\boldsymbol{s} \right\|^2 \left\| \boldsymbol{\phi} \right\|^2 - \sigma\hat{w}_m$$

and show that $\hat{w}_m(t) \ge 0$ if $\hat{w}_m(0) \ge 0$. The definition of all variables can be found in Eq. (8.63).

4. Refer to the Eqs. (8.71), and find the step-by-step solution of the following differential equation

$$\dot{V} \le -c_5V + [\alpha(t)\mathcal{N}(\omega) + 1]\,\dot{\omega} + c_4$$

where c_4, and c_5 are constants.

References

S.S. Ge, F. Hong, T.H. Lee, Adaptive neural control of nonlinear time-delay systems with unknown virtual control coefficients. IEEE Trans. Syst. Man Cybern. Part B (Cybern.) **34**(1), 499–516 (2004)

R.D. Nussbaum, Some remarks on a conjecture in parameter adaptive control. Syst. Control Lett. **3**(5), 243–246 (1983)

E.P. Ryan, A universal adaptive stabilizer for a class of nonlinear systems. Syst. Control Lett. **16**(3), 209–218 (1991)

K. Sachan, R. Padhi, Safety-constrained robust adaptive control for a class of MIMO non-linear systems. IET Control Theory Appl. **14**(7), 982–990 (2020)

J.-J.E. Slotine, W. Li, et al., *Applied Nonlinear Control*, vol. 199(1) (Prentice hall Englewood Cliffs, NJ, 1991)

W. Wang, D. Wang, Z. Peng, T. Li, Prescribed performance consensus of uncertain nonlinear strict-feedback systems with unknown control directions. IEEE Trans. Syst. Man Cybern. Syst. **46**(9), 1279–1286 (2016)

H. Xu, P. Ioannou, Robust adaptive control for a class of mimo nonlinear systems with guaranteed error bounds. IEEE Trans. Autom. Control **48**(5), 728–742 (2003)

K. Zhao, Y. Song, T. Ma, L. He, Prescribed performance control of uncertain Euler-Lagrange systems subject to full-state constraints. IEEE Trans. Neural Networks Learn. Syst. **29**(8), 3478–3489 (2018)

Chapter 9
Neural Networks and Evolving Systems

9.1 Introduction

In the era of abundant data, data-driven techniques have gained significant popularity. With the rapid advancement in artificial intelligence (AI), its usage in all fields is becoming inevitable. AI seeks to integrate intelligence into machines by leveraging data and learning algorithms to facilitate informed decision-making. Among the foundational paradigms in AI is the neural network (NN), a framework capable of learning to make decisions after being trained using the available data. NNs have demonstrated remarkable success in numerous dynamic system applications such as the estimation of unknown dynamics and adaptive control.

For many nonlinear control problems, designing a control system incorporates some form of function estimation. For instance, the estimation of the unmodeled dynamics, uncertainty and external disturbances can be achieved using well-established NN-based techniques. These estimations can be used to build a control system such as feedback linearization, SMC, or adaptive controllers.

This chapter builds a solid foundation of NNs to provide the readers with the required knowledge to utilize NNs in control systems. One of the most suitable types of feedforward neural networks (FFNNs) for online control problems is the radial basis functions neural networks (RBFNNs) since they are linearly parameterized, i.e., their output is linearly related to tunable weights, and they are suitable for online adaptation with a good learning speed (Ge et al. 2013). Thus, RBFNNs and general regression neural networks (GRNNs), a subclass of them, are detailed in the following subsections.

K. Sachan et al., *Intelligent Adaptive Control with System Constraints*,
https://doi.org/10.1007/978-981-95-8220-4_9

9.2 Radial Basis Functions Neural Networks

RBFNNs are single hidden layer FFNNs with Gaussian activation functions (Yu et al. 2011). The input vector $\boldsymbol{x} \in \Re^m$ is passed through this layer using the Gaussian activation function $\boldsymbol{\phi}(\boldsymbol{x}) \in \Re^n$. RBFNN diagram is demonstrated in Fig. 9.1. The output of RBFNN $\boldsymbol{y} \in \Re^l$ can be expressed as a linearly parameterized FFNN output as follows:

$$\boldsymbol{y}(\boldsymbol{x}) = \boldsymbol{W}^T \boldsymbol{\phi}(\boldsymbol{x}) + \boldsymbol{b} \tag{9.1}$$

where $\boldsymbol{W} \in \Re^{n \times l}$ is a matrix of the hidden-output weights, $\boldsymbol{\phi}(\boldsymbol{x}) \in \Re^n$ is a vector of the Gaussian basis functions (kernels) and $\boldsymbol{b} \in \Re^l$ is the bias vector.

$\boldsymbol{\phi}(\boldsymbol{x})$ is constructed as

$$\boldsymbol{\phi}(\boldsymbol{x}) = \begin{bmatrix} \phi_1(\boldsymbol{x}) \\ \phi_2(\boldsymbol{x}) \\ \vdots \\ \phi_n(\boldsymbol{x}) \end{bmatrix} \tag{9.2}$$

where $\phi_i(\boldsymbol{x})$ is calculated as

$$\phi_i(\boldsymbol{x}) = \exp\left(-\frac{||\boldsymbol{x} - \boldsymbol{c}_i||^2}{2\sigma^2}\right), \quad i = 1, 2, \ldots, n \tag{9.3}$$

where $\boldsymbol{c}_i \in \Re^m$ is the center of the basis function ϕ_i.

RBFNNs are trained in multiple phases, including using clustering of the input data into groups to calculate the centers of the basis function $\boldsymbol{c}$ and then utilizing gradient-based optimization to calculate the output weights $\boldsymbol{W}$. RBFNNs are frequently used in dynamic system modeling and control because of their simple structure, swift online learning capabilities, acceptable inference, and capacity to handle uncertainties (Yu et al. 2011).

Fig. 9.1 RBFNN

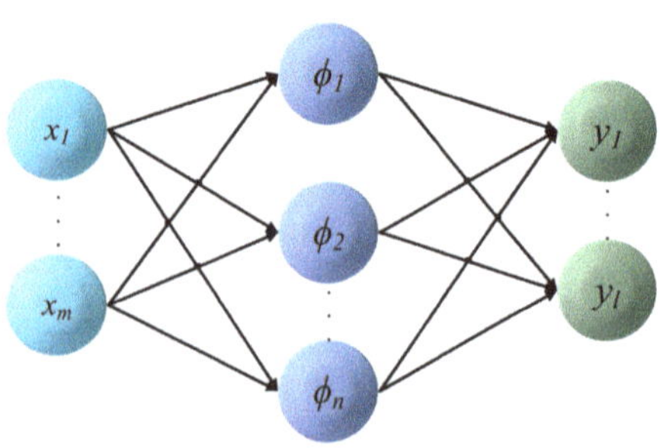

Table 9.1 RBFNN parameters

Centers	Weights and biases	σ
$c_1 = \begin{bmatrix} -1 \\ 3 \end{bmatrix}, c_2 = \begin{bmatrix} 2 \\ 4 \end{bmatrix}$ and $c_3 = \begin{bmatrix} 0 \\ 2 \end{bmatrix}$	$W = \begin{bmatrix} 4 & 8 \\ 2 & 4 \\ 3 & 9 \end{bmatrix}, b = \begin{bmatrix} 0.7 \\ 0.6 \end{bmatrix}$	$\sigma = 1$

Example 9.1 This example demonstrates how RBFNN output is calculated. Considering an RBFNN with three basis functions and having the following parameters (Table 9.1).

Using an input $\boldsymbol{x} = \begin{bmatrix} 1 \\ 2 \end{bmatrix}$ to calculate the network output.

Firstly, let us calculate the basis functions $\boldsymbol{\phi}$ as

$$\phi_1(x) = \exp\left(\frac{-\left\| \begin{bmatrix} 1 \\ 2 \end{bmatrix} - \begin{bmatrix} -1 \\ 3 \end{bmatrix} \right\|^2}{2}\right) = 0.082, \quad \phi_2(x) = \exp\left(\frac{-\left\| \begin{bmatrix} 1 \\ 2 \end{bmatrix} - \begin{bmatrix} 2 \\ 4 \end{bmatrix} \right\|^2}{2}\right) = 0.082$$

$$\phi_3(x) = \exp\left(\frac{-\left\| \begin{bmatrix} 1 \\ 2 \end{bmatrix} - \begin{bmatrix} 0 \\ 2 \end{bmatrix} \right\|^2}{2}\right) = 0.607.$$

$$\boldsymbol{\phi}(\boldsymbol{x}) = \begin{bmatrix} 0.082 \\ 0.082 \\ 0.607 \end{bmatrix}$$

Next, let us calculate the network output using (9.1) as

$$\boldsymbol{y} = \boldsymbol{W}^T \boldsymbol{\phi}(\boldsymbol{x}) + \boldsymbol{b} = \begin{bmatrix} 4 & 8 & 2 \\ 4 & 3 & 9 \end{bmatrix} \begin{bmatrix} 0.082 \\ 0.082 \\ 0.607 \end{bmatrix} + \begin{bmatrix} 0.7 \\ 0.6 \end{bmatrix} = \begin{bmatrix} 3.012 \\ 7.044 \end{bmatrix}$$

One special class of RBFNNs is the general regression neural network (GRNN), it inherits RBFNN properties but with improved training method making it even more suitable for control system applications.

9.3 General Regression Neural Networks

General regression neural networks (GRNNs), as shown in Fig. 9.2, have the same structure and activation functions as for RBFNN, and thus, they are considered a subclass of RBFNNs. Hence, GRNN falls under RBFNN in the NN chart as shown in Fig. 9.3.

GRNN can learn its parameters in one pass on the input/output data, deeming it precise and fast learner (Specht 1991). It is regarded as a suitable technique for dynamic systems modeling and control since quick and precise learning is vital in dynamic systems as it is concerned with decision making in real-time. In the general case, GRNN has multiple inputs and multiple outputs, and it is trained once on the whole training dataset.

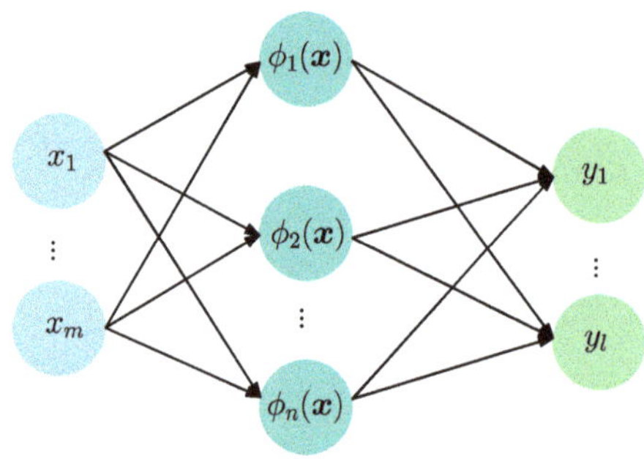

Fig. 9.2 GRNN structure

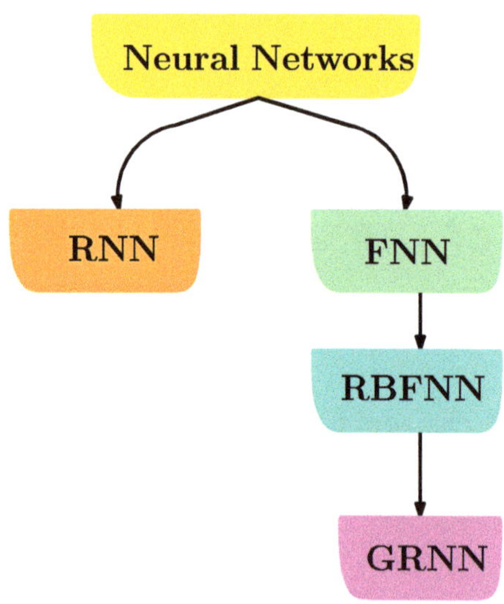

Fig. 9.3 GRNN location in NNs hierarchy

GRNN output can be written as

$$y(x) = \frac{W^T \phi(x)}{\sum_{i=1}^{n} \phi_i(x)} \tag{9.4}$$

where $y \in \Re^l$ is the output of the network, $x \in \Re^m$ is the input vector, $W \in \Re^{n \times l}$ is a matrix of the output weights, and $\phi(x) \in \Re^n$ is a vector of the hidden layer outputs, and it is calculated using (9.2) and (9.3).

9.3.1 Training of GRNN

GRNN training is the process of storing the training instances made of input data, which will be stored in the Gaussian activation function centers, and the output data, which will be saved in the output weights. Thus, GRNN hidden layer size will grow to the number of the training samples. For instance, if we are training GRNN with n samples then the hidden layer size will become n nodes after training. GRNN hidden layer growth is illustrated in Fig. 9.4.

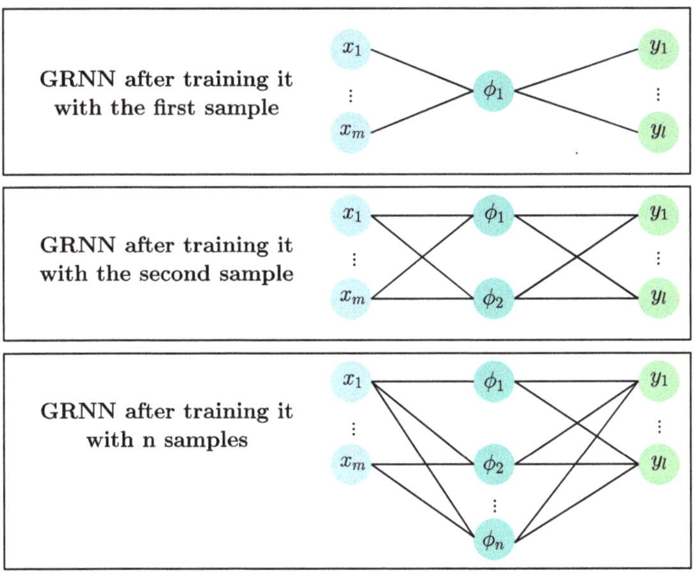

Fig. 9.4 GRNN hidden layer growth in the training process

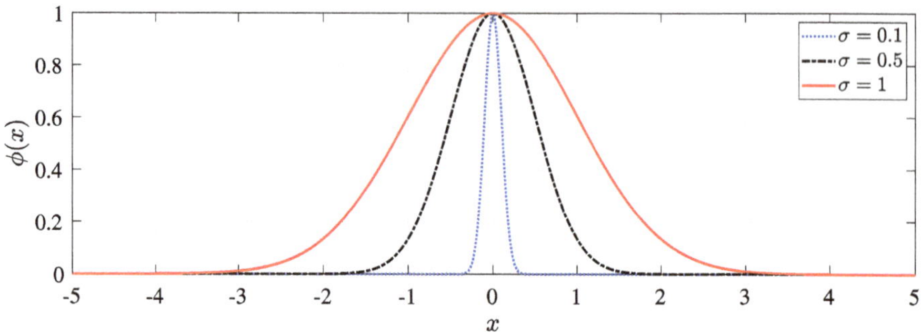

Fig. 9.5 $\phi(x)$ with different values of σ

Consider a dataset of n samples made of inputs $\boldsymbol{X} \in \Re^{m \times n} = \begin{bmatrix} \boldsymbol{x}_1 \\ \vdots \\ \boldsymbol{x}_n \end{bmatrix}$ and outputs $\boldsymbol{Y} \in \Re^{l \times n} = \begin{bmatrix} \boldsymbol{y}_1 \\ \vdots \\ \boldsymbol{y}_n \end{bmatrix}$, where m is the number of the input variables and l is the number of output variables; GRNN training is summarized as

For every training sample $(\boldsymbol{x}_i, \boldsymbol{y}_i)$:

- The center $\boldsymbol{c}_i$ of the activation function $\phi_i(\cdot)$ is set to $\boldsymbol{x}_i$.
- The output weights $\boldsymbol{w}_i$ are set to $\boldsymbol{y}_i$.
- $\sigma > 0$ is the smoothing parameter, and it determines the width of the Gaussian activation function $\phi(\cdot)$ as demonstrated in Fig. 9.5. Small values of σ leads to narrow activation function shape, while large values of σ leads to wider activation function shape. σ should be selected before starting the training. A good practice to select σ is setting it as the standard deviation of the input data ($std(\boldsymbol{x}) = \sqrt{\frac{1}{m}(\sum_{i=1}^{m}(x_i - \mu)^2}$, μ is the mean of the input data) to ensure the activation function covers the input space properly.

After the training concludes, we will have a GRNN with centers $\boldsymbol{C} = \begin{bmatrix} \boldsymbol{x}_1 \\ \vdots \\ \boldsymbol{x}_n \end{bmatrix}$ and output weights matrix $\boldsymbol{W} = \begin{bmatrix} \boldsymbol{y}_1 \\ \vdots \\ \boldsymbol{y}_n \end{bmatrix}$; these centers and weights are used to calculate GRNN output for any input $\boldsymbol{x}$ using (9.3) and (9.1). The following example demonstrates GRNN training and output calculation.

Table 9.2 GRNN training dataset

x_1	x_2	y
1	1	2
3	3	18
4	4	32

Example 9.2 This example depicts how GRNN output is calculated and how the smoothing parameter σ affects GRNN estimation accuracy. GRNN was trained with the dataset in Table 9.2 to approximate the function $y(\boldsymbol{x}) = x_1^2 + x_2^2$.

Find the output of the trained GRNN for the following input vector $\boldsymbol{x} = \begin{bmatrix} x_1 \\ x_2 \end{bmatrix} = \begin{bmatrix} 2.5 \\ 3.5 \end{bmatrix}$. Use $\sigma = 1$, $\sigma = 0.5$ and $\sigma = 0.1$. Use error $e(\boldsymbol{x}) = y(\boldsymbol{x}) - \hat{y}(\boldsymbol{x})$ to quantify the estimation error.

Applying the training procedure of GRNN detailed in Sect. 9.3.1 to three training samples given in Table 9.2, the following Gaussian activation function centers are determined $\boldsymbol{c}_1 = \begin{bmatrix} 1 \\ 1 \end{bmatrix}$, $\boldsymbol{c}_2 = \begin{bmatrix} 3 \\ 3 \end{bmatrix}$, and $\boldsymbol{c}_3 = \begin{bmatrix} 4 \\ 4 \end{bmatrix}$. Also, the output weights are assigned as $\boldsymbol{w} = \begin{bmatrix} 2 \\ 18 \\ 32 \end{bmatrix}$

1. For σ=1:

   The activation vector $\boldsymbol{\phi}(\boldsymbol{x})$ is calculated as

$$\phi_1(\boldsymbol{x}) = \exp\left(-\frac{\|\boldsymbol{x}-\boldsymbol{c}_1\|^2}{2\sigma^2}\right) = \exp\left(-\frac{\left\|\begin{bmatrix} 2.5 \\ 3.5 \end{bmatrix} - \begin{bmatrix} 1 \\ 1 \end{bmatrix}\right\|^2}{2}\right) = 0.0143$$

$$\phi_2(\boldsymbol{x}) = \exp\left(-\frac{\|\boldsymbol{x}-\boldsymbol{c}_2\|^2}{2\sigma^2}\right) = \exp\left(-\frac{\left\|\begin{bmatrix} 2.5 \\ 3.5 \end{bmatrix} - \begin{bmatrix} 3 \\ 3 \end{bmatrix}\right\|^2}{2}\right) = 0.779$$

$$\phi_3(\boldsymbol{x}) = \exp\left(-\frac{\|\boldsymbol{x}-\boldsymbol{c}_3\|^2}{2\sigma^2}\right) = \exp\left(-\frac{\left\|\begin{bmatrix} 2.5 \\ 3.5 \end{bmatrix} - \begin{bmatrix} 4 \\ 4 \end{bmatrix}\right\|^2}{2}\right) = 0.287$$

$$\boldsymbol{\phi}(\boldsymbol{x}) = \begin{bmatrix} \phi_1(\boldsymbol{x}) \\ \phi_2(\boldsymbol{x}) \\ \phi_3(\boldsymbol{x}) \end{bmatrix} = \begin{bmatrix} 0.0143 \\ 0.779 \\ 0.287 \end{bmatrix}, \sum_{i=1}^{3} \phi_i = 0.0143+0.779+0.287 = 1.080$$

(continued)

Example 9.2 (continued)
GRNN output is found as

$$\hat{y}(\boldsymbol{x}) = \frac{\boldsymbol{w}^T \boldsymbol{\phi}(\boldsymbol{x})}{\sum_{i=1}^{3} \phi_i} = \frac{\begin{bmatrix} 2 \ 18 \ 32 \end{bmatrix} \begin{bmatrix} 0.0143 \\ 0.779 \\ 0.287 \end{bmatrix}}{1.080}$$

$$= \frac{\left[2 \times 0.0143 + 18 \times 0.779 + 32 \times 0.287\right]}{1.080}$$

$$= \frac{23.235}{1.080} = 21.514$$

The actual output is $y(\boldsymbol{x}) = x_1^2 + x_2^2 = 2.5^2 + 3.5^2 = 18.5 \rightarrow e(\boldsymbol{x}) = y(\boldsymbol{x}) - \hat{y}(\boldsymbol{x}) = 18.500 - 21.514 = -3.014$

2. For σ=0.5:
For different values of σ, only the activation vector $\boldsymbol{\phi}(\boldsymbol{x})$ will be different
thus, $\boldsymbol{\phi}(\boldsymbol{x}) = \begin{bmatrix} 4.140e^{-8} \\ 0.368 \\ 0.007 \end{bmatrix}$
$\hat{y}(\boldsymbol{x}) = 18.252 \rightarrow e(\boldsymbol{x}) = 18.500 - 18.252 = 0.248$
3. For σ=0.1:
$\boldsymbol{\phi}(\boldsymbol{x}) = \begin{bmatrix} 2.660e^{-185} \\ 1.390e^{-11} \\ 5.1664e^{-55} \end{bmatrix}$
$\hat{y}(\boldsymbol{x}) = 18 \rightarrow e(\boldsymbol{x}) = y(\boldsymbol{x}) - \hat{y}(\boldsymbol{x}) = 18.5 - 18 = 0.500$ (Table 9.3)

For this problem, with only three data samples, GRNN can produce reasonable output for unseen data when $\sigma > 0$. Thus, the smoothing parameter of GRNN is a dominant parameter for the approximation error, i.e., the difference between the network estimated output $\hat{y}(\boldsymbol{x})$ and the actual output $y(\boldsymbol{x})$ as quantified by the error $e(t)$. However, to obtain accurate estimations, more training data will be needed.

Example 9.3 This example demonstrates using the standard deviation of the input data to find GRNN smoothing parameter. Considering a GRNN with the following input/output samples (Table 9.4)

(continued)

Example 9.3 (continued)

Find a good value of σ and test the accuracy of the network for $x = 6$. Calculating the standard deviation of the input data ($std(\boldsymbol{x})$) yields 3.162. So $\sigma = 3.162$ would be a good choice. Using the training method in Sect. 9.3.1, the centers are assigned as $c_1 = 1, c_2 = 3, c_3 = 5, c_4 = 7, c_5 = 9$ and the output weight vector are $\boldsymbol{w} = \begin{bmatrix} 1 \\ 9 \\ 25 \\ 49 \\ 81 \end{bmatrix}$. When $x = 6$, the activation vector is calculated as $\boldsymbol{\phi}(x) = \begin{bmatrix} \exp(-\frac{\|6-1\|^2}{2\times 3.1620^2}) \\ \exp(-\frac{\|6-3\|^2}{2\times 3.1620^2}) \\ \exp(-\frac{\|6-5\|^2}{2\times 3.1620^2}) \\ \exp(-\frac{\|6-7\|^2}{2\times 3.1620^2}) \\ \exp(-\frac{\|6-9\|^2}{2\times 3.1620^2}) \end{bmatrix} = \begin{bmatrix} 0.286 \\ 0.638 \\ 0.951 \\ 0.951 \\ 0.638 \end{bmatrix}$

The summation of the activation vector is $\sum_{i=1}^{5} \phi_i(x) = 3.464$.

The network output $\hat{y}$ is calculated as

$$\hat{y}(x) = \frac{\begin{bmatrix} 1\ 9\ 25\ 49\ 81 \end{bmatrix} \begin{bmatrix} 0.286 \\ 0.638 \\ 0.951 \\ 0.951 \\ 0.638 \end{bmatrix}}{3.464} = 36.970$$

The actual output $y = 36$ for an input $x = 6$. The estimation error is $e(x) = y(x) - \hat{y}(x) = 36.000 - 36.970 = -0.970$ which is low given the network is trained with only five samples.

Table 9.3 GRNN example results

σ	$\hat{y}(\boldsymbol{x})$	$e(\boldsymbol{x})$
1	21.514	−3.014
0.5	18.525	0.248
0.1	18	0.500

Table 9.4 Training data

x	y
1	1
3	9
5	25
7	49
9	81

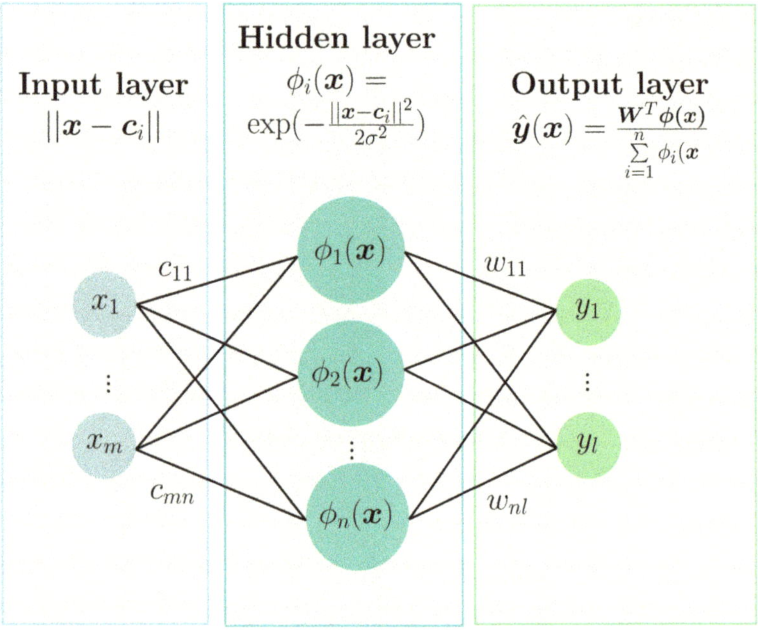

Fig. 9.6 GRNN layers

9.3.2 Layers of GRNN

After the training phase of GRNN, its output can be calculated using (9.4). GRNN layers are demonstrated in Fig. 9.6 and detailed here as:

- **The input layer**: Given GRNN has n basis function and each function has a center $\boldsymbol{c}_i$, $i = 1, \ldots, n$. The norm of the difference between the input $\boldsymbol{x}$ and the activation function center $\boldsymbol{c}_i$ is calculated in this layer as $\|\boldsymbol{x} - \boldsymbol{c}_i\|$. This norm is equivalent to the Euclidean distance between $\boldsymbol{x}$ and the center of the activation function $\boldsymbol{c}_i$. Thus, $\|\boldsymbol{x} - \boldsymbol{c}\| = \sqrt{(x_1 - c_{1,1})^2 + \cdots + (x_m - c_{m,n})^2}$. The aim of the distance calculation is to determine which centers of activation functions are closer to the input $\boldsymbol{x}$.
- **The hidden layer**: Gaussian activation function is applied to the norm calculated from the input layer $\phi_i(\boldsymbol{x}) = \exp(-\frac{\|\boldsymbol{x}-\boldsymbol{c}_i\|^2}{2\sigma^2})$, $i = 1, \ldots, n$. In this layer, the

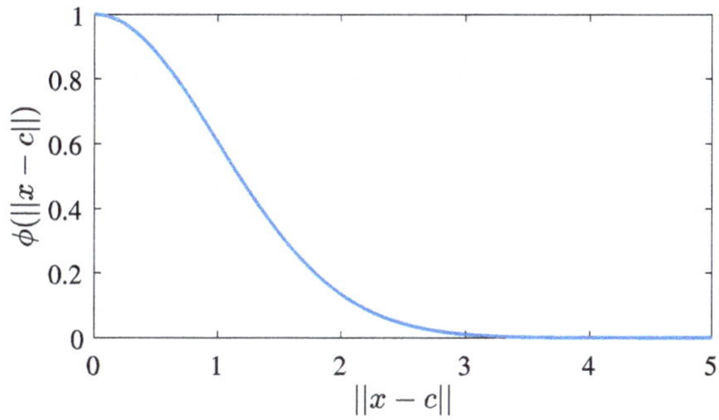

Fig. 9.7 Activation firing strength $\phi_i(\boldsymbol{x})$vs norm $||\boldsymbol{x} - \boldsymbol{c}_i||$

firing strength of the activation function $\phi(\cdot)$ is high when the function center is closer to the input $\boldsymbol{x}$ shown in Fig. 9.7. This figure explains that activation firing strength is inversely proportional to the distance between the input and activation functions centers. This layer is pivotal to associate the input data with the most similar basis functions.

- **The output layer**: The output of GRNN $\boldsymbol{y}(\boldsymbol{x})$ is the product of the output weights matrix transpose $\boldsymbol{W}^T$ by the activation vector $\boldsymbol{\phi}(\boldsymbol{x})$, and then this product is normalized by the summation of the activation vector $\sum_{i=1}^{n} \phi_i$ as follows: $\boldsymbol{y}(\boldsymbol{x}) = \frac{\boldsymbol{W}^T \boldsymbol{\phi}(\boldsymbol{x})}{\sum_{i=1}^{n} \phi_i(\boldsymbol{x})}$. Based on the above calculation, basis functions with large firing strengths impact GRNN output more significantly than the ones with small activation strengths.

9.3.3 *Merits and Shortcomings of GRNNs*

The primary merit of GRNN is the simple one pass non-gradient training. Its training encompasses a selection of only one hyperparameter, the smoothing parameter (σ), and thus, its training duration is significantly shorter than other FFNNs such as RBFNNs. In addition, it produces acceptable predictions even when trained with a small portion of the dataset. However, GRNN has some shortcomings that need to be dealt with before being utilized in dynamic systems applications. For instance, the size of the activation vector $\boldsymbol{\phi}(\boldsymbol{x})$ in GRNN might drastically increase during training with large datasets; this is because in GRNN training, it will create a basis function for every unique data sample. This would increase the time required to calculate its output after training. For example, when training GRNN with a dataset composed of 1000 unique data samples, the hidden layer size will grow to 1000 which is typically a large number that could potentially lead to the problem of overfitting. Overfitting happens when a model learns very well in the training phase but fails to generalize for new unseen data.

Moreover, there is no standard guide for selecting the smoothing parameter σ of GRNN in the one-pass training method. In addition, utilizing GRNN in online learning tasks such as adaptive control is still uncovered because it is designed for offline learning tasks. Data flow continuously in online systems, and thus, the size of the GRNN hidden layer evolves, which could consequently result in increased memory and computational costs. Finally, in unsupervised learning tasks, like control of dynamic systems, the training targets are missing, and thus, the offline supervised learning methods are deemed not useful. Therefore, online GRNN is to be developed to deal with online learning tasks.

9.3.4 *Online and Offline Learning in GRNN*

Understanding the difference between online and offline learning GRNN is essential for designing adaptive or learning-based controllers. Offline learning or batch learning is the standard approach for training GRNN from a given complete input/output dataset. This method is suitable to build an approximate model of a dynamic system. On the other hand, in the online or incremental learning, we do not have a stored dataset, but the network is updated once the data arrive. This approach is sometimes referred to as learning-on-the fly. Online learning fits adaptive control system design purposes more than offline learning. The output and parameters of an offline trained GRNN are static, while the output of an online trained GRNN is dynamic, i.e., it varies with time. Comparison between the offline and online learning methods used in GRNN training is depicted in Table 9.5

For example, to design a control system of a quadcopter using offline learning approach, one needs to collect datasets of the various variables of the system, such as the velocity, position, attitude, and altitude, and throttle commands which is time-consuming and prone to errors such as measurement errors, data transmission issues, storage issues, etc. and then train a GRNN to learn the mapping between the states and the control commands. In contrast, in the online learning framework, these offline data are not necessary and can be avoided if GRNN is designed to perform online weight updates, usually using adaptation laws and utilizing feedback in the form of error, such as the tracking error.

Table 9.5 Comparison of offline and online learning for GRNN

Aspect	Offline learning	Online learning
Training mode	Once on the complete dataset	Incrementally once the data arrive
Data access	Full dataset available for training	Data arrive continuously
Applications	Modeling dynamic systems.	Adaptive and learning-based control
Network output and parameters	Static after training	Dynamic

Example 9.4 This example illustrates the difference between online and offline learning in GRNN. GRNN is trained offline to predict the function $y(x_k) = \sin(x_k)$, $x_k = 0.3k$, $k = 0, 1, \ldots, 20$. Taking $x_k = \pi/2$ as test point. The actual output $y(\pi/2) = 1$.

The output of the offline trained GRNN is $\hat{y}(\pi/2) = 0.995$. On the other hand, the output of the online trained GRNN varies when a new samples arrive as depicted in Fig. 9.8. It starts around zero and then reaches to a steady state value of almost 1.

In the practical dynamical systems, the data flow sequentially; hence, for modeling and controlling such systems in real-time, online learning of GRNN should be considered. Based on that, GRNN weights should be adjusted once new data arrive to achieve a certain objective function, such as minimizing the modeling error or reducing the states' deviation from the reference states in the control problem.

9.3.5 GRNN Stability Analysis

Definition 9.1 Bounded-Input Bounded-Output (BIBO) Stability: A given system is BIBO stable if for every bounded input $x \in \Re$, the system produces a bounded output $y \in \Re$.

To examine the stability of GRNN according to the BIBO stability, let us consider the following:

1. GRNN hidden layer outputs for a bounded input vector $\boldsymbol{x} \in \Re^m$ are defined as $\phi_i(\boldsymbol{x}) = \exp(-\frac{\|\boldsymbol{x}-\boldsymbol{c}_i\|^2}{2\sigma^2})$, $i = 1, \ldots, n$,. The elements of the vector $\boldsymbol{\phi}(\boldsymbol{x})$ are

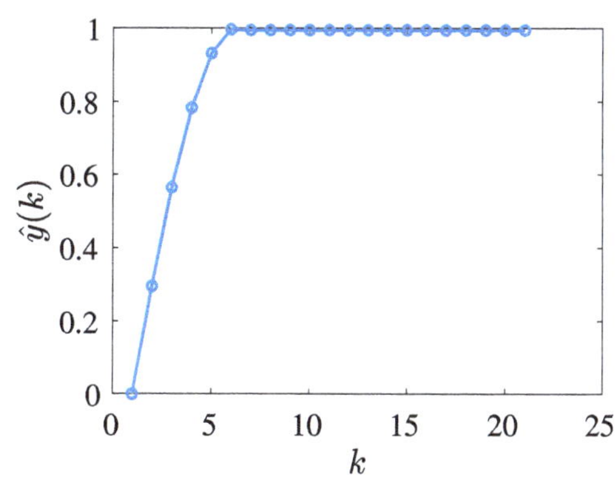

Fig. 9.8 Online trained GRNN output variation with the samples number

bounded $\in [0, 1]$ since Gaussian activation functions are also bounded $\in [0, 1]$. Thus, for GRNN activation vector of n elements the activation vector is bounded as $||\boldsymbol{\phi}(\boldsymbol{x})|| \leq n$, which leads to the boundness of GRNN hidden layer.

2. The output of GRNN for a bounded output weights matrix $\boldsymbol{W}$ is expressed as $\hat{y(\boldsymbol{x})} = \frac{\boldsymbol{W}^T \boldsymbol{\phi}(\boldsymbol{x})}{\sum_{i=1}^{n} \phi_i(\boldsymbol{x})}$. Since both $\boldsymbol{\phi}(\boldsymbol{x})$ and $\boldsymbol{W}$ are bounded, and $\sum_{i=1}^{n} \phi_i$ is used to normalize the activation vector elements between 0 and 1, $\hat{\boldsymbol{y}}(\boldsymbol{x})$ is also bounded as $\min\{\boldsymbol{W}^T \frac{\boldsymbol{\phi}(\boldsymbol{x})}{n}\} \leq \hat{y}(\boldsymbol{x}) \leq \max\{\boldsymbol{W}^T \frac{\boldsymbol{\phi}(\boldsymbol{x})}{n}\}$.

Based on the above analysis, if a bounded input $\boldsymbol{x}$ is applied to GRNN, its output $\hat{\boldsymbol{y}}(\boldsymbol{x})$ is also bounded, and thus, GRNN is BIBO stable.

Example 9.5 GRNN with ten hidden neurons given $\boldsymbol{\phi}(\boldsymbol{x}) = \begin{bmatrix} 0.1 \leq \phi_1(\boldsymbol{x}) \leq 1 \\ \vdots \\ 0.1 \leq \phi_1 0(\boldsymbol{x}) \leq 1 \end{bmatrix}$, the output weights matrix are bounded as

$$\boldsymbol{W} = \begin{bmatrix} -10 \leq w_{1,1} \leq 10 & -10 \leq w_{1,2} \leq 10 \\ \vdots & \vdots \\ -10 \leq w_{10,1} \leq 10 & -10 \leq w_{10,2} \leq 10 \end{bmatrix}$$

Find the minimum and the maximum of the GRNN output $\hat{\boldsymbol{y}}(\boldsymbol{x})$. The summation of the activation vector bounded as is $0.1 \times 10 \leq \sum_{i=1}^{10} \phi_i(\boldsymbol{x}) \leq 10 \times 1$, and thus, the minimum of GRNN output is $\hat{y}(\boldsymbol{x}) = \frac{0.1 \times -10 + \cdots + 0.1 \times -10}{1} = -10$ and the maximum of GRNN output is

$\hat{y}(\boldsymbol{x}) = \frac{1 \times 10 + \cdots + 1 \times 10}{10} = 10$

9.3.6 *Potential Singularities in GRNN*

To eliminate any potential singularities in GRNN when utilized in approximation tasks, the following should be considered:

1. Avoiding feeding NaN (not a number) in the input/output dataset of GRNN as it will lead to NaN outputs.
2. Avoiding division by zero when calculating GRNN hidden layer outputs by not allowing σ to be selected as zero.
3. Adding a small number τ such as 0.1 to the numerator of the GRNN output equation to avoid division by zero as follows: $\hat{\boldsymbol{y}} = \frac{\boldsymbol{W}^T \boldsymbol{\phi}}{(\sum_{i=1}^{n} \phi_i) + \tau}$

Online learning is more prominent in dynamic systems particularly in adaptive control systems. Since online learning incorporates learning from the incoming data stream, thus, network structure might be become redundant when incoming data distribution varies over time. To handle such issue, simultaneous parameters and structure, learning is considered. This leads to the concept of self-evolving neural networks (SENNs).

9.4 Self-Evolving Neural Networks (SENNs)

To further elaborate the learning abilities of NNs, self-evolving neural networks (SENNs) are considered as an extension to the typical adaptive NNs. They are characterized by multi-level adaptability, including structural and parametric adjustments. These systems begin learning with a simple structure and a few parameters which will be modified later according to the new data received by these models (Angelov 2002). The learning in SENNs will proceed continually as new data are received. Learning both the structure and parameters is beneficial when dealing with dynamic systems since it overcomes the issue of NNs structure design and provides higher degrees of uncertainty tolerance than the classical adaptive systems.

Definition 9.2 Node: Every neuron in the hidden layer with connections to the input and output layers is called hidden layer node. For simplification purpose, hidden layer node term is replaced with node.

SENNs have three major parts: addition of nodes, removal of nodes, and parameter adaptation (Lughofer 2011). Node addition and removal are demonstrated in Figs. 9.9 and 9.10. We are limiting the scope of the chapter here and after for self-evolving GRNNs (SEGRNNs) (Al-Mahasneh et al. 2020).

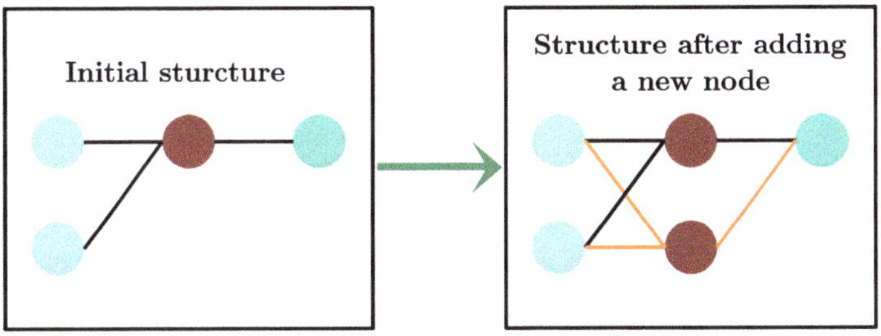

Fig. 9.9 Node addition in SENNs

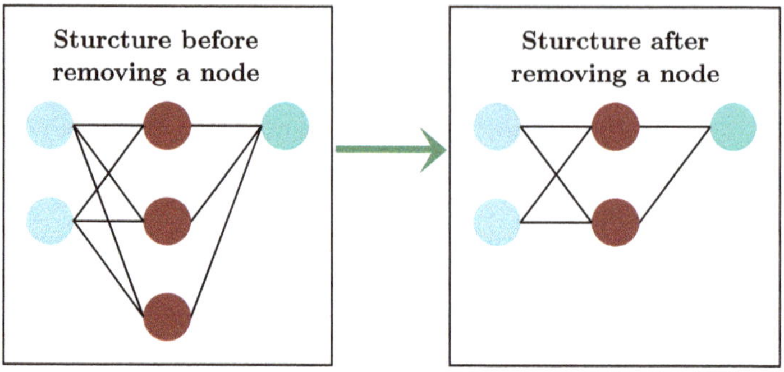

Fig. 9.10 Node removal in SENNs

9.4.1 Nodes Addition in SENNs

Node addition in SENNs aims to learn the new knowledge introduced by the recently received data and to reduce the redundancy in the network structure to the minimum. To determine if a new node is to be added to the network or not, the similarity score between the new potential node and the current nodes is calculated. Once this similarity is found low, a new node will be added. Considering a GRNN with an input vector $x \in \Re^m$ and n hidden nodes with Gaussian activation functions centered at $\boldsymbol{C} = \begin{bmatrix} \boldsymbol{c}_1 \\ \vdots \\ \boldsymbol{c}_n \end{bmatrix} \in \Re^{m \times n}$, one of the simplest methods to calculate the similarity between the center of each node and the input $\boldsymbol{x}$ is using distance calculation as

$$d_i(x) = \text{dist}(\boldsymbol{c}_i, \boldsymbol{x}) \tag{9.5}$$

$$= \sqrt{(c_{1,1} - x_1)^2 + \cdots + (c_{n,m} - x_m)^2},\ i = 1, \ldots, n \tag{9.6}$$

Algorithm 9.1 describes the nodes addition in SENNs.

Other approaches to quantify the similarity include density estimation (Pratama et al. 2016), mutual information (Lin et al. 2015) and entropy (Paeedeh et al. 2024). However, these methods require more computations than the distance-based similarity.

Algorithm 9.1 Nodes addition in SEGRNNs

Require: $\boldsymbol{C}, d_{th}, \boldsymbol{x}$
  for Every center $\boldsymbol{c}_i \in \boldsymbol{C}$ **do**
    Find $d_i(\boldsymbol{x}) = \text{dist}(\boldsymbol{c}_i, \boldsymbol{x})$
  end for
  Form the vector $\boldsymbol{d}(\boldsymbol{x}) = [d_1(\boldsymbol{x}), \ldots, d_n(\boldsymbol{x})]$
  if $\min\{\boldsymbol{d}(\boldsymbol{x})\} > d_{th}$ **then**
    Add a new node
  else
    No node will be added
  end if

Example 9.6 This example demonstrates how SEGRNN adds nodes to the hidden layer. Consider SEGRNN with three centers $\boldsymbol{C} = \begin{bmatrix} -1 & 1 \\ 1 & 1 \\ 1 & -1 \end{bmatrix}$, using the Euclidean distance similarity with a distance threshold $d_{th} = 0.5$, determine whether a new node needs to be added to SEGRNN for the following inputs: $\boldsymbol{x} = [0.8, -0.8]$, $\boldsymbol{x} = [0.8, 0.8]$, $\boldsymbol{x} = [-1.6, 0.8]$.

- For $\boldsymbol{x} = [0.8, -0.8]$:
  Calculating the distance vector as
  $\boldsymbol{d} = [\sqrt{(-1-0.8)^2 + (1+0.8)^2}, \sqrt{(1-0.8)^2 + (1+0.8)^2},$
  $\sqrt{(1-0.8)^2 + (-1+0.8)^2}] = [2.550, 1.811, 0.283]$ since $\min\{d\} < d_{th}$
  no node addition will take place.
- For $\boldsymbol{x} = [0.8, 0.8]$:
  Calculating the distance vector as
  $\boldsymbol{d} = [\sqrt{(-1-0.8)^2 + (1-0.8)^2}, \sqrt{(1-0.8)^2 + (1-0.8)^2},$
  $\sqrt{(1-0.8)^2 + (-1-0.8)^2}] = [1.811, 0.283, 1.811]$ since $\min\{d\} < d_{th}$
  no node addition will take place.
- For $\boldsymbol{x} = [-1.6, 0.8]$:
  Calculating the distance vector as
  $\boldsymbol{d} = [\sqrt{(-1+1.6)^2 + (1-0.8)^2}, \sqrt{(1+1.6)^2 + (1-0.8)^2},$
  $\sqrt{(1+1.6)^2 + (-1-0.8)^2}] = [0.633, 2.610, 3.162]$ since $\min\{d\} > d_{th}$
  a new node will be added.

9.4.2 Nodes Removal in SENNs

Node removal or pruning aims to maintain the overall system health by keeping only the best nodes and deleting the worst ones. One of the possible approaches to achieve this object is to calculate each node's contribution to the output of the hidden layer of the SEGRNN. Thus, nodes with the lowest contribution will be removed. This process is depicted in Algorithm 9.2.

Algorithm 9.2 Removing Nodes in SEGRNNs

Require: $\boldsymbol{C}, \phi_{th}, \sigma, \boldsymbol{x}$
  for Every center $\boldsymbol{c}_i \in \boldsymbol{C}$ **do**
    Find $\phi_i(\boldsymbol{x}) = \exp(-\frac{\|\boldsymbol{x}-\boldsymbol{c}_i\|^2}{2\sigma^2})$
    if $\phi_i(\boldsymbol{x}) < \phi_{th}$ **then**
      Remove node i
    else
      Do not remove i
    end if
  end for

Other possible methods to remove the nodes include their contribution to the overall network output (Nguyen et al. 2012), statistical contribution (Ferdaus et al. 2019), and distance-driven methods (Li et al. 2016).

Example 9.7 SEGRNN with $\boldsymbol{C} = \begin{bmatrix} -1 & 1 \\ 0 & 1 \\ 1 & 0 \\ 1 & -1 \end{bmatrix}$ $\sigma = 1$, $\phi_{th} = 0.2$. Determine whether a node will be removed or not for the following inputs $\boldsymbol{x} = [0, 0]$, $\boldsymbol{x} = [1, 1]$:

- For $\boldsymbol{x} = [0, 0]$:

$$\text{Calculating } \boldsymbol{\phi}(\boldsymbol{x}) = \begin{bmatrix} \phi_1(\boldsymbol{x}) = \exp(-\frac{\|\boldsymbol{x}-\boldsymbol{c}_1\|^2}{2\sigma^2}) = \exp(-\frac{\|[0,0]-[-1,1]\|^2}{2}) = 0.368 \\ \phi_2(\boldsymbol{x}) = \exp(-\frac{\|\boldsymbol{x}-\boldsymbol{c}_2\|^2}{2\sigma^2}) = \exp(-\frac{\|[0,0]-[0,1]\|^2}{2}) = 0.607 \\ \phi_3(\boldsymbol{x}) = \exp(-\frac{\|\boldsymbol{x}-\boldsymbol{c}_3\|^2}{2\sigma^2}) = \exp(-\frac{\|[0,0]-[1,0]\|^2}{2}) = 0.607 \\ \phi_4(\boldsymbol{x}) = \exp(-\frac{\|\boldsymbol{x}-\boldsymbol{c}_4\|^2}{2\sigma^2}) = \exp(-\frac{\|[0,0]-[1,-1]\|^2}{2}) = 0.368 \end{bmatrix}$$

  Since no nodes have $\phi(\boldsymbol{x}) < \phi_{th}$, no nodes will be removed.

- For $\boldsymbol{x} = [1, 1]$:

$$\text{Calculating } \boldsymbol{\phi}(\boldsymbol{x}) = \begin{bmatrix} \phi_1(\boldsymbol{x}) = \exp(-\frac{\|\boldsymbol{x}-\boldsymbol{c}_1\|^2}{2\sigma^2}) = \exp(-\frac{\|[1,1]-[-1,1]\|^2}{2}) = 0.135 \\ \phi_2(\boldsymbol{x}) = \exp(-\frac{\|\boldsymbol{x}-\boldsymbol{c}_2\|^2}{2\sigma^2}) = \exp(-\frac{\|[1,1]-[0,1]\|^2}{2}) = 0.607 \\ \phi_3(\boldsymbol{x}) = \exp(-\frac{\|\boldsymbol{x}-\boldsymbol{c}_3\|^2}{2\sigma^2}) = \exp(-\frac{\|[1,1]-[1,0]\|^2}{2}) = 0.607 \\ \phi_4(\boldsymbol{x}) = \exp(-\frac{\|\boldsymbol{x}-\boldsymbol{c}_4\|^2}{2\sigma^2}) = \exp(-\frac{\|[1,1]-[1,-1]\|^2}{2}) = 0.135 \end{bmatrix}$$

  Since node 1 and node 4 have $\phi(\boldsymbol{x}) < \phi_{th}$, these nodes will be removed.

9.4.3 Parameter Adaptation in SENNs

The last stage in SENNs operation is the parameters update. The goal of adapting the parameters in SENNs is to update their values according to the latest data arrival. Recursive update methods are usually used to achieve this. In SEGRNN-based control systems, which is a direct neural adaptive controller with evolving structure, the tracking error is used to update the network parameters. For example in an SEGRNN controller, the output weights $\boldsymbol{w}$ are updated recursively as follows:

$$\dot{w} = -\gamma e(t)\bar{\boldsymbol{\phi}}(\boldsymbol{x}(t)) \tag{9.7}$$

where γ is the adaptation gain, $e(t)$ is the tracking error, and $\bar{\boldsymbol{\phi}}(\boldsymbol{x}(t))$ is the normalized hidden layer output (hidden layer activation vector) written as

$$\bar{\boldsymbol{\phi}}(x) = \frac{\boldsymbol{\phi}(x)}{\sum_{i=1}^{n} \phi_i(x)}.$$

9.4.4 Stability Analysis of SEGRNN

For simplicity let us consider a single-input and single-output (SISO) nonlinear system as

$$\dot{x}(t) = f(x) + bu(t) \tag{9.8}$$

Consider an estimation of the control signal by SEGRNN as

$$\hat{u}(t) = \frac{\hat{\boldsymbol{w}}^T \boldsymbol{\phi}(x)}{\sum_{i=1}^{n} \phi_i(x)} \tag{9.9}$$

The above equation can be rewritten as

$$\hat{u}(t) = \hat{\boldsymbol{w}}^T \bar{\boldsymbol{\phi}}(x) \tag{9.10}$$

where $\bar{\boldsymbol{\phi}}(x) = \frac{\boldsymbol{\phi}(x)}{\sum_{i=1}^{n} \phi_i(x)}$

Assuming an ideal control that cancels out the nonlinearities and achieves asymptotic stability, i.e., drives the tracking error $e(t)$ to zero when $t = \infty$ as follows:

$$f(x) + bu^*(t) = \dot{x}_d(t) - ae(t) \tag{9.11}$$

where $a > 0$ is a feedback gain.

The ideal control signal can be written as

$$u^*(t) = \boldsymbol{w}^{*T}\bar{\boldsymbol{\phi}}(x) + \epsilon(x) \tag{9.12}$$

where $\epsilon(x)$ is a bounded approximation error as $|\epsilon(x)| \leq \bar{\epsilon}$ for all x.

The control error is defined as

$$\tilde{u}(t) = \hat{u}(t) - u^*(t) \tag{9.13}$$

$$= \hat{\boldsymbol{w}}^T\bar{\boldsymbol{\phi}}(x) - \boldsymbol{w}^{*T}\bar{\boldsymbol{\phi}}(x) - \epsilon(x) \tag{9.14}$$

Using the above equation, the weights error is written as

$$\tilde{\boldsymbol{w}} = \hat{\boldsymbol{w}} - \boldsymbol{w}^* \tag{9.15}$$

Defining the tracking error dynamics as

$$\begin{aligned}\dot{e}(t) &= \dot{x}(t) - \dot{x}_d(t)\\ &= -ae(t) + b(\hat{u}(t) - u(t)^*)\\ &= -ae(t) + b(\tilde{w}^T\bar{\boldsymbol{\phi}}(x) - \epsilon(x))\end{aligned} \tag{9.16}$$

Consider the following Lyapunov candidate function:

$$V = \frac{1}{2}e^2(t) + \frac{1}{2\gamma}\tilde{\boldsymbol{w}}^T\tilde{\boldsymbol{w}} \tag{9.17}$$

The time derivative of the Lyapunov function is written as

$$\dot{V} = \dot{e}(t)e(t) + \frac{1}{\gamma}\tilde{\boldsymbol{w}}^T\dot{\tilde{\boldsymbol{w}}} \tag{9.18}$$

Substituting $\dot{e}$ from (9.16) leads to

$$\dot{V} = e(t)(-ae(t) + b(\tilde{w}^T\bar{\boldsymbol{\phi}}(x) - \epsilon(x)) + \frac{1}{\gamma}\tilde{\boldsymbol{w}}^T\dot{\tilde{\boldsymbol{w}}} \tag{9.19}$$

Simplifying (9.19) leads to

$$\dot{V} = -ae^2(t) + b\tilde{w}^T\bar{\boldsymbol{\phi}}(x) - be(t)\epsilon(x) + \frac{1}{\gamma}\tilde{\boldsymbol{w}}^T\dot{\tilde{\boldsymbol{w}}} \tag{9.20}$$

Choosing the weight adaptation law as

$$\dot{\hat{\boldsymbol{w}}} = -\gamma\bar{\boldsymbol{\phi}}(x)be(t) \tag{9.21}$$

$\dot{V}$ becomes

$$\dot{V} = -ae^2(t) + be(t)\tilde{\boldsymbol{w}}^T\bar{\boldsymbol{\phi}}(x) - be(t)\epsilon(x) - be(t)\tilde{\boldsymbol{w}}^T\bar{\boldsymbol{\phi}}(x) \tag{9.22}$$

Simplifying (9.22) leads to

$$\dot{V} = -ae^2(t) - be(t)\epsilon(x) \tag{9.23}$$

Assuming $|b| \leq \bar{b}, |\epsilon(x)| \leq \bar{\epsilon}$, then $|be\epsilon(x)| \leq \bar{b}\bar{\epsilon}|e(t)|$, hence, $\dot{V}$ can be rewritten as

$$\dot{V} = -ae^2(t) - \bar{b}\bar{\epsilon}|e(t)| \tag{9.24}$$

Rewriting (9.24) as

$$\dot{V} \leq -a(e^2(t) - \frac{\bar{b}\bar{\epsilon}}{a}|e(t)|) \tag{9.25}$$

Completing the squares leads to

$$\dot{V} \leq -a\left(|e(t)| - \frac{\bar{b}\bar{\epsilon}}{2a}\right)^2 + \frac{(\bar{b}\bar{\epsilon})^2}{4a} \tag{9.26}$$

This leads to the ultimate boundness of the tracking error $e(t)$ as

$$\limsup_{t\to\infty} |e(t)| \leq \frac{\bar{b}\bar{\epsilon}}{2a} \tag{9.27}$$

where lim sup is the limit superior or the upper limit and the constant $\frac{\bar{b}\bar{\epsilon}}{2a}$ is a radius of a ball around the zero where the tracking error settles in.

Example 9.8 Let us consider using SEGRNN controller to control a nonlinear dynamic system with external disturbances. The system dynamics are given as

$$\dot{x}_1(t) = x_2(t)$$
$$\dot{x}_2(t) = -x_1(t)x_2(t) + u(t) + d(t)$$
$$y(t) = x_2(t)$$

(continued)

Example 9.8 (continued)
The aim of the control is to stabilize the output to zero; thus, $y_d(t) = 0$, and the external disturbance is given as $d(t) = 0.1\cos(t)$. The adaptation gain γ is selected as 1, d_{th} is 0.5, and ϕ_{th} is 0.1.

The SEGRNN estimated control law is designed as

$$\hat{u}(t) = \hat{\boldsymbol{w}}^T \phi(\boldsymbol{x}) \tag{9.28}$$

Defining the output tracking error $e(t)$ as

$$e(t) = y_d(t) - y(t) \tag{9.29}$$

The adaptation law is chosen as

$$\dot{\boldsymbol{w}} = -\gamma e(t)\boldsymbol{\phi}(\boldsymbol{x}(t)) \tag{9.30}$$

The output tracking results as depicted in Fig. 9.11 reveal that initial oscillations take place since the controller is learning the structure and parameter in this phase. After these oscillations, the output $y(t)$ accurately follows the desired output $y_d(t)$.

In addition, as the results depict in Fig. 9.12, the hidden layer size of SEGRNN will grow to five nodes, then shrinks to four, and stay fixed to four after that with no nodes addition, and removal takes place beyond that. Also, the norm of the SEGRNN output weights is bounded to almost 27; it starts from zero since there is only one node initially in the network and then stays fixed around 27. SEGRNN takes almost 2 seconds to learn the proper structure and parameters.

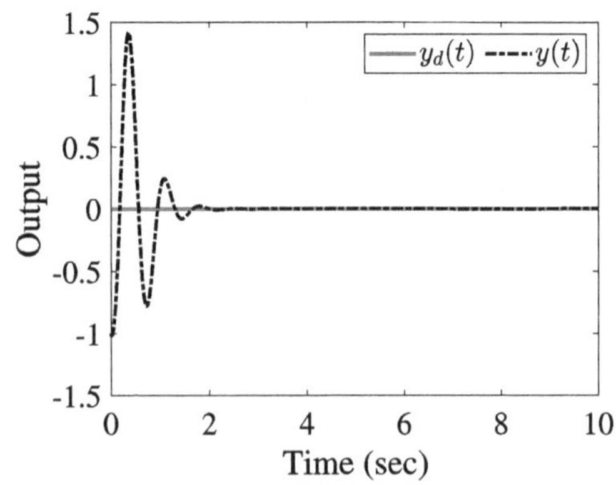

Fig. 9.11 Output history

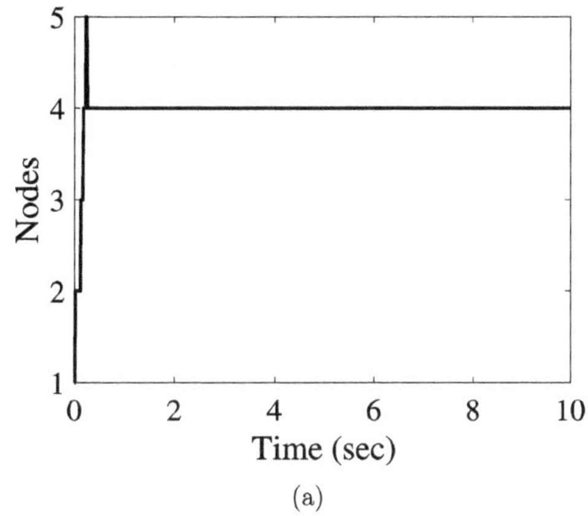

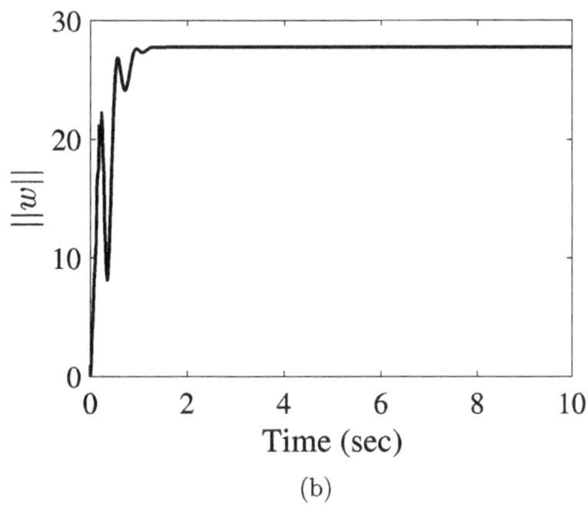

Fig. 9.12 (**a**)SEGRNN nodes evolution (**b**) Evolution of the Norm of SEGRNN output weights—Example 9.8

9.5 Summary

This chapter lays the foundation of neural networks while paying close attention to GRNNs including their training process, layers, mathematical background, properties, convergence, and stability analyses. It also introduce evolving systems properties, structure evolution, and parameter adaptation.

9.6 Exercise Problems

1. GRNN was trained with the following (input, output) data samples (1,1), (3,27), and (5,125). Find the output of trained GRNN for inputs 1.5, 3.5, 4.5. Use σ=1 and repeat the solution when σ=0.5 and σ=0.1. Which σ value of the three used above will produce the best approximation?

2. GRNN was trained with the following (input, target) data samples: (1,2.72), (3,20.085), and (5,148.413). Find the output of trained GRNN for inputs 2, 2.5, 3.5. Use σ=1. What is the upper threshold of σ for which the approximation will not change after increasing σ value.
3. Model the following functions using GRNN: $y = x^2$, $y = e^x$, $y = \sin(x)\cos(x)$ on the domain $x \in [-100, 100]$.
4. SEGRNN with empty centers and output weights. Determine whether a node will be added to SEGRNN for the inputs $\boldsymbol{x} = [1, 3, 4, 5, 6, 7]$. Use Euclidean distance similarity with $d_{th} = 0.01$ and $\sigma = 1$
5. SEGRNN with centers $\boldsymbol{c} = [-1, -0.5, 0, 0.5, 1]$ and output weights $\boldsymbol{w} = [1, 0.25, 0, 0.25, 1]$. Determine whether a node will be removed for the inputs $\boldsymbol{x} = [-0.75, 0.2, 0.75]$. Use firing strength threshold $\phi_{th} = 0.1$ and $\sigma = 1$
6. Which one of the following σ values could potentially lead to a singularity in GRNN output computation $\boldsymbol{\sigma} = [0.01, 0.0001, 0]$.
7. Write a Matlab code to train GRNN to model $\sin(x)$ for $x \in [-\pi, \pi]$ and change the values of σ from 0.01 to 1 and plot the results for each σ value.
8. Write a Matlab code for SEGRNN using Euclidean distance and test it with $y = \sin(x)\cos(x)$ function on $x \in [-1, 1]$. use $d_{th} = 0.5$, $\phi_{th} = 0.1$.

References

A.J. Al-Mahasneh, S.G. Anavatti, M.A. Garratt, Self-evolving neural control for a class of nonlinear discrete-time dynamic systems with unknown dynamics and unknown disturbances. IEEE Trans. Industr. Inform. **16**(10), 6518–6529 (2020)

P.P. Angelov, *On-Line Identification of Flexible TSK-Type Models* (Physica-Verlag HD, Heidelberg, 2002), pp. 87–109

M.M. Ferdaus, M. Pratama, S. Anavatti, M.A. Garratt, Y. Pan, Generic evolving self-organizing neuro-fuzzy control of bio-inspired unmanned aerial vehicles, in *IEEE Transactions on Fuzzy Systems* (2019), pp. 1–1

S. Ge, C. Hang, T. Lee, T. Zhang, *Stable Adaptive Neural Network Control*, ser. The International Series on Asian Studies in Computer and Information Science (Springer, London, 2013)

J. Li, Q. Xiong, K. Wang, X. Shi, S. Liang, A recurrent self-evolving fuzzy neural network predictive control for microwave drying process. Dry. Technol. **34**(12), 1434–1444 (2016)

C. Lin, N.R. Pal, S. Wu, Y. Liu, Y. Lin, An interval type-2 neural fuzzy system for online system identification and feature elimination. IEEE Trans. Neural Networks Learn. Syst. **26**(7), 1442–1455 (2015)

E. Lughofer, *EFS Approaches for Regression and Classification* (Springer, Berlin, 2011), pp. 93–164

N.N. Nguyen, C. Quek, E. Y. Cheu, Traffic prediction using a generic self-evolving takagi-sugeno-kang (GSETSK) fuzzy neural network, in *The 2012 International Joint Conference on Neural Networks (IJCNN)* (2012), pp. 1–7

N. Paeedeh, M. Pratama, M.A. Ma'sum, W. Mayer, Z. Cao, R. Kowlczyk, Cross-domain few-shot learning via adaptive transformer networks. Knowl.-Based Syst. **288**, 111458 (2024)

M. Pratama, E. Lughofer, M.J. Er, W. Rahayu, T. Dillon, Evolving type-2 recurrent fuzzy neural network, in *2016 International Joint Conference on Neural Networks (IJCNN)* (2016), pp. 1841–1848

D.F. Specht, A general regression neural network. IEEE Trans. Neural Netw. **2**(6), 568–576 (1991)

H. Yu, T. Xie, S. Paszczyñski, B.M. Wilamowski, Advantages of radial basis function networks for dynamic system design. IEEE Trans. Ind. Electron. **58**(12), 5438–5450 (2011)

Chapter 10
Adaptive Control Based on General Regression Neural Networks (GRNN)

10.1 Introduction

Data-driven controllers might be a good alternative to model-based controllers. In this chapter and the next chapter, neural data-driven controllers are thoroughly studied paying attention to their design, adaptation, and utilization in the control of nonlinear systems. NN controllers can be used in direct adaptive control, i.e., NN provides a direct control signal to the plant. GRNN-based adaptive controllers can be used to approximate the control signal to be sent to the plant. The philosophy of GRNN-based adaptive control as depicted in Fig. 10.1 consists of three main stages: designing an objective function j, which represents a cost function to be minimized by suitable control actions, deriving weights' update laws based on the objective function, and calculating the control signal based on the updated weights. The design of the objective function is problem-dependent. For example, in the output/states tracking problem, the network weights are iteratively adapted in time using gradient descent (GD), i.e., taking steps toward minimizing the gradient of the objective function, which is typically a function of the tracking error. The gradient of the objective function is used to update the network weights $\boldsymbol{W}$. These weight updates can be formulated as

$$\dot{\boldsymbol{W}} = -\gamma \frac{\partial j}{\partial \boldsymbol{W}} \tag{10.1}$$

where $\boldsymbol{W}$ is the network weight matrix, γ is called the adaptation gain, and j is the objective function. After achieving the weights update, the new control signal will be calculated.

The next sections will discuss two direct adaptive GRNN-based controllers: the improved GRNN (I_GRNN) and the hybrid GRNN and sliding mode control (GRNNSMC) controllers.

K. Sachan et al., *Intelligent Adaptive Control with System Constraints*,
https://doi.org/10.1007/978-981-95-8220-4_10

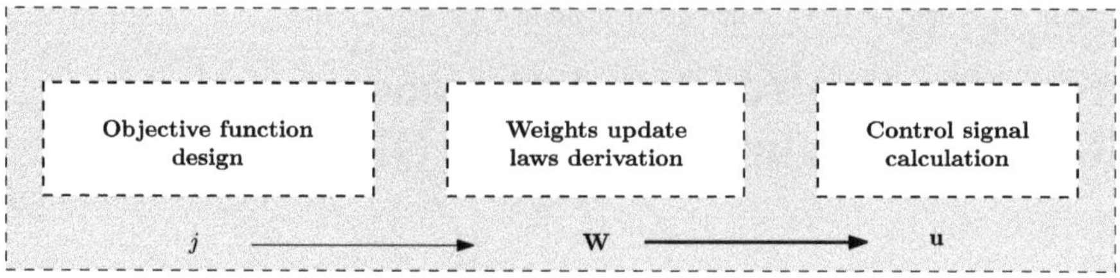

Fig. 10.1 Philosophy of GRNN-based adaptive control

10.2 Improved GRNN-Based Adaptive Control

I_GRNN[1] is an improved version of the original GRNN which has better potential than the original one in the adaptive control tasks (Al-Mahasneh and Anavatti 2023). I_GRNN is enhanced in the following ways:

1. The statistical means are used to calculate the input-hidden weights of the network. This will automatically adapt the input-hidden weights of I_GRNN and avoid poor initialization of these weights.
2. Another layer is added to the network with direct connection weights from the inputs to the new layer. This layer and its adaptation will improve the control accuracy.
3. The smoothing parameter of GRNN is identified as a range of numbers rather than a single number. Since selecting the GRNN smoothing parameter is a major issue and highly affects its accuracy, using a range of numbers would eliminate trying different values or using heuristic approaches to select it.

10.2.1 I_GRNN Layers Details

Detailed explanations of the I_GRNN layers are provided in this subsection. I_GRNN structure is shown in Fig. 10.2.

The I_GRNN is made of the following layers:

1. **Input layer**: The inputs are handled in this layer. The number of the network inputs is m, and it has h neurons in the first hidden layer, n is the number of the network outputs, and l is the length of the discrete range of the smoothing parameter σ. The hidden weights matrix $\boldsymbol{P} \in \Re^{m\times h}$ connects the inputs vector $\boldsymbol{x} \in \Re^{m}$ with the neurons in the first layer. It is made of the statistical means of

[1] Some contents of the current Sect. 10.2 are used from Publication "Novel General Regression Neural Networks for Improving Control Accuracy of Nonlinear MIMO Discrete-Time Systems," IEEE Transactions on Cybernetics, vol. 51, pp. 6122–6132, April, 2023, with permission from IEEE (License Number: 6160361390672).

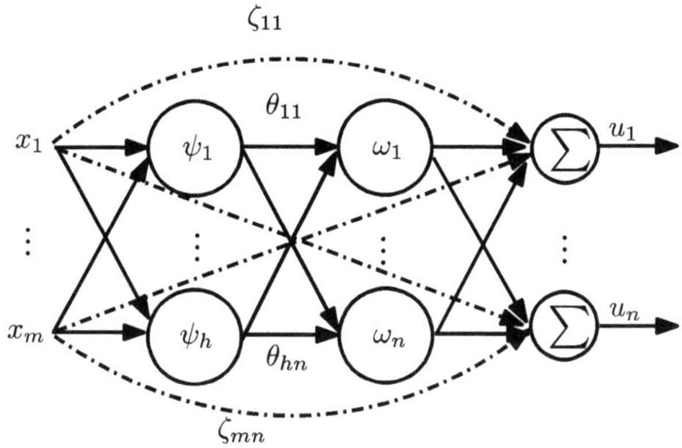

Fig. 10.2 I_GRNN structure, dashed-dotted lines indicate the new input-output weights ζ

the networks inputs $\boldsymbol{x}(k)$ updated recursively at each sample k as follows:

$$\boldsymbol{P} = \begin{bmatrix} \mu_{11}(k) \ \dots \ \mu_{1h}(k) \\ \vdots \\ \mu_{m1}(k) \ .. \ \mu_{mh}(k) \end{bmatrix} \tag{10.2}$$

The elements of the matrix $\boldsymbol{P}$ are adapted recursively once a new input is received as

$$\mu_{ij}(k) = \frac{(k-1)}{k}\mu_{ij}(k-1) + \frac{1}{k}x_i(k) \tag{10.3}$$

$i = 1, \dots, m$, i is an index for each network input. $j = 1, \dots, h$ j is an index for each hidden layer node.

Example 10.1 I_GRNN with two inputs x_1 and x_1 and five neurons in the first layer. If the inputs at the first sample $k = 1$ are $\boldsymbol{x}(1) = \begin{bmatrix} x_1 = 1 \\ x_2 = 0 \end{bmatrix}$ and they varied at the second sample $k = 2$ to $\boldsymbol{x}(2) = \begin{bmatrix} x_1 = 2 \\ x_2 = 1 \end{bmatrix}$, find the hidden weights matrix $\boldsymbol{P}$.

Since $\boldsymbol{P} \in \Re^{2\times 5}$ is made of the input means updated recursively, the $\boldsymbol{P}$ is at sample $k = 1$ is constructed as

$$\boldsymbol{P} = \begin{bmatrix} 1\ 1\ 1\ 1\ 1 \\ 0\ 0\ 0\ 0\ 0 \end{bmatrix}$$

(continued)

Example 10.1 (continued)

At sample $k = 2$, the matrix $\boldsymbol{P}$ is updated using as (10.3), as $\mu_{11}(2) = \frac{(2-1)}{2}(1)+\frac{1}{2}(2) = 0.5+1 = 1.5$, $\mu_{21}(2) = \frac{(2-1)}{2}(0)+\frac{1}{2}(1) = 0+0.5 = 0.5$; thus, the matrix $\boldsymbol{P}$ becomes after the update at sample $k = 2$

$$\boldsymbol{P} = \begin{bmatrix} 1.5 & 1.5 & 1.5 & 1.5 & 1.5 \\ 0.5 & 0.5 & 0.5 & 0.5 & 0.5 \end{bmatrix}.$$

The Euclidean distance vector $\boldsymbol{q} \in \Re^{1\times h}$ between the current inputs vector $\boldsymbol{x}(k) \in \Re^{m}$ and the hidden weight matrix $\boldsymbol{P}$ is computed as

$$\boldsymbol{q}(k) = \mathrm{dist}(\mathrm{P}, \boldsymbol{x}(k)) \tag{10.4}$$

where dist is the Euclidean distance function calculated as

$$\boldsymbol{q}(k) = \left\| \boldsymbol{p}_i - \boldsymbol{x}(k) \right\| \tag{10.5}$$

$i = 1, \ldots, h$. The size of distance vector $\boldsymbol{q}(k)$ depends on the number of neurons in the first hidden layer h.

Example 10.2 I_GRNN with hidden weights matrix given as $\boldsymbol{P} = \begin{bmatrix} 1.5 & 1.5 & 1.5 \\ 0.5 & 0.5 & 0.5 \end{bmatrix}$, if the input is given as $\boldsymbol{x}(k) = \begin{bmatrix} 3 \\ 4 \end{bmatrix}$, find the distance vector $\boldsymbol{q}(k)$.

Using (10.5), the distance vector $\boldsymbol{q}(k)$ is found as

$$\begin{aligned} \boldsymbol{q}(k) &= \left\| \boldsymbol{p}_i - \boldsymbol{x}(k) \right\| \\ &= \left[\left\| \begin{bmatrix} 1.5 \\ 0.5 \end{bmatrix} - \begin{bmatrix} 3 \\ 4 \end{bmatrix} \right\| \left\| \begin{bmatrix} 1.5 \\ 0.5 \end{bmatrix} - \begin{bmatrix} 3 \\ 4 \end{bmatrix} \right\| \left\| \begin{bmatrix} 1.5 \\ 0.5 \end{bmatrix} - \begin{bmatrix} 3 \\ 4 \end{bmatrix} \right\| \right] \\ &= \left[3.810 \ 3.810 \ 3.810 \right] \end{aligned}$$

2. **The first hidden layer**: In this layer, the activation vectors $\boldsymbol{\phi}(k) \in \Re^{l}$ are computed for every element in $\boldsymbol{\sigma}$ using

$$\boldsymbol{\phi}_j(k) = \exp\left(-\frac{\boldsymbol{q}^T(k)\boldsymbol{q}(k)}{2\sigma_j^2} \right), \ j = 1, \ldots, l \tag{10.6}$$

The statistical means of the activation vectors $\boldsymbol{\phi}(k) \in \Re^{1\times h}$ are computed as

$$\boldsymbol{\phi}(k) = \frac{\sum_{j=1}^{h} \boldsymbol{\phi}_j(k)}{h} \tag{10.7}$$

3. **The second hidden layer**: In this layer, the weighted average of the statistical means $\boldsymbol{\omega} \in \Re^n$ of the activation vectors from the previous layer is computed

$$\boldsymbol{\omega}(k) = \frac{\boldsymbol{\Theta}^T(k)\boldsymbol{\phi}(k)}{\sum_{i=1}^{n} \boldsymbol{\phi}_i} \tag{10.8}$$

where $\boldsymbol{\Theta}^T \in \Re^{h\times n} = \begin{bmatrix} \theta_{11} & \dots & \theta_{1h} \\ \vdots & \dots & \vdots \\ \theta_{1h} & \dots & \theta_{nh} \end{bmatrix}$ is an adaptable weight matrix.

Defining $\boldsymbol{\psi}(k)$ as

$$\boldsymbol{\psi}(k) = \frac{\boldsymbol{\phi}(k)}{\sum_{i=1}^{h} \boldsymbol{\phi}_i(k)} \tag{10.9}$$

Substituting (10.9) into (10.8) yields

$$\boldsymbol{\omega}(k) = \boldsymbol{\Theta}^T(k)\boldsymbol{\psi}(k) \tag{10.10}$$

4. **The output layer**: In this layer, the adaptable input-output weights $\boldsymbol{\zeta} \in \Re^{m\times n} = \begin{bmatrix} \zeta_{11} & \dots & \zeta_{1n} \\ \vdots & \vdots & \vdots \\ \zeta_{m1} & \dots & \zeta_{mn} \end{bmatrix}$ are defined, and the output of I_GRNN is computed as

$$\begin{aligned} \boldsymbol{u}(k) &= \boldsymbol{\zeta}^T(k)\boldsymbol{x}(k) + \boldsymbol{\omega}(k) \\ \boldsymbol{u}(k) &= \boldsymbol{\zeta}^T(k)\boldsymbol{x}(k) + \boldsymbol{\Theta}^T(k)\boldsymbol{\psi}(k) \end{aligned} \tag{10.11}$$

Example 10.3 For I_GRNN with two inputs given as $\boldsymbol{x} = \begin{bmatrix} 1 \\ 2 \end{bmatrix}$, the input-hidden weight matrix is given as $\boldsymbol{P} = \begin{bmatrix} 1\ 0 \\ 0\ 1 \\ 1\ 1 \end{bmatrix}$, $\boldsymbol{\sigma} = [1, 2, 3]$, $\boldsymbol{\varsigma} = \begin{bmatrix} 1\ 1 \\ 2\ 1 \end{bmatrix}$, $\boldsymbol{\Theta} = \begin{bmatrix} 3\ 6 \\ 4\ 7 \\ 5\ 8 \end{bmatrix}$. Calculate the output of the network.

In the first step, the distance $\boldsymbol{q}$ is calculated as $\boldsymbol{q} = \text{dist}(\boldsymbol{P}, \boldsymbol{x}) = \begin{bmatrix} 2 \\ 1.414 \\ 1 \end{bmatrix}$

In the second step, the activation of the inputs over the whole $\boldsymbol{\sigma}$ range is calculated as $\boldsymbol{\phi}_1 = \begin{bmatrix} 0.1353 \\ 0.606 \\ 0.800 \end{bmatrix}$ $\boldsymbol{\phi}_2 = \begin{bmatrix} 0.368 \\ 0.779 \\ 0.890 \end{bmatrix}$ $\boldsymbol{\phi}_3 = \begin{bmatrix} 0.607 \\ 0.880 \\ 0.946 \end{bmatrix}$

The normalized activation average is computed as $\boldsymbol{\psi} = \begin{bmatrix} 0.256 \\ 0.339 \\ 0.406 \end{bmatrix}$

The network output is calculated as

$$\boldsymbol{u} = \begin{bmatrix} 1\ 2 \\ 1\ 1 \end{bmatrix} \begin{bmatrix} 1 \\ 2 \end{bmatrix} + \begin{bmatrix} 3\ 4\ 5 \\ 6\ 7\ 8 \end{bmatrix} \begin{bmatrix} 0.256 \\ 0.339 \\ 0.406 \end{bmatrix} = \begin{bmatrix} 7.148 \\ 12.148 \end{bmatrix}$$

10.2.2 Design of I_GRNN Controller

The design of the I_GRNN controller is discussed next in detail. Consider a MIMO discrete-time dynamic system given as

$$\begin{aligned} \boldsymbol{X}(k+1) &= \boldsymbol{f}(\boldsymbol{X}(k), \boldsymbol{u}(k)) + \boldsymbol{G}(\boldsymbol{X}(k), \boldsymbol{u}(k)) \\ \boldsymbol{y}_i(k) &= \boldsymbol{x}_{1i}(k) \\ \boldsymbol{u}(k) &= [u_1(k), \ldots, u_n(k)] \\ \boldsymbol{X} &= \begin{bmatrix} x_{1,1}, \ldots, x_{1,n} \\ x_{2,1}, \ldots, x_{2,n} \\ \vdots \\ x_{n,1}, \ldots, x_{n,n} \end{bmatrix} \end{aligned} \tag{10.12}$$

where $\boldsymbol{X} \in \Re^{n\times n}$ is the state of the system, $\boldsymbol{f} \in \Re^{n\times n}$ is a nonlinear function of the system dynamics, $\boldsymbol{u} \in \Re^{n}$ is the control input, and $\boldsymbol{G} \in \Re^{n\times n}$ is the control input gain.

The output tracking error $\boldsymbol{e}(k) \in \Re^{n}$ is represented as

$$\boldsymbol{e}(k) = \boldsymbol{y}_d(k) - \boldsymbol{y}(k) \tag{10.13}$$

where $\boldsymbol{y}_d(k) \in \Re^{n}$ is the desired output trajectory and $\boldsymbol{y}(k) \in \Re^{n}$ is the current output trajectory. The controller objective is to ensure the system outputs $\boldsymbol{y}(k)$ track the desired outputs $\boldsymbol{y}_d(k)$ precisely.

Let there be weight matrices $\boldsymbol{\zeta}^*$ and $\boldsymbol{\Theta}^*$ that reconstruct the desired control vector $\boldsymbol{u}_d(k)$ with reconstruction error $\boldsymbol{\epsilon}(k)$ as

$$\boldsymbol{u}_d(k) = \boldsymbol{\zeta}^{*T}\boldsymbol{x}(k) + \boldsymbol{\Theta}^{*T}\boldsymbol{\psi}(k) + \boldsymbol{\epsilon}(k) \tag{10.14}$$

The control error $\tilde{\boldsymbol{u}}(k) = \boldsymbol{u}_d(k) - \boldsymbol{u}(k)$ is practically unknown because $\boldsymbol{u}_d(k)$ is also unknown; however, it affects the tracking error, which is known, based upon that the tracking error is used to update the I_GRNN weights as follows:

$$\begin{aligned} \boldsymbol{\Theta}(k+1) &= \boldsymbol{\Theta}(k) - \gamma_1 \boldsymbol{e}(k)\boldsymbol{\psi}(k) \\ \boldsymbol{\zeta}(k+1) &= \boldsymbol{\zeta}(k) - \gamma_2 \boldsymbol{e}(k)\boldsymbol{x}(k) \end{aligned} \tag{10.15}$$

where γ_1 and γ_2 are the adaptation gains.

The current estimation of the control signal $\boldsymbol{u}(k)$ using I_GRNN can be expressed as

$$\boldsymbol{u}(k) = \boldsymbol{\zeta}^T(k)\boldsymbol{x}(k) + \boldsymbol{\Theta}^T(k)\boldsymbol{\psi}(k) \tag{10.16}$$

The control error $\tilde{\boldsymbol{u}}(k)$ is constructed as

$$\begin{aligned} \tilde{\boldsymbol{u}}(k) = {} & \boldsymbol{\zeta}^{*T}\boldsymbol{x}(k) + \boldsymbol{\Theta}^{*T}\boldsymbol{\psi}(k) + \boldsymbol{\epsilon}(k) \\ & - \boldsymbol{\zeta}^T(k)\boldsymbol{x}(k) - \boldsymbol{\Theta}^T(k)\boldsymbol{\psi}(k) \end{aligned} \tag{10.17}$$

Define the construction errors $\tilde{\boldsymbol{\zeta}}(k)$ and $\tilde{\boldsymbol{\Theta}}(k)$ as

$$\begin{aligned} \tilde{\boldsymbol{\zeta}}(k) &= \boldsymbol{\zeta}^* - \boldsymbol{\zeta}(k) \\ \tilde{\boldsymbol{\Theta}}(k) &= \boldsymbol{\Theta}^* - \boldsymbol{\Theta}(k) \end{aligned} \tag{10.18}$$

Combining (10.18) and (10.17) leads to

$$\tilde{\boldsymbol{u}}(k) = \tilde{\boldsymbol{\zeta}}^T(k)\boldsymbol{x}(k) + \tilde{\boldsymbol{\Theta}}^T(k)\boldsymbol{\psi}(k) + \boldsymbol{\epsilon}(k) \tag{10.19}$$

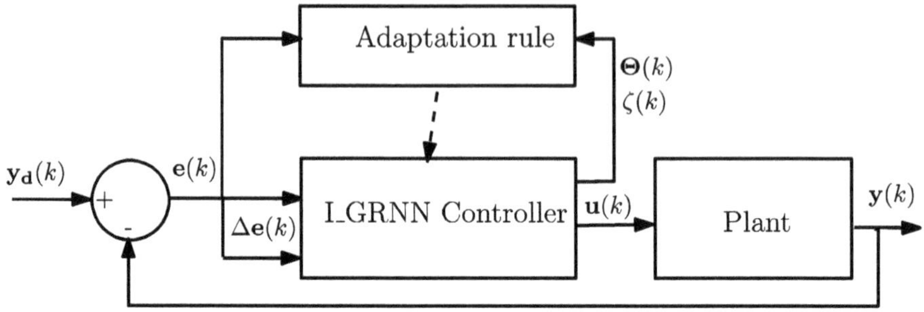

Fig. 10.3 Block diagram of the I_GRNN

I_GRNN block diagram in the closed-loop form is shown in Fig. 10.3.

The weights of I_GRNN are adapted online according to a specific objective function $j(k)$ as follows:

$$j(k) = \frac{1}{2}(\tilde{\boldsymbol{u}}(k)^T \tilde{\boldsymbol{u}}(k)) \tag{10.20}$$

I_GRNN weights are adapted online using the gradient-descent approach to minimize the above function. Thus, I_GRNN's weight adaptation rules are developed as

$$\begin{aligned}\boldsymbol{\Theta}(k+1) &= \boldsymbol{\Theta}(k) + \gamma_1 \frac{\Delta j}{\Delta \boldsymbol{\Theta}(k)} \\ \boldsymbol{\zeta}(k+1) &= \boldsymbol{\zeta}(k) + \gamma_2 \frac{\Delta j}{\Delta \boldsymbol{\zeta}(k)}\end{aligned} \tag{10.21}$$

$\frac{\Delta j}{\Delta \boldsymbol{\zeta}(k)}$ and $\frac{\Delta j}{\Delta \boldsymbol{\Theta}(k)}$ are calculated using the expression for the control input as

$$\begin{aligned}\frac{\Delta j(k)}{\Delta \boldsymbol{\Theta}(k)} &= \frac{\Delta j(k)}{\Delta \boldsymbol{u}(k)} \frac{\Delta \boldsymbol{u}(k)}{\Delta \boldsymbol{\Theta}(k)} \\ &= -\tilde{\boldsymbol{u}}(k)\boldsymbol{\psi}(k)\end{aligned} \tag{10.22}$$

$$\begin{aligned}\frac{\Delta j(k)}{\Delta \boldsymbol{\zeta}(k)} &= \frac{\Delta j(k)}{\Delta \boldsymbol{u}(k)} \frac{\Delta \boldsymbol{u}(k)}{\Delta \boldsymbol{\zeta}(k)} \\ &= -\tilde{\boldsymbol{u}}(k)\boldsymbol{x}(k)\end{aligned} \tag{10.23}$$

Considering the approximation that $\tilde{\boldsymbol{u}}(k) \approx \boldsymbol{e}(k)$

$$\begin{aligned}\boldsymbol{\Theta}(k+1) &= \boldsymbol{\Theta}(k) - \gamma_1 \boldsymbol{e}(k)\boldsymbol{\psi}(k) \\ \boldsymbol{\zeta}(k+1) &= \boldsymbol{\zeta}(k) - \gamma_2 \boldsymbol{e}(k)\boldsymbol{x}(k)\end{aligned} \tag{10.24}$$

10.2.3 Stability Analysis of I_GRNN

The stability of I_GRNN is investigated using Lyapunov's second method.

Firstly, a positive definite function $V(k)$ is chosen as

$$V(k) = \sum_{i=1}^{3} V_i(k) \tag{10.25}$$

The first Lyapunov function is

$$V_1(k) = \boldsymbol{e}(k)^T \boldsymbol{e}(k) \tag{10.26}$$

The first difference of V_1 is

$$\Delta V_1(k) = \boldsymbol{e}^T(k+1)\boldsymbol{e}(k+1) - \boldsymbol{e}^T(k)\boldsymbol{e}(k) \tag{10.27}$$

Considering the tracking error difference as $\Delta \boldsymbol{e}(k) = \boldsymbol{e}(k+1) - \boldsymbol{e}(k)$ and replacing $\Delta \boldsymbol{e}(k)$ with $-\boldsymbol{\beta} \in \Re^n$, the following can be written $\boldsymbol{e}(k+1) = \boldsymbol{e}(k) - \boldsymbol{\beta}$, and using the inequality of Cauchy-Schwarz (Bresch-Pietri et al. 2012), $\Delta V_1(k)$ is expressed as

$$\Delta V_1(k) \le (\boldsymbol{e}(k) - \boldsymbol{\beta})^T(\boldsymbol{e}(k) - \boldsymbol{\beta}) - \boldsymbol{e}^T(k)\boldsymbol{e}(k) \tag{10.28}$$

$$\begin{aligned} \Delta V_1(k) &\le 2\boldsymbol{e}^T(k)\boldsymbol{e}(k) - 2\boldsymbol{\beta}^T\boldsymbol{\beta} - \boldsymbol{e}(k)^T\boldsymbol{e}(k) \\ &\le \boldsymbol{e}^T(k)\boldsymbol{e}(k) - \boldsymbol{\beta}^T\boldsymbol{\beta} \end{aligned} \tag{10.29}$$

$$\Delta V_1(k) \le ||\boldsymbol{e}(k)||^2 - 2||\boldsymbol{\beta}||^2 \tag{10.30}$$

The function V_1 is stable if

$$||\boldsymbol{e}(k)|| \le \pm\sqrt{2}||\boldsymbol{\beta}|| \tag{10.31}$$

The second Lyapunov function is

$$V_2(k) = \tilde{\boldsymbol{\Theta}}(k)^T \tilde{\boldsymbol{\Theta}}(k) \tag{10.32}$$

The first difference of V_2 is defined as

$$\Delta V_2(k) = \tilde{\boldsymbol{\Theta}}(k+1)^T \tilde{\boldsymbol{\Theta}}(k+1) - \tilde{\boldsymbol{\Theta}}(k)^T \tilde{\boldsymbol{\Theta}}(k) \tag{10.33}$$

Substituting (10.24) and simplifying further leads

$$\Delta V_2(k) \leq (\tilde{\boldsymbol{\Theta}}(k) - \gamma_1 \boldsymbol{e}(k)\boldsymbol{\psi}(k))^T(\tilde{\boldsymbol{\Theta}}(k) - \gamma_1 \boldsymbol{e}(k)\boldsymbol{\psi}) - \tilde{\boldsymbol{\Theta}}(k)^T\tilde{\boldsymbol{\Theta}}(k) \tag{10.34}$$

Utilizing Cauchy-Schwarz inequality (Bresch-Pietri et al. 2012), (10.34) can be expressed as

$$\Delta V_2(k) \leq 2(\tilde{\boldsymbol{\Theta}}^T(k)\tilde{\boldsymbol{\Theta}}(k) - (\gamma_1 \boldsymbol{e}(k)\boldsymbol{\psi}(k))^T(\gamma_1 \boldsymbol{e}(k)\boldsymbol{\psi}(k))) - \tilde{\boldsymbol{\Theta}}(k)^T\tilde{\boldsymbol{\Theta}}(k) \tag{10.35}$$

Considering $||\boldsymbol{\psi}|| \leq \alpha,$ (10.35) becomes

$$\Delta V_2(k) \leq ||\tilde{\boldsymbol{\Theta}}(k)||^2 - 2\gamma_1^2\alpha^2||\boldsymbol{e}(k)||^2 \tag{10.36}$$

$\Delta V_2(k) \leq 0$ when the adaptation gain γ_1 is selected as

$$\gamma_1 \geq \pm\frac{||\tilde{\boldsymbol{\Theta}}(k)||}{\sqrt{2}\alpha||\boldsymbol{e}(k)||} \tag{10.37}$$

The third Lyapunov function $V_3(k)$ is chosen as

$$V_3(k) = \tilde{\boldsymbol{\xi}}(k)^T\tilde{\boldsymbol{\xi}}(k) \tag{10.38}$$

The first difference of V_3 is expressed as

$$\Delta V_3(k) = \tilde{\boldsymbol{\xi}}(k+1)^T\tilde{\boldsymbol{\xi}}(k+1) - \tilde{\boldsymbol{\xi}}(k)^T\tilde{\boldsymbol{\xi}}(k) \tag{10.39}$$

Substituting (10.24) and simplifying yields

$$\Delta V_3(k) \leq (\tilde{\boldsymbol{\xi}}(k) - \gamma_2 \boldsymbol{e}(k)\boldsymbol{x}(k))^T(\tilde{\boldsymbol{\xi}}(k) - \gamma_2 \boldsymbol{e}(k)\boldsymbol{e}(k)) - \tilde{\boldsymbol{\xi}}(k)^T\tilde{\boldsymbol{\xi}}(k) \tag{10.40}$$

Using Cauchy-Schwarz inequality and simplifying further yields

$$\Delta V_3(k) \leq ||\tilde{\boldsymbol{\xi}}(k)||^2 - 2\gamma_2^2||\boldsymbol{e}(k)\boldsymbol{x}(k)||^2 \tag{10.41}$$

$\Delta V_3(k) \leq 0$ if the adaptation gain γ_2 is chosen as

$$\gamma_2 \geq \pm\frac{||\tilde{\boldsymbol{\xi}}(k)||}{\sqrt{2}||\boldsymbol{e}(k)\boldsymbol{x}(k)||} \tag{10.42}$$

Based on the above analysis, if the conditions in (10.31), (10.37), and (10.42) are fulfilled, the closed-loop system is stable.

Let us consider an example of a MIMO nonaffine discrete-time system in which I_GRNN is used for reference tracking of simple and complex set points.

Example 10.4 Consider the MIMO nonaffine discrete-time nonlinear system (Ge et al. 2004). The system dynamics are expressed as

$$
\begin{aligned}
x_{1,1}(k+1) &= f_{1,1}(x_{1,1}(k)) + g_{1,1}(x_{1,1}(k))\\
x_{1,2}(k+1) &= f_{1,2}(x_{1,2}(k))\\
&\quad + g_{1,2}(x_{1,1}(k))u_1(k) + d_1(k)\\
x_{2,1}(k+1) &= f_{2,1}(x_{2,1}(k)) + g_{2,1}(x_{2,2}(k))\\
x_{2,2}(k+1) &= f_{2,2}(x_{2,2}(k), u_1(k))\\
&\quad + g_{2,2}(x_{2,2}(k))u_2(k) + d_2(k)\\
y_1(k+1) &= x_{1,1}(k)\\
y_2(k+1) &= x_{2,1}(k)\\
f_{1,1}(x_{1,1}(k)) &= \frac{x_{1,1}^2(k)}{1+x_{1,1}^2(k)}\\
g_{1,1}(x_{1,1}(k)) &= 0.3\\
f_{1,2}(x_{1,2}(k)) &= \frac{x_{1,1}^2(k)}{1+x_{1,2}^2(k)+x_{2,1}^2(k)+x_{2,2}^2(k)}\\
g_{1,2}(x_{1,1}(k)) &= 1\\
d_1(k) &= 0.01\cos(0.05k)\cos(x_{1,1}(k))\\
f_{2,1}(x_{2,1}(k)) &= \frac{x_{2,1}^2(k)}{1+x_{2,1}^2(k)}\\
g_{2,1}(x_{2,2}(k)) &= 0.2\\
f_{2,2}(x_{2,2}(k), u_2(k)) &= \frac{x_{2,1}^2(k)u_2^2(k)}{1+x_{1,1}^2(k)+x_{2,1}^2(k)+x_{2,2}^2(k)}\\
g_{2,2}(x_{2,2}(k)) &= 1\\
d_2(k) &= 0.01\cos(0.05k)\cos(x_{2,1}(k))
\end{aligned}
$$

(continued)

Example 10.4 (continued)
The initial states are $[x_{1,1}(0), x_{1,2}(0), x_{2,1}(0), x_{2,2}(0)] = [0,0,0,0]$.
Design the I_GRNN controller for the two cases, one with simple reference input and one with complete reference input.
Case One: Simple Reference
Firstly, the desired output references are selected as simple step inputs starting at step zero as follows:

$$y_{d1}(k) = \begin{Bmatrix} 0, & \text{for } k < 0 \\ 0.6, & \text{for } k \geq 0 \end{Bmatrix}$$

$$y_{d2}(k) = \begin{Bmatrix} 0, & \text{for } k < 0 \\ 0.5, & \text{for } k \geq 0 \end{Bmatrix}$$

I_GRNN controller is designed using (10.16) as

$$\boldsymbol{u}(k) = \boldsymbol{\zeta}^T(k)\boldsymbol{x}(k) + \boldsymbol{\Theta}^T(k)\boldsymbol{\psi}(k) \tag{10.43}$$

where $\boldsymbol{x}(k) = \begin{bmatrix} \boldsymbol{e}(k) \\ \Delta\boldsymbol{e}(k) \end{bmatrix}$. The learning rates are $\gamma_1 = 0.1$, $\gamma_2 = 0.3$, the hidden layer size is ten, and $\sigma = [0.01 : 0.1 : 1]$. For the original GRNN, the learning rates are the same as γ_1, the hidden layer size is ten, and $\sigma = 1$.
The simple reference tracking results, as depicted in Figs. 10.4 and 10.5, illustrate accurate tracking power of I_GRNN. Small oscillations occur when I_GRNN starts learning. After the initial transient response, I_GRNN response exhibit accurate set point following, while the original GRNN response keeps oscillating around the set point, mainly due to the external disturbances.
Case Two: Complex Reference
Secondly, more complex desired output references composed of the sum of sines are selected as

$$\begin{aligned} y_{d_1}(k) &= 0.35\cos(\pi Tk/3) + 0.25\sin(\pi Tk/3) \\ y_{d_2}(k) &= 0.35\sin(\pi Tk/3) + 0.25\sin(\pi Tk/3) \end{aligned} \tag{10.44}$$

For the second set of references, the learning rates are $\gamma_1 = 0.1$, $\gamma_2 = 0.3$, hidden layer size is ten, and $\sigma = [0.01 : 0.1 : 1]$. For the original GRNN, the learning rate $\gamma = 0.6$ and the hidden layer size is ten and $\sigma = 1$.
The tracking results of the complex reference as shown in Figs. 10.6 and 10.7 depict the accurate response of I_GRNN compared to the original GRNN. Thus, I_GRNN has much lower tracking errors compared to the original GRNN.

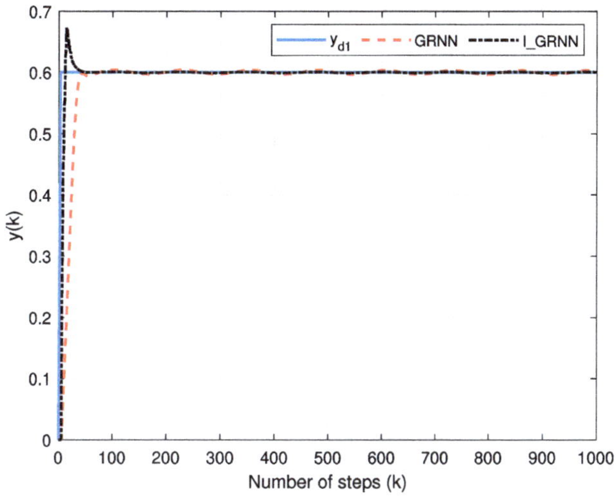

Fig. 10.4 Tracking of simple reference y_1

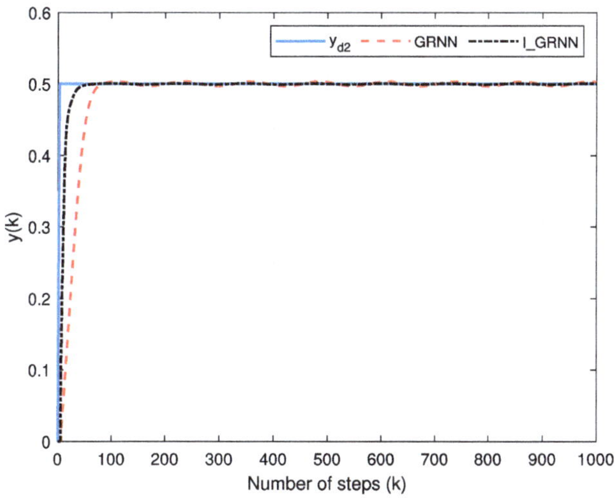

Fig. 10.5 Tracking of simple reference y_2

It is also possible to combine GRNN with other control methods such as SMC, using the robustness properties of SMC to improve the GRNN controller. In the next section, the combined GRNN and SMC controller is discussed in detail.

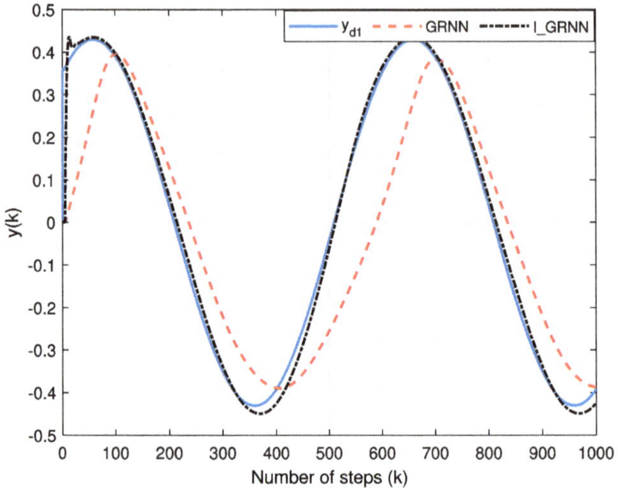

Fig. 10.6 Complex reference tracking response y_1

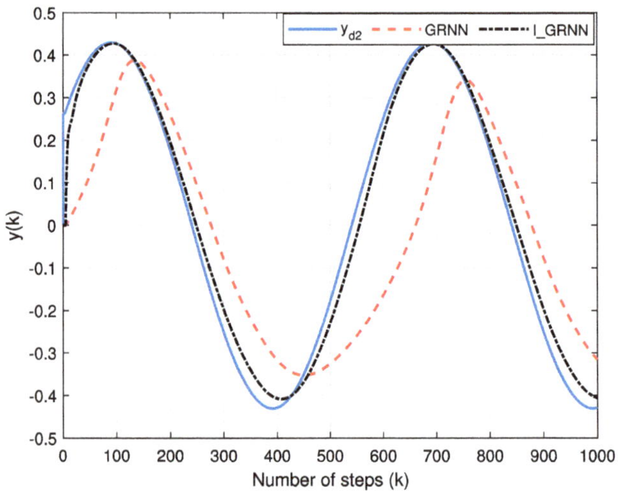

Fig. 10.7 Complex reference tracking response y_2

10.3 GRNNSMC-Based Adaptive Control

SMC is a popular robust controller for dynamic systems; however, it depends on the model of the system (Plestan et al. 2010). To design a learning-based SMC, the learning capability of GRNN is integrated with the robustness of SMC. In this framework, SMC is used to update GRNN learnable parameters (Al-Mahasneh et al.

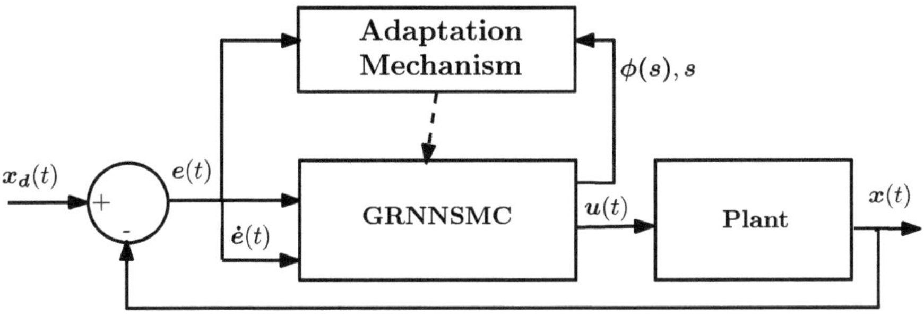

Fig. 10.8 GRNNSMC block diagram

2021). The closed-loop block diagram of GRNNSMC.[2] is shown in Fig. 10.8. Here s is the sliding surface as discussed in the next section.

10.3.1 GRNNSMC Controller Design

Consider a class of nonlinear MIMO dynamic systems (Ertugrul and Kaynak 2000) formulated as

$$\dot{\boldsymbol{x}}(t) = \boldsymbol{f}(\boldsymbol{x}(t)) + \boldsymbol{B}\boldsymbol{u}(t) \tag{10.45}$$

where $\boldsymbol{x} \in \Re^n$ is a vector of the system states, $\boldsymbol{f} \in \Re^n$ is a vector of nonlinear functions representing the drift dynamics, $\boldsymbol{B} \in \Re^{n\times n}$ is a matrix of the input gains, and $\boldsymbol{u} \in \Re^n$ is a vector of the control inputs.

The first step in SMC design is to design a sliding surface $\boldsymbol{s}$; the sliding surface $\boldsymbol{s}$ is designed as

$$\boldsymbol{s}(\boldsymbol{x}, t) = \boldsymbol{s}_a(t) - \boldsymbol{s}_b(\boldsymbol{x}) = \boldsymbol{R}(\boldsymbol{x}_d(t) - \boldsymbol{x}(t)) \tag{10.46}$$

where $\boldsymbol{s}_a(t) = \boldsymbol{R}(\boldsymbol{x}_d(t))$ is the first part of $\boldsymbol{s}$ and $\boldsymbol{s}_b(\boldsymbol{x}) = \boldsymbol{R}(\boldsymbol{x}(t))$ is the second part of $\boldsymbol{s}$, $\boldsymbol{x}_d(t)$ is the reference states vector, and $\boldsymbol{R} \in \Re^{n\times n}$ is the slope of the sliding surface. This slope is selected such that the tracking error $\boldsymbol{e}(t) = \boldsymbol{x}(t) - \boldsymbol{x}_d(t)$ converges to zero when $\boldsymbol{s} = \boldsymbol{0}$.

The sliding surface $\boldsymbol{s}$ is selected as $\boldsymbol{s}(t) = \dot{\boldsymbol{e}}(t) + \boldsymbol{\lambda}\boldsymbol{e}(t)$, all the elements of $\boldsymbol{\lambda}$ are positive, and it is a controller design parameter that influences the reaching speed to the sliding surface.

[2] Some contents of current Sect. 10.3 are used from Publication "Stable Adaptive Controller Based on Generalized Regression Neural Networks and Sliding Mode Control for a Class of Nonlinear Time-Varying Systems," IEEE Transactions on Systems, Man, and Cybernetics: Systems, vol. 51, pp. 2525–2535, April, 2021, with permission from IEEE (License Number: 6160201218650).

The next step in SMC design is to define a positive-definite Lyapunov function as

$$V(\boldsymbol{s}) = \frac{1}{2}\boldsymbol{s}^T\boldsymbol{s} \tag{10.47}$$

To guarantee asymptotic stability, $\dot{V}(\boldsymbol{s})$ is set as

$$\frac{dV(\boldsymbol{s})}{dt} = -\boldsymbol{s}^T\boldsymbol{Q}\text{sgn}(\boldsymbol{s}) \tag{10.48}$$

where $\boldsymbol{Q}$ is a positive definite matrix and sgn is the signum function.

Taking the derivative of (10.47) with time and substituting it back into (10.48) yields

$$\boldsymbol{s}^T\frac{d\boldsymbol{s}}{dt} = -\boldsymbol{s}^T\boldsymbol{Q}\text{sgn}(\boldsymbol{s}) \tag{10.49}$$

The term $\frac{d\boldsymbol{s}}{dt}$ can be obtained using (10.46) as

$$\frac{d\boldsymbol{s}}{dt} = \frac{d\boldsymbol{s}_a}{dt} - \frac{d\boldsymbol{s}_b}{d\boldsymbol{x}}\frac{d\boldsymbol{x}}{dt} = \frac{d\boldsymbol{s}_a}{dt} - \boldsymbol{R}(\boldsymbol{f}(\boldsymbol{x}(t)) + \boldsymbol{B}\boldsymbol{u}(t) \tag{10.50}$$

Substituting (10.50) in (10.49) yields

$$\boldsymbol{s}^T(\frac{d\boldsymbol{s}_a}{dt} - \boldsymbol{R}(\boldsymbol{f}(\boldsymbol{x}(t)) + \boldsymbol{B}\boldsymbol{u}(t)) = -\boldsymbol{s}^T\boldsymbol{Q}\text{sgn}(\boldsymbol{s}) \tag{10.51}$$

Simplification of (10.51) yields

$$\boldsymbol{u}(t) = (\boldsymbol{R}\boldsymbol{B})^{-1}(-\boldsymbol{Q}\text{sgn}(s) - \frac{d\boldsymbol{s}_a}{dt} + \boldsymbol{R}(\boldsymbol{f}(\boldsymbol{x}(t))) \tag{10.52}$$

The SMC control law $\boldsymbol{u}(t)$ is made of reaching $\boldsymbol{u}_r$ and sliding $\boldsymbol{u}_s$ control parts as follows:

$$\boldsymbol{u}(t) = \boldsymbol{u}_r + \boldsymbol{u}_s \tag{10.53}$$

where

$$\begin{aligned} \boldsymbol{u}_r &= (\boldsymbol{R}\boldsymbol{B})^{-1}\Big(-\boldsymbol{Q}\text{sgn}(s)\Big) \\ \boldsymbol{u}_s &= (\boldsymbol{R}\boldsymbol{B})^{-1}\Big(\boldsymbol{R}(\boldsymbol{f}(\boldsymbol{x}(t)) - \frac{d\boldsymbol{s}_a}{dt}\Big) \end{aligned} \tag{10.54}$$

The control term is not realizable if $(\boldsymbol{RB})$ is not a square matrix since its inverse $((\boldsymbol{RB})^{-1})$ cannot be computed. In addition, when the model of system is unknown or uncertain, the control law can be approximated using GRNNSMC $\hat{\boldsymbol{u}}$ as

$$\hat{\boldsymbol{u}} = \boldsymbol{W}^T \boldsymbol{\phi}(\boldsymbol{s}) \tag{10.55}$$

where $\boldsymbol{W} \in \Re^{m \times n}$ is an output weights matrix of GRNNSMC and $\boldsymbol{\phi}(\boldsymbol{s}) \in \Re^m$ is GRNNSMC hidden layer output vector and it is expressed as

$$h_i(\boldsymbol{s}) = \exp\left(-\frac{||\boldsymbol{s} - \boldsymbol{c}_i||^2}{2\sigma^2}\right), i = 1, \ldots, n \tag{10.56}$$

$$\boldsymbol{h}(\boldsymbol{s}) = \begin{bmatrix} h_1(\boldsymbol{s}) \\ h_2(\boldsymbol{s}) \\ \vdots \\ h_n(\boldsymbol{s}) \end{bmatrix} \tag{10.57}$$

$$\phi_i(\boldsymbol{s}) = \frac{h_i(\boldsymbol{s})}{\sum\limits_{i=1}^{n} h_i(\boldsymbol{s})}, i = 1, \ldots, n \tag{10.58}$$

$$\boldsymbol{\phi}(\boldsymbol{s}) = \begin{bmatrix} \phi_1(\boldsymbol{s}) \\ \phi_2(\boldsymbol{s}) \\ \vdots \\ \phi_n(\boldsymbol{s}) \end{bmatrix} \tag{10.59}$$

where $\boldsymbol{c}$ is the hidden weight and σ is the smoothing parameter.

The next step in GRNNSMC design is to select the GRNN structure. In this step, the input nodes are the sliding surfaces, and the number of input nodes depends on the number of the controlled states. The number of nodes in the hidden layer is at least equal to the number of the sliding surfaces. Also, the number of the nodes in the output layer equals the number of the control inputs. GRNNSMC structure is depicted in Fig. 10.9. GRNNSMC structure links the sliding variables $\boldsymbol{s}$ in the input layer with the hidden layer output of GRNNSMC $\boldsymbol{\phi}(s)$, while the estimated control signal $\hat{u}$ is calculated in the output layer.

GRNNSMC algorithm is detailed in Algorithm 10.1. The adaptation laws of GRNNSMC output weights are derived utilizing Lyapunov stability theory in the next section.

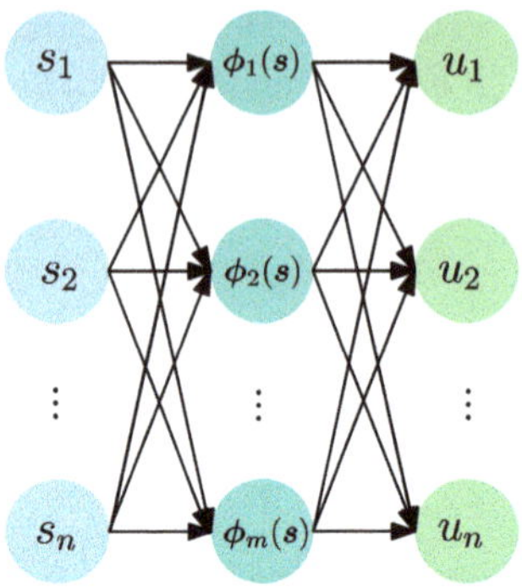

Fig. 10.9 GRNNSMC structure

10.3.2 GRNNSMC Weights Adaptation

GRNNSMC output weights are adapted at every time step following the processing of the inputs' sliding surfaces in the hidden layer. GRNNSMC output can be determined after the weights update.

In this section, Lyapunov's second method is utilized to derive the stable adaptation laws of GRNNSMC output weights and to guarantee Lyapunov stability of the controlled system.

GRNNSMC uses a normalized radial basis function to calculate the hidden layer output; hence, the maximum output of every node in the hidden output is 1. For a GRNNSMC with l nodes in the hidden layer, the output of the hidden layer is limited as

$$0 \leq \boldsymbol{\phi}^T(s)\boldsymbol{\phi}(s) \leq l \tag{10.60}$$

where $\boldsymbol{\phi}(s)$ is GRNNSMC hidden layer output.

Since (10.60) is always bounded and the output of GRNNSMC is a product of the hidden layer outputs and the output weights, the output weights are the main parameters of GRNNSMC stability.

Using Lyapunov direct approach, the derivative of (10.47) with time must be strictly negative. Based on that, to achieve the sliding mode reachability condition, the following can be expressed:

$$\boldsymbol{s}^T\dot{\boldsymbol{s}} < 0 \tag{10.61}$$

To derive the weights update law, $\boldsymbol{s}^T\dot{\boldsymbol{s}}$ should be minimized with respect to GRNNSMC weights $\hat{\boldsymbol{W}}$ as

$$\dot{\boldsymbol{W}} = -\gamma\frac{\partial \boldsymbol{s}^T(t)\dot{\boldsymbol{s}}(t)}{\partial \boldsymbol{W}} \tag{10.62}$$

where $\boldsymbol{\gamma}$ is a vector of the adaptation gains with only positive elements.

By utilizing the chain rule, (10.62) can expressed as

$$\dot{\boldsymbol{W}} = -\gamma \frac{\partial \boldsymbol{s}^T(t)\dot{\boldsymbol{s}}(t)}{\partial \hat{\boldsymbol{u}}(t)} \frac{\partial \hat{\boldsymbol{u}}(t)}{\partial \boldsymbol{W}} \tag{10.63}$$

To obtain $\boldsymbol{s}^T(t)\dot{\boldsymbol{s}}(t)$, (10.50) is multiplied by $\boldsymbol{s}^T$ as

$$\boldsymbol{s}^T(t)\dot{\boldsymbol{s}}(t) = \boldsymbol{s}^T(t)\frac{d\boldsymbol{s}_a}{dt} - \boldsymbol{s}^T(t)\boldsymbol{R}(\boldsymbol{f}(\boldsymbol{x}(t)) + \boldsymbol{s}^T(t)\boldsymbol{B}\boldsymbol{u}(t) \tag{10.64}$$

Using (10.64), $\dfrac{\partial \boldsymbol{s}^T(t)\dot{\boldsymbol{s}}(t)}{\partial \hat{\boldsymbol{u}}(t)}$ can be acquired as

$$\frac{\partial \boldsymbol{s}^T(t)\dot{\boldsymbol{s}}(t)}{\partial \hat{\boldsymbol{u}}(t)} = \boldsymbol{s}^T(t)\boldsymbol{B} \tag{10.65}$$

The term $\dfrac{\partial \hat{\boldsymbol{u}}(t)}{\partial \boldsymbol{W}}$ can be obtained by differentiating (10.55) with respect to $\boldsymbol{W}$ which yields

$$\frac{\partial \hat{\boldsymbol{u}}(t)}{\partial \boldsymbol{W}} = \boldsymbol{\phi}(\boldsymbol{s}) \tag{10.66}$$

Substituting (10.65) and (10.66) in (10.63) yields

$$\dot{\boldsymbol{W}} \leq \gamma \boldsymbol{s}^T(t)\boldsymbol{\phi}(\boldsymbol{s}) \tag{10.67}$$

Equation (10.67) is used to adapt GRNNSMC weights to achieve the SMC reachability condition. The weights are adapted as

$$\dot{\boldsymbol{W}} = -\gamma \boldsymbol{s}^T(t)\boldsymbol{\phi}(\boldsymbol{s}) \tag{10.68}$$

GRNNSMC algorithm is detailed in Algorithm 10.1. The algorithm begins with the tracking error $\boldsymbol{e}(t)$ and its derivative $\dot{\boldsymbol{e}}(t)$. These are utilized to obtain the sliding surface. In the next step, the sliding surfaces $\boldsymbol{s}$ are used to obtain GRNNSMC hidden layer output $\boldsymbol{\phi}(\boldsymbol{s})$. Then GRNNSMC output weights are adapted using (10.68). In the last step, GRNNSMC output is calculated.

Table 10.1 GRNNSMC's parameters for the MIMO system control

λ	γ
$\begin{bmatrix} 10.5 \\ 10.1 \end{bmatrix}$	$\begin{bmatrix} 7.1 \\ 7.9 \end{bmatrix}$

Algorithm 10.1 GRNNSMC

Input: $e(t), \dot{e}(t)$
Output: $x(t)$
Initialization :
Initialize the output weights to 0
Set the sliding surface parameters λ
Set the learning rates γ
Control Loop :
for $t = 0$ to t_{final} **do**
Find $s(t) = \dot{e}(t) + \lambda e(t)$
Find $\phi(s)$ based on (10.56)
Modify GRNNSMC's weights based on (10.67)
Find the output of GRNNSMC (u)
Obtain the plant states x
Find the tracking error and its time derivative using $e(t) = x_d(t) - x(t)$ and $\dot{e}(t) = \dot{x}_d(t) - \dot{x}(t)$
end for

Example 10.5 Design a GRNNSMC controller for the following MIMO nonlinear system using

$$\begin{bmatrix} \dot{x}_1(t) \\ \dot{x}_2(t) \end{bmatrix} = \begin{bmatrix} x_1(t)x_2(t) + x_2^2(t) \\ x_1^2(t) + x_2^2(t) \end{bmatrix} + \begin{bmatrix} 1 & 0 \\ 0 & 1 \end{bmatrix} u(t) \tag{10.69}$$

where $x \in \Re^2$ is the state of the system. $u \in \Re^2$ is the control signal and the initial states are $x(0) = [0.2, -0.2]$. $\sigma = 1$ The output weights of GRNNSMC are initialized to zero.

The desired states are selected as saturated sine wave which can be described as

$$x_d(t) = \begin{Bmatrix} -0.6, & \text{for } x_d(t) \leq -0.6 \\ \sin(t), & \text{for } x_d(t) < 0.6 \\ 0.6, & \text{for } x_d(t) \geq 0.6 \end{Bmatrix} \tag{10.70}$$

The parameters of the controller are given in Table 10.1.

The output of the system accurately follows the desired reference as shown in Fig. 10.10, even at the sharp edges, and the learning is quick.

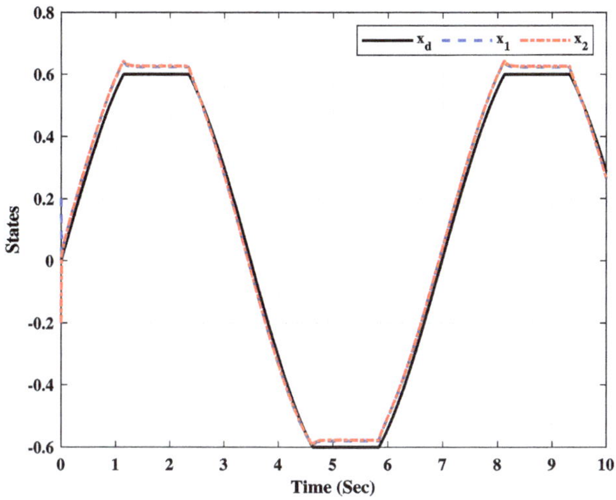

Fig. 10.10 Tracking of x_1 and x_2 using GRNNSMC

10.4 Summary

This chapter concentrates on GRNN-based model-free adaptive control methods by introducing to different approaches to address this problem, namely, the improved GRNN (I_GRNN) controller, which provides a modified version of GRNN to achieve higher control accuracy than the basic GRNN, and the combined GRNN and SMC adaptive controller (GRNNSMC), which aims to adapt GRNN using the robust sliding mode control framework.

10.5 Exercise Problems

1. For the following pendulum system, design a I_GRNN-based controller to control pendulum angle x_1:

$$\dot{x_1} = x_2$$

$$\dot{x_2} = \frac{-g}{l}\sin(x_1) - \frac{k}{m}x_2 + u$$

where x_1 is the pendulum angle θ. g is the gravitational acceleration, l is the length of the pendulum rope, k is the stiffness of the system, and m is the total mass of the system.

2. For the following pendulum system, design a GRNNSMC-based controller to control pendulum angle x_1:

$$\dot{x_1} = x_2$$

$$\dot{x_2} = \frac{-g}{l}\sin(x_1) - \frac{k}{m}x_2 + u$$

where x_1 is the pendulum angle θ. g is the gravitational acceleration, l is the length of the pendulum rope, k is the stiffness of the system, and m is the total mass of the system. Compare the results of I_GRNN and GRNNSMC.
3. What are the main developments in I_GRNN compared to the original GRNN?
4. Design I_GRNN and GRNNSMC controllers for a nonlinear two-tank system where tank 1 drains into tank 2. The system dynamics are governed by Torricelli's law, leading to the following equations:

$$\dot{h}_1(t) = -\frac{a_1}{A_1}\sqrt{2gh_1(t)} + \frac{1}{A_1}u_1(t)$$

$$\dot{h}_2(t) = \frac{a_1}{A_2}\sqrt{2gh_1(t)} - \frac{a_2}{A_2}\sqrt{2gh_2(t)} + \frac{1}{A_2}u_2(t)$$

where $a_1 = a_2 = 50\,\text{cm}^2$ are the areas of the outlet orifices of tank 1 and tank 2, respectively.

$A_1 = A_2 = 50\,\text{cm}^2$ are the cross-sectional areas of tank 1 and tank 2, respectively. $g = 981\,\text{cm/s}^2$. The set water heights are $h_{1d} = 15\,\text{cm}$ and $h_{2d} = 10\,\text{cm}$. $u_1(t)$ and $u_2(t)$ are inflow rates for tank 1 and tank 2 respectively.

Compare the performance of both controllers.
5. What are the benefits of making GRNN's smoothing parameter a range of numbers rather than a single number?
6. What is the significance of adding a direct connection from the input to the output layer of I_GRNN?

References

A.J. Al-Mahasneh, S.G. Anavatti, Novel general regression neural networks for improving control accuracy of nonlinear MIMO discrete-time systems. IEEE Trans. Cybern. **53**(10), 6122–6132 (2023)

A.J. Al-Mahasneh, S.G. Anavatti, M.A. Garratt, M. Pratama, Stable adaptive controller based on generalized regression neural networks and sliding mode control for a class of nonlinear time-varying systems. IEEE Trans. Syst. Man Cybern. Syst. 51(4), 2525–2535 (2021)

D. Bresch-Pietri, J. Chauvin, N. Petit, Adaptive control scheme for uncertain time-delay systems. Automatica **48**(8), 1536–1552 (2012)

M. Ertugrul, O. Kaynak, Neuro sliding mode control of robotic manipulators. Mechatronics **10**(1), 239–263 (2000)

S. Ge, J. Zhang, T. Lee, Adaptive neural network control for a class of MIMO nonlinear systems with disturbances in discrete-time. IEEE Trans. Syst. Man Cybern. Part B (Cybern.) **34**(4), 1630–1645 (2004)

F. Plestan, Y. Shtessel, V. Bregeault, A. Poznyak, New methodologies for adaptive sliding mode control. Int. J. Control **83**(9), 1907–1919 (2010)

Chapter 11
GRNN-Based Evolving Control

11.1 Introduction

Self-evolving neural networks (SENNs) have a great potential to perform better than adaptive neural networks in the control tasks with the addition of structure adaptation to the typical adaptive NN. Thus, adaptive NNs are enclosed in SENNs as shown Fig. 11.1. Adaptive controllers can only adapt their parameters while evolving controllers offer twofold of adaptation: on the structure and parameters levels. Simultaneous structure and parameter adaptations might lead to high control system accuracy particularly when dealing with systems that exhibit significant uncertainties and faults. These include the actuator faults, a commonly occurring phenomenon in dynamic systems.

In this chapter, two evolving GRNN-based controllers, namely, self-evolving GRNN (SEGRNN) and automatic configuration GRNNs (ARGRNNs) are discussed in detail. SEGRNN aims to control nonlinear systems subjected to unknown dynamics and unknown external disturbances. The ARGRNN is shown to handle different types of actuator faults, including complex faults.

11.2 Self-Evolving GRNN (SEGRNN) Controller

The details of the SEGRNN controller will be discussed in this section. SEGRNN[1] controller is developed for reference tracking in a class of unknown discrete-time dynamic systems with external disturbances.

---

[1] Some contents of current Sect. 10.3 are used from Publication "Self-Evolving Neural Control for a Class of Nonlinear Discrete-Time Dynamic Systems With Unknown Dynamics and Unknown Disturbances," IEEE Transactions on Industrial Informatics, vol. 16, pp. 6518–6529, October, 2020, with permission from IEEE (License Number: 6160371274178).

K. Sachan et al., *Intelligent Adaptive Control with System Constraints*,
https://doi.org/10.1007/978-981-95-8220-4_11

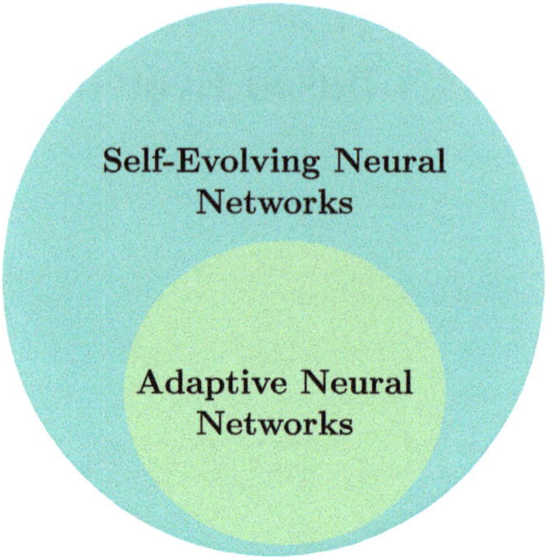

Fig. 11.1 Adaptive NNs and SENNs

SEGRNN begins learning from scratch and simultaneously adapts its structure and parameters to achieve the tracking control goals (Al-Mahasneh et al., 2020). SEGRNN has nodes addition, removal, and replacement to accomplish the control objectives and reject the external disturbances. A robustifying term is also added with SEGRNN output to improve external disturbances rejection. Also, SEGRNN data pool is used as temporary memory of the removed nodes. These nodes might be reused when they are found to be significant.

In addition, Lyapunov direct method is used to analyze the stability of SEGRNN controller and calculate the controller adaption gains.

11.2.1 Design of SEGRNN Controller

Considering a class of nonlinear SISO discrete-time dynamic systems represented as

$$\begin{cases} x_i\left(k+1\right) = f_i\left(x_i\left(k\right)\right), \quad i = 1, \ldots, m-1 \\ x_m\left(k+1\right) = f_m\left(x_m\left(k\right)\right) + g_m\left(x_m\left(k\right)\right) u\left(k\right) \\ \qquad\qquad + d\left(k\right) \\ y\left(k\right) = x_1\left(k\right) \end{cases} \tag{11.1}$$

where $\boldsymbol{x}(k) \in \Re^m$ is a vector of the system states, $u(k) \in \Re$ is the control input, $y(k) \in \Re$ is the system output, $f_i\left(x_i\left(k\right)\right) \in \Re$, $f_m\left(x_m\left(k\right)\right) \in \Re$ and $g_m\left(x_m\left(k\right)\right) \in \Re$ are unknown nonlinear continuous functions, $g_m\left(x_m\left(k\right)\right)$ is bounded as $\left|g_m\left(x_m\left(k\right)\right)\right| < \bar{g}_m$, and $d(k) \in \Re$ is an unknown bounded external disturbance with $|d(k)| \leq \bar{d}$ where $\bar{d}$ is a positive constant.

The control aims to develop an evolving tracking controller using SEGRNN to achieve:

- Boundness of closed-loop tracking error $e(k)$.
- Tracking of the desired reference output $y_d(k)$ in the presence of unknown system dynamics and external disturbances.

Define the filtered tracking error as $s(k) = \Delta e(k) + \lambda e(k)$ with $\Delta e(k)$ is the tracking error difference and $\lambda \in (0, \infty)$ is a positive design parameter and $e(k) = y_d(k) - y(k)$ is the output tracking error.

The control law of the system represented in (11.1) consists of two parts: tracking part $u_{tr}(k)$ and disturbance rejection part $u_{dr}(k)$. SEGRNN is used to estimate the tracking control part $u_{tr}(k)$ while robustifying control term $u_{dr}(k)$ is employed for the disturbance rejection part. Based on that, the control law $u(k)$ can be expressed as

$$u(k) = u_{tr}(k) + u_{dr}(k) \tag{11.2}$$

where u_{tr} is the tracking control term of SEGRNN and u_{dr} is the robustifying control term (Yang and Calise, 2007).

11.2.2 Robustifying Control Design

The robustifying control term is designed as

$$u_{dr}(k) = \tau \operatorname{sat}(s(k)) \tag{11.3}$$

where τ is a design parameter. sat is the saturation function and it is expressed as

$$\operatorname{sat}(s(k)) = \left\{ \begin{array}{ll} \operatorname{sgn}(s(k)) & \text{if } |s(k)| > a \\ \frac{s(k)}{a} & \text{if } |s(k)| \leq a \end{array} \right\}$$

where sgn is the signum function and $a > 0$ is a design parameter depicting a small boundary layer around the filtered tracking error. One can start a with 0.01 and then increase it when the control signal starts chattering.

The signum operator is used to determine the sign of the input and it is expressed as

$$\operatorname{sgn}(s(k)) = \left\{ \begin{array}{ll} 1 & \text{if } s(k) > 0 \\ 0 & \text{if } s(k) = 0 \\ -1 & \text{if } s(k) < 0 \end{array} \right\}$$

Remark 11.1 Since the signum operator is a non-smooth function and discontinuous at $s = 0$, it is usually replaced with other smooth functions such as

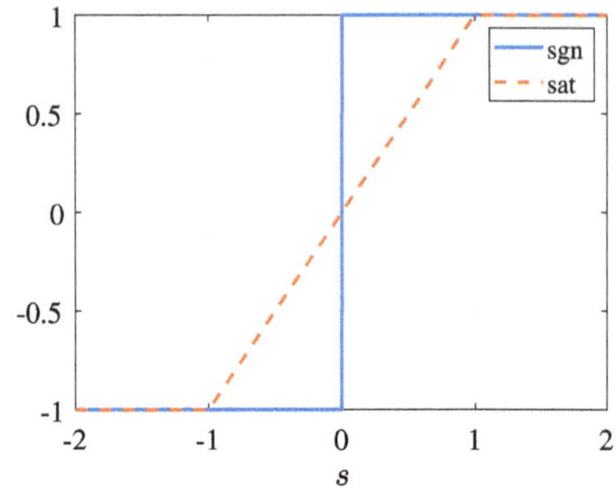

Fig. 11.2 Signum and saturation functions

saturation or tan sigmoid functions to avoid chattering phenomena in the real-time implementation (Sahamijoo et al., 2016). Figure 11.2 demonstrates the difference between the signum and saturation functions.

11.2.3 Tracking Control Design

An ideal tracking control law $u_{trd}(k)$ can be written as an output of SEGRNN as

$$u_{trd}(k) = \boldsymbol{\theta}^{*T}\boldsymbol{\phi}(k) + \epsilon(k) \tag{11.4}$$

where $\boldsymbol{\theta}^* \in \Re^n$ is a constant vector of the ideal SEGRNN output weights, $\boldsymbol{\phi}(k) \in \Re^n$ is SEGRNN hidden layer output vector, and $\epsilon(k) \in \Re$ is SEGRNN estimation error.

The approximation of the tracking control signal $\hat{u}_{tr}(k)$ is expressed as

$$\hat{u}_{tr}(k) = \hat{\boldsymbol{\theta}}(k)^T\boldsymbol{\phi}(k) \tag{11.5}$$

where $\hat{\boldsymbol{\theta}}(k)$ is a vector of SEGRNN output weights.

The approximation error of SEGRNN $\tilde{u}_{tr}(k)$ is the difference between the ideal and the estimated control signals and is expressed as

$$\tilde{u}_{tr}(k) = u_{trd}(k) - \hat{u}_{tr}(k) \tag{11.6}$$

Substituting (11.4) and (11.5) into (11.6) yields

$$\tilde{u}_{tr}(k) = \tilde{\boldsymbol{\theta}}^T(k)\boldsymbol{\phi}(k) - \epsilon(k) \tag{11.7}$$

where $\tilde{\boldsymbol{\theta}}(k) = \hat{\boldsymbol{\theta}}(k) - \boldsymbol{\theta}^*$ is the estimation error of SEGRNN weights.

11.2.4 Structure Adaptation of SEGRNN

The SEGRNN structure adaptation algorithm is composed of the addition, removal, and replacement of the nodes.

- Nodes addition: Adding new nodes to SEGRNN is considered as the backbone of its learning as it allows it to build up its knowledge in a healthy way. It is crucial to check the similarity between the current network input and its current nodes using distance similarity calculations. Euclidean distance $\boldsymbol{d} \in \Re^n$ between the current inputs and SEGRNN hidden weights $\boldsymbol{P}$ is calculated and compared with an addition threshold $d_{th} \geq 0$ as

$$\boldsymbol{d}(k) = \text{dist}(\boldsymbol{P}, \boldsymbol{x}(k)) \tag{11.8}$$

  Based on that, if $\min\{\boldsymbol{d}(k)\} \geq d_{th}$, a new node will be added; otherwise, no node addition will take place.
- Node removal: The contribution of each hidden node in the hidden layer, which is bounded as $0 \leq \phi \leq 1$, is calculated and compared to the contribution threshold $0 \leq \phi_{th} \leq 1$ as

$$\boldsymbol{\phi}(k) = \exp(-(\boldsymbol{d}(k)^2/2\sigma^2)) \tag{11.9}$$

  If any node in the hidden layer has $\phi < \phi_{th}$, it will be removed. Every node removed in the node removal process is stored in a temporary memory for possible reuse in the future. Every node stored in the temporary have hidden weight $\boldsymbol{p}_{temp}$, activation value ϕ_{temp} and output weight $\hat{\theta}_{temp}$. This concept is similar to experience reply in reinforcement learning (Rolnick et al., 2019) but it is limited to the deleted nodes only.
- Nodes replacement: Current nodes might be replaced with nodes from the temporary data memory. Two distance similarity vectors are calculated: the first one between the current input $\boldsymbol{x}(k)$ and the nodes in the temporary memory $\boldsymbol{P}_{temp}$ and the second one between the current input $\boldsymbol{x}(k)$ and the current network nodes $\boldsymbol{P}$. If any node in the temporary memory has less similarity than a current node, it will replace it. Both d_{th} and ϕ_{th} are design parameters, the selection of these parameters affect the control accuracy and SEGRNN hidden layer size. Small d_{th} might reduce the control error but increase the SEGRNN hidden layer size. On the hand, large values of ϕ_{th} will decrease SEGRNN hidden layer size and might reduce the control error.

11.2.5 Weights Adaptation of SEGRNN

After node evolution in the previous stage, the weights of SEGRNN are adapted. To develop the adaptation laws of SEGRNN weights, an objective function j is defined as squared control error

$$j = \frac{1}{2}\tilde{u}_{tr}^2(k) \tag{11.10}$$

The weights are adapted using a gradient-descent approach as follows:

$$\hat{\boldsymbol{\theta}}(k+1) = \hat{\boldsymbol{\theta}}(k) - \gamma \frac{\partial j}{\partial \hat{\boldsymbol{\theta}}(k)} \tag{11.11}$$

where γ is a weight adaptation gain that controls SEGRNN learning speed and stability. Using the chain rule, the adaptation laws can be expressed as

$$\hat{\boldsymbol{\theta}}(k+1) = \hat{\boldsymbol{\theta}}(k) - \gamma \tilde{u}_{tr}(k)\boldsymbol{\phi}(k) \tag{11.12}$$

By assuming that $g_m(x_m(k))$ is bounded away from zero and the tracking error $e(k)$ and $\Delta e(k)$ are also bounded, then the control error is bounded by the tracking error as $\tilde{u}_{tr}(k) \leq e(k)$, (11.12) can be rewritten as

$$\hat{\boldsymbol{\theta}}(k+1) = \hat{\boldsymbol{\theta}}(k) - \gamma e(k)\boldsymbol{\phi}(k) \tag{11.13}$$

SEGRNN block diagram is shown in Fig. 11.3. In the first step, the tracking error $e(k) = y_d(k) - y(k)$ and its difference $\Delta e(k) = e(k) - e(k-1)$ are computed. After that, the evolution algorithm takes the tracking error $e(k) = y_d(k) - y(k)$ and its difference $\Delta e(k) = e(k) - e(k-1)$ and determine which addition, removal, or replacement actions will take place. Following, the adaptation algorithm updates

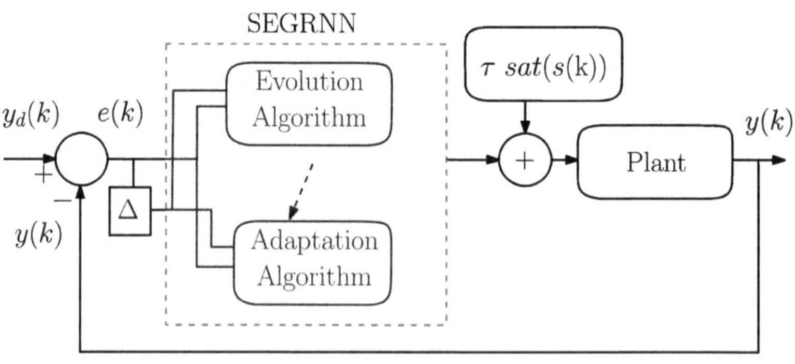

Fig. 11.3 SEGRNN block diagram

Algorithm 11.1 SEGRNN Algorithm

for $k = 1$ to $k =$ end **do**
  Require: $\boldsymbol{x}(k)$, tracking error $\boldsymbol{e}(k)$, d_{th} and ϕ_{th}
  Find the distance similarity vector $\boldsymbol{d}(k)$ using (11.8)
  if $\min\{d(k)\} \geq d_{th}$ **then**
    Add a new node
  end if
  Find nodes contribution $\boldsymbol{\phi}(k)$ using (11.9)
  for Each node i in the hidden layer **do**
    if $\phi_i < \phi_{th}$ **then**
      Remove node i
    end if
  end for
  for Every node in the temporary memory **do**
    Find the similarity between the network input $\boldsymbol{x}(k)$ and the input weights of the temporary nodes $\boldsymbol{P}_{temp}$ using (11.8)
    Find the similarity between the network input $\boldsymbol{x}(k)$ and the current network input weights $\boldsymbol{P}$ using (11.8)
    if Any node r in the temporary memory has less similarity to the network input than any network node i **then**
      Replace temporary node r withe the network node i
    end if
  end for
  Adapt SEGRNN output weights $\hat{\boldsymbol{\theta}}$ using (11.13)
end for

the SEGRNN weights. Also, the control signal $u_{tr}(k)$ is computed. In addition, a robustifying control term $u_{dr}(k)$ is computed. Eventually, the control signals of SEGRNN $u_{tr}(k)$ and the robustifying term $u_{dr}(k)$ are summed up to generate the whole control law $u(k)$.

11.2.6 Stability Analysis of SEGRNN

Lyapunov direct method is utilized to study the stability of the system.

Start with a candidate Lyapunov function $V(k)$ defined as

$$V(k) = e^2(k) + \frac{1}{\gamma}\tilde{\boldsymbol{\theta}}^T(k)\tilde{\boldsymbol{\theta}}(k) \tag{11.14}$$

The first difference of the Lyapunov candidate is

$$\begin{aligned}\Delta V(k) = e^2(k+1) - e^2(k) + \frac{1}{\gamma}(\tilde{\boldsymbol{\theta}}^T(k+1)\tilde{\boldsymbol{\theta}}(k+1) \\ - \tilde{\boldsymbol{\theta}}^T(k)\tilde{\boldsymbol{\theta}}(k))\end{aligned} \tag{11.15}$$

Using (11.13), Cauchy-Schwarz inequality, assuming $e(k+1) = e(k) + \mu$ and $||\boldsymbol{\phi}(k)|| \leq \psi$, (11.15) can be written as

$$\begin{aligned}
\Delta V(k) &< 2(e^2(k) + \rho^2) + \frac{2}{\gamma}(\tilde{\boldsymbol{\theta}}^T(k)\tilde{\boldsymbol{\theta}}(k) - \gamma^2 e^2(k)\psi^2) \\
&\quad - \frac{1}{\gamma}\tilde{\boldsymbol{\theta}}^T(k)\tilde{\boldsymbol{\theta}}(k) \\
&< (e^2(k) + \mu^2) + \frac{1}{\gamma}\tilde{\boldsymbol{\theta}}^T(k)\tilde{\boldsymbol{\theta}}(k) - 2\gamma e^2(k)\psi^2 \\
&< (e^2(k) + \mu^2) + \frac{1}{\gamma}||\tilde{\boldsymbol{\theta}}(k)||^2 - 2\gamma e^2(k)\psi^2
\end{aligned} \tag{11.16}$$

To guarantee Lyapunov stability, $\Delta V(k) < 0$, the adaptation gain γ must be chosen as $\gamma > \dfrac{-||\tilde{\boldsymbol{\theta}}^T(k)||^2}{2(e^2(k) + \mu^2)}$.

Example 11.1 This example deals with the control of continuous stirred tank reactor (CSTR) (Wang et al., 2021) using SEGRNN. Design an SEGRNN controller for the discrete-time CSTR system mathematically expressed as

$$\begin{aligned}
x_1(k+1) &= x_1(k) + f(x_1(k)) \\
x_2(k+1) &= x_2(k) + f(x_2(k)) + g_2(x_2(k))u + d(k) \\
y(k) &= x_1(k)
\end{aligned} \tag{11.17}$$

the dynamics functions $f(x_1(k))$ and $f(x_2(k))$ are expressed as

$$\begin{aligned}
f(x_1(k)) &= \left(-x_1(k) + 0.036(1 - x_1(k))\exp\left(\frac{28.5x_2(k)}{28.5 + x_2(k)}\right)\right) T_s \\
f(x_2(k)) &= \left(-x_2(k) + 0.7740(1 - x_1(k))\exp\left(\frac{28.5x_2(k)}{28.5 + x_2(k)}\right)\right. \\
&\quad \left. -25.2x_2(k)\right) T_s \\
g_2(x_2(k)) &= 28.5T_s
\end{aligned} \tag{11.18}$$

$T_s = 0.01$ seconds is the sampling time. The desired output is expressed as

$$y_d(k) = 0.008\cos(\pi) + \left(\frac{0.1k\pi}{20}\right) + 0.04 \tag{11.19}$$

(continued)

Example 11.1 (continued)
For this problem, the sampling time step T_s is taken as 0.01 seconds, the simulation time is 10 seconds, and the values of the addition and removal thresholds are chosen as $d_{th} = \phi_{th} = 0.1$; these are chosen so SEGRNN would add more nodes than the removed ones to achieve better tracking accuracy. The adaptation gain γ is chosen as 6 to allow fast tracking of the desired output. The robustifying term gain τ is chosen as 0.8 to improve disturbance rejection and the boundary layer is chosen as $a = 0.1$ to achieve sat smoothness.

The initial conditions are $x_1(0) = 0.1$ and $x_2(0) = -0.1$. The disturbance $d(k)$ is given as

$$d(k) = 0.1 \sin(0.05k) \sin(x_1(k)) \tag{11.20}$$

The output tracking history, as shown in Fig. 11.4, demonstrates that oscillatory behavior is observed in the initial learning of SEGRNN, and then the output precisely tracks the desired output. These results are supported by the tracking error history plot in Fig. 11.5 where initial oscillations in the tracking error occur and then the tracking error converges to a region around the zero. The control history is shown in Fig. 11.6. SEGRNN nodes evolution as shown in Fig. 11.7 depicts that SEGRNN nodes increase to five in the initial learning and then reduce to four once accurate tracking is achieved. Also, the evolution of SEGRNN weights norm $||\boldsymbol{\theta}||$ as shown in Fig. 11.8 reveals that the norm initially increases to more than nine and then reduces to an average of 0.5 after the tracking is achieved.

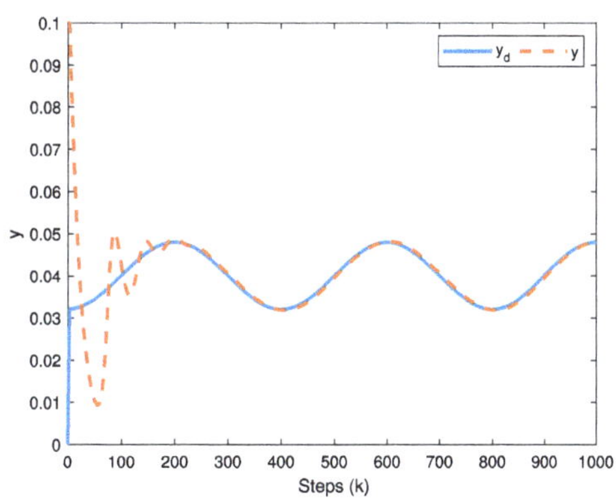

Fig. 11.4 CSTR output tracking history

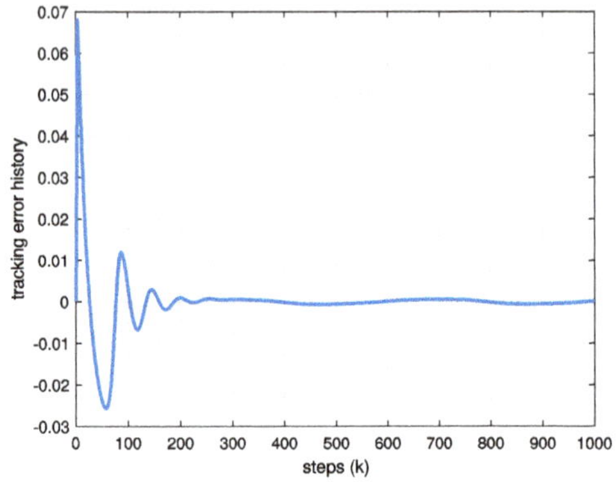

Fig. 11.5 Tracking error history

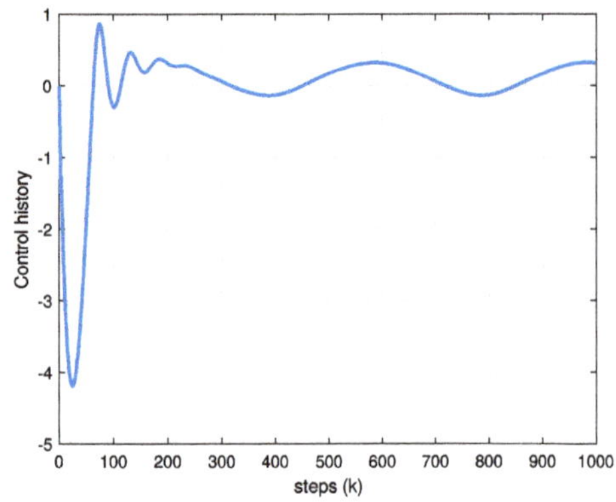

Fig. 11.6 CSTR control history

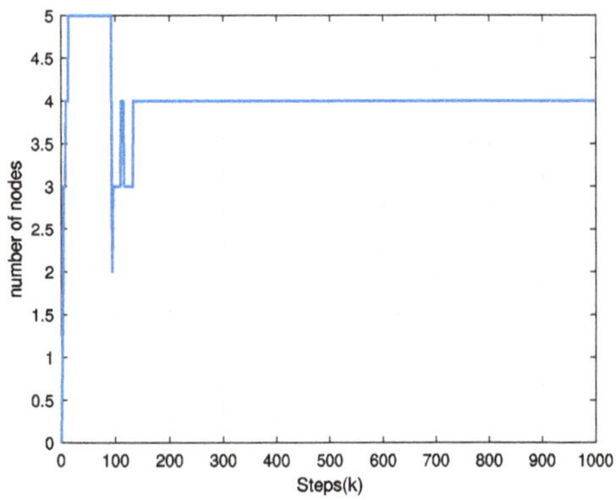

Fig. 11.7 Evolution SEGRNN nodes CSTR

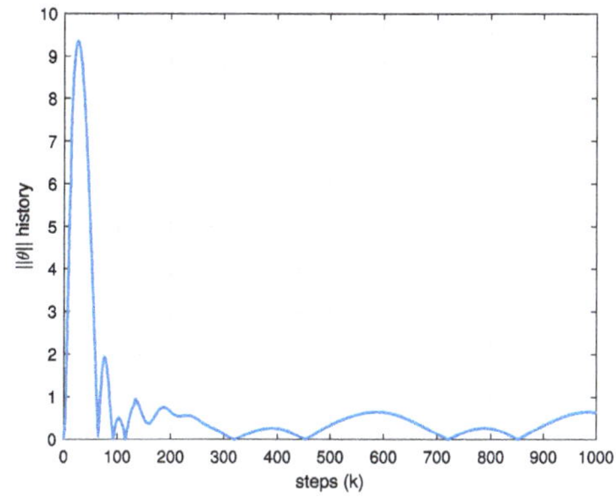

Fig. 11.8 $||\boldsymbol{\theta}||$ history

11.3 Automatic Reconfiguration GRNNs (ARGRNNs) Controller

11.3.1 Introduction

Automatic reconfiguration of controllers is a potential solution to handle actuator faults in dynamic systems (Tu et al., 2022). Actuator faults might affect the safety and efficiency of systems, and thus, they should be handled in real-time. Fault-tolerant control (FTC) attempts to handle potential faults in systems using fault diagnosis and accommodation to sustain control aims (Noura et al., 2009). Faults in dynamic systems might occur in different components such as the sensors or the actuators. A particular type of fault is the complex actuator faults that are considered in this section. Complex actuator faults include concurrent loss of effectiveness and bias faults (Xia et al., 2025). Automatic reconfiguration general regression neural networks (ARGRNNs) are designed to handle this type of fault by detecting anomalies in the input data and reconfiguring the controller structure to attain the control goals (Al-Mahasneh, 2020). ARNNs have two major algorithms, fault detection using the nearest neighbor distance similarity index and controller reconfiguration by evolving GRNN structure.

11.3.2 ARGRNN Controller Design

Consider a MIMO discrete-time dynamic system expressed as

$$\begin{aligned} \boldsymbol{x}(k+1) &= \boldsymbol{F}(\boldsymbol{x}(k)) + \boldsymbol{G}(\boldsymbol{x}(k))\boldsymbol{u}_f(k) + \boldsymbol{d}(k) \\ \boldsymbol{y}(k) &= \boldsymbol{h}(\boldsymbol{x}(k)) \end{aligned} \tag{11.21}$$

where $\boldsymbol{x}(k) = [x_1(k), x_2(k), ..., x_n] \in \Re^n$ is a vector of the system states, $\boldsymbol{f}(\boldsymbol{x}(k)) \in \Re^n$ and $\boldsymbol{G}(\boldsymbol{x}(k)) \in \Re^{n\times n}$ are unknown nonlinear functions, $\boldsymbol{u}_f(k) = [u_{f1}, ..., u_{fm}] \in \Re^m$ is the control input, $\boldsymbol{d}(k) \in \Re^n$ is an external disturbance bounded as $\boldsymbol{d}^T(k)\boldsymbol{d}(k) \leq \bar{d}$, where $\bar{d} > 0$, $y(k) \in \Re^m$ is the system output and $\boldsymbol{h}(\boldsymbol{x}(k)) \in \Re^n$ is the output function.

The control design aims to drive the system output $\boldsymbol{y}(k)$ to a reference output $\boldsymbol{y}_d(k)$ under complex actuator faults. The tracking error is expressed as $\boldsymbol{e}(k) = \boldsymbol{y}_d(k) - \boldsymbol{y}(k)$.

Remark 11.2 Complex actuator faults are defined as coincident loss of effectiveness and bias faults.

The outputs of faulty actuators under complex faults are defined as

$$\boldsymbol{u}_f(k) = \begin{bmatrix} \rho_1(k)u_{f1}(k) + \beta_1(k) \\ \vdots \\ \rho_m(k)u_{fm}(k) + \beta_m(k) \end{bmatrix} \tag{11.22}$$

where $0 < \rho_1(k), \ldots, \rho_m(k) \leq 1$ are the actuators efficiency factors and $\beta_1(k), \ldots, \beta_m(k) \in \Re$ are the actuators bias factors. When $\rho_1(k), \ldots, \rho_m(k) = 1, \ldots, 1$ and $\beta_1(k), \ldots, \beta_m(k) = 0, \ldots, 0$, the actuators are healthy.

Let the output of a non-faulty actuator be defined as $\boldsymbol{u}$ and the output of a faulty actuator be $\boldsymbol{u}_f$; hence, three classes of faults are defined as:

- Loss of effectiveness: $\boldsymbol{u}_f = \boldsymbol{\rho u}$ when any element of the $\boldsymbol{\rho} < 1$, loss of effectiveness faults exist.
- Bias faults: $\boldsymbol{u}_f = \boldsymbol{u} + \boldsymbol{\beta}$ when any element of the vector $\boldsymbol{\beta} \neq 0$, bias faults take place.
- Complex actuator faults: $\boldsymbol{u}_f = \boldsymbol{\rho u} + \boldsymbol{\beta}$ when loss of effectiveness and bias faults are coincident.

11.3.3 Fault Detection and Accommodation Using ARGRNN

ARGRNN is used to detect faults from the incoming data of $\boldsymbol{y}_d(k)$ and $\boldsymbol{u}(k)$ employing the Euclidean distance similarity index as

$$\boldsymbol{d}(k) = \text{dist}(\boldsymbol{P}, \boldsymbol{z}(k)) \tag{11.23}$$

where $\boldsymbol{P}$ is ARGRNN hidden weights matrix and $\boldsymbol{z}(k) = [\boldsymbol{y}_d(k), \boldsymbol{u}(k)]$ is the current network input. The nearest neighbor distance $r(k)$ is the minimum of the distance vector $\boldsymbol{d}(k)$ expressed as

$$r(k) = \min\{\boldsymbol{d}(k)\} \tag{11.24}$$

Faults are detected when $r(k)$ is higher than threshold parameter r_{th}.

After faults are detected, ARGRNN adjusts its structure to accommodate them by adding and removing its nodes. Node addition handles the loss of effectiveness faults while node removal handles bias faults.

- Node addition: If the nearest neighbor distance r is more than the threshold parameter r_{th}, a new node will be added to ARGRNN.
- Node removal: The importance of each node in the hidden layer is calculated as

$$\boldsymbol{m} = \exp\left(-\frac{\boldsymbol{d}^2(k)}{2\sigma^2}\right) \tag{11.25}$$

  where $\sigma > 0$ is the smoothing parameter of ARGRNN. Any node with importance less than an importance parameter m_{th} should be removed from ARGRNN to avoid having weak nodes in the network that do not significantly contribute to the network output.

11.3.4 ARGRNN Weight Adaptation Laws

The ideal control law $\boldsymbol{u}_d$ approximation by ARGRNNS is expressed as

$$\boldsymbol{u}_d(k) = \boldsymbol{\Theta}^{T*}\boldsymbol{\phi}(k) + \boldsymbol{\epsilon} \tag{11.26}$$

where $\boldsymbol{\Theta}^*$ is a constant matrix of the ideal ARGRNNs outputs weights, $\boldsymbol{\phi}(k)$ is ARGRNN hidden layer output, and ϵ is a bounded control approximation error expressed as $\epsilon^T\epsilon \leq \bar{\epsilon}$, where $\bar{\epsilon} > 0$.

The current estimation of the control law $\boldsymbol{u}(k)$ by ARGRNNs is expressed as

$$\boldsymbol{u}(k) = \boldsymbol{\Theta}^T\boldsymbol{\phi}(k) \tag{11.27}$$

The control estimation error $\tilde{\boldsymbol{u}}(k)$ is written as

$$\begin{aligned} \tilde{\boldsymbol{u}}(k) &= u_d(k) - u(k) \\ &= \boldsymbol{\Theta}^{T*}\boldsymbol{\phi}(k) + \boldsymbol{\epsilon} - \boldsymbol{\Theta}^T\boldsymbol{\phi}(k) \\ &= \tilde{\boldsymbol{\Theta}}^T\boldsymbol{\phi}(k) + \boldsymbol{\epsilon} \end{aligned} \tag{11.28}$$

The gradient-descent approach is used to adapt ARGRNN output weights using an objective function j as follows:

$$\boldsymbol{\Theta}(k+1) = \boldsymbol{\Theta}(k) - \gamma\frac{\partial j(k)}{\partial\boldsymbol{\Theta}(k)} \tag{11.29}$$

where γ is the adaptation gain and it affects the closed-loop stability.

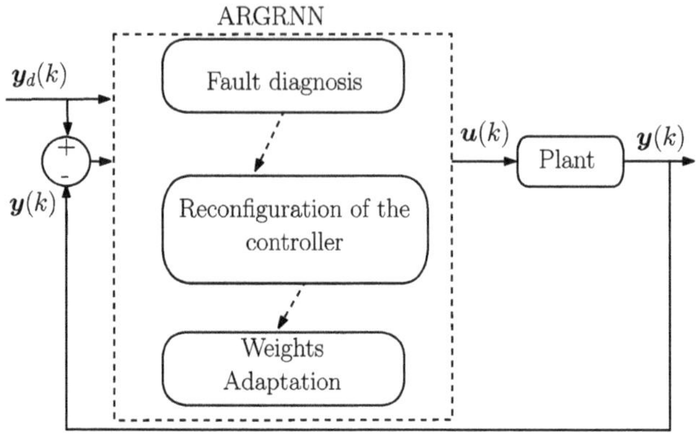

Fig. 11.9 ARGRNNs block diagram

The cost function j is selected to minimize the control estimation error $\tilde{\boldsymbol{u}}$ as follows:

$$j(k) = \frac{1}{2}(\tilde{\boldsymbol{u}}^T(k)\tilde{\boldsymbol{u}}(k)) \tag{11.30}$$

After finding the derivative $\frac{\partial j(k)}{\partial \boldsymbol{\Theta}(k)}$, the adaptation law in (11.29) becomes

$$\boldsymbol{\Theta}(k+1) = \boldsymbol{\Theta}(k) - \gamma \tilde{\boldsymbol{u}}(k)^T \boldsymbol{\phi}(k) \tag{11.31}$$

Assuming that the control error $\tilde{\boldsymbol{u}}(k)$ is bounded by the tracking error as $\tilde{\boldsymbol{u}}(k) \leq \boldsymbol{e}(k) + \boldsymbol{\mu}$, (11.31) becomes

$$\boldsymbol{\Theta}(k+1) = \boldsymbol{\Theta}(k) - \gamma(\boldsymbol{e}(k) + \boldsymbol{\mu})^T \boldsymbol{\phi}(k) \tag{11.32}$$

The ARGRNNs, as shown in Fig. 11.9, contain three major components: fault detection, automatic reconfiguration, and weights adaptation law. ARGRNNs algorithm is detailed in Algorithm 11.2.

11.3.5 Stability Analysis of ARGRNNs

Given $\tilde{\boldsymbol{u}}(k)$ is bounded as $\tilde{\boldsymbol{u}}(k) \leq \boldsymbol{e}(k) + \boldsymbol{\mu}$, Lyapunov direct method is used for stability analysis.

Consider the following Lyapunov function candidate $V(k)$ as

$$V(k) = \tilde{\boldsymbol{u}}^T(k)\tilde{\boldsymbol{u}}(k) \tag{11.33}$$

Algorithm 11.2 ARGRNNs Algorithm

for $k = 1$ to $k =$ end **do**
  Require: current input $z(k) = [\boldsymbol{y}_d(k), \boldsymbol{u}(k))]$, tracking error $\boldsymbol{e}(k)$
  Find the nearest neighbor distance r between the current input and hidden weights
  if $r \geq r_{th}$ **then**
    Add a new node
  end if
  Find the importance of each node Θ_i
  for $i = 1$ to $i =$ End **do**
    if $m_i \leq m_{th}$ **then**
      Remove node i
    end if
  end for
  Adapt ARGRNN output weights $\boldsymbol{\Theta}$ using (11.32)
end for

The first difference of $V(k)$ is expressed as

$$\begin{aligned}\Delta V(k) &= \tilde{\boldsymbol{u}}^T(k+1)\tilde{\boldsymbol{u}}(k+1) - \tilde{\boldsymbol{u}}^T(k)\tilde{\boldsymbol{u}}(k)\\ &= (\tilde{\boldsymbol{\Theta}}^T(k+1)\boldsymbol{\phi}(k+1) + \boldsymbol{\epsilon})^T\\ &\times (\tilde{\boldsymbol{\Theta}}^T(k+1)\boldsymbol{\phi}(k+1) + \boldsymbol{\epsilon})\\ &- (\tilde{\boldsymbol{\Theta}}^T(k)\boldsymbol{\phi}(k) + \boldsymbol{\epsilon})^T(\tilde{\boldsymbol{\Theta}}^T(k)\boldsymbol{\phi}(k) + \boldsymbol{\epsilon})\end{aligned} \tag{11.34}$$

Substituting (11.31) into (11.34) and since the output ARGRNN hidden layer is bounded as $||\boldsymbol{\phi}|| \leq q$ yields

$$\begin{aligned}\Delta V(k) = q^2(\tilde{\boldsymbol{\Theta}}(k) &- \gamma(\boldsymbol{e}(k) + \boldsymbol{\mu})^T + \boldsymbol{\epsilon})^T\\ &\times (\tilde{\boldsymbol{\Theta}}(k) - \gamma(\boldsymbol{e}(k) + \boldsymbol{\mu})^T + \boldsymbol{\epsilon}))\\ &- a^2(\tilde{\boldsymbol{\Theta}}^T(k) + \boldsymbol{\epsilon})^T(\tilde{\boldsymbol{\Theta}}^T(k) + \boldsymbol{\epsilon})\end{aligned} \tag{11.35}$$

Using Cauchy-Schwarz inequality, (11.35) can be rewritten as

$$\Delta V(k) \leq 3q^2||\tilde{\boldsymbol{\Theta}}(k)||^2 + 2q^2\boldsymbol{\epsilon} - 3\gamma^2 q^4||(\boldsymbol{e}(k) + \boldsymbol{\mu})||^2 \tag{11.36}$$

To achieve Lyapunov stability, γ should be selected as

$$\gamma \geq \sqrt{\frac{3||\tilde{\boldsymbol{\Theta}}(k)||^2 + 2\boldsymbol{\epsilon}}{3a^2||(\boldsymbol{e}(k) + \boldsymbol{\mu})||^2}} \tag{11.37}$$

Based on the above analysis, the closed-loop system is stable.

In the next section, numerical examples of ARGRNNs are provided to demonstrate their fault-tolerant control capabilities.

Example 11.2 ARGRNNs for SISO discrete-time system.

Consider the dynamic system given by

$$x_1(k+1) = \frac{1.4x_1^2(k)}{(1+x_1^2(k))} + (0.1 + 0.05\cos(x_1(k)))x_2(k)$$

$$x_2(k+1) = \frac{x_2^2(k)}{1+x_1^2(k)+x_2^2(k)} + u_f(k)$$

$$y(k) = x_1(k)$$

Two actuator faults are introduced: the first one takes place at step number 100 with parameters $\rho_1 = 0.5$ and $\beta_1 = 1.6$, and the second fault takes place at step number 500 with parameters $\rho_1 = 0.5$ and $\beta_1 = -0.5$. The initial states are $x_1(0) = 0.05$ and $x_2(0) = 0.01$. The adaptation gain γ is 1.5. r_{th} and m_{th} are selected as 0.1 and 0.5, respectively. The reference output trajectory $y_d(k)$ is

$$y_d(k) = 0.026\cos(kT_s\pi/10) + 0.035 \tag{11.38}$$

where the sampling time T_s is 0.01 seconds.

The output tracking results, as shown in Fig. 11.10, depict minor oscillation in the initial adaptation stage and more significant when faults take place. However, soon after the faults, ARGRNN was able to restore the system output to the desired output. The nearest neighbor distance evolution results, as shown in Fig. 11.11, depict that the distance substantially rose from zero to more than two when the first fault occurred and rose from zero to almost 0.5 when the second fault occurred. These increments in the distance triggered the fault detection algorithm which activated the node evolution in ARGRNN to accommodate the faults as shown in Fig. 11.12. When the first fault took place, the number of nodes decreased to two and then increased to 13 to accommodate the fault. In the second fault, ARGRNN nodes decreased to 10 and then increased to 15.

In the next example, ARGRNN and adaptive GRNN are compared for a MIMO system.

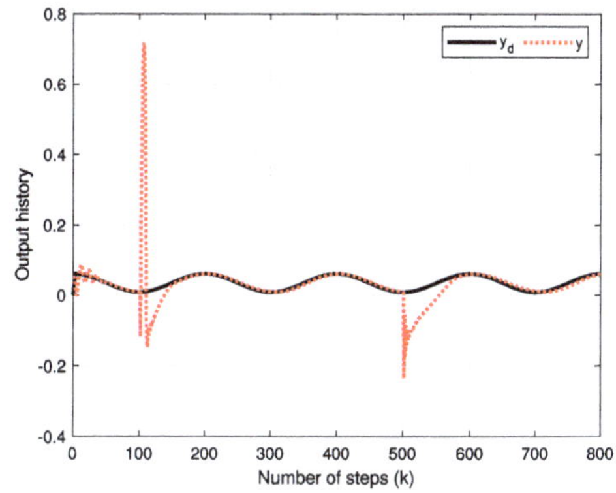

Fig. 11.10 Output tracking history

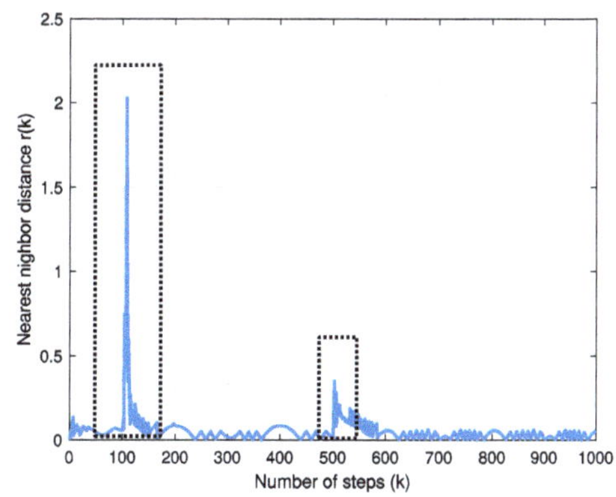

Fig. 11.11 Nearest neighbor distance evolution

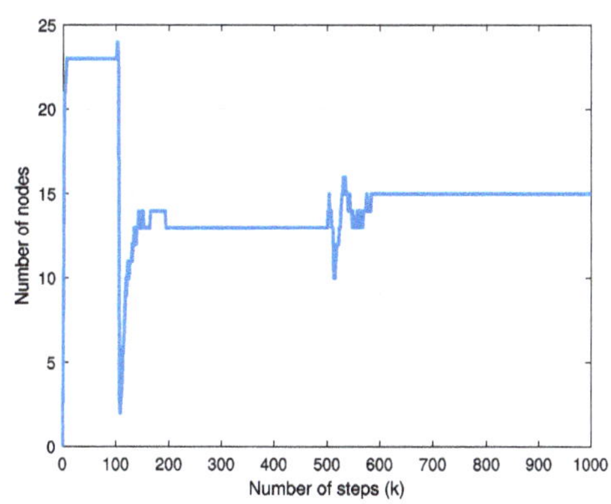

Fig. 11.12 ARGRNN nodes evolution

Example 11.3 Comparison of ARGRNNs with adaptive GRNNs for MIMO discrete-time system.

Consider the dynamic model of the system given as

$$\begin{aligned} x_{11}(k+1) &= x_{21}(k) \\ x_{12}(k+1) &= x_{22}(k) \\ x_{21}(k+1) &= \frac{0.4x_{11}(k)}{(1+x_{21}^2(k))} + 2u_{1f}(k) + d_1(k) \\ x_{22}(k+1) &= (0.1 + 0.05\cos(x_{12}(k)))x_{22}(k) \\ &\quad + 2u_{2f}(k) + d_2(k) \\ y_1(k) &= x_{11}(k) \\ y_2(k) &= x_{12}(k) \end{aligned} \tag{11.39}$$

The parameters of actuator fault are $\rho_1, \rho_2 = 0.2$ and $\beta_1, \beta_2 = -0.5$. The fault happens at step number 500. The initial states are $x_{11}(0) = -1, x_{12}(0) = 1, x_{21}(0) = -0.5$, and $x_{22}(0) = 1$. The adaptation gains are $\gamma = [0.18, 0.18]$. $r_{th} = 0.1$ and $m_{th} = 0.5$.

The reference outputs are expressed as

$$\begin{aligned} y_{d1}(k) &= 0.3\sin((0.5k\pi/50) + (\pi/4)) \\ y_{d2}(k) &= 0.2\cos((0.5k\pi/50) - (\pi/4)) \end{aligned} \tag{11.40}$$

The external disturbances are

$$\begin{aligned} d_1(k) &= 0.05\cos(0.01k)\sin(x_{11}(k)) \\ d_2(k) &= 0.05\sin(0.01k)\cos(x_{12}(k)) \end{aligned} \tag{11.41}$$

The output tracking results as shown in Figs. 11.13 and 11.14 demonstrate the existence of some oscillations in the initial adaptation for both the adaptive GRNN and ARGRNN. In addition, both controllers achieve accurate tracking after the initial learning. On the other hand, when the error took place at step number 500, ARGRNN was able to sustain acceptable tracking accuracy while the adaptive GRNN could not. Also, ARGRNN has a much faster response to the faults than the adaptive GRNN.

The nearest neighbor distance evolution $r(k)$, as shown in Fig. 11.15, demonstrates a significant increase in the distance to almost 0.8 which triggered the fault detection. After the fault was detected, ARGRNN evolved its nodes by increasing them to 68 and then decreasing them to 5 to

(continued)

Example 11.3 (continued)
accommodate the fault. The evolution of ARGRNNs' nodes in example 2, as depicted in Fig. 11.16, demonstrates the ability of the ARGRNNs to reconfigure themselves to accommodate for the actuator faults. The number of nodes initially increased to 43. When the faults happened at step number 500, the controller eliminated the redundant nodes and started gradually adding more meaningful nodes to accommodate the faults.

The evolution of the control signals u_{tr} and u_{dr} for both controllers, as shown in Fig. 11.17, depicts that both controllers have similar control signals until a fault occurs. When the faults occur, ARGRNN raises its control signals to achieve the desired output tracking while the adaptive GRNN could not reach a suitable control level. These results confirm the ability of ARGRNN to better handle faults than the adaptive GRNN.

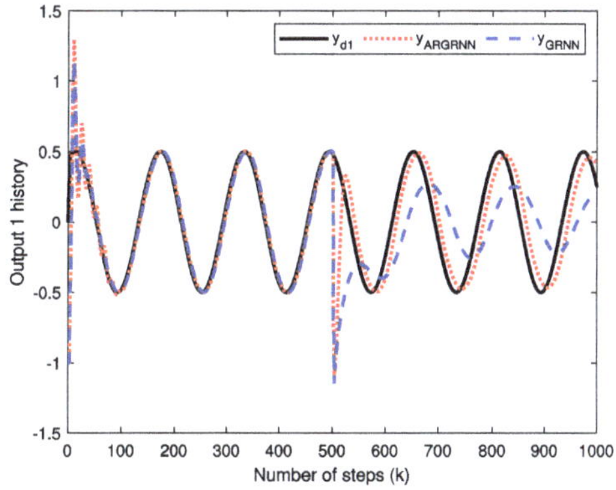

Fig. 11.13 Output 1 tracking adaptive Vs. ARGRNN

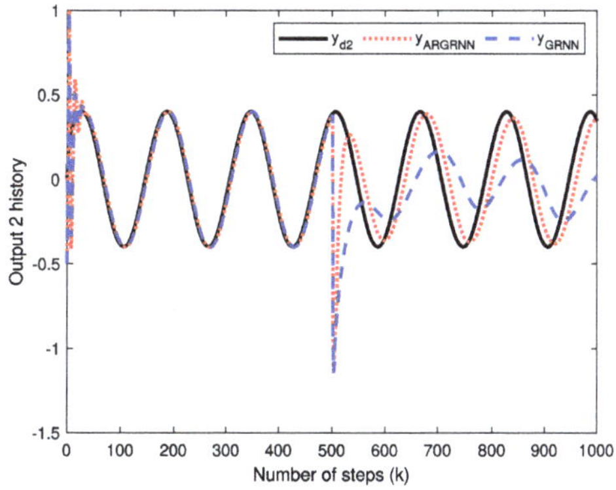

Fig. 11.14 Output 2 tracking adaptive Vs. ARGRNN

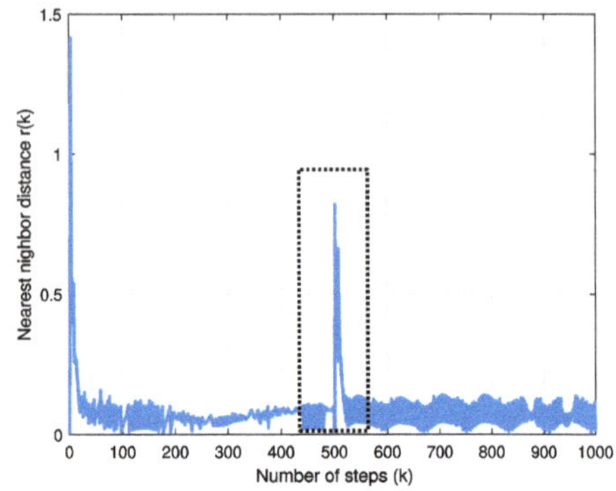

Fig. 11.15 Nearest neighbor distance evolution

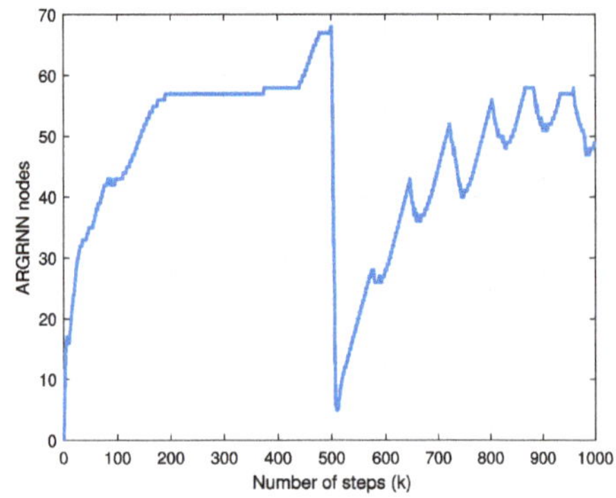

Fig. 11.16 Evolution of ARGRNNs nodes

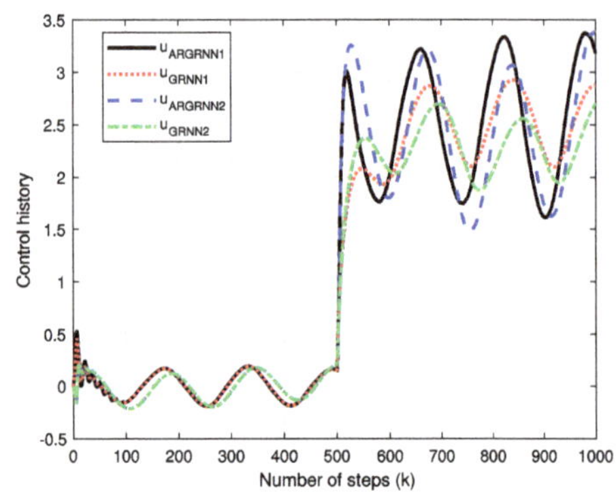

Fig. 11.17 Control history

11.4 Summary

This chapter develop two major evolving GRNN-based control systems, namely, SEGRNN and ARGRNN. SEGRNN is utilized to control systems with unknown dynamics and disturbances, while ARGRNN is used to control unknown systems with complex actuator faults.

11.5 Exercise Problems

1. For the following pendulum system, design an ARGRNN-based controller to control pendulum angle x_1.

$$\dot{x_1} = x_2$$

$$\dot{x_2} = \frac{-g}{l}\sin x_1 - \frac{k}{m}x_2 + u$$

   where x_1 is the pendulum angle θ. g is the gravitational acceleration, l is the length of the pendulum rope, k is the stiffness of the system, and m is the total mass of the system. The actuator fault can be expressed as $u = 0.5u + 0.1$ and happens at time = 2 seconds.
2. For the above pendulum system, design a SEGRNN-based controller to control pendulum angle x_1 and compare between ARGRNNs and SEGRNNs.
3. For the following system, design ARGRNN-based and SEGRNN-based controllers to drive the states x_1 and x_2 to $x_{1d} = 0.1\sin(t)$ and $x_{2d} = 0.2\cos(t)$. The actuator faults are $u_{tr} = 0.4u_{tr} + 0.5$ and $u_{dr} = 0.6u_{dr} + 0.7$ and both faults occur at time = 5 seconds.

$$\dot{x_1} = x_2 + u_{tr}$$

$$\dot{x_2} = x_1 + x_1 x_2 + u_{dr}$$

4. Design ARGRNN-based and SEGRNN-based controllers for the following system. Consider controlling the state x_1 to the desired state $x_{1d} = \sin(t)\cos(t)$. The actuator fault is expressed as $u = a(t)u + b(t)$, where $a(t) = 0.5|\sin(t)|$ and $b(t) = 0.4|\cos(t)|$. The fault occurs at time = 3 seconds.

$$\dot{x_1} = x_2$$

$$\dot{x_2} = x_1^2 + x_2^2 + u$$

5. What could be the effect of dynamic neural structure on the fault tolerance behavior in control systems?

References

A.J. Al-Mahasneh, Evolving neural networks for modeling and control of dynamic systems. Ph.D. dissertation, UNSW Sydney (2020)

A.J. Al-Mahasneh, S.G. Anavatti, M.A. Garratt, Self-evolving neural control for a class of nonlinear discrete-time dynamic systems with unknown dynamics and unknown disturbances. IEEE Trans. Ind. Inf. **16**(10), 6518–6529 (2020)

H. Noura, D. Theilliol, J.-C. Ponsart, A. Chamseddine, *Fault-Tolerant Control Systems: Design and Practical Applications* (Springer Science & Business Media, Berlin, 2009)
D. Rolnick, A. Ahuja, J. Schwarz, T. Lillicrap, G. Wayne, Experience replay for continual learning. Adv. Neural Inf. Proces. Syst. **32**, 350–360 (2019)
A. Sahamijoo, F. Piltan, M.H. Mazloom, M.R. Avazpour, H. Ghiasi, N.B. Sulaiman, Methodologies of chattering attenuation in sliding mode controller. Int. J. Hybrid Inf. Technol. **9**(2), 11–36 (2016)
Y. Tu, D. Wang, S.X. Ding, F. Fu, W. Li, A reconfiguration-based fault-tolerant control method for nonlinear uncertain systems. IEEE Trans. Autom. Control **67**(11), 6060–6067 (2022)
G. Wang, Q.-S. Jia, J. Qiao, J. Bi, M. Zhou, Deep learning-based model predictive control for continuous stirred-tank reactor system. IEEE Trans. Neural Networks Learn. Syst. **32**(8), 3643–3652 (2021)
Y. Xia, Y. Liu, W. Sun, A new fault tolerant strategy using adaptive time delay estimation for robot manipulators with actuator faults. Nonlin. Dyn. 1–13 (2025)
B.-J. Yang, A.J. Calise, Adaptive control of a class of nonaffine systems using neural networks. IEEE Trans. Neural Networks **18**(4), 1149–1159 (2007)

Chapter 12
Case Studies

This chapter presents several practical applications of the control design theories discussed in the book, illustrated through case studies. These example problems not only enrich the quality of the book but also strengthen the confidence of the reader in the applicability of nonlinear design techniques.

12.1 Neuro-adaptive Control of a Laser Beam for Jitter Correction

A laser beam is projected onto a distant target, aiming to consistently fall on the desired portion until the task is completed. This system can be applied in various fields, such as high-precision cutting, welding, drilling, and target range estimation in aerospace applications. Laser beams are highly coherent, maintaining a fixed phase relationship, which allows them to be narrow, focused, and collimated with minimal spread. These properties are valid in a uniform, ideal propagation medium. However, when the laser beam travels through the atmosphere, as in applications like micro-satellite range estimation or directed energy weapons for disabling drones, it experiences deviations due to atmospheric effects, known as jitter.

This pointing error is measured by a narrow field-of-view sensor and used to actuate a fast steering mirror (FSM), which adjusts horizontally and vertically to correct the beam deviation. The FSM system is often modeled as a spring-mass-damper system, with a transfer function that takes voltage as the control input and the angle of rotation as the output. Due to the challenges in perfectly modeling the FSM system, a neuro-adaptive controller is necessary, capable of adapting to uncertainties and working alongside nominal control designs. In this case

K. Sachan et al., *Intelligent Adaptive Control with System Constraints*,
https://doi.org/10.1007/978-981-95-8220-4_12

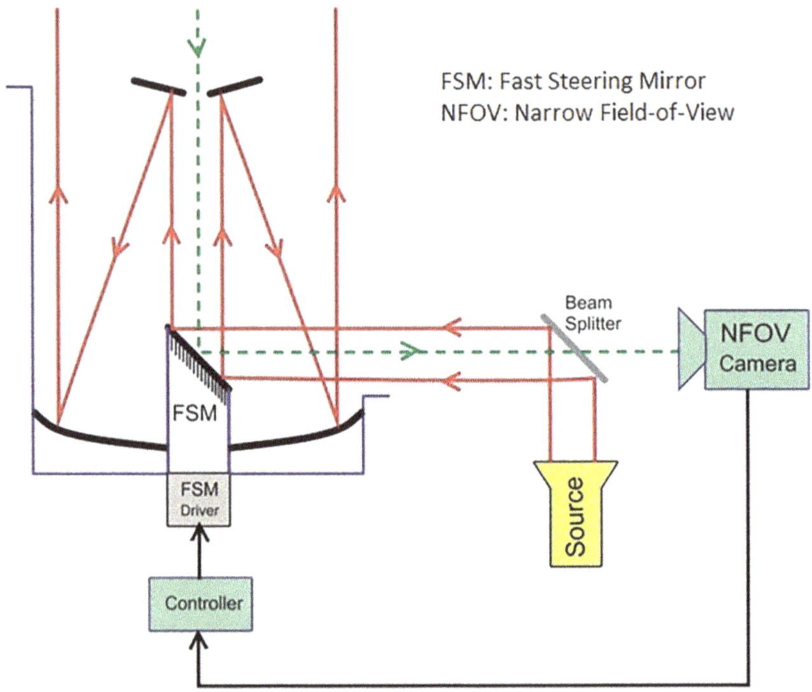

Fig. 12.1 Overall schematic of the FSM system

study,[1] a model-following neuro-adaptive control system is designed to enhance the performance of FSM used for tracking a reference signal in the presence of jitter.

12.1.1 Problem Formulation

The schematic diagram of the FSM system is shown in Fig. 12.1. The source emits a laser beam toward the target, and the reflected beam is observed by the narrow field-of-view (NFOV) sensor. The reference signal $\boldsymbol{y}^*$ represents the angular deviation of the beam along the horizontal and vertical axes from the target, expressed as $\boldsymbol{y}^* = [\psi^*, \gamma^*]^T$. This reference signal is filtered with a cutoff frequency corresponding to the actuator bandwidth, resulting in the filtered signal $\boldsymbol{y}_f^*$. Using the range measurement recorded by the laser range finder, the tracking error is estimated. The error signal is then fed to the controller, which generates the appropriate voltage

[1] The simulation results have been generated by Mr. Anand N. Sortur, a graduate student at the Indian Institute of Science, Bangalore, India. The authors express their sincere gratitude to him for his valuable contribution in providing the relevant results.

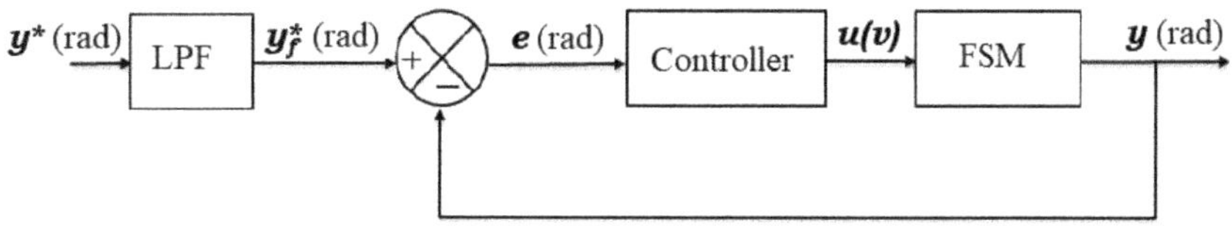

Fig. 12.2 Block diagram of FSM control system

input to the FSM, as illustrated in Fig. 12.2. The objective is to achieve a tracking root mean square (RMS) error on the order of 10^{-6} radians.

The transfer function of the FSM system is given by

$$G(s) = \frac{k\omega_n^2}{s^2 + 2\xi\omega_n s + \omega_n^2} \tag{12.1}$$

where $\frac{\psi(s)}{V_\psi(s)} = G(s)$, $\frac{\gamma(s)}{V_\gamma(s)} = G(s)$, $\psi(s)$, and $\gamma(s)$ are the horizontal and vertical angles, respectively, and $V_\psi(s)$, $V_\gamma(s)$ are their corresponding input voltages. From the system identification the parameters obtained are $k = 1.046$, $\omega_n = 1.554 \times 10^3$ and $\xi = 0.57$.

12.1.2 Adaptive Control Design

In this subsection, a model-following neuro-adaptive control system is designed to maintain reference signal tracking performance in the presence of jitter.

12.1.2.1 Low Pass Filter Design

The smoothness of the filtered signal and the delay have an inverse relationship. Therefore, a trade-off must be made between smoothness and delay when determining the filter order. In this case, a second-order low-pass filter is considered as follows (Samantaray and Padhi, 2017):

$$\frac{y_f^*(s)}{y^*(s)} = \frac{\bar{\omega}_n^2}{s^2 + 2\bar{\xi}\bar{\omega}_n s + \bar{\omega}_n^2} \tag{12.2}$$

where $\bar{\omega}_n$ and $\bar{\xi}$ are the undamped frequency and damping ratio of the low-pass filter. The 3 dB bandwidth of this filter is

$$\omega_{\mathrm{BW}} = \omega_n \sqrt{1 - 2\xi^2 + \sqrt{2 - 4\xi^2 + 4\xi^4}} \tag{12.3}$$

In the low-pass filter, the damping ratio $\xi = \frac{1}{\sqrt{2}}$ is chosen to ensure there is no overshoot in the frequency response and to equally weight all frequency components of the signal below the cutoff frequency. This results in $\omega_{\mathrm{BW}} = \bar{\omega}_n$, meaning $\bar{\omega}_n$ should match the actuator bandwidth of the FSM system. The actuator bandwidth is determined by substituting the corresponding ξ and ω_n from the FSM model into Eq. (12.3), yielding approximately 1500 rad/s (or a frequency of $\approx$300 Hz), which is equal to the ω_n of the low-pass filter. Therefore, the transfer function of the low-pass filter is given by Samantaray and Padhi (2017)

$$\frac{\boldsymbol{y}_f^*(s)}{\boldsymbol{y}^*(s)} = \frac{2.25 \times 10^6}{s^2 + 2121s + 2.25 \times 10^6} \tag{12.4}$$

For simulation purposes, the reference signal is filtered with a frequency of approximately 100 Hz, which is one-third (1/3) of the actuator specified frequency. This ensures that the actuator operates at one-third of its specified frequency.

12.1.2.2 Dynamic Inversion Control for Nominal System

A reference model is determined from the nominal model. The nominal plant model is defined by

$$\begin{aligned} \dot{\boldsymbol{x}}_m &= \boldsymbol{A}_1 \boldsymbol{x}_m + \boldsymbol{B}_1 \boldsymbol{u} \\ \boldsymbol{y}_m &= \boldsymbol{C}_1 \boldsymbol{x}_m \end{aligned} \tag{12.5}$$

where

$$\boldsymbol{x}_m = \left[\, \psi \ \dot{\psi} \ \gamma \ \dot{\gamma} \,\right]^T$$

$$\boldsymbol{u} = \left[\, V_\psi \ V_\gamma \,\right]^T.$$

$$\boldsymbol{A}_1 = \begin{bmatrix} 0 & 1 & 0 & 0 \\ -1.957 \times 10^6 & -1760 & 0 & 0 \\ 0 & 0 & 0 & 1 \\ 0 & 0 & -1.957 \times 10^6 & -1760 \end{bmatrix}$$

$$\boldsymbol{B}_1 = \begin{bmatrix} 0 & 0 \\ 2.047 \times 10^6 & 0 \\ 0 & 0 \\ 0 & 2.047 \times 10^6 \end{bmatrix}$$

$$\boldsymbol{C}_1 = \begin{bmatrix} 1\,0\,0\,0 \\ 0\,0\,1\,0 \end{bmatrix}$$

The objective is $\boldsymbol{y}_m$ should track $\boldsymbol{y}_f^*$ asymptotically. Hence, let us consider the error as

$$\boldsymbol{e}_m = \boldsymbol{y}_m - \boldsymbol{y}_f^* \tag{12.6}$$

The error dynamic equation can be written as

$$\ddot{\boldsymbol{e}}_m + \boldsymbol{K}_1\dot{\boldsymbol{e}}_m + \boldsymbol{K}_2\boldsymbol{e}_m = \boldsymbol{0} \tag{12.7}$$

where $\boldsymbol{K}_1$ and $\boldsymbol{K}_2$ are positive definite matrices. The order of the error dynamic equation is selected as second order since the relative degree of the open-loop model is two. This leads to the control equation as (assuming $(\dot{\boldsymbol{y}}_f^* = \boldsymbol{0}, \ddot{\boldsymbol{y}}_f^* = \boldsymbol{0})$)

$$\boldsymbol{u} = (\boldsymbol{C}_1\boldsymbol{A}_1\boldsymbol{B}_1)^{-1}\left[-\boldsymbol{K}_m\boldsymbol{x}_m + \boldsymbol{K}_2\boldsymbol{y}_f^*\right] \tag{12.8}$$

where $\boldsymbol{K}_m = \left(\boldsymbol{C}_1\boldsymbol{A}_1^2 + \boldsymbol{K}_1\boldsymbol{C}_1\boldsymbol{A}_1 + \boldsymbol{K}_2\boldsymbol{C}_1\right)$. Rearranging the terms, the reference model is obtained as

$$\dot{\boldsymbol{x}}_{\text{ref}} = \boldsymbol{A}_{\text{ref}}\boldsymbol{x}_{\text{ref}} + \boldsymbol{B}_{\text{ref}}\boldsymbol{y}_f^* \tag{12.9}$$

where $\boldsymbol{A}_{\text{ref}} = \left[\boldsymbol{A}_1 - \boldsymbol{B}_1(\boldsymbol{C}_1\boldsymbol{A}_1\boldsymbol{B}_1)^{-1}\boldsymbol{K}_m\right]$, $\boldsymbol{B}_{\text{ref}} = \left[\boldsymbol{B}_1(\boldsymbol{C}_1\boldsymbol{A}_1\boldsymbol{B}_1)^{-1}\boldsymbol{K}_2\right]$.

12.1.2.3 Neuro Adaptive Control Design

The perturbed model is constructed as

$$\dot{\boldsymbol{x}} = \boldsymbol{A}\boldsymbol{x} + \boldsymbol{B}\boldsymbol{u}$$

where $\boldsymbol{A} = \boldsymbol{A}_{ref} + \Delta\boldsymbol{A}_m$ and $\boldsymbol{B} = \boldsymbol{B}_1 + \Delta\boldsymbol{B}_m$ and $\Delta\boldsymbol{A}_m$ and $\Delta\boldsymbol{B}_m$ are matched uncertainties in $\boldsymbol{A}$ and $\boldsymbol{B}$, respectively.

The transfer function of FSM is given by

$$G(s) = \frac{k\omega_n^2}{s^2 + 2\zeta\omega_n s + \omega_n^2} = \frac{2.527 \times 10^6}{s^2 + 1760s + 2.416 \times 10^6}$$

For simulations, $\boldsymbol{A}$ and $\boldsymbol{B}$ are assumed to be

$$\boldsymbol{A} = \begin{bmatrix} 0 & 1 & 0 & 0 \\ -\omega_n^2 p_{\omega_n}^2 & -2\zeta\omega_n p_\zeta p_{\omega_n} & \omega_n^2 p_{c1} & 2\zeta\omega_n p_{c2} \\ 0 & 0 & 0 & 1 \\ \omega_n^2 p_{c1} & 2\zeta\omega_n p_{c2} & -\omega_n^2 p_{\omega_n}^2 & -2\zeta\omega_n p_\zeta p_{\omega_n} \end{bmatrix}$$

$$B = \begin{bmatrix} 0 & 0 \\ k\omega_n^2 p_{\omega_n}^2 p_k & 0 \\ 0 & 0 \\ 0 & k\omega_n^2 p_{\omega_n}^2 p_k \end{bmatrix}$$

where p_k, p_ζ, p_{ω_n}, p_{c1}, and p_{c2} are multiplicative perturbation uncertainties considered in the model parameters k, ω_n, and ζ. In ideal case $p_k = 1$, $p_{\zeta=1}$, $p_{\omega_n} = 1$, $p_{c1} = 0$, and $p_{c2} = 0$.

Since the system is non-square and affine, but the dimension of the state is greater than the dimension of control input ($n > m$), a similar control design strategy presented in Sect. 6.6.1.2 is followed.

12.1.3 Simulation Results

This section presents the simulation results for tracking the jitter of the laser beam in horizontal and vertical directions. The horizontal jitter tracking and the corresponding tracking error are shown in Figs. 12.3 and 12.4, respectively. The horizontal tracking achieved an RMS of 7.95 μrad. The vertical jitter tracking and its tracking error are shown in Figs. 12.5 and 12.6. The vertical tracking achieved an RMS of 7.78 μrad.

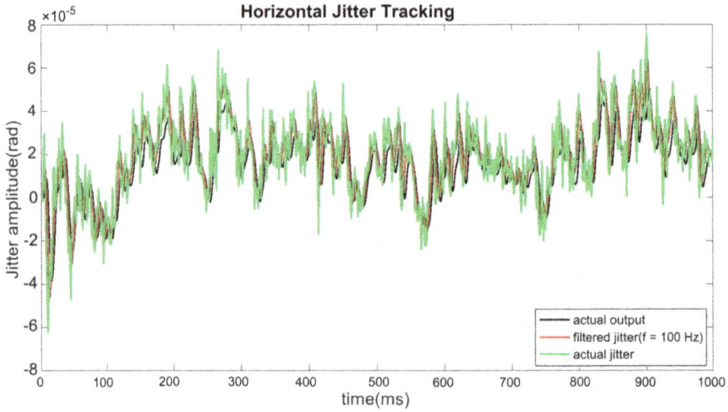

Fig. 12.3 Tracking of horizontal jitter

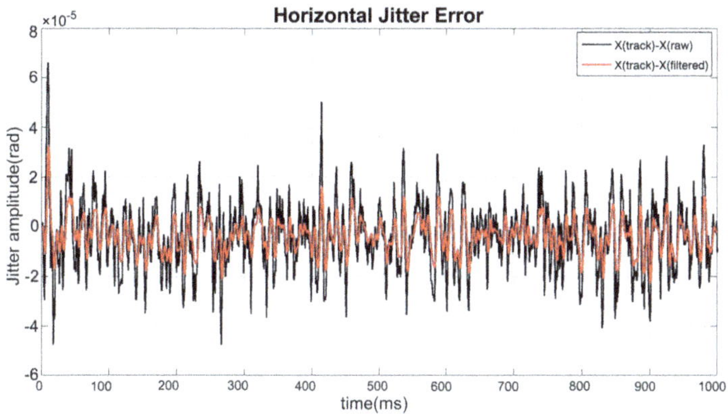

Fig. 12.4 Error in horizontal jitter

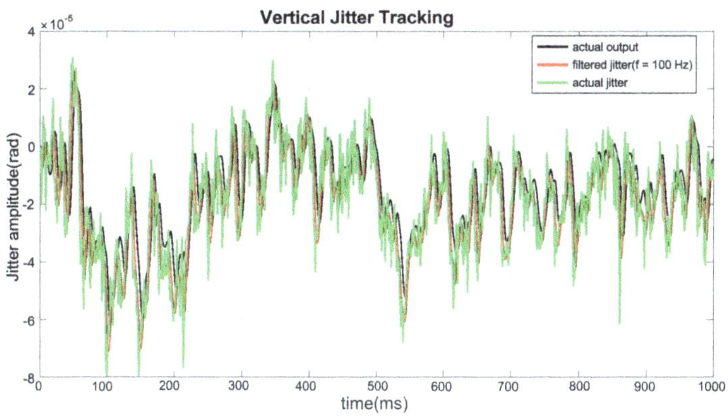

Fig. 12.5 Tracking of vertical jitter

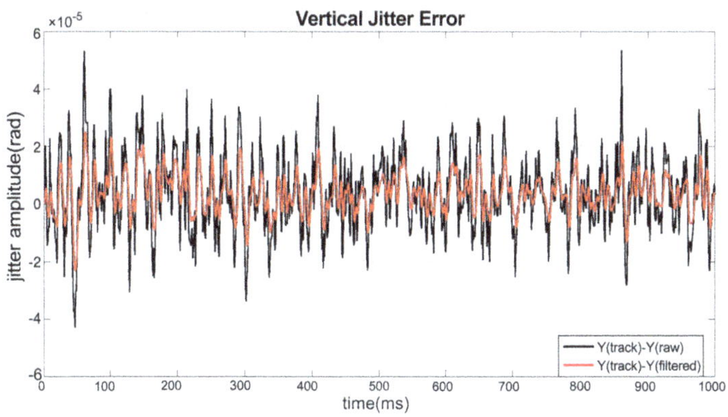

Fig. 12.6 Error in vertical jitter

12.2 State-Constrained Neuro-adaptive Control of Hypersonic Vehicles

The scramjet-powered air-breathing hypersonic vehicle (AHV), which travels at speeds exceeding five times that of sound, offers notable advantages in transportation, including rapid transit and reduced air traffic congestion. By utilizing atmospheric oxygen for propulsion, AHV can provide a cost-effective alternative to rocket-based systems that require onboard oxidizers. However, realizing these benefits depends on advancements in critical areas, particularly in sophisticated control system design (Keshmiri et al., 2006).

AHV control design is challenging due to its unique aerodynamic properties, which differ significantly from conventional vehicles.[2] At hypersonic speeds, AHVs exhibit nonlinear, coupled behaviors that are sensitive to flight conditions, leading to potential instability. The integration of scramjet engines further complicates control, as it introduces aero-propulsion coupling, limiting angle variations to prevent inlet unstart. This case study focuses on developing an adaptive control design using the winged-cone AHV model, the only openly available Six-DOF model. By employing a state-constrained adaptive controller, the attitude of the vehicle, body rates, and velocity are maintained within prescribed limits, ensuring stability and preventing inlet unstart (Rollins et al., 2013; Sachan and Padhi, 2018b; Mirmirani et al., 2005).

12.2.1 Description of Six-DOF Model

The nonlinear Six-DOF dynamics of the winged-cone AHV model (Shaughnessy et al., 1990; Sachan and Padhi, 2018a), shown in Fig. 12.7, is given by

$$\begin{bmatrix} \dot{u} \\ \dot{v} \\ \dot{w} \end{bmatrix} = \begin{bmatrix} Rv - Qw - g\sin\theta + \frac{F_{ax}+F_{Tx}}{M_v} \\ wP - uR + g\cos\theta\sin\phi + \frac{F_{ay}}{M_v} \\ uQ - vP + g\cos\theta\cos\phi + \frac{F_{az}}{M_v} \end{bmatrix} \tag{12.10}$$

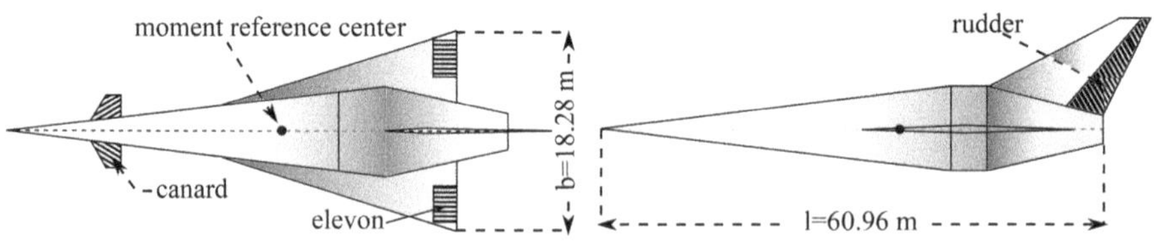

Fig. 12.7 Winged-cone model geometry of an air-breathing hypersonic vehicle

[2] Contents of current Sect. 12.2 are used from Publication "Nonlinear robust neuro-adaptive flight control for hypersonic vehicles with state constraints," Control Engineering Practice, vol. 102, p. 104526, Sep 2020, with permission from Elsevier (License Number: 6158230332176).

$$\begin{bmatrix} \dot{P} \\ \dot{Q} \\ \dot{R} \end{bmatrix} = \begin{bmatrix} c_1 QR + c_2 PQ + c_3 L_{am} + c_4 N_a \\ c_5 PR - c_6 \left(P^2 - R^2\right) + c_7 M_a \\ c_8 PQ - c_2 QR + c_4 L_{am} + c_9 N_a \end{bmatrix} \tag{12.11}$$

$$\begin{bmatrix} \dot{\phi} \\ \dot{\theta} \\ \dot{\psi} \end{bmatrix} = \begin{bmatrix} P + Q \sin\phi \tan\theta + R \cos\phi \tan\theta \\ Q \cos\phi - R \sin\phi \\ Q \sin\phi \sec\theta + R \cos\phi \sec\theta \end{bmatrix} \tag{12.12}$$

$$\begin{bmatrix} \dot{x} \\ \dot{y} \\ \dot{h} \end{bmatrix} = \begin{bmatrix} \begin{pmatrix} u \cos\theta \cos\psi + (\sin\phi \sin\theta \cos\psi - \cos\phi \sin\psi)\, v \\ + (\cos\phi \sin\theta \cos\psi + \sin\phi \sin\psi)\, w \end{pmatrix} \\ \begin{pmatrix} u \cos\theta \sin\psi + (\sin\phi \sin\theta \sin\psi + \cos\phi \cos\psi) v \\ +(\cos\phi \sin\theta \sin\psi - \sin\phi \cos\psi) w \end{pmatrix} \\ u \sin\theta - v \sin\phi \cos\theta - w \cos\phi \cos\theta \end{bmatrix} \tag{12.13}$$

where the variables used in AHV dynamics and important parameters are tabulated in Tables 12.1 and 12.2, respectively. Using transnational dynamics (12.10), V_t, α, and β of the vehicle can be calculated as $V_t = \sqrt{u^2 + v^2 + w^2}$, $\alpha = \tan^{-1}\left(\frac{w}{u}\right)$, and $\beta = \sin^{-1}\left(\frac{v}{V_t}\right)$.

Table 12.1 Parameters used in winged-cone AHV model

Description	Symbol	Value
Initial mass of the vehicle	$M_v(0)$	136,080 kg
Wing span of the vehicle	b	18.29 m
Wing reference area of the vehicle	S	334.73 m^2
Mean aerodynamics cord	c	24.38 m
Moment reference center	x_{mrc}	37.79 m
Maximum generated thrust	T_{max}	1,467,900 N
Engine specific impulse	I_{sp}	1000 sec

Table 12.2 Variables used in winged-cone AHV dynamics

Variables	Description
x, y	Distance covered along inertial north and east directions (m)
h	Altitude of the vehicle (m)
V_t, M	Total velocity (m/sec) and Mach number
u,v,w	Velocities along body x_b, y_b, and z_b directions (m/sec)
ϕ, θ, ψ	Roll, pitch, and yaw angle (deg)
α, β	Angle-of-attack (deg) and sideslip angle (rad)
P, Q, R	Roll, pitch, and yaw rate (deg/sec)
M_v	Mass of the vehicle (kg)
F_{ax}, F_{ay}, F_{az}	Aerodynamic forces along body x_b, y_b, and z_b directions
L_{am}, M_a, N_a	Rolling, pitching and yawing moments around body axis
F_{Tx}	Force generated by thrust along body x_b direction (N)
I_{xx}, I_{yy}, I_{zz}	Roll, pitch, and yaw moments of inertia, respectively (kg-m^2)
I_{xz}	Off-diagonal element of moment of inertia matrix (kg-m^2)

The coefficients $c_1, c_2, \ldots, c_9$, used in the rotational dynamics in (12.11), are defined as

$$\begin{bmatrix} c_1 \\ c_2 \\ c_3 \\ c_4 \\ c_8 \\ c_9 \end{bmatrix} = \frac{1}{\left(I_{xx}I_{zz} - I_{xz}^2\right)} \begin{bmatrix} I_{zz}(I_{yy} - I_{zz}) - I_{xz}^2 \\ I_{xz}(I_{zz} + I_{xx} - I_{yy}) \\ I_{zz} \\ I_{xz} \\ I_{xx}(I_{xx} - I_{yy}) + I_{xz}^2 \\ I_{xx} \end{bmatrix}, \begin{bmatrix} c_5 \\ c_6 \\ c_7 \end{bmatrix} = \begin{bmatrix} \frac{(I_{zz}-I_{xx})}{I_{yy}} \\ \frac{I_{xz}}{I_{yy}} \\ \frac{1}{I_{yy}} \end{bmatrix} \tag{12.14}$$

where the moment of inertia expressions for the vehicle (Zhang et al., 2010) is given by

$$\begin{bmatrix} I_{xx} \\ I_{yy} \\ I_{zz} \\ I_{xz} \end{bmatrix} = \begin{bmatrix} -7.8809 \times 10^{-5}\, M_v^2 + 25.88576\, M_v - 6.9683 \times 10^5 \\ -8.2890 \times 10^{-4}\, M_v^2 + 265.9889\, M_v - 7.3048 \times 10^6 \\ I_{yy} \\ 0 \end{bmatrix} \tag{12.15}$$

Since mass consumption rate of AHV is significant, along with the abovementioned traditional Six-DOF equations, the mass consumption dynamics of the AHV is also considered as a state equation. The mass consumption dynamics is given by $\dot{M}_v = -\frac{\sigma_T T_{max}}{I_{sp} g}$, where the control variable σ_T is defined as $\sigma_T = T/T_{max}$. Note that T is the required thrust magnitude and T_{max} is the maximum thrust generated by the AHV engine. Additionally, the hypersonic speed in terms of Mach number is calculated as $M = \frac{V_t}{a}$, where a is the speed of sound, $a = 298$ m/s for 24,000 m cruise altitude.

The aerodynamic and thrust forces and aerodynamic moments, given in the translation (12.10) and rotational dynamics (12.11), respectively, are calculated in the body frame as

$$F_{ax} = L \sin\alpha - D \cos\alpha \tag{12.16}$$

$$F_{ay} = Y \tag{12.17}$$

$$F_{az} = -D \sin\alpha - L \cos\alpha \tag{12.18}$$

$$F_{Tx} = \sigma T_{\max} \tag{12.19}$$

$$L_{am} = L_{Ac} \tag{12.20}$$

$$M_a = M_{Ac} - X_{cg} F_{az} \tag{12.21}$$

$$N_a = N_{Ac} + X_{cg} F_{ay} \tag{12.22}$$

where σ_T is the throttle utilized to maintain the thrust requirement for the speed regulation of the vehicle and X_{cg} is the center of gravity, given by

$$X_{cg} = 1.6210 \times 10^{-10} M_v^2 - 5.5226 \times 10^{-5} M_v + 7.3578. \qquad (12.23)$$

The forces (drag force (D), lift force (L), and side force (Y) in the stability frame) and the moments (rolling moment L_{Ac}, pitching moment M_{Ac}, and yawing moment N_{Ac}, relative to the moment reference center) of the vehicle are expressed as

$$D = \overline{q}SC_D, \qquad L = \overline{q}SC_L, \qquad Y = \overline{q}SC_Y \qquad (12.24)$$

$$L_{Ac} = \overline{q}SbC_{lm}, \qquad M_{Ac} = \overline{q}ScC_m, \qquad N_{Ac} = \overline{q}SbC_n \qquad (12.25)$$

where $\overline{q}$ is defined as $\overline{q} = \frac{1}{2}\rho V_t^2$ and ρ represents the atmospheric pressure. The terms $C_D, C_L, C_Y, C_{lm}, C_m$, and C_n are defined as

$$\begin{aligned}
C_D &= C_{Da} + C_{D,\delta_e} + C_{D,\delta_a} + C_{D,\delta_r}\\
C_L &= C_{La} + C_{L,\delta_e} + C_{L,\delta_a} + C_{L,\delta_r}\\
C_Y &= C_{Y_\beta}\beta + C_{Y,\delta_e} + C_{Y,\delta_a} + C_{Y,\delta_r}\\
C_{lm} &= C_{l_\beta}\beta + C_{l,\delta_e} + C_{l,\delta_a} + C_{l,\delta_r} + C_{l_p}\left(\frac{Pb}{2V}\right) + C_{l_r}\left(\frac{Rb}{2V}\right)\\
C_m &= C_{ma} + C_{m,\delta_e} + C_{m,\delta_a} + C_{m,\delta_r} + C_{m_q}\left(\frac{Qc}{2V}\right)\\
C_n &= C_{n_\beta}\beta + C_{n,\delta_e} + C_{n,\delta_a} + C_{n,\delta_r} + C_{n_p}\left(\frac{Pb}{2V}\right) + C_{n_r}\left(\frac{Rb}{2V}\right)
\end{aligned}$$

where δ_e, δ_a, and δ_r represent surface deflection of the right elevon, left elevon, and rudder, respectively. The expressions for each coefficient are given as a nonlinear function of M, α (in degree), and control surface deflections (in degrees). The expressions of these coefficients are given as follows (Keshmiri et al., 2005) (truncated up to three-digit precision):

$$\begin{aligned}
C_{Da} &= 8.717 \times 10^{-2} - 3.307 \times 10^{-2}M + 3.179 \times 10^{-2}\alpha + 5.036 \times 10^{-3}M^2\\
C_{D,\delta_e} &= 0\\
C_{D,\delta_a} &= 0\\
C_{D,\delta_r} &= 0\\
C_{La} &= -8.19 \times 10^{-2} + 4.7 \times 10^{-2}M + 1.86 \times 10^{-2}\alpha - 9.19 \times 10^{-3}M^2\\
C_{L,\delta_e} &= -4.14 \times 10^{-4}\delta_e\\
C_{L,\delta_a} &= -4.14 \times 10^{-4}\delta_a
\end{aligned}$$

$$
\begin{aligned}
C_{L,\delta_r} &= 0\\
C_{Y_\beta} &= -0.292M + 5.48 \times 10^{-2}M^2 - 4.32 \times 10^{-3}M^3\\
C_{Y,\delta_e} &= 0\\
C_{Y,\delta_a} &= 0\\
C_{Y,\delta_r} &= 3.84 \times 10^{-4}\delta_r\\
C_{l_\beta} &= -0.14 + 3.32 \times 10^{-2}M - 7.59 \times 10^{-4}\alpha - 3.79 \times 10^{-3}M^2\\
C_{l,\delta_e} &= -1.17 \times 10^{-4}\delta_e\\
C_{l,\delta_a} &= 1.17 \times 10^{-4}\delta_a\\
C_{l,\delta_r} &= 1.144 \times 10^{-4}\delta_r\\
C_{l_p} &= -0.299 + 7.47 \times 10^{-2}M + 1.38 \times 10^{-3}\alpha - 9.13 \times 10^{-3}M^2\\
C_{l_r} &= 0.382 - 0.106M + 1.94 \times 10^{-3}\alpha + 1.45 \times 10^{-2}M^2 - 1.02 \times 10^{-3}M^3\\
C_{ma} &= -2.19 \times 10^{-2} + 7.73 \times 10^{-3}M - 2.26 \times 10^{-3}\alpha\\
C_{m,\delta_e} &= 2.89 \times 10^{-4}\delta_e\\
C_{m,\delta_a} &= 2.89 \times 10^{-4}\delta_a\\
C_{m,\delta_r} &= 1.58 \times 10^{-3}M^2\\
C_{m_q} &= -1.36 + 0.38M - 5.42 \times 10^{-2}M^2 + 3.8 \times 10^{-3}M^3\\
C_{n_\beta} &= 5.91 \times 10^{-2}M - 1.48 \times 10^{-2}M^2 + 1.27 \times 10^{-3}M^3\\
C_{n,\delta_e} &= 0\\
C_{n,\delta_a} &= 0\\
C_{n,\delta_r} &= -5.28 \times 10^{-4}\delta_r\\
C_{n_p} &= 0.368 - 9.79 \times 10^{-2}M + 1.24 \times 10^{-2}M^2\\
C_{n_r} &= -2.41 + 0.596M - 2.74 \times 10^{-3}\alpha - 7.57 \times 10^{-2}M^2 + 4.9 \times 10^{-3}M^3
\end{aligned}
$$

12.2.2 Nominal Three-Loop Control Design

The nominal controller design leverages the inherent timescale separation between the slow and fast dynamics of the vehicle. As illustrated in Fig. 12.8, the three-loop control design comprises outer, intermediate, and inner loops. The outer loop receives reference signals and outputs a desired signal for the intermediate loop. The intermediate loop then provides input to the inner loop, which generates the necessary control deflections to track angular rates. Additionally, a separate velocity

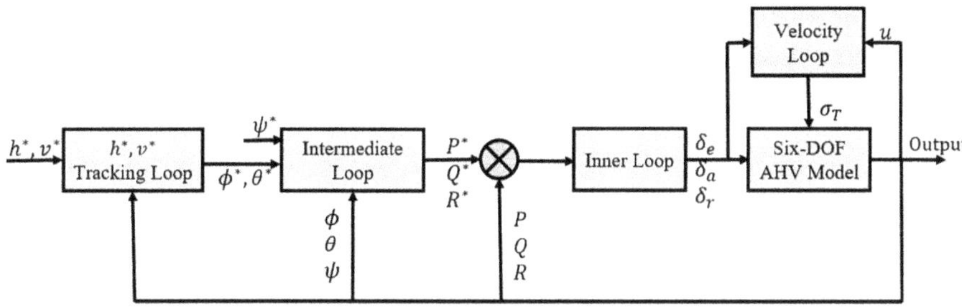

Fig. 12.8 Block diagram representation of three-loop control design

control loop ensures the AHV maintains the desired velocity. Each loop assumes continuous and smooth desired input signals, with their time derivatives set to zero. Symmetric constraints are also assumed on the vehicle states, and the control formulation for each loop is detailed as follows.

12.2.2.1 Outer Loop Control Formulation

The outer loop is designed to track the reference inputs h^* and v^* by generating the desired θ^* and ϕ^*. To track reference h^*, the following closed-loop stable error dynamics is enforced in h^* tracking loop:

$$\dot{e}_h + k_{h_B} e_h = 0 \tag{12.26}$$

where $e_h = h^* - h$ is the error between the reference altitude and the actual altitude and $k_{h_B} = \frac{k_h}{\left(k_{e_h}^2 - e_h^2\right)}$. The constant k_{e_h} is the symmetric limit on the error e_h and k_h is a positive constant gain. Since h^* is assumed to be constant, $\dot{h}^* = 0$. Thus, using (12.26) the following is obtained:

$$\dot{h} = k_{h_B}(h^* - h) \tag{12.27}$$

Using (12.13), $\dot{h}$ is substituted into (12.27), which gives

$$u \sin\theta - (v \sin\phi + w \cos\phi)\cos\theta = k_{h_B}(h^* - h) \tag{12.28}$$

To obtain desired θ^*, Eq. (12.28) is divided by $\sqrt{u^2 + [v \sin\phi + w \cos\phi]^2}$. By performing mathematical manipulations, the desired pitch angle ($\theta = \theta^*$) is obtained as

$$\theta^* = \tan^{-1}\left(\frac{v \sin\phi + w \cos\phi}{u}\right) + \sin^{-1}\left(\frac{k_{h_B}(h^* - h)}{\sqrt{u^2 + [v \sin\phi + w \cos\phi]^2}}\right) \tag{12.29}$$

To ensure zero sideslip angle β while turning, the coordinated turn philosophy is used. In this, the side velocity in the body frame is enforced to be zero. To achieve this, the following stable closed-loop error dynamics is chosen:

$$\left(\dot{v}-\dot{v}^{*}\right)+k_{vB}\left(v-v^{*}\right)=0 \tag{12.30}$$

where k_{vB} is defined as $k_{vB}=\frac{k_v}{(k_{e_v}^2-e_v^2)}$ and $e_v=v-v^*$. The constant k_{e_v} is the limit on the variation of error e_v, and k_v is the positive constant gain. By substituting $\dot{v}$ from (12.10) into (12.30), and setting $v^*=0$, the desired ϕ^* can be obtained as

$$\phi^{*}=\sin^{-1}\left(\frac{-k_{vB}V_t-\left(wP-uR+\frac{F_{ay}}{M_v}\right)}{g\cos\theta}\right) \tag{12.31}$$

12.2.2.2 Intermediate Loop Control Formulation

The output of the outer loop is used as the reference input for the intermediate loop. The desired roll rate P^*, pitch rate Q^*, and yaw rate R^* are obtained by solving the following nonlinear error dynamics:

$$\begin{bmatrix}\dot{e}_\phi\\ \dot{e}_\theta\\ \dot{e}_\psi\end{bmatrix}+\begin{bmatrix}k_{\phi B}&0&0\\ 0&k_{\theta B}&0\\ 0&0&k_{\psi B}\end{bmatrix}\begin{bmatrix}e_\phi\\ e_\theta\\ e_\psi\end{bmatrix}=0 \tag{12.32}$$

where $e_\phi=\phi^*-\phi$ is the error between the desired and actual roll angles. Similarly, $e_\theta=\theta^*-\theta$ is the error between the desired and actual pitch angles, and $e_\psi=\psi^*-\psi$ is the error between the desired and actual yaw angles. The parameters $k_{\phi B}$, $k_{\theta B}$, and $k_{\psi B}$ are defined as $\frac{k_\phi}{\left(k_{e_\phi}^2-e_\phi^2\right)}$, $\frac{k_\theta}{\left(k_{e_\theta}^2-e_\theta^2\right)}$, and $\frac{k_\psi}{\left(k_{e_\psi}^2-e_\psi^2\right)}$, respectively. Moreover, the terms k_ϕ, k_θ, k_ψ are positive constant gains. The constants, k_{e_ϕ}, k_{e_θ}, and k_{e_ψ} are positive constants and symmetric limits on the errors e_ϕ, e_θ, and e_ψ, respectively.

By substituting $\dot{\phi}$, $\dot{\theta}$, and $\dot{\psi}$ from (12.12) into the closed-loop error dynamics (12.32) of the intermediate loop, the following desired body rates are obtained:

$$\begin{bmatrix}P^*\\ Q^*\\ R^*\end{bmatrix}=\begin{bmatrix}1&\tan\theta\sin\phi&\tan\theta\cos\phi\\ 0&\cos\phi&-\sin\phi\\ 0&\sin\phi\sec\theta&\cos\phi\sec\theta\end{bmatrix}^{-1}\begin{bmatrix}k_{\phi B}\left(\phi^*-\phi\right)\\ k_{\theta B}\left(\theta^*-\theta\right)\\ k_{\psi B}\left(\psi^*-\psi\right)\end{bmatrix} \tag{12.33}$$

Note that the use of Euler angles in the control formulation usually suffers from the singularity problem. However, due to the imposed constraints, the singularity points are outside of the range of operation.

12.2.2.3 Inner Loop Control Formulation

In the inner loop control formulation, the inputs are obtained from the output of the intermediate loop. The required control surface deflections to track the desired angular rates are obtained by enforcing following stable error dynamics

$$\begin{bmatrix} \dot{e}_P \\ \dot{e}_Q \\ \dot{e}_R \end{bmatrix} + \begin{bmatrix} k_{pB} & 0 & 0 \\ 0 & k_{qB} & 0 \\ 0 & 0 & k_{rB} \end{bmatrix} \begin{bmatrix} e_P \\ e_Q \\ e_R \end{bmatrix} = 0 \tag{12.34}$$

where $e_P = P^* - P$ is the error between the desired roll rate and the actual roll rate. Similarly, $e_Q = Q^* - Q$ is the error between desired and actual pitch rates, and $e_R = R^* - R$ is the error between the desired and actual yaw rates. The terms k_{pB}, k_{qB}, and k_{rB} are defined as $\frac{k_p}{\left(k_{e_p}^2 - e_P^2\right)}$, $\frac{k_q}{\left(k_{e_q}^2 - e_Q^2\right)}$, and $\frac{k_r}{\left(k_{e_r}^2 - e_R^2\right)}$, respectively. The terms k_p, k_q, k_r are positive constant gains. The constants, k_{e_p}, k_{e_q}, and k_{e_r} are positive constants and provide limits on the errors e_P, e_Q, and e_R, respectively. By substituting the angular rates from (12.11) into (12.34), the following closed-loop error dynamics is obtained

$$\begin{bmatrix} -c_3 L_{am} \\ -c_5 PR - c_7 M_a \\ -c_8 PQ - c_9 N_a \end{bmatrix} + \begin{bmatrix} k_{pB} & 0 & 0 \\ 0 & k_{qB} & 0 \\ 0 & 0 & k_{rB} \end{bmatrix} \begin{bmatrix} (P^* - P) \\ (Q^* - Q) \\ (R^* - R) \end{bmatrix} = 0 \tag{12.35}$$

where, for the nominal case, the inertia values are substituted as $c_1 = c_2 = c_4 = c_6 = 0$, as defined in (12.14). However, in the actual system, these values are uncertain and nonzero. In (12.35), the aerodynamic moments are nonlinear functions of control deflections; thus, the control solution cannot be obtained explicitly from the above system of equations. Therefore, the control surface deflections δ_e, δ_a, and δ_r are obtained numerically using a fast Newton-Raphson algorithm.

12.2.2.4 Velocity Loop Control Formulation

The integration of the velocity control loop is required to maintain the constant desired velocity during flight. Since the forward velocity u is a major component of the total velocity, the required thrust magnitude is obtained by using forward velocity dynamics in the velocity control loop. The closed-loop error dynamics for designing the velocity control in the velocity loop is given as

$$\dot{e}_u + k_{uB} e_u = 0 \tag{12.36}$$

where $e_u = u^* - u$ is the error between the desired forward velocity and actual forward velocity and $k_{uB} = \frac{k_u}{(k_{eu}^2 - e_u^2)}$. The constant k_{e_u} is the limit on the error e_u, and k_u is a positive constant gain. Substituting $\dot{u}$ from (12.10) into (12.36) and solving for σ_T results in

$$\sigma_T = \left(\frac{M_v}{T_{max}}\right)\left[-\left(vR - QW - g\sin\theta + \frac{F_{ax}}{M_v}\right) + k_{u_B}(u^* - u)\right] \tag{12.37}$$

It can be observed that the nominal AHV controller is derived assuming a known vehicle model. In practice, however, a perfectly known model is impossible to design. Therefore, in the next subsection, an adaptive control design is presented to capture the model uncertainties while ensuring the stability of the vehicle during flight.

12.2.3 Neuro-adaptive Control Design

This section presents the adaptive control design for an AHV (Sachan and Padhi, 2020). The adaptive controller is augmented with the inner and velocity control loops of the control architecture. In the Six-DOF AHV model, structural uncertainties are assumed in the aerodynamic coefficients and inertia coefficients. The aerodynamic force coefficients related to the actual model are defined as

$$C_{D_a} = C_D\left(1 + \Delta C_D\right), \qquad C_{L_a} = C_L\left(1 + \Delta C_L\right), \qquad C_{Y_a} = C_Y\left(1 + \Delta C_Y\right)$$

where C_D, C_L, and C_Y are nominal aerodynamic force coefficients. The terms ΔC_D, ΔC_L, and ΔC_Y are used to calculate the uncertainties associated with the nominal aerodynamic force coefficients in percentage. For an example, if $\Delta C_D = 0.1$, then C_{D_a} has 10% uncertainty with respect to nominal coefficient. The aerodynamic moment coefficients of the actual model are defined as

$$C_{lm_a} = C_{lm}\left(1 + \Delta C_{lm}\right), \qquad C_{m_a} = C_m\left(1 + \Delta C_m\right), \qquad C_{n_a} = C_n\left(1 + \Delta C_n\right)$$

where C_{lm}, C_m, and C_n are nominal aerodynamic moment coefficients. The terms ΔC_{lm}, ΔC_m, and ΔC_n are used to calculate the uncertainties associated with the nominal aerodynamic moment coefficients in percentage. The values of the inertia components for the actual model are defined as

$$I_{xx_a} = I_{xx}\left(1 + \Delta I_{xx}\right), \qquad I_{yy_a} = I_{yy}\left(1 + \Delta I_{yy}\right), \qquad I_{zz_a} = I_{zz}\left(1 + \Delta I_{zz}\right)$$
$$I_{xz_a} = I_{xx}\Delta I_{xz}$$

where I_{xx}, I_{yy}, and I_{zz} are the nominal inertia values and ΔI_{xx}, ΔI_{yy}, and ΔI_{zz} are used to calculate the uncertainties associated with the nominal inertia values, in percentage. The uncertainty in cross moment of inertia term is assumed as ±5% of principle moment of inertia, i.e., $\Delta I_{xz} = \pm 0.05$. Next, the adaptive control formulation for the inner and velocity control loop is presented.

12.2.3.1 Inner Loop Adaptive Control Formulation

The actual rotational dynamics of the AHV can be expressed as

$$\dot{P}_a = c_{1_a} Q_a R_a + c_{2_a} P_a Q_a + c_{3_a} L_{am_a} + c_{4_a} N_{a_a} + d_1 \tag{12.38a}$$

$$\dot{Q}_a = c_{5_a} P_a R_a - c_{6_a} \left(P_a^2 - R_a^2\right) + c_{7_a} M_{a_a} + d_2 \tag{12.38b}$$

$$\dot{R}_a = c_{8_a} P_a Q_a - c_{2_a} Q_a R_a + c_{4_a} L_{am_a} + c_{9_a} N_{a_a} + d_3 \tag{12.38c}$$

where d_1, d_2, and d_3 are the model uncertainties associated with the system. The approximated rotational dynamics (12.38) can be expressed as

$$\dot{P}_{ap} = c_{1_a} Q_a R_a + c_{2_a} P_a Q_a + c_{3_a} L_{am_a} + c_{4_a} N_{a_a} + \hat{d}_1 + k_{ap_1} \left(P_a - P_{ap}\right) \tag{12.39a}$$

$$\dot{Q}_{ap} = c_{5_a} P_a R_a - c_{6_a} \left(P_a^2 - R_a^2\right) + c_{7_a} M_{a_a} + \hat{d}_2 + k_{ap_2} \left(Q_a - Q_{ap}\right) \tag{12.39b}$$

$$\dot{R}_{ap} = c_{8_a} P_a Q_a - c_{2_a} Q_a R_a + c_{4_a} L_{am_a} + c_{9_a} N_{a_a} + \hat{d}_3 + k_{ap_3} \left(R_a - R_{ap}\right) \tag{12.39c}$$

where $\hat{d}_1$, $\hat{d}_2$, and $\hat{d}_3$ are the approximation of the uncertainties. All variables and gains have their usual meaning as defined in generic neuro-adaptive control design theory in Sect. 7.1 of Chap. 7. Note that $\hat{d}_1$, $\hat{d}_2$, and $\hat{d}_3$ are obtained using a neural network, and the weights of the neural network are trained using weight update rule, which is defined in (12.40).

To approximate the model uncertainties, proper basis functions are required to be chosen. Since structured uncertainties are assumed in the model, the basis function vector is obtained from the uncertainty-perturbed model analysis. In this analysis, uncertainty perturbations are assumed in the rotational dynamics in (12.38), and then by comparing with the approximated dynamics (12.39), $\hat{d}_1$, $\hat{d}_2$, and $\hat{d}_3$ are obtained as a function of inner loop variables and aerodynamic model parameters, which serve as desired basis functions. It has been observed that the dominant aerodynamic parameters (up to second order) are sufficient to approximate the model uncertainties. Thus, the basis functions to approximate $\hat{d}_1$, $\hat{d}_2$, and $\hat{d}_3$ are selected as

$$\Phi_j = [\,1,\ P_a^2,\ Q_a^2,\ R_a^2,\ P_a Q_a,\ Q_a R_a,\ P_a R_a,\ \alpha,\ M,\ \alpha^2,\ M^2,\ \alpha M,\ \delta_e,\ \delta_e M,\ \delta_e \alpha,\ \delta_a,\ \delta_a M,\ \delta_a \alpha,\ \delta_r,\ \delta_r M,\ \delta_r \alpha]^T$$

where the indexing j correspond to the model uncertainties $\hat{d}_j$, $j = 1, 2, 3$. The Jacobian matrices of the basis functions $\left[\frac{\partial \Phi_j}{\partial X}\right]$ can be obtained by differentiating Φ_j with respect to variables X, where $X = [P_a, Q_a, R_a, \alpha, M, \delta_e, \delta_a, \delta_r]^T$.

Following similar steps as in Sect. 7.1, the weight update rule $\hat{W}_j$ is obtained as

$$\dot{\hat{W}}_j = \left[\frac{I_P}{\gamma_j} + \left[\frac{\partial \Phi_j}{\partial X}\right] \Theta_j \left[\frac{\partial \Phi_j}{\partial X}\right]^T\right]^{-1} \frac{\beta_j e_{a_j} \Phi_j}{k_{\beta_j}^2 - e_{a_j}^2}, \qquad j = 1, 2, 3. \quad (12.40)$$

where e_{a_j} are the errors between actual states and approximated states, and k_{β_j} are limits on e_{a_j}. To propagate (12.40), the initial conditions are selected as zero vector. Finally, the $\hat{d}_j$ can be obtained as $\hat{d}_j = \hat{W}_j^T \Phi_j$, $j = 1, 2, 3$.

After capturing the uncertainties, the next step in the adaptive controller design is to track the states of the nominal system. Hence, the following stable error dynamics is used to achieve the tracking performance:

$$\begin{bmatrix} \dot{P}_{ap} - \dot{P} \\ \dot{Q}_{ap} - \dot{Q} \\ \dot{R}_{ap} - \dot{R} \end{bmatrix} + \begin{bmatrix} k_{ad_1} & 0 & 0 \\ 0 & k_{ad_2} & 0 \\ 0 & 0 & k_{ad_3} \end{bmatrix} \begin{bmatrix} (P_{ap} - P) \\ (Q_{ap} - Q) \\ (R_{ap} - R) \end{bmatrix} = 0 \qquad (12.41)$$

where k_{ad_1}, k_{ad_2}, and k_{ad_3} are defined as $k_{ad_1} = \frac{k_{a_1}}{\left(k_{e_{a1}}^2 - (P_{ap}-P)^2\right)}$, $k_{ad_2} = \frac{k_{a_2}}{\left(k_{e_{a2}}^2 - (Q_{ap}-Q)^2\right)}$, and $k_{ad_3} = \frac{k_{a_3}}{\left(k_{e_{a3}}^2 - (R_{ap}-R)^2\right)}$, respectively. The terms, k_{a_1}, k_{a_2}, and k_{a_3} are positive controller gains, and $k_{e_{aj}}, \forall j$ are the symmetric limits on the corresponding errors. By substituting $\dot{P}_{ap}$, $\dot{Q}_{ap}$, $\dot{R}_{ap}$ from (12.39), and $\dot{P}$, $\dot{Q}$, $\dot{R}$ from (12.11), the following is obtained:

$$\left[c_{1_a} Q_a R_a + c_{2_a} P_a Q_a + c_{3_a} L_{am_a} + c_{4_a} N_{a_a} + \hat{d}_1 + k_{ap_1} \left(P_a - P_{ap}\right)\right] - \left[c_1 Q R + c_2 P Q + c_3 L_{am} + c_4 N_a\right] + \left[k_{ad_1}\right] (P_{ap} - P) = 0 \qquad (12.42a)$$

$$\left[c_{5_a} P_a R_a - c_{6_a} \left(P_a^2 - R_a^2\right) + c_{7_a} M_{a_a} + \hat{d}_2 + k_{ap_2} \left(Q_a - Q_{ap}\right)\right] - \left[c_5 P R - c_6 \left(P^2 - R^2\right) + c_7 M_a\right] + \left[k_{ad_2}\right] (Q_{ap} - Q) = 0 \qquad (12.42b)$$

$$\left[c_{8_a} P_a Q_a - c_{2_a} Q_a R_a + c_{4_a} L_{am_a} + c_{9_a} N_{a_a} + \hat{d}_3 + k_{ap_3} \left(R_a - R_{ap}\right)\right] - \left[c_8 P Q - c_2 Q R + c_4 L_{am} + c_9 N_a\right] + \left[k_{ad_3}\right] (R_{ap} - R) = 0 \qquad (12.42c)$$

where $k_{ap_j}, \forall j$ are positive observer gain constants (defined in (12.39)) and (7.1)). Since control surface deflections are nonlinear functions of aerodynamic coefficients, the systems of equations in (12.42) are non-affine in nature. Therefore, the Newton-Raphson method is used to extract the required control surface deflections numerically.

12.2.3.2 Adaptive Velocity Control Formulation

In this subsection, the adaptive control formulation for the velocity control loop is presented. The forward velocity equation with model uncertainties can be expressed as

$$\dot{u}_a = R_a v_a - Q_a w_a - g \sin\theta_a + \frac{F_{ax_a} + F_{Tx_a}}{M_{v_a}} + d_4 \tag{12.43}$$

where d_4 represents the uncertainty associated with the forward velocity equation. Further, (12.43) can be approximated as

$$\dot{u}_{ap} = R_a v_a - Q_a w_a - g \sin\theta_a + \frac{F_{ax_a} + F_{Tx_a}}{M_{v_a}} + \hat{d}_4 + k_{ap4}\left(u_a - u_{ap}\right) \tag{12.44}$$

where $\hat{d}_4$ is the approximation of the uncertainties d_4. The remaining variables have their usual meaning. The basis functions for the adaptive velocity control loop are chosen from the perturbation analysis and obtained as

$$\Phi_v = \begin{bmatrix} 1, & \alpha, & M, & \alpha^2, & M^2, & \alpha M, & \delta_e, & \delta_e M, & \delta_e\alpha, & \delta_a, & \delta_a M, & \delta_a\alpha, & \delta_r, & \delta_r M, & \delta_r\alpha \end{bmatrix}^T$$

The Jacobian matrix of the basis functions $\left[\frac{\partial \Phi_v}{\partial X_v}\right]$ is obtained by differentiating Φ_v with respect to variables $[\alpha, M, \delta_e, \delta_a, \delta_r]^T$. To approximate model uncertainties, the following weight update rule is used:

$$\dot{\hat{W}}_v = \left[\frac{I_P}{\gamma_4} + \left[\frac{\partial \Phi_v}{\partial X_v}\right] \Theta_4 \left[\frac{\partial \Phi_v}{\partial X_v}\right]^T\right]^{-1} \frac{\beta_4 e_{a_v} \Phi_v}{k_{\beta_4}^2 - e_{a_v}^2}, \tag{12.45}$$

where the error e_{a_v} is defined as $e_{a_v} = u_a - u_{ap}$ and k_{β_4} is the limit on error e_{a_v}. The initial condition of $\hat{W}_v(0)$ is selected as a zero vector. Finally, the $\hat{d}_4$ can be obtained as $\hat{d}_4 = \hat{W}_v^T \Phi_v$.

After capturing the uncertainties $\hat{d}_4$, the following stable error dynamics is used to achieve the tracking performance:

$$\left(\dot{u}_{ap} - \dot{u}\right) + k_{ad_4}\left(u_{ap} - u\right) = 0 \tag{12.46}$$

where k_{ad_4} is defined as $k_{ad_4} = \frac{k_{a4}}{\left(k_{e_{a4}}^2 - e_{u_{ap}}^2\right)}$ and $e_{u_{ap}} = u_{ap} - u$. The constant $k_{e_{a4}}$ is the constraint on the variation of $e_{u_{ap}}$, and k_{a_4} is a positive controller gain.

By substituting $\dot{u}_{ap}$ from (12.44) and $\dot{u}$ from (12.10), the following is obtained:

$$\left[R_a v_a - Q_a w_a - g \sin\theta_a + \frac{F_{ax_a} + F_{Tx_a}}{M_{v_a}} + \hat{d}_4 + k_{ap_4}\left(u_a - u_{ap}\right)\right] \\ - \left[Rv - Qw - g\sin\theta + \frac{F_{ax} + F_{Tx}}{M_v}\right] + k_{ad_4}(u_{ap} - u) = 0 \quad (12.47)$$

where k_{ap_4} is positive observer gain constant. By solving (12.47), the following closed-form expression for the adaptive throttle control is obtained:

$$\sigma_{T_{ad}} = -\left(\frac{M_v}{T_{max}}\right)\left(R_a v_a - Q_a w_a - g\sin\theta_a + \frac{F_{ax_a}}{M_{v_a}} + \hat{d}_4 + k_{ap_4}\left(u_a - u_{ap}\right)\right) \\ + \left(\frac{M_v}{T_{max}}\right)\left[\left(Rv - Qw - g\sin\theta + \frac{F_{ax} + F_{Tx}}{M_v}\right) - k_4(u_{ap} - u)\right] \quad (12.48)$$

Using $\sigma_{T_{ad}}$, the velocity of the vehicle is regulated within the desired limits.

12.2.4 Simulation Results

This subsection presents simulation results demonstrating the effectiveness of the proposed adaptive control formulation. Trim conditions are first calculated to provide initial values for the simulations. Trim conditions represent the state and control values where the vehicle maintains equilibrium, assuming constant altitude, roll angle, zero angular body rates, and constant forward velocity. The remaining state variables in the Six-DOF model, $[v, w, \theta, \delta_e, \delta_a, \delta_r, \sigma_T]^T$, are determined by solving $[\dot{u}, \dot{v}, \dot{w}, \dot{P}, \dot{Q}, \dot{R}, \dot{h}]^T = 0$. The trimmed states for the winged-cone AHV model are determined as $u = 1786.7$ m/s, $v = 0$, $w = 91.86$ m/s, $P = Q = R = 0$, $\phi = 0°$, $\theta = 2.94°$, $\psi = 0°$, $x = y = 0$, and $h = 24000$ m. The trimmed control values are $\delta_e = \delta_a = -7.2°$, $\delta_r = 0$, and $\sigma_T = 0.194$. To validate the control design under modeling uncertainties, the uncertain parameters are randomly selected from the interval $[-0.30, 0.30]$. A second-order actuator dynamics model with $\zeta = 0.7$ and $\omega_n = 20$ Hz is applied to each control variable. Symmetric state constraints are assumed, with control surface limits set to $[-30°, 20°]$ and throttle range $\sigma_T \in [0, 1]$. Gain values and other parameters for the nominal and adaptive control designs are detailed in Tables 12.3 and 12.4.

The simulation results using adaptive control under 30% model uncertainties are presented in Figs. 12.9, 12.10, 12.11, 12.12, 12.13, and 12.14. The angle-of-attack

Table 12.3 Parameters used in the nominal control design

Outer loop		Intermediate loop		Inner loop		Velocity loop	
k_h	0.001	k_ϕ	0.2	k_p	0.3	k_u	1
k_v	2.5	k_θ	0.001	k_q	0.5	k_{e_u}	± 1
k_{e_h}	± 200	k_ψ	0.004	k_r	0.012		
k_{e_v}	± 15	k_{e_ϕ}	$\pm 30°$	k_{e_p}	± 30		
		k_{e_θ}	$\pm 2°$	k_{e_q}	± 6		
		k_{e_ψ}	$\pm 5°$	k_{e_r}	± 6		

Table 12.4 Parameter values used in adaptive control design

P loop		Q loop		R loop		u loop	
$k_{e_{a1}}$	± 5	$k_{e_{a2}}$	± 2	$k_{e_{a3}}$	± 2	$k_{e_{a4}}$	± 0.2
k_{a_1}	0.001	k_{a_2}	0.001	k_{a_3}	0.001	k_{a_4}	0.004
k_{ap_1}	50	k_{ap_2}	12	k_{ap_3}	50	k_{ap_4}	35
k_{β_1}	$\pm 5°$	k_{β_2}	$\pm 2°$	k_{β_3}	$\pm 2°$	k_{β_4}	± 0.2
γ_1	20	γ_2	20	γ_3	20	γ_4	20
β_1	0.005	β_2	0.002	β_3	0.005	β_4	0.008
Θ_1	10	Θ_2	10	Θ_3	10	Θ_4	10

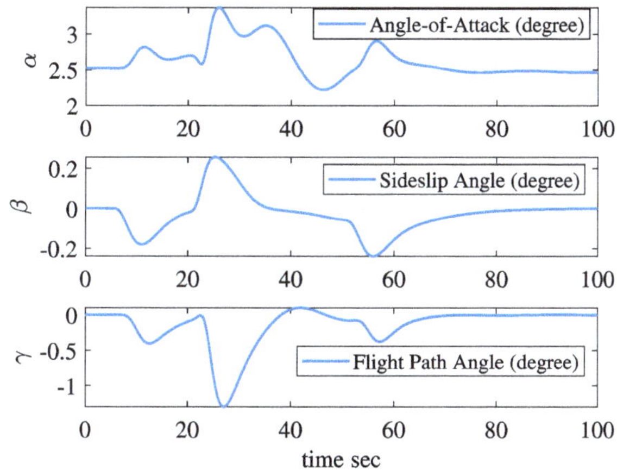

Fig. 12.9 Vehicle angle trajectories

(α) and sideslip angle (β) trajectories in Fig. 12.9 demonstrate successful prevention of inlet unstart by maintaining α within $[0°, 6°]$ and β within $[-2°, 2°]$. The flight path angle γ stabilizes rapidly with minimal transients. Figure 12.10 shows smooth attitude trajectories within imposed limits. The large roll angle bounds facilitate turn coordination without significant β variation. The lateral control capability is illustrated through yaw angle reversals triggered at $t = 5, 20$ and 50 seconds, transitioning from $0° \to 5°, 5° \to -5°$, and $-5° \to 0°$.

Angular body rate profiles in Fig. 12.11 remain stable and quickly converge after reversals. The adaptive controller effectively suppresses high-frequency transients via gradient-based weight updates. Figure 12.12 shows smooth side velocity v with minimal variation, confirming precise turn coordination. Constraints on forward velocity u and side velocity maintain a nearly constant Mach number, ensuring the

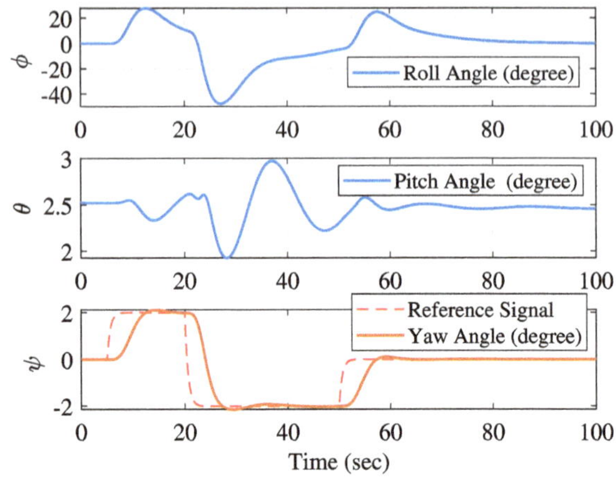

Fig. 12.10 Attitude trajectories

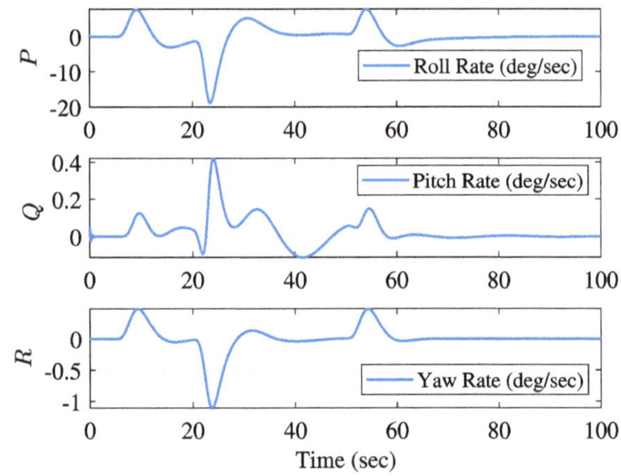

Fig. 12.11 Angular body rate trajectories

AHV remains in the hypersonic regime. The altitude profile indicates consistent cruise flight. Control deflections and throttle profiles in Figs. 12.13 and 12.14 are smooth, illustrating the effectiveness of proposed adaptive control. Minor transients during yaw reversals can be reduced by smoothing the reference signal.

To validate the effectiveness of the proposed adaptive control design thoroughly, extensive Monte Carlo-like high-fidelity simulations were conducted. A total of 2000 times simulations were performed for 100 seconds with 30% random model uncertainties. The controller demonstrated successful performance with a 100% and 98% confidence level for up to 25% and 30% random model uncertainties, respectively.

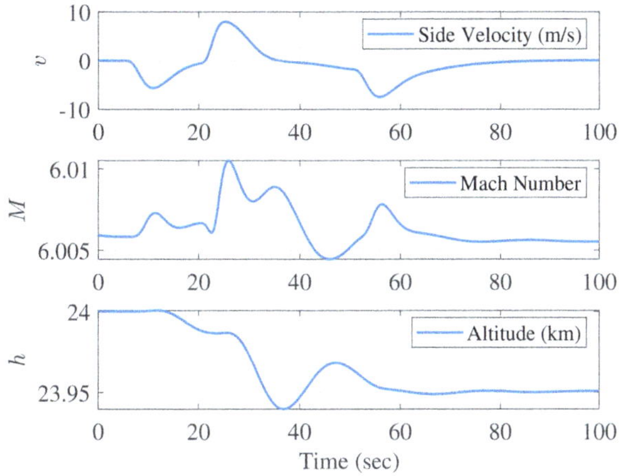

Fig. 12.12 Trajectories of vehicle velocity and altitude

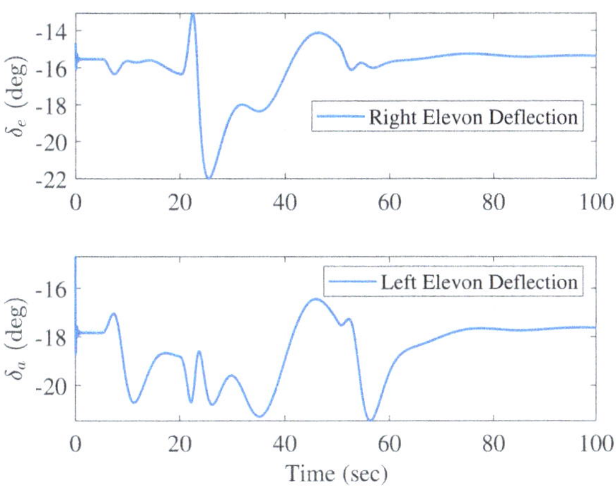

Fig. 12.13 Control deflection trajectories

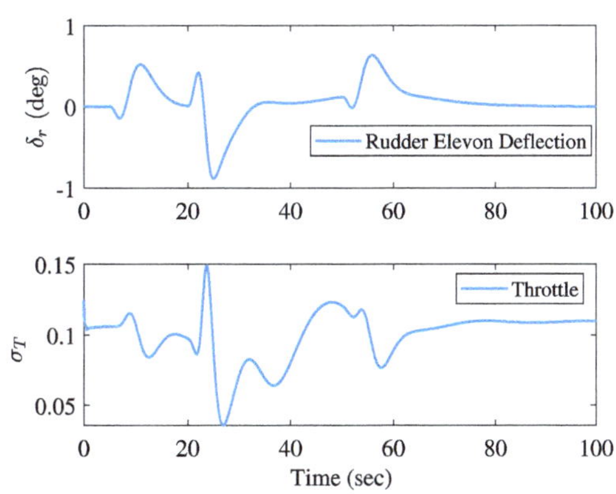

Fig. 12.14 Control deflection trajectories

12.3 Velocity Control of Hexacopter Using GRNNSMC

In this case study, GRNNSMC[3] is employed to control the velocity of a hexacopter as shown in Fig. 12.15.

12.3.1 Dynamic Model

The dynamic model of the hexacopter (Al-Mahasneh et al., 2019) can be expressed as

$$
\begin{aligned}
F_x &= m(\dot{v}_x + \omega_y v_z - \omega_z v_y) \\
F_y &= m(\dot{v}_y + \omega_z v_x - \omega_x v_z) \\
F_z &= m(\dot{v}_z + \omega_x v_y - \omega_y v_x) \\
T_x &= I_{xx}\dot{\omega}_x - I_{xz}\dot{\omega}_z + \omega_y\omega_z(I_{zz} - I_{yy}) - I_{xz}\omega_x\omega_y \\
T_y &= I_{yy}\dot{\omega}_y + \omega_z\omega_x(I_{xx} - I_{zz}) + I_{xz}(\omega_x^2 - \omega_z^2) \\
T_z &= -I_{xz}\dot{\omega}_x + I_{zz}\dot{\omega}_z + \omega_x\omega_y(I_{yy} - I_{xx}) + I_{xz}\omega_y\omega_z
\end{aligned}
\tag{12.49}
$$

where m is the mass, v is the translation speed in the body frame, ω is the rotational speed in the body frame, and I is the mass moment of inertia. The model of the hexacopter consists of altitude, yawing, and attitude controllers.

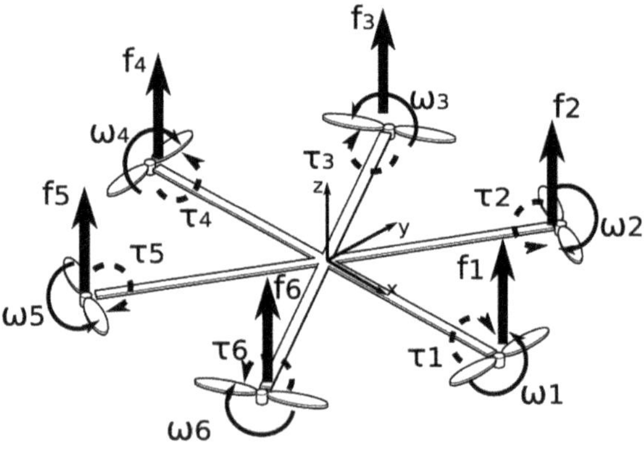

Fig. 12.15 The model of hexacopter

[3] Some contents of the current Sect. 12.3 are used from Publication "Stable Adaptive Controller Based on Generalized Regression Neural Networks and Sliding Mode Control for a Class of Nonlinear Time-Varying Systems," IEEE Transactions on Systems, Man, and Cybernetics: Systems, vol. 53, pp. 2525–2535, April, 2021, with permission from IEEE (License Number: 6160201218650).

12.3.2 Simulation Results

GRNNSMC is used to control the velocity of the hexacopter in the x and y directions. GRNNSMC's inputs are the sliding surfaces of the velocities in the x and y directions, and its outputs are the control signals for the velocities. The parameters of the controller are shown in Table 12.5.

The desired velocity trajectory for both x and y directions is $v_d(t) = 0.5\sin(0.1t)$. The velocity tracking performance of GRNNSMC is very accurate after the initial learning stage as shown in Fig. 12.16. Also, the control signals of the GRNNSMC are depicted in Fig. 12.17. The evolution of the norm of the

Table 12.5 GRNNSMC's parameters for the hexacopter velocity control

γ	λ
$\begin{bmatrix} 0.001 \\ 0.001 \end{bmatrix}$	$\begin{bmatrix} 1.0 \\ 0.8 \end{bmatrix}$

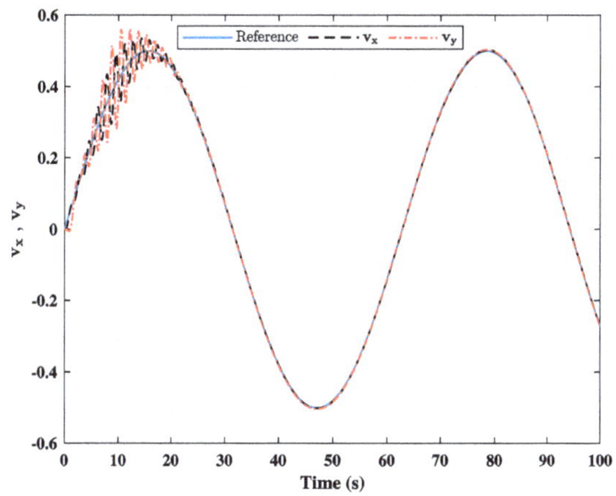

Fig. 12.16 Velocity tracking control using GRNNSMC

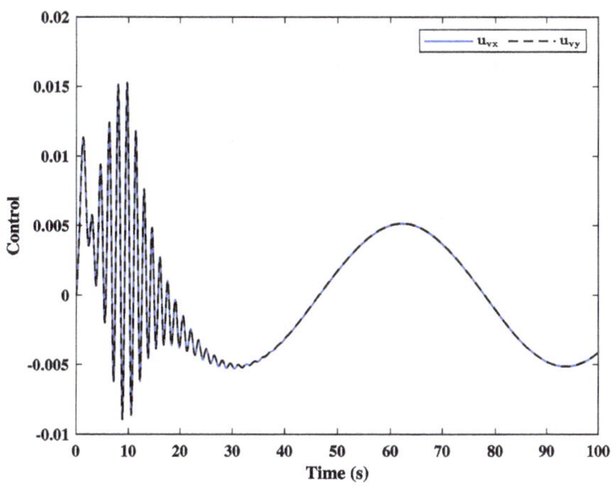

Fig. 12.17 GRNNSMC control signals

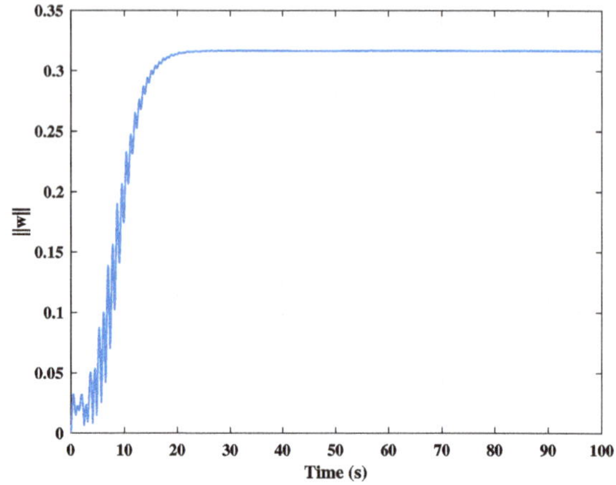

Fig. 12.18 Evolution of GRNNSMC weights norm

Table 12.6 GRNNSMC RMSE for different disturbance levels

Disturbance %	RMSE_vx	RMSE_vy
0%	0.0061	0.0068
25%	0.0101	0.0113
50%	0.0141	0.0158
75%	0.0154	0.0168
100%	0.0192	0.0208
200%	0.0324	0.0341

output weights $\boldsymbol{W}$ of GRNNSMC is also demonstrated in Fig. 12.18, after the initial learning phase, the norm of the weights converges to a fixed value around 0.32. Also, the accuracy of the controller is tested using the RMSE for different external disturbance values measured as a percentage of the magnitude of the reference signal as shown in Table 12.6. The results reveal that GRNNSMC can handle large values of disturbance without losing much accuracy.

12.4 Spacecraft Attitude Control Using Automatic Reconfiguration GRNNs (ARGRNNs)

In this case study, the ARGRNNs are used to control the attitude of a rigid spacecraft as in Zou et al. (2011).

12.4.1 Dynamic Model

The attitude dynamics of the spacecraft can be expressed as

$$\begin{aligned} \boldsymbol{I}\dot{\boldsymbol{\omega}}(t) &= -\boldsymbol{\omega}(t) \times \boldsymbol{I}\boldsymbol{\omega}(t) + \boldsymbol{u}(t) + \boldsymbol{d}(t) \\ \boldsymbol{y}(t) &= \boldsymbol{\omega}(t) \end{aligned} \tag{12.50}$$

where $\boldsymbol{\omega} \in \Re^3$ is a vector of the angular velocity of the spacecraft in the fixed-body frame. $\boldsymbol{I} \in \Re^{3\times3}$ is the inertia matrix of the spacecraft. $\boldsymbol{u} \in \Re^3$ is the control torque. $\boldsymbol{d} \in \Re^3$ is a vector of the external disturbances, and $\times$ is the cross product operator.

The inertia matrix is given as

$$\boldsymbol{I} = \begin{bmatrix} 1.5 & 0.1 & 0.3 \\ 0.1 & 1.7 & 0.4 \\ 0.3 & 0.4 & 0.7 \end{bmatrix} \text{kg.m}^2 \tag{12.51}$$

The desired output trajectories vector is selected as

$$\boldsymbol{\omega_d}(t) = \left[0.2\sin(0.1\pi t), 0.2\sin(0.15\pi t), 0.2\sin(0.25\pi t)\right] \tag{12.52}$$

The external disturbances vector is

$$\boldsymbol{d}(t) = \left[0.1\sin(t), 0.1\cos(t), 0.1\sin(2t)\right] \tag{12.53}$$

The initial values of the states are $\omega(0) = (0.1, -0.1, 0.2)$. The faults take place at 5 seconds. The parameters of the faults are $\rho_1, \rho_2, \rho_3 = 0.2$ and $\beta_1, \beta_2, \beta_3 = -8$, which corresponds to losing 80% of each actuator effectiveness and biasing it with -8. The adaptation gains are chosen as $\gamma = [40, 40, 40]$. The addition and removal thresholds are chosen as $r_{th} = 0.1$ and $m_{th} = 0.5$.

12.4.2 Simulation Results

The attitude tracking results as shown in Fig. 12.19 depicts that ARGRNN controller can drive the attitude to the desired values, after the fault took place by adjusting the its structure as shown in Fig. 12.20 soon after the fault is detected as depicted in Fig. 12.21. Fault detection is triggered by significant increase in the nearest neighbor distance from zero to 0.15. Also, ARGRNN needs six nodes to control the attitude for healthy actuators and 15 nodes to accommodate the faulty actuators.

The control history u_1, u_2, and u_3 for the three actuators, as depicted in Fig. 12.22, reveal small increments in the initial learning stage followed by a substantial increments from zero to 78 to accommodate the actuator faults.

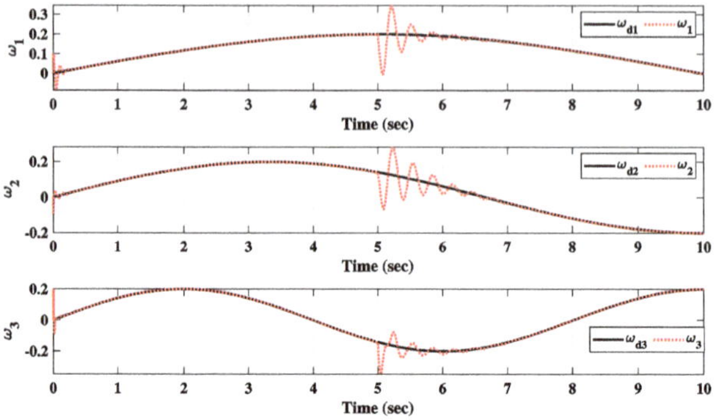

Fig. 12.19 Attitude tracking using ARGRNNs

Fig. 12.20 Evolution of ARGRNNs nodes

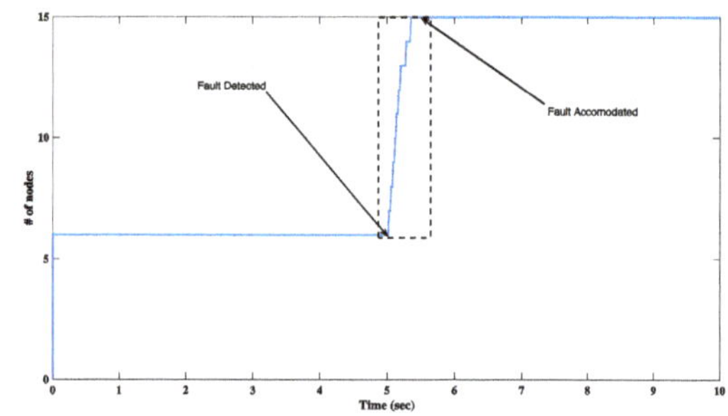

Fig. 12.21 The nearest neighbor distance evolution

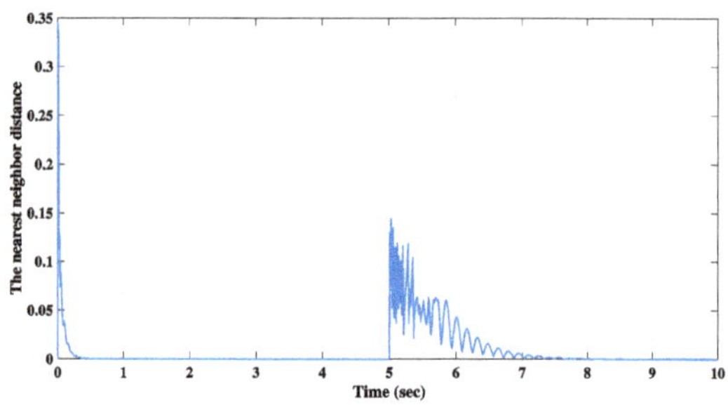

Fig. 12.22 Control history

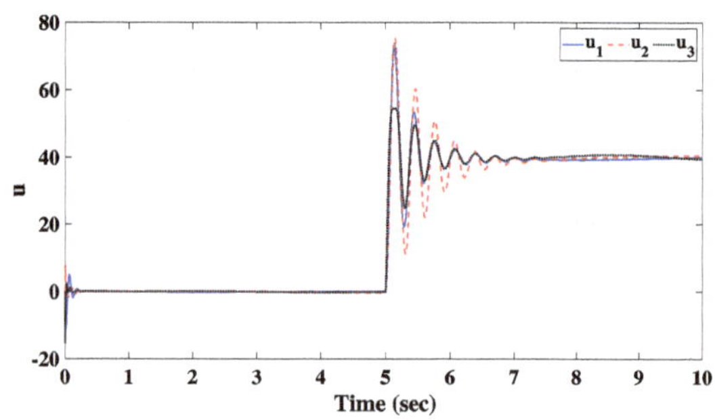

12.5 Two-Degree-of-Freedom (DOF) Planar Robotic Arm Control Using I_GRNN

Two-degrees-of-freedom (DOF) planar robotic arm as depicted in Fig. 12.23 is used to test I_GRNN performance and compare it with the original GRNN performance. This system is a nonlinear second-order system. The control aims to track the desired angles by varying joint torques.

12.5.1 Mathematic Model of 2-Dof Robotic Arm

Defining $\boldsymbol{q}$ as the joint variables vector expressed as $\boldsymbol{q} = \begin{bmatrix} q_1 \\ q_2 \end{bmatrix}$, then the dynamics of the robot arm can be written in the state-space format as

$$\boldsymbol{M}(\boldsymbol{q})\ddot{q} + \boldsymbol{v}(\dot{\boldsymbol{q}}, \boldsymbol{q}) + \boldsymbol{g}(\boldsymbol{q}) = \boldsymbol{\tau} \tag{12.54}$$

where $\boldsymbol{M}(\boldsymbol{q})$ is the mass matrix, $\boldsymbol{v}(\dot{\boldsymbol{q}}, \boldsymbol{q}$ is the centrifugal and Coriolis vector, $\boldsymbol{g}(\boldsymbol{q})$ is the gravity vector, and $\boldsymbol{\tau}$ is the torque vector and they are expressed as

$$\boldsymbol{M}(\boldsymbol{q}) = \begin{bmatrix} (m_1 + m_2)l_1^2 + m_2 l_2^2 + 2m_2 l_1 l_2 \cos(\theta_2) & m_2 l_2^2 + m_2 l_1 l_2 \cos(\theta_2) \\ m_2 l_2^2 + m_2 l_1 l_2 \cos(\theta_2) & m_2 l_2^2 \end{bmatrix} \tag{12.55}$$

$$\boldsymbol{v}(\dot{\boldsymbol{q}}, \boldsymbol{q}) = \begin{bmatrix} -m_2 l_1 l_2 \sin(\theta_2)(2\dot{\theta}_1\dot{\theta}_2 + \dot{\theta}_2{}^2) \\ -m_2 l_1 l_2 \sin(\theta_2)\dot{\theta}_1\dot{\theta}_2 \end{bmatrix} \tag{12.56}$$

$$\boldsymbol{g}(\boldsymbol{q}) = \begin{bmatrix} -(m_1 + m_2) g l_1 \sin(\theta_1) - m_2 g l_2 \sin(\theta_1 + \theta_2) \\ -m_2 g l_2 \sin(\theta_1 + \theta_2) \end{bmatrix} \tag{12.57}$$

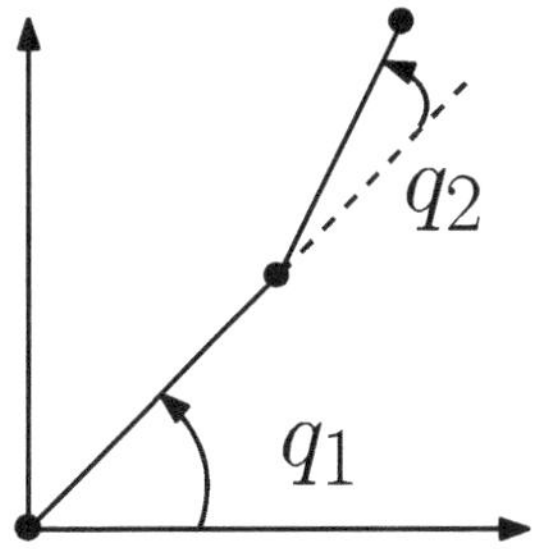

Fig. 12.23 Two-DOF planar robotic arm

$$\boldsymbol{\tau} = \begin{bmatrix} u_1 \\ u_2 \end{bmatrix} \tag{12.58}$$

where $l_1 = 0.7\,\text{m}$, $l_2 = 0.8\,\text{m}$ are the links lengths, $m_1 = 0.5\,\text{kg}$, $m_2 = 0.4\,\text{kg}$ are the links masses and $g = 9.81\,\text{m/s}^2$. The desired angles q_{d1} and q_{d2} are $[q_{d1}, q_{d2}] = [\pi/4, \pi/4]$ between 1 and 5 seconds and zero at all the other times.

12.5.2 Simulation Results

The initial values are zeros. The learning rates for I_GRNN are $\gamma_1 = 2.5$, $\gamma_2 = 0.5$, the hidden layer size is ten, and $\sigma = [0.01 : 0.1 : 1]$. For the original GRNN, the learning rate is $\gamma_1 = 2.5$, the hidden layer size is ten, and $\sigma = 1$.

The output evolution of q_1 and q_2 as depicted in Figs. 12.24 and 12.25 illustrates that the performance of I_GRNN is better than the original GRNN with less overshoot and oscillations for both controlled variables.

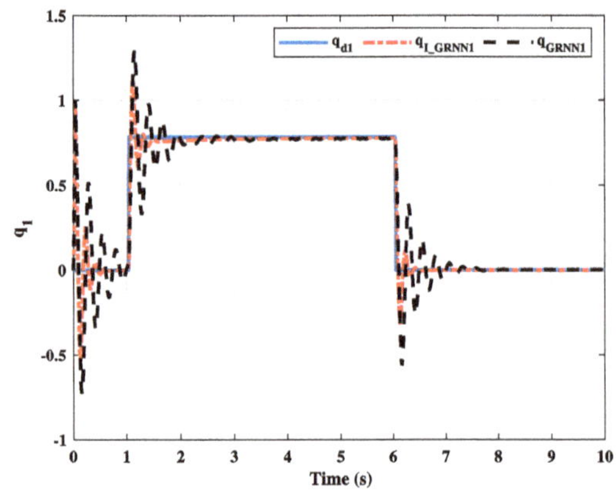

Fig. 12.24 q_1 tracking for robotic arm

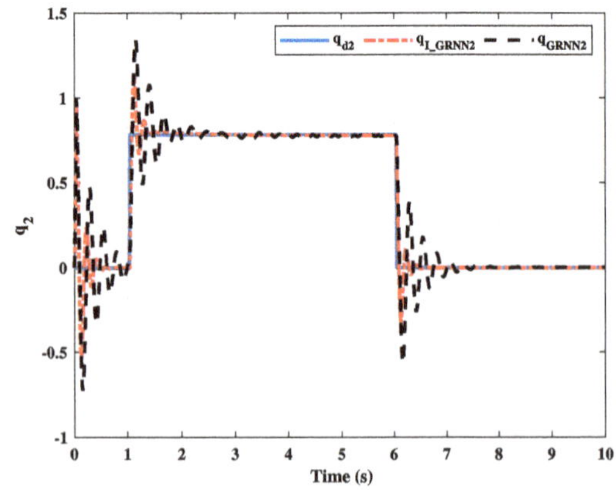

Fig. 12.25 q_2 tracking for robotic arm

References

A.J. Al-Mahasneh, S.G. Anavatti, M. Ferdaus, M.A. Garratt, Adaptive neural altitude control and attitude stabilization of a hexacopter with uncertain dynamics, in *2019 IEEE International Conference on Industry 4.0, Artificial Intelligence, and Communications Technology (IAICT)* (2019), pp. 44–49

S. Keshmiri, R. Colgren, M. Mirmirani, Development of an aerodynamic database for a generic hypersonic air vehicle, in *AIAA Guidance, Navigation, and Control Conference and Exhibit* (2005)

S. Keshmiri, R. Colgren, M. Mirmirani, Six-dof modeling and simulation of a generic hypersonic vehicle, for control and navigation purposes, in *AIAA Guidance, Navigation, and Control Conference and Exhibit* (2006)

M. Mirmirani, C. Wu, A. Clark, S. Choi, B. Fidan, Airbreathing hypersonic flight vehicle modeling and control, review, challenges, and a CFD-based example, in *Workshop on Modeling and Control of Complex Systems* (2005), pp. 1–15

E. Rollins, J. Valasek, J.A. Muse, M.A. Bolender, Nonlinear adaptive dynamic inversion applied to a generic hypersonic vehicle, in *AIAA Guidance, Navigation, and Control (GNC) Conference* (2013)

K. Sachan, R. Padhi, State-constrained robust adaptive cruise control design for air-breathing hypersonic vehicles, in *2018 AIAA Guidance, Navigation, and Control Conference* (2018a)

K. Sachan, R. Padhi, A brief survey on six-degree-of-freedom modeling for air-breathing hypersonic vehicles. IFAC-PapersOnLine **51**(1), 492–497 (2018b)

K. Sachan, R. Padhi, Nonlinear robust neuro-adaptive flight control for hypersonic vehicles with state constraints. Control Eng. Pract. **102**, 104526 (2020)

S. Samantaray, R. Padhi, Model reference adaptive sliding mode control of a laser beam for jitter correction, in *25th AIAA/AHS Adaptive Structures Conference* (2017)

J.D. Shaughnessy, S.Z. Pinckney, J.D. McMinn, C.I. Cruz, M.-L. Kelley, Hypersonic vehicle simulation model: winged-cone configuration (1990)

Z. Zhang, L. Yang, Y. Zhong, G. Shen, Modeling and analysis for a generic hypersonic vehicle, in *2010 8th World Congress on Intelligent Control and Automation* (2010), pp. 152–158

A. Zou, K.D. Kumar, Z. Hou, X. Liu, Finite-time attitude tracking control for spacecraft using terminal sliding mode and chebyshev neural network. IEEE Trans. Syst. Man Cybernet. Part B (Cybernet.) **41**(4), 950–963 (2011)

Index

K. Sachan et al., *Intelligent Adaptive Control with System Constraints*,
https://doi.org/10.1007/978-981-95-8220-4

www.ingramcontent.com/pod-product-compliance
Ingram Content Group UK Ltd.
Pitfield, Milton Keynes, MK11 3LW, UK
UKHW020631260926
475829UK00003BA/21
* 9 7 8 9 8 1 9 5 8 2 1 9 8 *